Periodic Table of the Elements with the Gmelin System Numbers

1	2	3	4	5	6	7	8	9	10	11	12	13	14	15	16	17	18
1 H 2																	2 He 1
3 Li 20	4 Be 26											5 B 13	6 C 14	7 N 4	8 O 3	9 F 5	10 Ne 1
11 Na 21	12 Mg 27											13 Al 35	14 Si 15	15 P 16	16 S 9	17 Cl 6	18 Ar 1
19* K 22	20 Ca 28	21 Sc 39	22 Ti 41	23 V 48	24 Cr 52	25 Mn 56	26 Fe 59	27 Co 58	28 Ni 57	29 Cu 60	30 Zn 32	31 Ga 36	32 Ge 45	33 As 17	34 Se 10	35 Br 7	36 Kr 1
37 Rb 24	38 Sr 29	39 Y 39	40 Zr 42	41 Nb 49	42 Mo 53	43 Tc 69	44 Ru 63	45 Rh 65	46 Pd 65	47 Ag 61	48 Cd 33	49 In 37	50 Sn 46	51 Sb 18	52 Te 11	53 I 8	54 Xe 1
55 Cs 25	56 Ba 30	57** La 39	72 Hf 43	73 Ta 50	74 W 54	75 Re 70	76 Os 70	77 Ir 67	78 Pt 68	79 Au 62	80 Hg 34	81 Tl 38	82 Pb 47	83 Bi 19	84 Po 12	85 At 8a	86 Rn 1
87 Fr 25a	88 Ra 31	89*** Ac 40	104 71	105 71													

* NH₄ 23 → NH_4 23

Lanthanides 39

58 Ce	59 Pr	60 Nd	61 Pm	62 Sm	63 Eu	64 Gd	65 Tb	66 Dy	67 Ho	68 Er	69 Tm	70 Yb	71 Lu

***Actinides**

90 Th 44	91 Pa 51	92 U 55	93 Np 71	94 Pu 71	95 Am 71	96 Cm 71	97 Bk 71	98 Cf 71	99 Es 71	100 Fm 71	101 Md 71	102 No 71	103 Lr 71

A Key to the Gmelin System is given on the Inside Back Cover

Gmelin Handbook of Inorganic Chemistry

8th Edition

Gmelin Handbook
of Inorganic Chemistry

8th Edition

Gmelin Handbuch der Anorganischen Chemie

Achte, völlig neu bearbeitete Auflage

Prepared
and issued by

Gmelin-Institut für Anorganische Chemie
der Max-Planck-Gesellschaft
zur Förderung der Wissenschaften
Director: Ekkehard Fluck

Founded by

Leopold Gmelin

8th Edition

8th Edition begun under the auspices of the
Deutsche Chemische Gesellschaft by R. J. Meyer

Continued by

E.H.E. Pietsch and A. Kotowski, and by
Margot Becke-Goehring

Springer-Verlag Berlin Heidelberg GmbH   1988

Organometallic Compounds in the Gmelin Handbook

The following listing indicates in which volumes these compounds are discussed or are referred to:

Ag   Silber B 5 (1975)

Au   Organogold Compounds (1980)

Be   Organoberyllium Compounds 1 (1987)

Bi   Bismut-Organische Verbindungen (1977)

Co   Kobalt-Organische Verbindungen 1 (1973), 2 (1973), Kobalt Erg.-Bd. A (1961), B 1 (1963), B 2 (1964)

Cr   Chrom-Organische Verbindungen (1971)

Cu   Organocopper Compounds 1 (1985), 2 (1983), 3 (1986), 4 (1987), Index (1987)

Fe   Eisen-Organische Verbindungen A 1 (1974), A 2 (1977), A 3 (1978), A 4 (1980), A 5 (1981), A 6 (1977), A 7 (1980), A 8 (1985), B 1 (partly in English; 1976), Organoiron Compounds B 2 (1978), Eisen-Organische Verbindungen B 3 (partly in English; 1979), B 4 (1978), B 5 (1978), Organoiron Compounds B 6 (1981), B 7 (1981), B 8 to B 10 (1985), B 11 (1983), B 12 (1984), B 13 (1988) **present volume**, Eisen-Organische Verbindungen C 1 (1979), C 2 (1979), Organoiron Compounds C 3 (1980), C 4 (1981), C 5 (1981), C 7 (1985), and Eisen B (1929-1932)

Ga   Organogallium Compounds 1 (1986)

Ge   Organogermanium Compounds 1 (1988)

Hf   Organohafnium Compounds (1973)

Nb   Niob B 4 (1973)

Ni   Nickel-Organische Verbindungen 1 (1975), 2 (1974), Register (1975), Nickel B 3 (1966), and C 1 (1968), C 2 (1969)

Np, Pu   Transurane C (partly in English; 1972)

Pb   Organolead Compounds 1 (1987)

Pt   Platin C (1939) and D (1957)

Ru   Ruthenium Erg.-Bd. (1970)

Sb   Organoantimony Compounds 1 (1981), 2 (1981), 3 (1982), 4 (1986)

Sc, Y, La to Lu   D 6 (1983)

Sn   Zinn-Organische Verbindungen 1 (1975), 2 (1975), 3 (1976), 4 (1976), 5 (1978), 6 (1979), Organotin Compounds 7 (1980), 8 (1981), 9 (1982), 10 (1983), 11 (1984), 12 (1985), 13 (1986), 14 (1987), 15 (1988)

Ta   Tantal B 2 (1971)

Ti   Titan-Organische Verbindungen 1 (1977), 2 (1980), Organotitanium Compounds 3 (1984), 4 and Register (1984)

U   Uranium Suppl. Vol. E 2 (1980)

V   Vanadium-Organische Verbindungen (1971), Vanadium B (1967)

Zr   Organozirconium Compounds (1973)

Gmelin Handbook
of Inorganic Chemistry

8th Edition

Fe
Organoiron Compounds

Part B 13

Mononuclear Compounds 13

With 5 illustrations

AUTHOR   Christa Siebert (Maintal)

EDITOR   Jürgen Faust

CHIEF EDITOR   Johannes Füssel

Springer-Verlag Berlin Heidelberg GmbH   1988

LITERATURE CLOSING DATE: 1986

Library of Congress Catalog Card Number: Agr 25-1383

ISBN 978-3-662-06920-2   ISBN 978-3-662-06918-9 (eBook)
DOI 10.1007/978-3-662-06918-9

© by Springer-Verlag Berlin Heidelberg 1988

Originally published by Springer-Verlag, Berlin · Heidelberg · New York · Paris · London · Tokyo in 1988.

Softcover reprint of the hardcover 8th edition 1988

Preface

"Organoiron Compounds" B 13 (present volume) systematically covers the literature through the end of 1986, and includes additional references up to 1988.

The volume continues Series B (volumes B 1 to B 12 already published) on the mononuclear organoiron compounds, whereas Series A (volumes A 1 to A 8 already published) is devoted to the ferrocenes and Series C (volumes C 1 to C 5 and C 7 already published) treats organoiron compounds with two or more Fe atoms in the molecule.

Series B so far includes mononuclear organoiron compounds as follows:

"Eisen-Organische Verbindungen" B 1 (1976), B 2 (1978), B 3 (1979)
Sections 1 to 1.1.4.8 on σ-compounds and carbonyl compounds.

"Eisen-Organische Verbindungen" B 4 (1978)
Sections 1.1.5 to 1.2.3.2.3 on isonitrile and carbene compounds and on compounds with ligands bonded by two C atoms (^{2}L ligands) to the Fe atom.

"Eisen-Organische Verbindungen" B 5 (1978)
Sections 1.3 to 1.3.6 on compounds with ligands bonded by three C atoms (^{3}L ligands) to the Fe atom.

"Organoiron Compounds" B 6 (1981), B 7 (1981), B 8 (1985), B 9 (1985), B 10 (1986)
Sections 1.4 to 1.4.3.4 on compounds with ligands bonded by four C atoms (^{4}L ligands) to the Fe atom.

"Organoiron Compounds" B 11 (1983), B 12 (1984)
Sections 1.5 to 1.5.2.3.16.1.5 on compounds with ligands bonded by five C atoms (^{5}L ligands) to the Fe atom up to the first subsections concerning compounds of the type $C_5H_5Fe(CO)_2R$.

"Organoiron Compounds" B 13 continues the description of $C_5H_5Fe(CO)_2R$ compounds, which will be finished shortly in "Organoiron Compounds" B 14. It deals with compounds in which the $C_5H_5Fe(CO)_2$ group (abbreviated as Fp) is bonded to alkyl groups substituted by heterocycles, to a C=O, C=N, or C=S unit, and to alkenyl, alkynyl, or carbocycles. More than 200 compounds with Fp bonded to carbocycles other than aryls cover a substantial part of this volume. An index covering both B 13 and B 14 will be included in Part B 14.

For abbreviations used throughout this volume, see p.X.

Frankfurt am Main, October 1988   Johannes Füssel

Remarks on Abbreviations and Dimensions

Most compounds and reagents in this volume are presented in tables. For the sake of conciseness, some abbreviations are used and some dimensions are omitted in the tables. This necessitates the following clarification.

Geometric isomers are designated according to the IUPAC rules. Structural labels are missing when authors fail to report structural details.

Temperatures are given in °C, otherwise K stands for Kelvin. Abbreviations used with temperatures are m.p. for melting point, b.p. for boiling point, subl. for sublimes, and dec. for decomposition.

Nuclear magnetic resonance is abbreviated as NMR, noise decoupling is indicated by braces { }. Chemical shifts are given as δ values in ppm with the positive sign for downfield shifts. Reference substances are $Si(CH_3)_4$ for ^1H and ^{13}C NMR, $CFCl_3$ for ^{19}F NMR, and H_3PO_4 for ^{31}P NMR.

Multiplicities of the signals are abbreviated as s, d, t, q (singlet to quartet), quint, sext, sept (quintet to septet), and m (multiplet); terms like dd (double doublet) and t's (triplets) are also used. Assignments referring to labeled structural formulas are given in the form C-4, H-3,5. Coupling constants ^nJ in Hz are given as J(A,B) or as J(1,3) referring to labeled structural formulas, n is the number of bonds between the coupled nuclei.

Optical spectra are labeled as IR (infrared), R (Raman), and UV (electronic spectrum including the visible region). IR bands and Raman lines are given in cm^{-1}, as far as necessary the assigned bands are labeled with the symbols ν for stretching vibration and δ for deformation vibration. IR bands resulting from the terminal CO groups (two strong bands usually in the range 1920 to 2060 cm^{-1}) are presented first, followed by further bands given in the respective publication. The CO stretching force constant and CO,CO interaction constant (in mdyn/Å) are denoted as k and k', respectively. Intensities are indicated by the common qualitative terms (vs, s, m, w, vw) or as numerical relative intensities in parentheses. Multiplicities are abbreviated as with NMR. The UV absorption maxima, λ_{max}, are given in nm followed by the extinction coefficient ϵ ($L \cdot cm^{-1} \cdot mol^{-1}$) or log ϵ in parentheses; sh means shoulder, br means broad. If reported, solvents or the physical state are given in parentheses immediately after the spectral symbol.

Electron paramagnetic resonance and **electron spin resonance** are abbreviated as EPR and ESR, hyperfine coupling constants are given as a(X).

Mössbauer spectra ($^{57}Fe-\gamma$). Both the isomer shift δ (vs. $Na_2[Fe(CN)_5NO]$ at room temperature) and the quadrupole splitting Δ are given in mm/s; the experimental error has generally been omitted. Other reference substances for δ are indicated after the numerical value, e.g., $\delta = 0.23$ (Fe).

Mass spectral data are given as the ions (e.g., $[M]^+$) or the m/e values followed by the relative intensities in parentheses.

Further abbreviations:

d_c	calculated density	SCE	saturated calomel electrode
d_m	experimental density	soln.	solution
$[M]^+$	molecular ion in mass spectroscopy	THF	tetrahydrofuran
aq.	aqueous	$i-C_3H_7$	isopropyl $CH(CH_3)_2$
conc.	concentrated	$s-C_4H_9$	sec-butyl $CH(CH_3)C_2H_5$
emf	electromotive force	$t-C_4H_9$	tert-butyl $C(CH_3)_3$
NQR	nuclear quadrupole resonance		

Table of Contents

Organoiron Compounds, Part B

Mononuclear Compounds B 13

1.5.2.3.16 Compounds of the $C_5H_5Fe(CO)_2{}^1L$ Type (continued)

1.5.2.3.16.1 $C_5H_5Fe(CO)_2R$ Compounds with Aliphatic R Groups (continued)

1.5.2.3.16.1.6 $C_5H_5Fe(CO)_2R$ Compounds with Alkyl Groups Substituted by Heterocycles

The compounds listed in Table 1 have been prepared by the following methods. Other procedures used for some of the compounds are described under "Further information"; see also under "remarks" in Table 1.

Method I: Reaction of Na[Fp] with the corresponding halogenoalkylthiophene, –piperidine, or –pyridine (halogen = Cl, Br; 1:1 mole ratio) in THF for 16 to 20 h. After solvent removal under reduced pressure the residue is extracted with hexane (Nos. 2 and 3) or CH_2Cl_2 (Nos. 9 and 10). The extracts are filtered and the solvent is again removed. Workup of the residue is carried out by chromatography on Al_2O_3 with hexane (Nos. 2 and 3), ether (No. 10), or CH_2Cl_2 (Nos. 11 and 12) as eluents. The products are obtained either by evaporation of the solvent (Nos. 2 and 3) [4] and further purified by three low-temperature recrystallizations from pentane (No. 10) [1] or by addition of hexane to the eluate and concentration of the solution (Nos. 11 and 12) [8]. In the case of No. 9 the liquid reaction residue is extracted with pentane and the filtered extracts are cooled at −78 °C for several hours to give the product. The recrystallization from pentane is repeated twice [1].

Method II: Successive deprotonation of $[C_5H_5Fe(CO)_2CH_2=CH(CH_2)_nNH_3][BF_4]_2$, prepared by exchange reaction from $[C_5H_5Fe(CO)_2CH_2=C(CH_3)_2]BF_4$ and $[CH_2=CH(CH_2)_nNH_3]BF_4$ (n = 3 or 4), with $N(C_4H_9-n)_3$ followed by $KOC(CH_3)_3$. Further details were not given [12].

Compound No. 8 is converted into the lactam I by treatment with Ag_2O in THF at 65 °C for 20 h [12]. Similar oxidation of compounds No. 4 and 5 has led only to a polyamide [12] and a trace of lactam II [14], respectively. However, heating compound No. 4 in THF for 4 h in the presence of 10 mol% of $P(C_6H_5)_3$ [12] or of No. 5 in CH_3CN at 65 °C for 7.5 h in the presence of a small amount of $P(C_4H_9-n)_3$ [14] gives the chelate III ($R^1 = H$, CH_3) which on subsequent treatment with Ag_2O in THF at 25 °C for 5 min (No. 4) [12] or 2 h (No. 5) [14] affords the lactam II. For the stereochemistry of II ($R^1 = CH_3$), see [14].

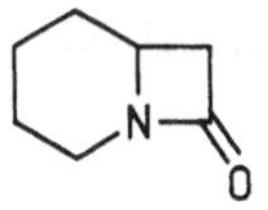

I 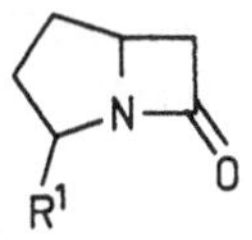 II 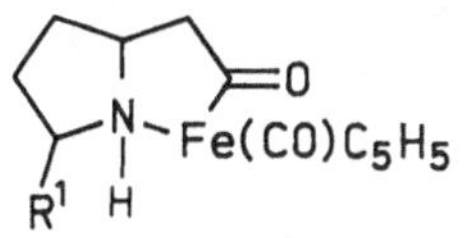 III

References on p. 6

Table 1

$C_5H_5Fe(CO)_2R$ Compounds with R = Alkyl Substituted by Heterocycles.

Further information on compounds with numbers preceded by an asterisk is given at the end of the table.

For abbreviations and dimensions, see p. X.

No.	compound method of preparation (yield)	properties and remarks
*1		cis- and trans-isomers
*2	 I (64%) [4]	red-brown liquid [4] ^1H NMR (CDCl$_3$): ~2.9 (s, CH$_2$), ~4.6 (C$_5$H$_5$), ~6.8 (complex, C$_4$H$_3$S) [4] IR: 1966s, 2005s (both CO) in C$_6$H$_{12}$, 2920, 2960, 3100, 3125 (all CH) in KBr, other bands reported [4]
*3	 I (55%) [4]	red-brown liquid [4] ^1H NMR (CDCl$_3$): 2.85 (s, CH$_2$), ~4.0 (C$_5$H$_5$), ~6.8 (complex, C$_4$H$_3$S) [4] IR: 1963s, 2004s (both CO) in C$_6$H$_{12}$, 2925, 2960sh, 3065, 3110sh, 3115 (all CH) in KBr, other bands reported [4]
4	 II [12]	for reactions, see p. 1
*5		1:1 mixture of diastereomers [14] orange oil [14] ^1H NMR (CDCl$_3$): 0.99, 1.16 (2d, CH$_3$, J=7), 1.2, 1.9, 3.1 (3m, CH$_2$CH$_2$, CH), 4.74, 4.79 (2s, C$_5$H$_5$) [14] IR (CH$_2$Cl$_2$): 1942, 2010 (both CO) [14] for reactions, see p. 1
*6		red-orange oil [14] ^1H NMR (CDCl$_3$): 1.1 to 1.6 (m, FeCH$_2$), 1.99 (s, CH$_3$), 1.9 to 2.6 (m, CH$_2$CH$_2$), 3.88 (m, CH), 4.79 (s, C$_5$H$_5$) [14] ^{13}C NMR (CDCl$_3$): 8.8 (FeCH$_2$), 19.9 (CH$_3$), 32.6, 39.2 (CH$_2$CH$_2$), 79.4 (C–N), 85.4 (C$_5$H$_5$), 171.7 (C=N), 217.6 (CO) [14] IR (CH$_2$Cl$_2$): 1937, 2010 (both CO), 1640 [14]
7		obtained from $C_5H_5Fe(CO)_2CH_2C(CH_2Cl)=CH_2$ and $4-CH_3C_6H_4SO_2NCO$ [13]

Table 1 (continued)

No.	compound method of preparation (yield)	properties and remarks
8	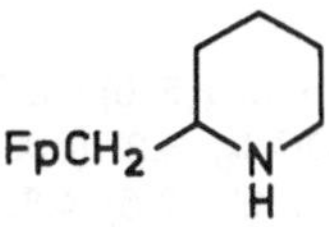 FpCH$_2$ — (piperidine ring, N–H) II [12]	for reaction, see p. 1
*9	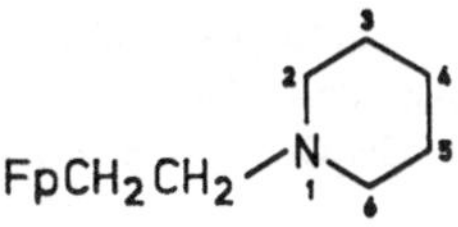 FpCH$_2$CH$_2$ — (piperidine ring) I (39%) [1]	yellow–orange crystals, m.p. 38 to 39° [1] ^{1}H NMR (CS$_2$): 1.43 (br apparent s, FeCH$_2$, CH$_2$-3, 5), 1.56 (apparent s, CH$_2$-4), 2.27 (complex with 3 peaks, CH$_2$N, CH$_2$-2, 6), 4.62 (s, C$_5$H$_5$), relative intensities 6:2:6:5, only tentative assignment [1] IR: 1960s, 2015s (both CO) in C$_6$H$_{12}$, 2740, 2775, 2840, 2910, 3020 (all CH) in KBr, other bands reported [1] UV (C$_6$H$_{12}$): 348 (ε = 1410) [1]
*10	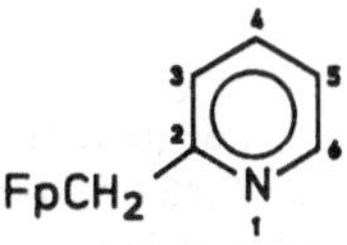 FpCH$_2$ — (pyridine ring) I (25%) [1]	yellow–orange, m.p. 42 to 44° [1] ^{1}H NMR (CS$_2$): 2.55 (s, CH$_2$), 4.71 (s, C$_5$H$_5$), 6.90 (H–3, 5), 7.32 (apparent t with additional fine structure, H–4, J ~ 8), 8.28 (apparent d with additional fine structure, H–6, J = 4) [1] IR: 1933w, 1970vs, 2020sh, 2025s (all CO) in C$_6$H$_{12}$, 2900, 2950, 3040 (all CH), other bands reported in KBr [1] UV (C$_6$H$_{12}$): 315 (ε = 10650) [1]
*11	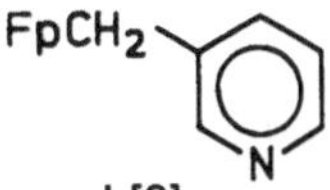 FpCH$_2$ — (pyridine ring) I [8]	—
*12	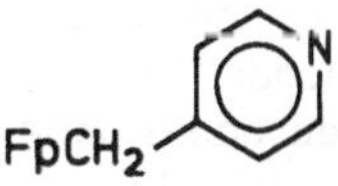 FpCH$_2$ — (pyridine ring) I [8]	—
13	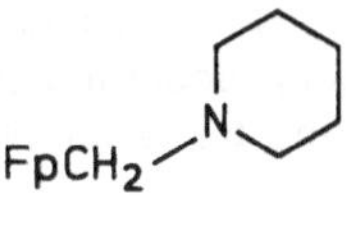 FpCH$_2$ —N (piperidine ring) I (77%) [16]	red–brown oil [16] ^{1}H NMR (C$_6$D$_6$): 1.48 (complex m, 6H, CCH$_2$C), 2.37 (complex m, 4H, CCH$_2$N), 4.22 (FeCH$_2$), 4.27 (C$_5$H$_5$) [16] IR (C$_6$H$_{12}$): 1927, 1933sh, 1949, 1985, 1991sh, 2003 (all CO) [16]
14	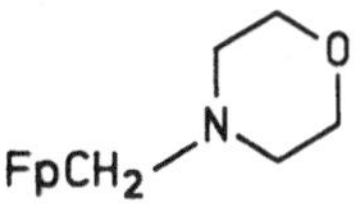 FpCH$_2$ —N (morpholine ring, O) I (94%) [16]	dark red oil [16] ^{1}H NMR (C$_6$D$_6$): 1.48, 2.37 (both: complex t, NCH$_2$CH$_2$O), 3.87 (FeCH$_2$), 4.25 (C$_5$H$_5$) [16] IR (C$_6$H$_{12}$): 1933, 1937 sh, 1952, 1988, 1993 sh, 2008 (all CO) [16]

References on p. 6

Table 1 (continued)

No.	compound method of preparation (yield)	properties and remarks

15 FpCH₂ ... R= OCOCH₃ I (92%) [17]

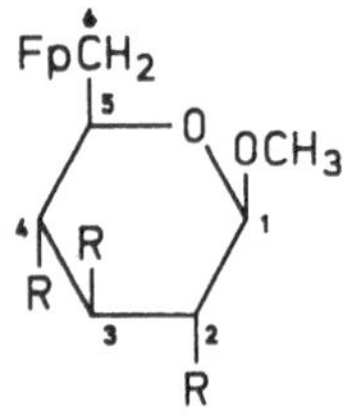

yellow, rather unstable [17]
^{1}H NMR (CDCl$_3$): 1.24 (m, H-6'), 1.62 (dd, H-6, J(5, 6) = 2.7, J(6, 6') = 11), 2.00, 2.04, 2.07 (3s, 9H, COCH$_3$), 3.27 (oct, H-5), 3.53 (s, 3H, OCH$_3$), 4.35 (d, H-1, J(1, 2) = 7.5), 4.80 (s superposed on m, 6H, C$_5$H$_5$ and H-2), 4.95 (m, H-4), 5.13 (t, H-3, J = 9.3) [17]
^{13}C NMR (CDCl$_3$): 20.7, 20.9 (both: CH$_3$ of CH$_3$CO), 57.1 (OCH$_3$), 71.9 to 79.4 (5C, C-2 to C-6), 85.4 (C$_5$H$_5$), 101.9 (C-1), 169.5, 169.9, 170.4 (all CO of CH$_3$CO), 216.8, 217.4 (both CO) [17]
IR (neat): 1940, 2015 (both CO), 1745 (C=O) [17]
UV (CH$_3$OH): 412 [17]
mass spectrum: [M]$^+$ (10), [M−CO]$^+$ (15), [M−2CO]$^+$ (70), [M−2CO−C$_5$H$_5$]$^+$ (100), [M−2CO−C$_3$H$_6$O$_2$]$^+$ (30), [M−Fp+H]$^+$ (50) [17]

16 FpCH₂ ... C₆H₅OCO ... R= OCOCH₃ I [17]

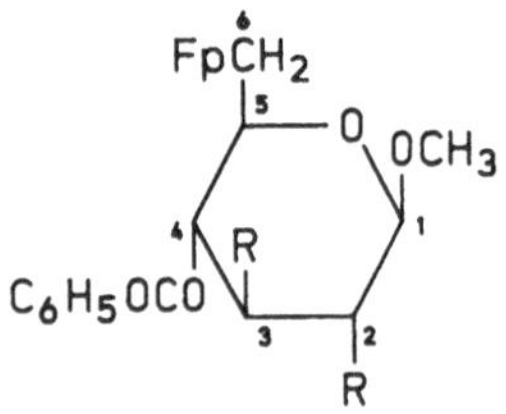

yellow syrup, darkens within 1d [17]
^{1}H NMR(CDCl$_3$): 1.34 (center of m, 1H, H-6), 1.68 (center of m, 1H, H-6'), 1.86, 2.04 (2s, 6H, COCH$_3$), 3.43 (m, H-5), 3.55 (s, 3H, OCH$_3$), 4.43 (d, H-1, J(1, 2) = 7.8), 4.74 (s, C$_5$H$_5$), 5.03 (dd, H-2, J(1, 2) = 7.8, J(2, 3) = 9), 5.06, 5.34 (2t, H-3, H-4, J ~ 9), 7.7 (m, C$_6$H$_5$) [17]

* Further information:

C$_5$H$_5$Fe(CO)$_2$CH$_2$(C$_8$H$_{10}$O$_2$)OC$_2$H$_5$ (Table. 1, No. 1) has been obtained in 93% yield as a mixture of the diastereomers (m.p. 84 to 88 °C) of the cis-lactone IV by the reduction of C$_5$H$_5$Fe(CO)$_2$CH$_2$C(OC$_2$H$_5$)(COOC$_2$H$_5$)C$_6$H$_9$O (see "Organoiron Compounds" B 12, 1984, p. 271) with Li[BH(C$_4$H$_9$-s)$_3$] in THF at −78 °C (1.5 h). If, however, the reduction is performed with NaBH$_4$, the main product of the mixture is the trans-lactone V, 70% yield [15].

Treatment of the cis- or trans-lactones thus obtained with HPF$_6$ · O(C$_2$H$_5$)$_2$ in CH$_2$Cl$_2$ at −78 °C (0.5 h) followed by demetalation of the intermediate VI with NaI in acetone or with [N(C$_2$H$_5$)$_4$]Br in CH$_2$Cl$_2$ at room temperature (15 min) gives as main product the corresponding free ligands of VI (the cis-lactone in the former case and the trans-lactone in the latter case) [15].

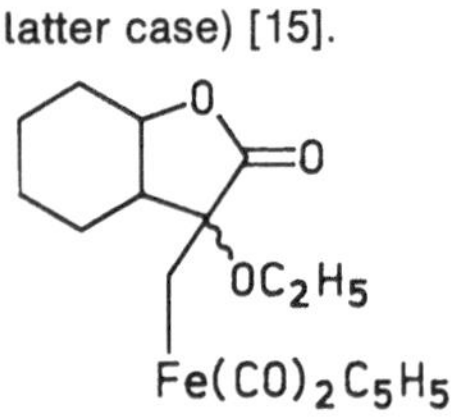

IV 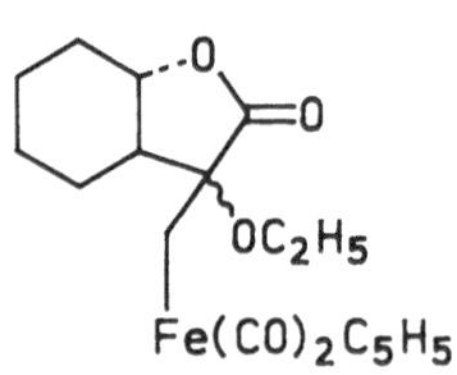   V 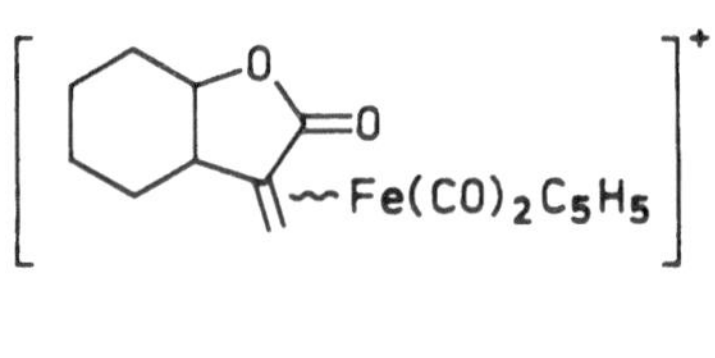   VI

References on p. 6

$C_5H_5Fe(CO)_2CH_2SC_4H_3$ (Table 1, Nos. **2** and **3**). The ^{1}H NMR spectra are broad and poorly resolved due to decomposition giving paramagnetic material [4].

Attempts to obtain mass spectra failed also, because of decomposition within the mass spectrometer [4].

UV irradiation of No. 3 in hexane for 36 h results in decarbonylation to give the π-thenyl derivative VII in small amounts (~2%) besides Fp_2 [4, 5]. Attempts to repeat this preparation have been unsuccessful [4]. Cleavage of the Fe-then-2-yl bond occurs when compound No. 2 is either refluxed in methylcyclohexane or vacuum distilled at 0.1 Torr, the products are Fp_2 in the former case and Fp_2, ferrocene, and an unidentified dirt-brown material in the latter case [4].

VII

$C_5H_5Fe(CO)_2CH_2(NC_4H_7)CH_3$ (Table 1, No. **5**) has been obtained by reduction of compound No. 6 with $NaBH_4$ (1:1 mole ratio) in ethanol at room temperature for 1 h. After cooling to 0 °C the mixture is treated with water, concentrated under reduced pressure, neutralized with HCl (after addition of water), and extracted with CH_2Cl_2. Removal of the solvent leaves a red oil which is chromatographed on basic Al_2O_3 with CH_2Cl_2/CH_3OH (9:1) as eluent. The product (52% yield) is shown by NMR spectroscopic analysis to be a 1:1 mixture of two diastereomers [14].

$C_5H_5Fe(CO)_2CH_2(NC_4H_5)CH_3$ (Table 1, No. **6**) has been prepared by passing NH_3 over a CH_2Cl_2 solution of $[C_5H_5Fe(CO)_2CH_2=CHCH_2CH_2COCH_3]BF_4$, formed by exchange reaction from $[C_5H_5Fe(CO)_2CH_2=C(CH_3)_2]BF_4$ and $CH_2=CHCH_2CH_2COCH_3$, at room temperature for 2 h. Addition of ether, filtration through Celite, and concentration under reduced pressure gives a viscous oil which is chromatographed on basic Al_2O_3 and eluted with ether/CH_2Cl_2 (1:1), 68% yield [14].

On heating in THF at 60 °C for 19 h, the compound is readily transformed into the chelate complex VIII as an almost equal mixture of diastereomers [14]. Reduction to compound No. 5 takes place on treatment with $NaBH_4$ in ethanol at room temperature for 1 h [14].

VIII

$C_5H_5Fe(CO)_2CH_2CH_2NC_5H_{10}$ (Table 1, No. **9**). The mass spectrum shows the molecular ion $[M]^+$ and the fragments $[M-nCO]^+$ (n=1, 2). The resulting $[C_5H_5FeCH_2CH_2NC_5H_{10}]^+$ eliminates then up to five hydrogen atoms, giving the ions $[C_5H_5FeCH_2CH_2NC_5H_n]^+$ (n=5, 6, and 8). All can lose a C_2H_4 fragment, forming $[C_5H_5FeNC_5H_n]^+$ (n=5, 6, 8, and 10). Other ions observed are: $[C_5H_5Fe(CO)_n]^+$, $[C_5H_5Fe(CO)_nH]^+$, $[C_5H_nNCH_2CH_2]^+$ (n=2, 1, and 0), and fragmentation products of the $C_5H_{10}NCH_2CH_2$ group. For the complete mass spectrum, see the figure in [3].

                     References on p. 6

UV irradiation leads only to the recovery of unchanged starting material. Reactions with I_2, CH_3I, or $C_5H_5Fe(CO)_2I$ in pentane or ether give complex mixtures of yellow to yellow-brown products, which have not been investigated further [1].

$C_5H_5Fe(CO)_2CH_2NC_5H_4$ (Table 1, No. **10**). A characteristic feature of the mass spectrum, which is otherwise very similar to those of related cyclopentadienyl metal carbonyl compounds, is the elimination of HCN fragments from the ion $[C_5H_5FeCH_2NC_5H_4]^+$ to give $[(C_5H_5)_2Fe]^+$. For the spectrum, see [3].

$C_5H_5Fe(CO)_2CH_2NC_5H_4$ (Table 1, Nos. **11** and **12**). Compound No. 12 is reported to be soluble in water and stable for long periods, even in the presence of oxygen [6]. Also, no decomposition takes place in dilute mineral acid at room temperature in the dark for several days [2]. However, at 65 °C and in the presence of concentrated mineral acid the compound is rapidly protonated at the pyridine nitrogen [2, 6].

Yellow crystals of $[C_5H_5Fe(CO)_2CH_2C_5H_5N]BF_4$ are formed on acidification of methanolic solutions of compounds No. 11 and 12 with HBF_4 [8].

For pK_a measurements of the conjugated cations of Nos. 11 and 12 [8], the kinetics of their decomposition in aqueous acid [7], and in nitrous acid [2, 6] as well as the kinetics of their reactions with metallic electrophiles such as Hg^+ [11], Hg^{2+}, or Tl^{3+} [9, 10], see $[C_5H_5Fe(CO)_2CH_2C_5H_5N]BF_4$, Section 1.5.2.3.20 in "Organoiron Compounds" B 14.

References:

[1] King, R.B.; Bisnette, M.B. (Inorg. Chem. **5** [1966] 293/300).
[2] Johnson, M.D.; Winterton, N.; Shortland, A.C. (New Aspects Chem. Metal Carbonyls Deriv. Proc. 1st Intern. Symp., Venice 1968, Abstr. C4, pp. 1/5).
[3] King, R.B. (Org. Mass Spectrom. **2** [1969] 387/99).
[4] King, R.B.; Kapoor, R.N. (Inorg. Chem. **8** [1969] 2535/9).
[5] King, R.B.; Kapoor, R.N. (4th Intern. Conf. Organometal. Chem., Bristol 1969, Abstr. K 3).
[6] Johnson, M.D. (Record Chem. Progr. **31** [1970] 143/54).
[7] Johnson, M.D.; Winterton, N. (J. Chem. Soc. A **1970** 511/6).
[8] Johnson, M.D.; Winterton, N. (J. Chem. Soc. A **1970** 507/11).
[9] Johnson, M.D.; Dodd, D. (J. Chem. Soc. B **1971** 662/7).
[10] Dodd, D.; Johnson, M.D.; Winterton, N. (J. Chem. Soc. A **1971** 910/5).

[11] Dodd, D.; Johnson, M.D. (J. Chem. Soc. Perkin Trans. II **1974** 219/23).
[12] Wong, P.K.; Madhavarao, M.; Marten, D.F.; Rosenblum, M. (J. Am. Chem. Soc. **99** [1977] 2823/4).
[13] Waterman, P.S. (Diss. Boston Univ. 1979, pp. 1/154; Diss. Abstr. Intern. B **40** [1979] 2200).
[14] Berryhill, S.R.; Rosenblum, M. (J. Org. Chem. **45** [1980] 1984/6).
[15] Chang, T.C.T.; Rosenblum, M. (J. Org. Chem. **46** [1981] 4626/7).
[16] Barefield, E.K.; Sepelak, D.J. (J. Am. Chem. Soc. **101** [1979] 6542/9).
[17] Baer, H.H.; Hanna, H.R. (Carbohydr. Res. **102** [1982] 169/83).

1.5.2.3.16.1.7 Oxo Compounds of the Type $C_5H_5Fe(CO)_2COR'$

Table 2 (pp. 13/27) summarizes compounds where an Fp moiety is bonded to a COR' group, the ligand R' represents aliphatic (Nos. 2 to 49), alicyclic (Nos. 50 to 66), aromatic (Nos. 67 to 71), heterocyclic groups (Nos. 72 to 75), and substituted derivatives thereof. Also included in the table are the aldehyde, FpCHO (No. 1), the carboxylic acid, FpCOOH

(No. 76), the alkoxycarbonyl, $FpCOOR^1$ (Nos. 77 and 78), the carboxamido, $FpCONR^1R^2$ (Nos. 79 to 89), and the carbazoyl compounds $FpCONR^1NHR^2$ (Nos. 90 to 92).

The complexes $FpCOCH(R^1)CH_2CCl_3$ ($R^1 = H$, CH_3, C_2H_5, $n-C_6H_{13}$) are not listed in the table because they are only supposed to be formed as intermediates in the addition of CCl_4 and CO to the olefins $CH_2=CHR^1$ in the presence of Fp_2 to give $Cl_3CCH_2CH(R^1)COCl$ and $Cl_3CCH_2CH(R^1)Cl$. They have not been further characterized [31].

The methods used to prepare the compounds in Table 2 (see pp. 13/27) are described below.

Method I: Reaction of Na[Fp'] or K[Fp] with R'COCl (1:1 mole ratio) in THF at room tempera-
            ture [2 to 5, 22, 33, 48, 50, 54, 60, 63, 81, 86], the reaction times range from
            0.5 to 16 h. In some cases the reactants are mixed at 0 °C [56, 110, 114, 145]
            or −78 to −70 °C [8, 32, 35, 55, 57, 65, 90, 110, 116, 133, 142, 150, 161] before
            warming to room temperature. For the preparation of $FpCOC_3H_5$-cyclo, Na[Fp]
            and cyclo-C_3H_5COCl (1:2 mole ratio) are allowed to react in refluxing THF for
            3 h (56% yield) [25] or at −78 °C (65% yield) [140], or at 0 °C (1.5 h) to room
            temperature (45 min, 20% yield) [110]. Because the workup of the residue ob-
            tained after solvent removal varies so much for the individual compounds, they
            are described under further information.

Method II: A suspension of $[C_5H_5Fe(CO)_3]PF_6$ in THF or C_6H_6 is treated with $(CH_3)_2S(O)=CH_2$
            at 0 °C for 30 min or with $(C_6H_5)_3P=CHR^1$ (1:2 mole ratio) for 2 h. The solid
            ($[(CH_3)_3SO]PF_6$ and $[(C_6H_5)_3PCH_2R^1]PF_6$, respectively) is filtered off and the solu-
            tion evaporated to dryness. In the first case the orange–red residue is stirred
            in toluene for 20 min, filtered, washed with petroleum ether, dried, and recrystal-
            lized from THF/petroleum ether [64]. In the second case the remaining oil is
            eluted with pentane through a column and the yellow residue recrystallized
            twice from benzene/pentane [73]. For the preparation of $FpCOCH=P(CH_3)_3$,
            $(CH_3)_3P=CHSi(CH_3)_3$ is added to $[C_5H_5Fe(CO)_3]BF_4$ (3:1 mole ratio) in THF at
            −30 °C and the mixture is stirred at −20 °C for 2 h before warming to room
            temperature. Workup of the oily residue remaining after evaporation of the sol-
            vent from the filtrate is performed by extraction with toluene, addition of pentane,
            and cooling to −30 °C. The mechanism of formation is discussed [84].

Method III: Gaseous R^1NH_2 is bubbled into a mixture of $[C_5H_5Fe(CO)_3]PF_6$ in N_2-saturated
            ether for 30 min. $[R^1NH_3]PF_6$ formed is filtered off and the filtrate is evaporated
            until crystallization begins. Addition of pentane, cooling in a dry ice–acetone
            bath, and filtration afford the products $FpCONHR^1$. In the case of liquid R^1NH_2
            a tenfold excess of the amine is added to the ether solution and after filtration
            the solvent is evaporated to dryness. The residue is purified by dissolution
            in ether, filtration, and addition of pentane. The compounds $FpCONC_4H_8$-cyclo
            and $FpCONC_5H_{10}$-cyclo have been prepared similarly from pyrrolidine and piper-
            idine. Both complexes are purified by recrystallization from benzene/heptane
            [20]. The oily residue remaining after solvent removal from the reaction mixture
            of $[C_5H_5Fe(CO)_3]PF_6$ and cyclohexylamine is worked up by washing with water,
            extraction of the powder formed with warm ether, and cooling the extracts to
            −20 °C [40]. $FpCONH_2$ is obtained by the reaction of $[C_5H_5Fe(CO)_3]PF_6$ with liquid
            NH_3 in a cooled Schlenk tube at −40 °C for 10 min, followed by distillation of
            excess NH_3 at ∼ −50 °C, and separation of the product from $[NH_4]PF_6$ by dissolu-
            tion in C_6H_6 or CH_2Cl_2 [44].

Method IV: Reaction of FpR with $[C(C_6H_5)_3]BF_4$ (1:1 mole ratio) in CH_2Cl_2 under excess
CO (1 atm) at $-78\,°C$ for 1 h and at $-20\,°C$ for 16 h followed by addition of
ether to the concentrated solution at $0\,°C$ and chromatography of the concentrated filtrate on Al_2O_3 at $-10\,°C$, elution with 2% ether in pentane. The product
resulting from $FpCH_2CH_2CH(CH_3)_2$ is contaminated with a substance of unknown
structure [131]. For the preparation of $FpCOC_7H_{11}$ (Nos. 58 and 61) the reaction
is carried out at $25\,°C$ for 5 h and at $-20\,°C$ for 2 h, respectively, then the
mixture is cooled to $-78\,°C$, treated with ether and further worked up as above.
The exo-compound No. 61 is mixed with a trace of benzophenone [146].

Method V: Addition of $FpCOCH_3$ in THF to $LiN(Si(CH_3)_3)_2$ ($\sim$1:1 mole ratio) in the same
solvent at $-63\,°C$ (formation of $FpC(OLi)=CH_2$) followed after 15 min and cooling
to $-78\,°C$ by addition of R^1X ($R^1 = CH_3$, $CH(CH_3)_2$, C_4H_9-n, $CH_2CH=CH_2$,
$CH_2C(CH_3)=CH_2$, $CH_2C_6H_5$, $CH_2C_6H_4OCH_3$-4; $X = Br$, I, CF_3SO_3), $C_2H_5C_2H_3O$
($C_2H_4O = oxiran$; in the presence of $BF_3 \cdot O(C_2H_5)_2$), or C_2H_5CHO (1:1.2 mole ratio). The mixture is stirred at $-63\,°C$ for 3 h and after warming to $-20\,°C$ over
a period of 0.5 h quenched by addition of saturated aqueous NH_4Cl solution
and concentrated. The residue is eluted through Al_2O_3 with CH_2Cl_2, concentrated,
and purified by radial chromatography on SiO_2. The products are $FpCOCH_2R^1$
from R^1X, $FpCOCH_2CH_2CH(OH)C_2H_5$ from the oxirane, and $FpCOCH_2CH(OH)C_2H_5$
from C_2H_5CHO. Similarly, $FpCOCH(CH_3)_2$ has been obtained from $FpCOC_2H_5$ by
treatment with $LiN(Si(CH_3)_3)_2$ and CH_3I [141].

Method VI: A mixture of $FpCOCR^1=CR^2R^3$ and three to five equivalents of the corresponding
amine or thiol is stirred without solvent at $25\,°C$ for 48 h. Details of workup
or purification of the products (Nos. 94 to 105) were not given [150]; see also
[156].

FpCOR' and FpH are assumed to be intermediates during hydroformylation of propene
and pent-1-ene in toluene at 100 to $150\,°C$ under 70 to 140 atm (1000 to 2000 lb/in.^{2}) syngas
pressure ($CO/H_2 = 1:1$) in the presence of Fp_2, giving branched and linear aldehydes [134].

Addition of triphenylmethyl radical to the reaction mixtures obtained according to Method
IV inhibits the insertion of CO into FpR in the cases where $R = C_3H_7$-n and endo-norbornyl
and affords only a 13% yield of FpCOR' ($R = R'$) for $R = CH_2CH_2CH(CH_3)_2$. It is assumed
that the CO insertion (Method IV) follows a catalytic radical chain process initiated by a
one-electron transfer from FpR to the cation $[C(C_6H_5)_3]^+$ [131]. This assumption could be
confirmed by EPR studies and electrochemical measurements [146].

A linear relationship has been found between ^1HNMR chemical shifts of C_5H_5 and the
force constants k(CO) of FpCOR' and other cyclopentadienyl metal carbonyls [19].

^{13}C NMR studies of the compounds FpCOR' ($R' = NHCH_3$, NHC_6H_{11}-cyclo, and piperid-1-
yl) reveal that the carbamoyl chemical shift is affected only modestly by changes in the
amido group. However, it lies at higher field than the corresponding acetyl carbon, for
example, in $FpCOCH_3$ [120].

Compounds No. 10, 11, and 18 are moderately soluble in pentane, but readily soluble
in benzene and polar solvents [73].

Solutions of the compounds No. 25, 34, and 35 in CH_2Cl_2 are stable at $-20\,°C$ under
CO for extended periods [131].

Compounds No. 72 to 75 are stable as solids but less stable in solution. They are soluble
in the common organic solvents [63].

References on pp. 47/51

In the solid state the carboxamido complexes No. 79 to 82, 84, 85, 88, and 89 slowly decompose in air but can be stored for several weeks in the dark under vacuum. Their stability in solution decreases as the size of the substituents in the amide group increases. Upon melting considerable decomposition occurs yielding Fp$_2$. The complexes react with water to give yellow, basic (pH ~9 to 10) solutions that in a few minutes turn dark, producing a brown precipitate [20]. The presence of the free amines $(CH_3)_2NH$ or $(C_2H_5)_2NH$ in the aqueous solutions of FpCON(CH$_3$)$_2$ and FpCON(C$_2$H$_5$)$_2$ could be evidenced by their precipitation as tetraphenylborates. The iron appeared to be present in a brown, insoluble residue [4].

The cyclic voltammograms of FpCOR′ (Pt electrode, CH$_3$CN/0.1 M [N(C$_2$H$_5$)$_4$]ClO$_4$, 25 °C) show only an anodic peak at E$_p$ = 1.20 V (vs. NaCl/SCE) for R′ = CH$_3$ and 1.22 V for R′ = C$_6$H$_5$. The oxidation was irreversible, even at sweep rates as high as 10 V/s. The resultant acyliron cation radicals [C$_5$H$_5$Fe(CO)$_2$COR′]$^{+\cdot}$ are highly labile. They may be trapped by ethanol to yield organic esters [88]. The electrochemical oxidation of FpCOR′ (R′ = CH$_3$, C$_6$H$_5$) is reversible ([123]; see also [87]) giving the corresponding radical cations which slowly decompose into [C$_5$H$_5$Fe(CO)(CH$_3$CN)COR′]$^+$, whereas the oxidations of FpCOR′ (R′ = C$_2$H$_5$, CH$_2$CH(CH$_3$)$_2$, CH$_2$C$_6$H$_5$) are irreversible at all accessible scan rates and temperatures [123]. The cyclic voltammogram of FpCOCH$_3$ is reported to exhibit scan rate and external CO dependencies, the CO dependence suggests that the ligand substitution process in the cation radical proceeds via a coordinatively unsaturated intermediate [124].

Chemical oxidation of FpCOR′ (R′ = CH$_3$, C$_6$H$_5$) by AgPF$_6$ or AgBF$_4$ in CH$_2$Cl$_2$ gives the carbene complexes [C$_5$H$_5$(CO)$_2$Fe=C(R′)OFe(CO)$_2$C$_5$H$_5$]PF$_6$, whereas from FpCOR′ (R′ = C$_2$H$_5$, CH$_2$CH(CH$_3$)$_2$, CH$_2$C$_2$H$_5$) the complexes [Fp=C(OR′)R′]PF$_6$ and [C$_5$H$_5$Fe(CO)$_3$]PF$_6$ are obtained. Only [C$_5$H$_5$Fe(CO)$_3$]PF$_6$ has been isolated from the oxidation of FpCOC$_3$H$_5$–cyclo or FpCOC$_3$H$_7$–i [123]. Reduction of the red–orange oil resulting from solid–state oxidation of FpCOCH$_3$ by AgPF$_6$, with zinc powder affords FpCOCH$_3$ (20%) and the carboxonium salt [FpCOCH$_3$Fp]$^+$ (13%) [124].

The compounds FpCOR′ are decarbonylated by treatment with Rh(P(C$_6$H$_5$)$_3$)$_3$Cl in CH$_2$Cl$_2$ (Nos. 2, 66, and 70 [78]), benzene (Nos. 23 [22, 33] and 65 [60]), or CH$_2$Cl$_2$/benzene at 25 to 30 °C for 1 to 5 h (Nos. 2, 23, and 66 [21, 48]), or in THF overnight (Nos. 39 and 41 [90]) to give the compounds FpR′. Other isolated products are C$_5$H$_5$Fe(CO)(P(C$_6$H$_5$)$_3$)COR′ and Rh(P(C$_6$H$_5$)$_3$)$_2$(CO)Cl. Only traces of FpC(CH$_3$)$_3$ in addition to ferrocene, C$_5$H$_5$Fe-(CO)(P(C$_6$H$_5$)$_3$)COC(CH$_3$)$_3$, and Rh(P(C$_6$H$_5$)$_3$)$_2$(CO)Cl have been obtained from the reaction of FpCOC(CH$_3$)$_3$ with the rhodium complex in refluxing benzene for 0.5 h [60]. No detectable decarbonylation occurred by reacting FpCOC$_3$H$_5$–cyclo with Rh(P(C$_6$H$_5$)$_3$)$_3$Cl in CH$_2$Cl$_2$ for 24 h [50], see also [25], and Rh(P(C$_6$H$_5$)$_3$)$_2$(CO)Cl is mentioned in [48] as the only product which could be isolated from the reactions of FpCOR′ (R′ = C(C$_6$H$_5$)$_2$CH$_3$, C(C$_6$H$_5$)$_3$, (E)–CH= CHC$_6$H$_5$) with Rh(P(C$_6$H$_5$)$_3$)$_3$Cl in benzene/CH$_2$Cl$_2$ (but see above the reaction of (E)–FpCOCH= CHC$_6$H$_5$ (No. 39) in THF [90]). It has also not been possible to convert FpCOCF$_3$ (No. 3) into FpCF$_3$ [21, 48]. To avoid the formation of the P(C$_6$H$_5$)$_3$–substituted acyliron complexes, [Rh(P(C$_6$H$_5$)$_3$)$_2$Cl]$_2$ in CH$_3$CN (20 h) has been employed as the decarbonylation agent for FpCOR′ (R′ = C$_6$H$_5$, C$_6$H$_4$OCH$_3$–4). However, attempts to abstract CO from FpCOCF$_3$ by [Rh(P(C$_6$H$_5$)$_3$)$_2$Cl]$_2$ have also been unsuccessful [78]. The observed pattern of ^{13}CO distribution in the products (characterized by IR spectroscopy) from the reaction of Fp^{13}COCH$_3$ (No. 2a) with Rh(P(C$_6$H$_5$)$_3$)$_3$Cl reveals a reaction mechanism similar to that for the photochemical decarbonylation, that is, formation of the coordinatively unsaturated C$_5$H$_5$Fe(CO)^{13}COCH$_3$ which then rearranges via a methyl group migration to C$_5$H$_5$Fe(CO)(^{13}CO)CH$_3$ or, alternatively, combines with P(C$_6$H$_5$)$_3$ yielding C$_5$H$_5$Fe(CO)(P(C$_6$H$_5$)$_3$)^{13}COCH$_3$. The Rh(P(C$_6$H$_5$)$_3$)$_2$(CO)Cl formed contains only the natural abundance of ^{13}CO thus indicating that the abstracted

CO must have been a terminal one [21, 48]. A comparison shows that the yields of the decarbonylated and $P(C_6H_5)_3$-substituted products depend on the nature of R′ in FpCOR′, reflecting the relative rates of migration and substitution, respectively. Thus, the CH_3 group migrates rather readily and only a small amount of the substitution product C_5H_5Fe-(CO)($P(C_6H_5)_3$)COCH$_3$ is observed, whereas both substitution and migration occur at comparable rates in FpCOC$_6$H$_5$ [48]. The stereospecificity of the reaction has been investigated on $(-)_{546}$-FpCOCH(C_6H_5)CH$_3$ ([α]$_{546}^{27}$ = $-64.3°$ in CHCl$_3$) which reacts with the rhodium complex to give $(+)_{546}$-FpCH(C_6H_5)CH$_3$ ([α]$_{546}^{27}$ = $+78.4°$ in CHCl$_3$) in 54% yield. Because the carbonylation reaction of threo-FpCHDCHDC(CH$_3$)$_3$ with $P(C_6H_5)_3$ leads to the threo-acyl complex with >95% retention of configuration [29], it is concluded that $(-)_{546}$-FpCOCH(C_6H_5)CH$_3$, with the R configuration, retains the same absolute configuration at the α-carbon on decarbonylation [48].

Giving low yields of FpR, the CO abstraction from FpCOR′ (R = R′ = CH$_3$, C$_6$H$_5$) also takes place with Ir($P(C_6H_5)_3$)$_3$Cl in benzene at room temperature or with Ru($P(C_6H_5)_3$)$_3$Cl$_2$ in refluxing benzene (8 to 12 h). The complexes Ir($P(C_6H_5)_3$)$_2$(N_2)Cl, Pt($P(C_6H_5)_3$)$_4$, and Pt($P(C_6H_5)_3$)$_3$, however, failed to show any decarbonylation activity [78].

UV irradiation of FpCOR′ in the presence of $P(C_6H_5)_3$ in cyclohexane for 60 h (No. 50 [25]) or 16 h (No. 66 [13, 16]) results in substitution of one CO group and formation of the acyl compounds C_5H_5Fe(CO)($P(C_6H_5)_3$)COR′. In contrast, photolysis of $P(C_6H_5)_3$ and FpCOC(CH$_3$)=CH$_2$ (No. 42) in hexane/benzene (19:1) for 2.5 h [114, 145] or FpCOCH$_2$CH(CH$_3$)-C$_6$H$_5$ (No. 29) in benzene for 1 h in a Pyrex tube gives the (ar)alkyl derivatives C_5H_5Fe-(CO)($P(C_6H_5)_3$)R′. The product with R′ = CH$_2$CH(CH$_3$)C$_6$H$_5$ is obtained as an optically active compound when $(-)_D$-FpCOCH$_2$CH(CH$_3$)C$_6$H$_5$ is used as the starting material [57]. C_5H_5Fe-(CO)($P(OCH_3)_3$)R′ (R′ = (E)-CH=CHC(CH$_3$)=CH$_2$, (E,E)-CH=CHCH=CHCH$_3$) result from photolysis of the respective acyl complexes and $P(OCH_3)_3$ in ether/benzene at -5 °C [133]. Both C_5H_5Fe(CO)($P(C_6H_5)_3$)COR′ and C_5H_5Fe(CO)($P(C_6H_5)_3$)R′ form on irradiation of equimolar amounts of FpCOR′ and $P(C_6H_5)_3$ in hexane (Nos. 2a [58] and 3 [28]) or in ether for 1.5 h (Nos. 72 to 75 [63]). On the basis of the nature and extent of labeling of the products in the reaction of Fp^{13}COCH$_3$ (No. 2a) with $P(C_6H_5)_3$ it was demonstrated that the intermediate C_5H_5Fe(CO)^{13}COCH$_3$, formed by irradiation, undergoes nucleophilic attack by $P(C_6H_5)_3$ faster than methyl migration [58].

The reaction of FpCOR′ with $P(C_6H_5)_3$ carried out by heating without UV irradiation either does not occur (No. 50 [25]) or proceeds only very slowly (No. 65; 6 days in refluxing benzene [60]) to give C_5H_5Fe(CO)($P(C_6H_5)_3$)COR′.

UV irradiation of FpCOR′ leads to decarbonylation and formation of FpR′ in hexane (Nos. 2 [8], 3 [28], 28 [111]), pentane (Nos. 58 to 61 [146], 65 [60]), hexane/toluene (No. 42 [114, 145]), acetone (Nos. 39, 41 [90]), acetone-d$_6$ at 0 °C (Nos. 64 [142], 38, 40, 42 [113]), ether (Nos. 72 to 75 [63]), benzene (No. 66 [66]), ether/benzene at -5 °C (Nos. 46, 49 [133]), THF (Nos. 39 [36, 37, 86], 68 [32]), or petroleum ether (b.p. 30 to 60 °C; No. 56 [86]).

The reaction times range from 15 min to 16 h. In the cases where R′ is CH=CHC$_6$H$_5$ (No. 39) [36, 37] and adamant-1-yl (No. 65) [60], Fp$_2$ has been formed in addition to FpR′. For attempts to decarbonylate FpCOC$_3$H$_5$-cyclo, see No. 50, under further information [25, 50, 110]. The photoinduced decarbonylation of the latter and related compounds (Nos. 52, 53, and 57) gives fair to excellent yields of the respective FpR′ complexes when acetone-d$_6$ is used as the solvent. However, the yields are significantly lower (16% FpC$_3$H$_5$-cyclo) for petroleum ether (solvent used in [25]) and lie between acetone-d$_6$ and petroleum ether for benzene-d$_6$ [110].

Attempts at photochemical decarbonylation of $FpCOCH(CH_3)C_6H_5$ (No. 23) led only to decomposition [48]. Photolysis of $Fp^{13}COCH_3$ in hexane, giving $C_5H_5Fe(CO)(^{13}CO)CH_3$ indicates that the decarbonylation proceeds via loss of a terminal CO group. A mechanism involving formation of an unsaturated intermediate, that is, $C_5H_5Fe(CO)^{13}COCH_3$, followed by methyl migration was proposed [58]. Evidence for radical formation on photolysis of FpCOR' ($R' = CH_3$, $CH_2C_6H_5$) has been noted in ESR studies carried out in CH_2Cl_2 at $-30\,°C$ with R^1NO ($R^1 = C_6H(CH_3)_4$-2,3,5,6) as the spin trap. ESR signals assignable to adducts derived from both R'' ($R^1NO^.R'$, weak signal) and $Fp^.$ ($R^1NO^.Fp$, strong signal) are observed for $R' = CH_2C_6H_5$. However, when $R' = CH_3$, the signal due to $R^1NO^.Fp$ is weak and that from $R^1NO^.R'$ could not be observed. Spin-trapped acyl radicals have never been detected [106].

NMR studies on (E)-$FpCOCH=CHR^1$ ($R^1 = CH_3$, C_6H_5) have shown that the photochemical decarbonylation proceeds with retention of the olefin geometry [90]. Also, the irradiation of compounds No. 58 to 61 gives the corresponding norbornyl complexes of retained configuration [146]. For photolysis in frozen gas matrices and polyvinyl chloride film matrices, see "Further information" on $FpCOCH_3$ (No. 2).

Irradiation of FpCOR' (Nos. 67, 73, 74, and 75) and $C_6H_5C\equiv CC_6H_5$ (1:1 mole ratio) in benzene for 4 h gives in addition to the corresponding decarbonylation products FpR (R = R') the compounds I to IV. Under similar conditions compound No. 72 does not react with $C_6H_5C\equiv CC_6H_5$. The only product obtained was the decarbonylated derivative. A possible mechanism leading to formation of the products I to IV was discussed [159]. Studies (^{13}C NMR and mass spectra) on compound No. 75 labeled with ^{13}C at the acyl group revealed that only the acyl CO group and none of the terminal CO groups participates in the cyclization of the thenoyl ligand with $C_6H_5C\equiv CC_6H_5$ [117].

The reaction of $FpCONHR^1$ ($R^1 = CH_3$, C_6H_{11}-cyclo) with halogens ($X_2 = Cl_2$, Br_2, or I_2) in pentane or hexane proceeds with cleavage of the Fe-carboxamido bond according to the equation: $3C_5H_5Fe(CO)_2CONHR^1 + 2X_2 \rightarrow C_5H_5Fe(CO)_2X + [C_5H_5Fe(CO)_3]X + 2R^1NCO + [R^1NH_3]X$ [34, 40]. A mechanism like Hofmann's rearrangement of primary carbonic acid amides is proposed for the reaction of $FpCONH_2$ with I_2 in benzene: $2C_5H_5Fe(CO)_2CONH_2 + I_2 \rightarrow C_5H_5Fe(CO)_2NCO + [C_5H_5Fe(CO)_3]I + NH_4I$. Because NH_4I reacts further with FpNCO, the compounds FpI and NH_4NCO are always formed as byproducts [44].

Treatment of $FpCONHR^1$ ($R^1 = CH_3$, C_2H_5, i-C_3H_7, n-C_4H_9) in ether [20] or of $FpCONH_2$ in benzene [44] with gaseous HCl gives a pale yellow precipitate of $[C_5H_5Fe(CO)_3]Cl$ and $[R^1NH_3]Cl$ or NH_4Cl, respectively.

References on pp. 47/51

Addition of CH_3Li in ether to FpCOR' ($R' = CH_3$, $CH(C_6H_5)_2$ [65], $i-C_3H_7$ [100]) in the same solvent at -55 to -40 °C/0.5 h or to $FpCOCH_3$ in THF at -50 to -30 °C/1 h [121] (see also [45]) leads to formation of the corresponding β-diketonate anions V and VI.

$$
\left[\begin{array}{c} R' \\ \backslash \\ C{=}O \\ / \\ C_5H_5(CO)Fe{=}C \\ \backslash \\ C{=}O \\ / \\ R^1 \end{array} \right]^{-} Li^{+}
$$

$$
\begin{array}{l} V \quad R^1 = CH_3 \\ VI \quad R^1 = H \end{array}
$$

A mechanism involving deprotonation of the β-diketonate anion by lithium tetramethylpiperidide with concomitant interligand C–C bond coupling to give the dianion $C_5H_5Fe(CO)(\eta^3\text{-}CH_2C(O^-)C(O^-)CH_3)$ as intermediate was proposed for the latter reaction [121].

The reactions of FpCOR' ($R' = CH_3$, C_6H_5, $C_6H_4OCH_3$) with 1.2 equivalent of $Li[BH(C_2H_5)_3]$ in THF at -50 °C afford anionic metal formyl complexes VI, which are unstable at room temperature; VI ($R' = C_6H_5$) decomposes, producing Fp_2, $[Fp]^-$, and $C_6H_5CH_2OH$. The rates of thermal decomposition were monitored and found to be first-order. A possible reaction mechanism was discussed [75].

Additions take place on treatment of $FpCOCR^1{=}CR^2R^3$ (Nos. 41, 42, and 93) with amines or thiols to give the compounds No. 94 to 105 (see Method VI). Lewis acids or bases did not accelerate the reaction, but rather promoted the Fp_2 formation [150], see also [156].

Alkylation of FpCOR' ($R' = CH_3$, C_2H_5, $i-C_3H_7$) with $CH_3OSO_2CF_3$ in CH_2Cl_2 at 25 °C (24 h) generates the carbene complexes $[Fp{=}C(OCH_3)R']^+$, which on subsequent reduction with $NaBH_4$ in $CH_3OH/NaOCH_3$ give the ethers $FpCH(OCH_3)R'$ [151]; see also [98].

Acyliron phosphorus ylides should be able to add electrophiles on the ylid carbon as well as on the acyl oxygen. Both reaction types are observed when the ylides $FpCOCR^1{=}PR_3^2$ are treated with CH_3OSO_2F in benzene at 8 to 10 °C/2 h. Whereas $FpCOCH{=}P(CH_3)_3$ yields the phosphonium salt $[FpCOCH(CH_3)P(CH_3)_3]SO_3F$, $FpCOC(CH_3){=}P(C_2H_5)_3$ undergoes exclusively O-alkylation to give the phosphoniovinyl salt $[FpC(OCH_3){=}C(CH_3)P(C_2H_5)_3]SO_3F$. A mixture of both products is obtained from $FpCOC(CH_3){=}P(CH_3)_3$ and $FpCOCH{=}P(C_2H_5)_3$, the C-methylated component predominates in the former case and the O-methylated product in the latter case. The difference in the reaction products is explained by the increasing inductive donor effect of the ligands at the P atom and the ylid carbon in the series $FpCOCH{=}P(CH_3)_3$, $FpCOC(CH_3){=}P(CH_3)_3$, $FpCOCH{=}P(C_2H_5)_3$, and $FpCOC(CH_3){=}P(C_2H_5)_3$, favoring the phosphonioenolate structure (cf. Formula X on p. 34) [155].

The compounds FpCOR' ($R' = CH_3$, C_6H_5) are effective antiknock agents and can be used as additives to petroleum lubricants and as vapor-phase metal plating agents. They also find utility in biocidal applications, for example, as fungicides, herbicides, and pesticides [3].

Remarks Concerning Table 2 on pp. 13/27: Up to four mostly strong IR bands in the range 1920 to 2060 cm^{-1} result from the terminal CO groups and from the acyl group generally one medium-intensity band in the range 1450 to 1650 cm^{-1} results. Force constants (in mdyn/Å) are designated as k for the stretching constant k(CO) and k' for the interaction constant k(CO,CO). They were calculated by the Cotton–Kraihanzel method.

References on pp. 47/51

Table 2

Oxo Compounds of the Type $C_5H_5Fe(CO)_2COR'$.

Further information on compounds with numbers preceded by an asterisk is given at the end of the table.

For abbreviations and dimensions, see p. X.

No.	compound method of preparation (yield)	properties and remarks
*1	FpCHO	^1H NMR (CD_3COCD_3, $-80°$): 5.38 (C_5H_5), 14.25 (CHO) [136]
*2	FpCOCH$_3$ I (60%) [2, 3], (63%) [4], (78%) [151]	orange crystals, m.p. 53 to 56° [82], 56 to 59° [3, 4, 9, 12, 94], subl. at 80°/0.01 Torr [82] ^1H NMR: 2.03 (CH_3), 3.77 (C_5H_5) [82], 2.42 (s, CH_3), 4.88 (s, C_5H_5) [151] in C_6D_6, 2.35 (s, CH_3), 4.13 (s, C_5H_5) in $C_6D_5CD_3$ [95], 2.44, 4.25 in CD_3COCD_3 [61], 2.42, 4.77 [15], 2.57, 4.87 [12] in CS_2, 2.50, 4.93 [102], 1.66 (?), 4.97 [61] in C_4D_8O, 2.58, 5.03 in $CDCl_3$ [127], for data measured in $CHCl_3$, see [4, 43] ^{13}C NMR: 52.1 (CH_3), 86.7 (C_5H_5), 215.0 (CO), 253.3 (C=O) in $CDCl_3$ [79], similar in $CHCl_3$ [38], 87.1 (C_5H_5), 215.4 (CO) in CH_2Cl_2 [46] ^{57}Fe-γ (78 K): $\delta = 0.019$ (^{57}Co/Cr), $\Delta = 0.168$ [6] IR: 1954, 2014 (both CO), 1659 (C=O) in THF [45], 1958, 2017 (both CO), 1661 (C=O) in $C_6H_5CH_3$ [95], 1961, 2009 (both CO), 1667 (C=O) in $C_{14}H_{30}$ [76], 1962, 2010 (both CO), 1663 (C=O) in Nujol [127], 1963, 2018 (both CO), 1655 (C=O) in halocarbon oil mull [4], 1963, 2015 (both CO) [46], 1960, 2010 (both CO), 1638, 1680 (both C=O) [151] in CH_2Cl_2, 1964, 2022 (both CO), 1669 (C=O) in C_6H_{14} [45], similar in $CHCl_3$ [47], 1969, 2035 (both CO), 1673 (C=O) in C_6H_{12} [12], 2900, 2940, 3050 (all CH), other bands at 828, 840, 885, 910, 930, 945, 1004, 1018, 1065, 1318, 1352, 1415, 1430 in KBr [4], 1919, 1954, 2013 (all CO), 1657 (C=O) in polystyrene [162], 1936, 2015 (both CO), 1655 (C=O) [79] $k = 16.01$ [46, 47], $k' = 0.386$ [46], 0.48 [47], $k = 16.19$, $k' = 0.53$ [19] UV (CH_3CN): 322 [87]
*2a	Fp^{13}COCH$_3$ I (65%) [48]	yellow [48] IR (C_6H_{14}): 1968, 2021 (both CO), 1636 (^{13}C=O), 1671 (^{12}C=O) [58], 1973, 2016 (both CO), 1629, 1668 (both C=O) [48]
*3	FpCOCF$_3$	orange crystals, m.p. 33 to 34° [8] ^1H NMR ($CHCl_3$, containing Freon 112 and $(CH_3)_3SiO$-$Si(CH_3)_3$): 4.91 (C_5H_5) [8] ^{13}C NMR ($CDCl_3$): 86.8 (C_5H_5), 114.6 (CF_3), 212.1 (CO), 243.6 (C=O) [79] ^{19}F NMR ($CHCl_3$, containing Freon 112 and $(CH_3)_3SiO$-$Si(CH_3)_3$): 80.1 [8]

References on pp. 47/51

Table 2 (continued)

No.	compound method of prepara- tion (yield)	properties and remarks
*3 (continued)		IR: 1994, 2045 in CH_2Cl_2 [48], 1995, 2046 (all CO), 1654 (C=O) in halocarbon oil mull, 3090 (CH), 1125, 1175, 1220 (all CF_3), other bands at 686, 716, 840 to 860 br, 940 br, 1000, 1015, 1060, 1415, 1430 in KBr [8] k = 16.48, k' = 0.42 [19] UV (C_6H_{12}): 242 (ε = 10 100), 321 (ε = 2900) [8]
*4	$FpCOCH_2OCH_3$ I (100%) [55]	amber oil [55] ^1H NMR ($CDCl_3$): 3.38 (s, CH_3), 4.0 (s, CH_2), 4.91 (s, C_5H_5) [55] IR (KBr): 1960, 2040 (both CO), 1660 (C=O) [55]
5	$FpCOCH_2OC_2H_5$	acid–induced isomerization to $FpCH_2COOC_2H_5$ is mentioned in [147]; no other data given
*6	$FpCOCHS(O)(CH_3)_2$ II (78%) [64]	yellow, m.p. 76 to 78° [64] ^1H NMR: 3.20 (s, 6H, CH_3), 4.90 (s, 1H, CH), 4.90 (s, 5H, C_5H_5) in CD_3CN, 3.36, 4.81, 4.94 in CD_3COCD_3 [64] ^{57}Fe-γ (78 K): δ = 0.320, Δ = 1.66 [64] IR: 1943, 2010 (both CO), 1525 (C=O), other bands at 1058 sh, 1070, 1175 br in CH_2Cl_2, 1923, 1990, 1996 (all CO), 1500, 1528 (both C=O), other bands at 1072, 1090, 1148 in Nujol [64]
*7	$FpC^1OC^2H=P(CH_3)_3$ II (67%) [84]	amber "ashlars", m.p. 93° (dec.) [84, 109] ^1H NMR (C_6H_6): 0.97 (d, 9H, PCH_3, ^2J (H, P) = 14), 3.55 (d, 1H, HC=P, ^2J (H, P) = 37), 4.74 (s, 5H, C_5H_5) [84] ^{13}C {^1H} NMR (C_6D_6): 12.6 (d, PCH_3, J (C, P) = 57.4), 86.6 (s, C_5H_5), 218.9 (s, CO) [84] ^{31}P {^1H} NMR (C_6D_6): −10.2 [84] IR (C_6H_6): 1930, 1995 (both CO), 1510 (C=O) [84]
8	$FpCOCH=P(C_2H_5)_3$ II [155]	
*9	$FpCOCH=P(CH_3)_2C_2H_5$	yellow-orange crystals, m.p. 72° (dec.) [109] ^1H NMR (C_6H_6): 0.73 (dt, 3H, PCH_2CH_3, ^3J (H, P) = 16.4), 0.97 (d, 6H, PCH_3, ^2J (H, P) = 13.3), 1.45 (dq, 2H, PCH_2, ^2J (H, P) = 12.2, ^3J (H, H) = 7.5), 3.45 (d, 1H, P=CH, ^2J (H, P) = 32.4), 4.72 (s, 5H, C_5H_5) [109] ^{31}P NMR (C_6D_6): −1.8 [109] IR (C_6H_6): 1995 (CO), 1505 (C=O) [109]
10	$FpCOCH=P(C_6H_5)_3$ II (56%) [73]	yellow, m.p. 125 to 126° [73] ^1H NMR (C_6D_6): 4.03 (d, P=CH, J (H, P) = 33.87), 4.40 (s, C_5H_5), 6.8 to 7.8 (m, C_6H_5) [73] ^{31}P NMR (C_6H_6): 7.4 [73] IR (KBr): 1926, 1992 (both CO), 1498 (C=O) [73]

References on pp. 47/51

Table 2 (continued)

No.	compound method of prepara- tion (yield)	properties and remarks
11	FpCOC(C_6H_5)=P(C_6H_5)$_3$ II (55%) [73]	yellow, m.p. 122 to 123° [73] ^{1}H NMR (C_6D_6): 4.22 (s, C_5H_5), 6.7 to 7.6 (m, C_6H_5) [73] ^{31}P NMR (C_6H_6): +13.0 [73] IR (KBr): 1932, 1993 (both CO), 1442 (C=O) [73]
*12	FpCOCH$_2C_6H_5$ I (36%) [54]	yellow, m.p. 77.5 to 78° [54] ^{1}H NMR: 4.10 (s, CH$_2$), 4.75 (s, C_5H_5), 7.22 (m, C_6H_5) [54] IR (C_6H_{12}): 1962, 2022 (both CO) [54]
*13	FpCOCH(C_6H_5)$_2$ I (57%) [65]	golden brown, m.p. 115 to 116° [65] ^{1}H NMR (CS$_2$): 4.61 (s, C_5H_5), 5.54 (s, CH), 7.11 (m, 10H, C_6H_5) [65] IR (C_6H_{12}): 1958, 2005 (both CO), 1655 (C=O) [65]
*14	FpCOC(C_6H_5)$_3$ I [48]	IR (CH$_2$Cl$_2$): 1958, 2038 (both CO) [48]
*15	FpCOC$_2H_5$ I [5], (86%) [151] V (84%) [141]	orange-red oil [5] ^{1}H NMR: 0.84 (CH$_3$), 2.72 (CH$_2$), 4.44 (C_5H_5) in C_6H_6 [5], 1.05 (t, CH$_3$, J=8), 2.78 (q, CH$_2$, J=8), 4.25 (s, C_5H_5) in C_6D_6 [151] IR: 1955, 2005 (both CO), 1637 (C=O) in CH$_2$Cl$_2$ [151], 1960, 2026 (both CO), 1657 (C=O) in KBr (?) [5]
*16	FpCOC$_2F_5$ I (12%) [8]	orange crystals, m.p. 53 to 57° [7, 8, 39] somewhat air-sensitive [7] ^{1}H NMR (CDCl$_3$): 5.02 (C_5H_5) [39]; see also [8] ^{19}F NMR (CHCl$_3$, containing Freon 112 and (CH$_3$)$_3$SiO- Si(CH$_3$)$_3$): +80.3 (CF$_3$), +113.6 (CF$_2$) [8] ^{57}Fe–γ (25°): δ=0.13 (Fe/Cr), Δ=1.78 [30] IR: 1996, 2044 in C_6H_{12} [39], 1998, 2048 (all CO), 1659 (C=O) in halocarbon oil mull, 3120 (CH) in KBr, other bands 714 to 1570 in KBr [8] k=16.53, k'=0.41 [19] UV (C_6H_{12}): 246 (ε=11600), 324 (ε=3230) [8] attempts to decarbonylate failed [7]
*17	FpCOC(CH$_3$)=P(CH$_3$)$_3$	yellow-orange crystals, m.p. 94° (dec.) [109] ^{1}H NMR (C_6H_6): 1.04 (d, 9H, PCH$_3$, ^{2}J (H,P)=13.2), 2.02 (d, 3H, P=C(CH$_3$), ^{3}J (H,P)=15), 4.70 (s, 5H, C_5H_5) [109] ^{31}P NMR (C_6D_6): −3.4 [109] IR (C_6H_6): 1932, 1995 (both CO), 1505 (C=O) [109]
18	FpCOC(CH$_3$)=P(C_6H_5)$_3$ II (50%) [73]	yellow, m.p. 120 to 121° [73] ^{1}H NMR (C_6D_6): 1.91 (d, =CCH$_3$, ^{3}J (H,P)=16.0), 4.33 (s, C_5H_5), 6.8 to 7.7 (m, C_6H_5) [73] ^{31}P NMR (C_6H_6): 15.4 [73] IR (KBr): 1924, 1993 (both CO), 1461 (C=O) [73]

References on pp. 47/51

Table 2 (continued)

No.	compound method of prepara- tion (yield)	properties and remarks
*19	FpCOCH$_2$CH$_2$—	yellow oil, 1:1 mixture with No. 20 [118] spectra of the mixture: ^{1}H NMR (C$_6$D$_6$): 2.52 to 2.80 (m, 4H, CCH$_2$C= and ring–CH$_2$), 2.91 to 3.16 (m, 2H, CH$_2$CO), in both cases 2 groups of signals in a 1:1 ratio, 4.12 (m, 5H, C$_5$H$_5$), 5.90 to 6.50 (m, 3H, –CH=) [118] IR (neat): 1950, 2015 (both CO), 1640 (C=O) [118]
*20	FpCOCH$_2$CH$_2$—	yellow oil, 1:1 mixture with No. 19 [118] spectra given under No. 19
21	FpCOCH$_2$CH$_2$C$_6$H$_5$ V (56%) [141]	—
22	FpCOCH$_2$CH$_2$C$_6$H$_4$OCH$_3$–4 V (96%) [141]	—
*23	FpCOCH(CH$_3$)C$_6$H$_5$ I (32%) [33]	yellow crystals, m.p. 57 to 59° [33] ^{1}H NMR (CDCl$_3$): 1.32 (d, CH$_3$, J=7), 4.35 (q, CH, J=7), 4.72 (s, C$_5$H$_5$), 7.36 (s, C$_6$H$_5$) [22, 33] IR (CH$_2$Cl$_2$): 1960, 2015 (both CO), 1623, 1645 (both C=O) [33, 48]
	(−)$_{546}$-compound I [33]	yellow crystals at −78°, yellow-brown oil at room tempera- ture [33] $[\alpha]_{546}^{27} = -64.3 \pm 0.4°$ (CHCl$_3$) [33, 48], −130° (CHCl$_3$) [22]
24	FpCOC(C$_6$H$_5$)$_2$CH$_3$ I [48]	IR (CH$_2$Cl$_2$): 1960, 2040 [48] workup as No. 2a [48]
*25	FpCOC$_3$H$_7$-n IV (13%) [131]; see also [152]	^{1}H NMR (CD$_2$Cl$_2$): 0.77 (t, CH$_3$, J=7.1), 1.34 (~sext, CCH$_2$C, J≈7.2), 2.87 (t, COCH$_2$, J=7.2), 4.85 (s, 5H, C$_5$H$_5$) [131] ^{13}C {^{1}H} NMR (CDCl$_3$): 13.59 (CH$_3$), 18.67 (CCH$_2$C), 68.63 (COCH$_2$), 86.38 (C$_5$H$_5$), 214.5 (CO), 259.52 (C=O) [131] IR (CH$_2$Cl$_2$): 1957, 2012 (both CO), 1639 (C=O) [131] mass spectrum: [M]$^+$, [M−CO]$^+$, [M−2CO]$^+$, [M−C$_3$H$_7$]$^+$, [M−COC$_3$H$_7$]$^+$ [131]
*26	FpCOC$_3$H$_7$-i I (56%) [65], (91%) [151] V (90%) [141]	orange needles, m.p. 13 to 14° [65] ^{1}H NMR: 0.91 (d, CH$_3$, J=6.9), 2.94 (sept, CH, J=6.9), 4.83 (s, C$_5$H$_5$) in CS$_2$ [65], 1.03 (d, CH$_3$, J=7), 3.01 (sept, CH, J=7), 4.37 (s, C$_5$H$_5$) in C$_6$D$_6$ [151] IR: 1950, 2003 (both CO), 1662 (C=O) in C$_6$H$_{12}$ [65], 1950, 2010 (both CO), 1618, 1665 (both C=O) in CH$_2$Cl$_2$ [151]
*27	FpCOC$_3$F$_7$-n I (11%) [8]	orange liquid [8, 39], solidifies at −15° [8] ^{1}H NMR (CHCl$_3$, containing Freon 112 and (CH$_3$)$_3$SiO- Si(CH$_3$)$_3$): 4.98 (C$_5$H$_5$) [8]

References on pp. 47/51

Table 2 (continued)

No.	compound method of prepara- tion (yield)	properties and remarks
		^{19}F NMR (CHCl$_3$, containing Freon 112 and (CH$_3$)$_3$SiO- Si(CH$_3$)$_3$): 81.0 (CF$_3$), 110.7 (COCF$_2$), 126.2 (CCF$_2$C) [8] IR: 1980, 2030 (both CO), 1645 (C=O), 3100 (CH), other bands 688 to 1430 as liquid film [8], 1997, 2044 (both CO) in C$_6$H$_{12}$ [39] k = 16.23, k' = 0.40 [19]
*28	FpCO(CH$_2$)$_3$Cl I (71%) [56]	red oil [56] ^{1}H NMR (C$_6$D$_6$): 1.74 ("q", 2H, CH$_2$, J = 7.0, 7.0), 2.85 (t, 2H, CH$_2$), 3.25 (t, 2H, COCH$_2$), 4.25 (s, 5H, C$_5$H$_5$) [56] IR (C$_6$H$_{14}$): 1967, 2026 (both CO), 1656 (C=O) [56]
*29	FpCOCH$_2$CH(CH$_3$)C$_6$H$_5$ I (72%) [57]	yellow crystals, m.p. 106 to 107° [57] ^{1}H NMR (CDCl$_3$): 1.19 (d, CH$_3$, J = 7), 3.19 (m, complex, CH$_2$, CH), 4.65 (s, C$_5$H$_5$), 7.20 (s, C$_6$H$_5$) [57] IR (CH$_2$Cl$_2$): 1954, 2017 (both CO), 1648 (C=O) [57]
	(−)$_D$-compound I [57]	m.p. 73.5 to 75° [57] $[\alpha]_D^{23}$ = −30.6° (CHCl$_3$) [57]
30	FpCOC$_4$H$_9$-n	from [FpC(OC$_2$H$_5$)C$_4$H$_9$-n]BF$_4$ and H$_2$O [104]
31	FpCOCH$_2$CH(OH)CH$_2$CH$_3$ V (62%) [141]	—
*32	FpCOC$_4$H$_9$-t I (61%) [60]	orange-yellow needles, m.p. 75 to 76° [60]
33	FpCOCH$_2$CH(CH$_3$)$_2$ V (60%) [141]	for electrochemical and chemical oxidation, see p. 9 [123]
*34	FpCOC^{1}H$_2$C^{2}H$_2$CH(CH$_3$)$_2$ IV (78%) [131]	^{1}H NMR (CDCl$_3$): 0.90 (d, CH$_3$, J = 5.4), 1.40 (m, 3H, CH$_2$CH), 2.86 (t, COCH$_2$, J = 7.2), 4.90 (s, C$_5$H$_5$) [131] ^{13}C {^{1}H} NMR (CDCl$_3$): 22.32 (CH$_3$), 27.5 (C^2?), 33.87 (CH?), 64.66 (C^1), 86.31 (C$_5$H$_5$), 214.44 (CO), 256.38 (C=O) [131] IR (CH$_2$Cl$_2$): 1960, 2022 (both CO), 1641 (C=O) [131] mass spectrum: [M]$^+$, [M−CO]$^+$, [M−2 CO]$^+$, [M−C$_5$H$_{11}$]$^+$, [M−C$_5$H$_{11}$CO]$^+$, [M−C$_5$H$_{11}$CO−2 CO]$^+$ [131]
35	FpCOC^{1}H$_2$C^{2}H$_2$C^3(CH$_3$)$_3$ IV (92%) [131]	m.p. 47 to 49° [131] ^{1}H NMR (CDCl$_3$): 0.90 (s, CH$_3$), 1.40 (t, C^{2}H$_2$, J = 7.5), 2.84 (t, COCH$_2$, J = 7.5), 4.94 (s, C$_5$H$_5$) [131] ^{13}C {^{1}H} NMR (CDCl$_3$): 29.14 (CH$_3$), 29.92 (C-2), 38.32 (C-3), 62.39 (C-1), 86.30 (C$_5$H$_5$), 214.40 (CO), 256.78 (C=O) [131] IR (CH$_2$Cl$_2$): 1960, 2022 (both CO), 1647 (C=O) [131] mass spectrum: [M]$^+$, [M−CO]$^+$, [M−2 CO]$^+$, [M−C$_6$H$_{13}$]$^+$, [M−C$_6$H$_{13}$CO]$^+$, [M−C$_6$H$_{13}$CO−2 CO]$^+$ [131]
36	FpCOC$_5$H$_{11}$-n V (85%) [141]	from [FpC(OC$_2$H$_5$)C$_5$H$_{11}$-n]BF$_4$ and H$_2$O [104]

　　　　References on pp. 47/51

Table 2 (continued)

No.	compound method of preparation (yield)	properties and remarks

37 FpCOCH$_2$CH$_2$CH(OH)CH$_2$CH$_3$
    V (68%) [141]   —

*38

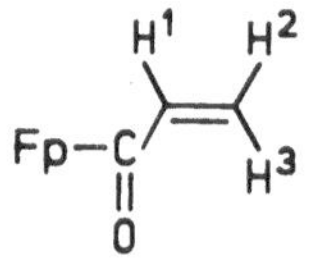

    I (5%) [8], (72%) [150]

orange liquid, solidifies at $\sim$ $-15°$, yellow crystals at $-78°$ [8]
^{1}H NMR: 4.84 (C$_5$H$_5$), 4.78 (H-2), 5.20 (H-3), 6.45 (H-1), J(H-1, H-3) = 17, J(H-1, H-2) = 10, J(H-2, H-3) = 2 in CS$_2$ [402], 4.87 (C$_5$H$_5$), 4.89 (H-2), 5.33 (H-3), 6.59 (H-1), J(H-1, H-3) = 17, J(H-1, H-2) = 10.0, J(H-2, H-3) = 1.5 in CDCl$_3$ [37]
IR: 1960, 2010 (both CO), 1622 (C=O), 1587 (C=C), 3080 (CH), other bands 690 to 1430 as liquid film [8], 1966, 2025 (both CO) in C$_6$H$_{12}$, 1610 (C=O), 1586 (C=C) in KBr [37]
k = 15.91, k' = 0.40 [19]

*39 (E)-FpCOCH^1=CH^{2}C$_6$H$_5$
    I (80%) [8],
    (25%) [86]

orange, m.p. 90 to 92° [8, 90]
^{1}H NMR: 4.84 (s, C$_5$H$_5$), coupled d's centered at 6.48, 6.84 (H-1, H-2, J = 15), 7.29 (C$_6$H$_5$) in CS$_2$ [8], 4.87 (C$_5$H$_5$), 6.68 (H-2), 6.96 (H-1), 7.18 to 7.68 (C$_6$H$_5$), J(H-1, H-2) = 16.0 in CDCl$_3$ [36, 37], further data (in CDCl$_3$) are given in [86, 90]
^{13}C NMR (CDCl$_3$): 86.6 (C$_5$H$_5$), 128.3, 128.9, 129.6, 135.2, 138.7, 214.3, 250.4, acyl carbon not observed [80]
IR: 1958, 2018 (both CO), 1633 (C=O) in THF [80], similar in CHCl$_3$ [48], 1960, 2025 (both CO), 1635 (C=O) in petroleum ether (bp. 30 to 60°) [86], 1963, 2009, 2031 (all CO), 1634 (C=O) in halocarbon oil mull [8], 1968, 2016 (sh), 2025 (all CO), 1643 (C=O) in C$_6$H$_{14}$ [90], similar in C$_6$H$_{12}$ [36, 37], 1622, 1581 (both C=C) [36, 37], 3020, 3050, 3080 (all CH), other bands 693 to 1582 in KBr [8]
k = 16.01, k' = 0.46 [19]
UV (C$_6$H$_{12}$): 221 (ε = 25900), 298 (ε = 29600) [8]

40 FpCOC($\dot{C}_6$H$_5$)=CH$_2$
    I [113]   —

*41 FpCOCH=CHCH$_3$
    I [90], (75%) [150]

oil [90]
^{1}H NMR (CDCl$_3$): 1.70 (d, CH$_3$, J(CH$_3$, H) = 6), 4.75 (s, C$_5$H$_5$), 6.00 (m, CH=CH, J(H, H) = 15) [90]
IR (C$_6$H$_{14}$): 1968, 2021 sh, 2027 (all CO), 1658 (C=O) [90]

*42 FpCOC^1(CH$_3$)=C^{2}H$_2$
    I (64%) [114],
    (70%) [145, 150]

m.p. 24 to 28° [114, 145]
^{1}H NMR (CDCl$_3$): 1.75 (dd, CH$_3$, J = 1.4, 0.9), 4.86 (s, C$_5$H$_5$), 5.26 (m, 1H, C=CH$_2$), 5.34 (m, 1H, C=CH$_2$) [113], 1.75 (s, CH$_3$), 4.83 (s, C$_5$H$_5$), 5.28 (s, 1H, C=CH$_2$), 5.37 (br s, 1H, C=CH$_2$) [114, 145]
^{13}C NMR (CDCl$_3$): 18.56 (CH$_3$), 86.13 (C$_5$H$_5$), 118.76 (C-2), 157.77 (C-1), 214.18 (CO), 255.79 (C=O) [113]; the assign-

References on pp. 47/51

Table 2 (continued)

No.	compound method of prepara- tion (yield)	properties and remarks
		ment for C–1 and C–2 differs from that given in the follow- ing ^{13}C {^1H} NMR (C_6D_6): 18.9 (CH_3), 86.4 (C_5H_5), 119.1 (C–1), 157.8 (C–2), 215.0 (CO), 251.7 (C=O) [114, 145] IR ($CHCl_3$): 1967, 2007 (both CO), 1597, 1624 (both C=O) [114, 145]
*43	$FpCOCH_2CH=CH_2$ I (20%) [35]	oil [35]
*44	$FpCOCH_2CH_2CH=CH_2$ I (20%) [35], V (54%) [141]	oil [35]
45	$FpCOCH_2CH_2C(CH_3)=CH_2$ V (92%) [141]	—
46	(E)-$FpCOCH=CHC(CH_3)=CH_2$ I (91%) [133]; see also [153]	^1H NMR ($CDCl_3$): 1.85 (br s, 3H), 4.88 (s, 5H), 5.33 (br s, 2H), 6.38 (s, 2H) [133] IR (Nujol): 1960, 2008 (both CO), 1643 (C=O) [133]
*47	$FpCO(CH_2)_3CH=CH_2$ I (20%) [35]	oil [35]
*48	$FpCOCH_2CH_2CH=CHCH_3$	oil [35]
49	(E,E)-$FpCOCH=CHCH=CHCH_3$ I (80%) [133]; see also [153]	^1H NMR ($CDCl_3$): 1.82 (d, 3H, J = 5.5), 4.86 (s, 5H), 5.74 to 6.55 (m, 4H) [133] IR (Nujol): 1960, 2000 (both CO), 1648 (C=O) [133]
*50	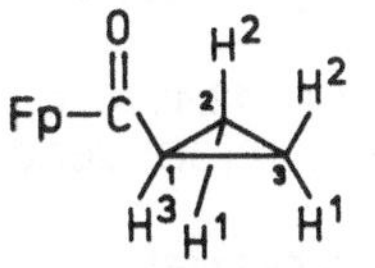 I (56%) [25], (65%) [140], (20%) [110]	bright orange-yellow oil [25], orange-brown liquid [110] ^1H NMR: 0.39 (m, 2H), 1.12 (m, 2H), 2.33 (m, 1H), 4.22 (s, 5H) in C_6D_6 [140], 0.53 (m, H–1), 0.86 (m, H–2), 2.38 (t of t's, H–3, J–1, 3 = 4.7, J–2, 3 = 7.5), 4.72 (s, C_5H_5) in CS_2 [25], 0.70 (m, 4H, CH_2), 2.75 (m, CH), 5.0 (s, C_5H_5) in $CDCl_3$ [110] ^{13}C {^1H} NMR: 10.5 (t), 39.4 (d), 86.2 (d), 215.1 (s), 250.0 (s) in C_6D_6 [140], 10.5 (C–2, C–3), 39.2 (C–1), 85.9 (C_5H_5), 214.0 (CO), 255.7 (C=O) in $CDCl_3$ [110] IR: 1958, 2023 (both CO), 1639, 1654 (both C=O), other bands at 848, 983, 1060, 1173, 1350, 1470 in C_6H_{12} [25], 1955, 2015 (both CO), 1630 (C=O) in CH_2Cl_2 [140], other bands 570 to 3220 (neat) [110] mass spectrum: $[M-CO]^+$ [110]

References on pp. 47/51

Table 2 (continued)

No.	compound method of prepara- tion (yield)	properties and remarks

*51

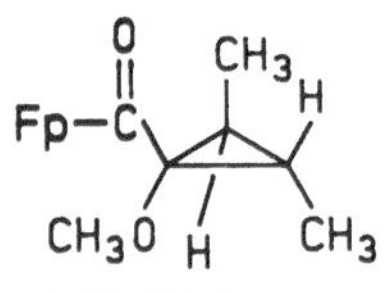

I (35%, based on ROH)

m.p. 61 to 62° [144]

^{1}H NMR (C_6D_6): 0.90 (q?, CH), 1.00 (d, CH_3), 1.08 (d, CH_3), 1.43 (q?, CH), 3.16 (s, OCH_3), 4.37 (s, C_5H_5) [144]

^{13}C NMR (C_6D_6): 12.80, 13.68 (both CH_3), 26.93, 30.83 (both CH), 57.54 (OCH_3), 86.58 (C_5H_5), 88.53 (OC≮), 215.52, 216.01 (both CO), 255.88 (C=O) [144]

IR ($CDCl_3$): 1950, 2000 (both CO), 1650 (C=O) [144]

mass spectrum: $[M-CO]^+$, $[M-2CO]^+$, $[C_5H_5Fe(CO)_3]^+$, $[M-FpCO]^+$ [144]

*52

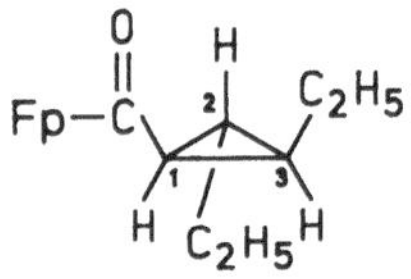

I (63%) [110]

orange-brown, air-sensitive oil [110]

^{1}H NMR ($CDCl_3$): 1.20 (m, 12H), 2.50 (m, CH), 4.80 (s, C_5H_5) [110]

^{13}C {^{1}H} NMR (CD_3COCD_3): 13.1, 14.1 (C-2, C-3), 26.5 (CH_2), 28.1 (CH_2), 31.8 (CH_3), 35.8 (CH_3), 53.7 (C-1), 86.0 (C_5H_5), 214.1, 214.4 (CO), 253.0 (C=O) [110]

IR (neat): 580, 660, 620(?), 735, 830, 950, 985, 1020, 1070, 1140, 1205, 1240, 1320, 1380, 1420, 1430, 1470, 1640, 1965, 2005, 2880, 2940, 2970 [110]

mass spectrum: $[M-CO]^+$ [110]

*53

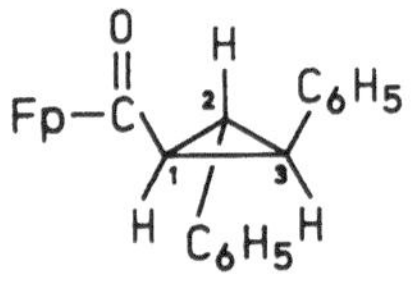

I (77%) [110]

orange-brown, air-sensitive oil [110]

^{1}H NMR (CD_3COCD_3): 3.12 (m, 1H), 3.50 (m, 2H), 4.85 (s, C_5H_5), 7.25 (s, 10H, C_6H_5) [110]

^{13}C {^{1}H} NMR (CD_3COCD_3): 29.7, 36.1 (C-2, C-3), 58.3 (C-1), 85.9 (C_5H_5), 125.8, 126.1, 128.4, 129.1, 135.6, 140.0 (C_6H_5), 213.8, 213.9 (CO), 249.8 (C=O) [110]

IR (CD_3COCD_3): 515, 580, 695, 735, 750, 790, 830, 910, 950, 1020, 1080, 1155, 1250, 1310, 1420, 1450, 1500, 1635, 1705, 1955, 2010, 2255, 3040, 3060, 3090, 3110 [110]

mass spectrum: $[M-3CO]^+$ [110]

*54

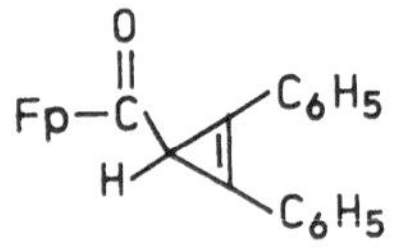

I (70%) [116]

yellow crystals, m.p. 85 to 87° [116]

^{1}H NMR ($CDCl_3$): 3.45 (s, CH), 4.76 (s, C_5H_5), 7.2 to 7.8 (m, C_6H_5) [116]

^{13}C NMR ($CDCl_3$): 47.56 (CH), 86.36 (C_5H_5), 112.41 (C=C), 127.91, 128.64, 128.96, 129.70 (C_6H_5), 214.7 (CO), 262.8 (C=O) [116]

IR (C_6H_{14}): 1960, 2016 (both CO), 1638 (C=O) [116]

*55

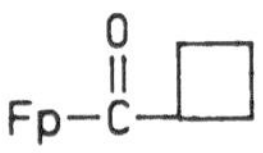

I [50]

red oil [50]

IR (C_6H_{12}): 1961, 2025 (both CO), 1635, 1650 (both C=O), other bands at 850, 988, 1075, 1158 [50]

*56

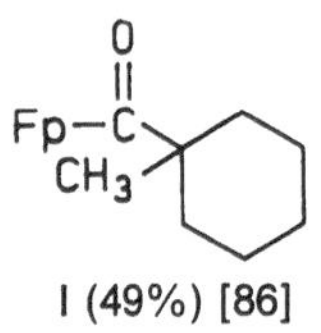

I (49%) [86]

crystals, m.p. 50°, very stable [86]

^{1}H NMR ($CDCl_3$): 1.00 (s, CH_3), 1.2 to 2.3 (m, CH_2), 4.80 (s, C_5H_5) [86]

IR (petroleum ether, b.p. 30 to 60°): 1954, 2022 (both CO), 1654 (C=O) [86]

References on pp. 47/51

Table 2 (continued)

No.	compound method of prepara- tion (yield)	properties and remarks
*57	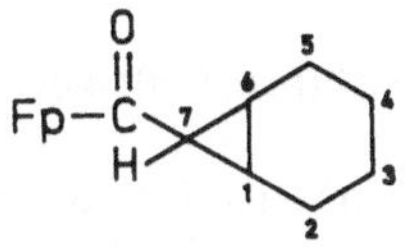 I (18%) [110]	brown, air-sensitive solid, m.p. 56 to 57° [110] ^{1}H NMR (CDCl$_3$): 1.25 (m, 4H), 1.70 (m, 6H), 2.42 (m, 1H, H-7), 4.88 (s, C$_5$H$_5$) [110] ^{13}C {^{1}H} NMR (CDCl$_3$): 20.8, 22.2 (C-2 to C-5), 24.7 (C-1, C-6), 54.3 (C-7), 85.9 (C$_5$H$_5$), 215.0 (CO), 253.0 (C=O) [110] IR (CDCl$_3$): 570, 650, 730, 870, 905, 1005, 1090, 1205, 1270, 1285, 1395, 1450, 1630, 1970, 2015, 2865, 2940 [110] mass spectrum: [M−CO]$^+$ [110]
*58	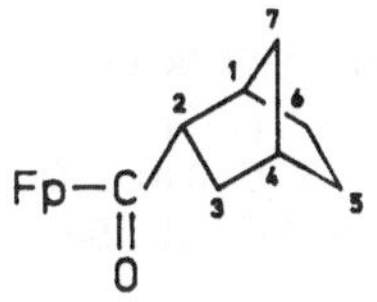 IV (69%) [146]	yellow, m.p. 111.5 to 113.5° [146] ^{1}H NMR (CCl$_4$): 1.31 (m, 8H), 2.18, 2.97 (both br s, bridgehead CH), 3.30 (m, CHCO), 4.83 (s, C$_5$H$_5$) [146] ^{13}C {^{1}H} NMR (CD$_2$Cl$_2$/CH$_2$Cl$_2$ = 1:5): 24.4 (C-6), 29.3 (C-5), 31.2 (C-3), 37.9 (C-4), 39.6 (C-7), 46.6 (C-1), 76.2 (C-2), 86.6 (C$_5$H$_5$), 214.7, 215.1 (both CO), 254.9 (C=O) [146] IR (CH$_2$Cl$_2$): 1958, 2018 (both CO), 1652 (C=O) [146] mass spectrum: [M]$^+$, [M−CO]$^+$, [M−2CO]$^+$, [M−Fp]$^+$ [146]
*59	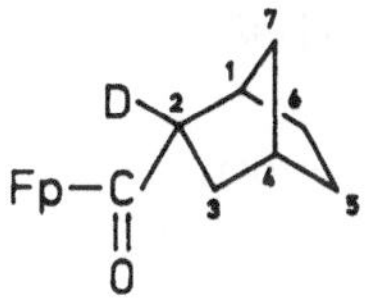	—
*60	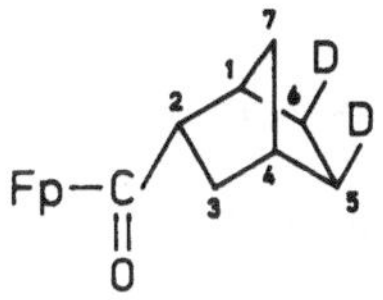	yellow solid [146]
*61	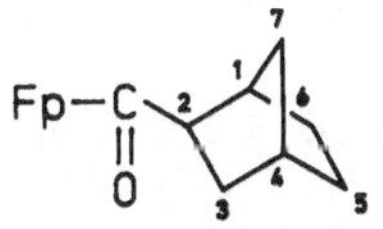 IV (73%) [146]	yellow crystals, m.p. 91 to 92° [146] ^{1}H NMR (CCl$_4$): 1.21 (m, 8H), 2.18, 2.41 (both br s, bridgehead CH), 2.76 (m, CHCO), 4.77 (s, C$_5$H$_5$) [146] ^{13}C {^{1}H} NMR (CD$_2$Cl$_2$/CH$_2$Cl$_2$ = 1:5): 29.3 (C-6), 29.7 (C-5), 34.7 (C-3), 36.2 (C-4), 36.9 (C-7), 40.2 (C-1), 76.5 (C-2), 86.9 (C$_5$H$_5$), 215.0, 215.5 (both CO), 256.7 (C=O) [146] IR (CH$_2$Cl$_2$): 1958, 2015 (both CO), 1652 (C=O) [146] mass spectrum: [M]$^+$, [M−CO]$^+$, [M−2CO]$^+$, [M−Fp]$^+$ [146]
*62	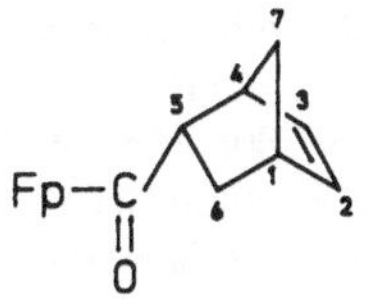	yellow crystals, m.p. 74 to 76° [146] ^{1}H NMR (CCl$_4$): 1.3 (m, 4H, CH$_2$), 2.79, 3.25 (both br s, bridgehead CH), 3.40 (m, CHCO), 4.77 (s, C$_5$H$_5$), 5.85 (m, 2H, CH=CH) [146] ^{13}C {^{1}H} NMR (CDCl$_3$): 29.5 (C-6), 42.6 (C-1), 46.0 (C-7), 48.7 (C-4), 73.5 (C-5), 86.6 (C$_5$H$_5$), 132.6 (C-3), 136.4 (C-2), 214.9 (CO), 255.5 (C=O) [146] IR (CH$_2$Cl$_2$): 1960, 2017 (both CO), 1660 (C=O) [146] mass spectrum: [M]$^+$, [M−CO]$^+$, [M−2CO]$^+$, [M−Fp]$^+$ [146]

References on pp. 47/51

Table 2 (continued)

No.	compound method of preparation (yield)	properties and remarks

*63

Fp—C=O

m.p. 97 to 99° [132]

^{1}H NMR (CDCl$_3$): 1.49 (m, 10H, CH$_2$), 2.21 (br s, CH), 4.80 (s, C$_5$H$_5$) [132]

^{13}C {^{1}H} NMR (CDCl$_3$): 29.6 (C-3, C-5), 32.3 (C-2, C-6), 35.8 (C-7), 41.9 (C-4), 75.3 (C-1), 86.1 (C$_5$H$_5$), 215.0 (CO), 261.6 (C=O) [132]

IR (CH$_2$Cl$_2$): 1955, 2005 (both CO), 1640 (C=O) [132]

*64

Fp—C=O

CH$_3$O

I (27%) [142]

brown, air-sensitive oil [142]

^{1}H NMR (CDCl$_3$): 1.00 to 2.90 (m, 8H, CH$_2$), 3.20 (s, OCH$_3$), 4.20 (m, H-7), 4.90 (s, C$_5$H$_5$), 6.45 (m, H-2) [142]

^{13}C {^{1}H} NMR (CDCl$_3$): 24.2, 26.1, 26.7, 29.7 (C-3 to C-6), 55.8 (OCH$_3$), 75.7 (C-7), 85.8 (C$_5$H$_5$), 144.4 (C-2), 158.1 (C-1), 214.2 (CO), 252.7 (C=O) [142]

IR (neat): 510, 570, 630, 680, 710, 830, 885, 970, 1000, 1025, 1090, 1120, 1165, 1200, 1220, 1315, 1340, 1370, 1440, 1450, 1600, 1960, 2010, 2820, 2860, 2930, 3110 [142]

mass spectrum: [M$-$CO]$^+$ [142]

*65 FpCOC$_{10}$H$_{15}$-1
(C$_{10}$H$_{15}$=adamantyl)
I (91%) [60]

bright yellow needles, m.p. 149.5 to 150.5° [60]

IR: 1954, 2015 (both CO), 1633 sh, 1652 (both C=O) [60]

*66 FpCOC$_{10}$H$_{15}$-2
(C$_{10}$H$_{15}$=adamantyl)

pale yellow, m.p. 97 to 99° [60]

IR: 1955, 2014 (both CO), 1651 sh, 1668 (both C=O) [60]

*67 FpCOC$_6$H$_5$
I (75%) [8]

yellow-orange crystals, m.p. 59 to 62° [8], 60 to 63° [94], 63 to 65° [9, 12, 80]

^{1}H NMR: 4.78 (s, C$_5$H$_5$), 7.27 (m, C$_6$H$_5$) in CS$_2$ [67]; see also [8], 4.88 (C$_5$H$_5$) in CH$_2$Cl$_2$ [46]

^{13}C NMR: 86.2 (C$_5$H$_5$), 125.7, 127.9 (C$_6$H$_5$, C-3, 5, C-2, 6), 130.1 (C$_6$H$_5$, C-4), 150.9 (C$_6$H$_5$, C-1), 213.9 (CO), 254.6 (C=O) [135], 86.3 (C$_5$H$_5$), 125.9, 128.1, 130.3, 151.0 (C$_6$H$_5$), 214.0 (CO), 254.8 (C=O) [67] in CDCl$_3$, 86.7, 125.9, 128.1, 130.3, 150.9, 214.0, 263.5 in CDCl$_3$ containing Cr(acac)$_3$ [80], 87.1, 215.1 in CH$_2$Cl$_2$ [46]

^{57}Fe-γ (78 K): $\delta = 0.032$ (α-Fe), $\Delta = 1.844$ [135]

IR: 1943w, 1972, 2029 (all CO), 1603 (C=O) in halocarbon oil mull [8], 1945, 2030 (both CO), 1605 (C=O) in Nujol [67], 1956, 2019 (both CO), 1613 (C=O) in THF [80], 1969, 2029 in CCl$_4$ [135], 1968, 2018 [46], 1981, 2036 [48] in CH$_2$Cl$_2$, 1976, 2033 (all CO), 1633 (C=O) in C$_6$H$_{12}$ [12], 3080 (CH), other bands 694 to 1582 in KBr [8]

k = 16.03 [19, 46], 16.14 [135], k' = 0.402 [46], 0.57 [19], 0.484 [135]

UV (C$_6$H$_{12}$): 232 ($\epsilon = 21600$), 332 ($\epsilon = 4570$) [8]

References on pp. 47/51

Table 2 (continued)

No.	compound method of preparation (yield)	properties and remarks
*68	FpCOC$_6$H$_4$F-4 I (77%) [32]	—
*69	FpCOC$_6$F$_5$	yellow crystals, m.p. 86 to 87° [14, 18]
*70	FpCOC$_6$H$_4$OCH$_3$-4 I (70%) [161]	orange or yellow crystals [161] m.p. 108 to 110° (from benzene/petroleum ether) [161] ^{1}H NMR (CDCl$_3$): 3.76 (s, CH$_3$), 4.86 (s, C$_5$H$_5$), 6.82, 7.55 (both: complex, 2H, C$_6$H$_4$) [161] ^{13}C NMR (CDCl$_3$): 55.4 (CH$_3$), 86.4 (C$_5$H$_5$), 130.0, 133.4, 144.1, 161.7 (all C$_6$H$_4$), 214.1 (CO), 250.0 (C=O) [79] IR: 1958, 2017 (both CO), 1608 (C=O) [79] IR (mulls): 1940, 1990sh, 2000 (all CO), 1249 (C–O), 1585 (C=O), 3030, 3095, 3120 (all CH), further bands given [161]
*71	FpCOC$_6$H$_4$COOH-4	yellow crystals, dec. ~160° [161] ^{1}H NMR (HCOOH): 5.19 (s, C$_5$H$_5$), 7.55, 8.18 (both: complex, 2H, C$_6$H$_4$) [161] IR (mulls): 1940, 1990sh, 2010 (all CO), 1610 (FeC=O), 1675 (C=O of COOH), 2560, 2675 (both: weak and broad, OH), 3115, 3130, 3140 (all CH), further bands given [161]
*72	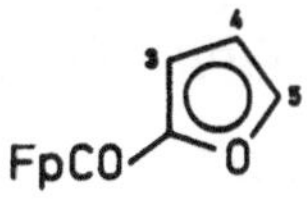 (FpCO–furanyl, positions 3,4,5) I (80%) [63]	lemon-yellow, thin needles, m.p. 127 to 128° [63] ^{1}H NMR (CDCl$_3$): 4.91 (C$_5$H$_5$), 6.42 (H-3), 6.70 (H-4), 7.50 (H-5) [63] IR: 1980, 2026 (both CO) in C$_6$H$_6$, 1576 (C=O) in KBr [63]
*73	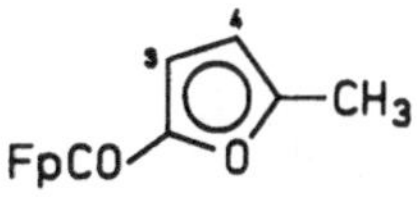 (FpCO–methylfuranyl–CH$_3$, positions 3,4) I (63%) [63]	lemon-yellow, thin needles, m.p. 102 to 103° [63] ^{1}H NMR (CDCl$_3$): 2.30 (CH$_3$), 4.88 (C$_5$H$_5$), 6.06 (H-3), 6.67 (H-4) [63] IR: 1975, 2028 (both CO) in C$_6$H$_6$, 1574 (C=O) in KBr [63]
*74	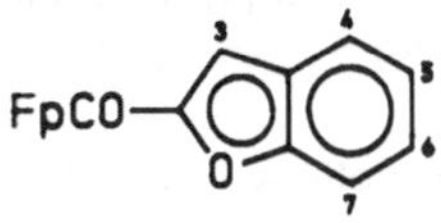 (FpCO–benzofuranyl, positions 3,4,5,6,7) I (51%) [63]	lemon-yellow, thin needles, m.p. 127 to 128° [63] ^{1}H NMR (CDCl$_3$): 4.96 (C$_5$H$_5$), 6.80 (H-3), 7.20 to 7.70 (H-4 to H-7) [63] IR: 1987, 2028 (both CO) in C$_6$H$_6$, 1578 (C=O) in KBr [63]
*75	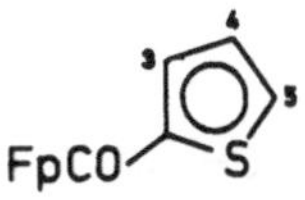 (FpCO–thienyl, positions 3,4,5) I (59%) [63]	lemon-yellow, thin needles, m.p. 111 to 112° (from C$_6$H$_{14}$/CH$_2$Cl$_2$ = 5:1) [63] ^{1}H NMR (CDCl$_3$): 4.98 (C$_5$H$_5$), 6.68 (H-3), 6.92 (H-4), 7.25 (H-5) [63] ^{13}C NMR (CH$_2$Cl$_2$): 86.1 (C$_5$H$_5$), 127.0 (C-5), 130.4 (C-4), 132.0 (C-3), 156.3 (C-2), 214.6 (CO), 237.6 (C=O) [117] IR: 1975, 2030 (both CO) in C$_6$H$_6$, 1570 (C=O) in KBr [63]
*76	FpCOOH	yellow [83]

References on pp. 47/51

Table 2 (continued)

No.	compound method of preparation (yield)	properties and remarks
*77	FpCOOCH$_3$	yellow-orange, m.p. 34 to 36° [9, 12] ^{1}H NMR: 3.39 (s, CH$_3$), 4.86 (s, C$_5$H$_5$) in CS$_2$ [9, 12], 3.52 (s, CH$_3$), 4.30 (s, C$_5$H$_5$) in C$_6$D$_6$ [103] ^{13}C NMR (C$_6$D$_6$): 51.7 (CH$_3$), 86.17 (C$_5$H$_5$), 197.6 (C=O), 214.5 (CO) [103] IR: 1945, 2020 (both CO), 1615 (C=O) [9], 1965 br, 2025 (both CO), 1632 (C=O), 1034 (O-CH$_3$) [103], 2900, 3060 (both CH), other bands at 768, 835 br, 915, 1025, 1165, 1410, 1425 [12] in KBr, 1974, 2012 (both CO), 1674 (C=O) in THF [139], 1995, 2046 (both CO), 1665 (C=O) in C$_6$H$_{12}$ [42] k = 16.48, k' = 0.42 [19]
*78	FpCOOC$_6$H$_{11}$-cyclo	—
*79	FpCONH$_2$ III (73%) [44]	light yellow crystals [44] IR: 1963, 2015 (both CO), 1593 (C=O) [20], 1970 br, 2031 (both CO), other bands at 360, 398, 460, 490, 517, 535, 579, 594, 633, 669, 1556, 1570, 1608 in CH$_2$Cl$_2$; the complete IR spectrum (KBr, Nujol) and tentative assignments of all bands were also given [44]
*80	FpCONHCH$_3$ III (60%) [20]	yellow, m.p. 118° (dec.) [20] ^{1}H NMR: 2.57 (d, CH$_3$, J(HC, NH) = 5), 4.87 (C$_5$H$_5$), 5.8 (br, NH) in CD$_3$CN [40], 2.88 (d, CH$_3$), 4.94 (s, C$_5$H$_5$) in CDCl$_3$, 6.80 (br, NH) in CD$_3$SOCD$_3$ [20] ^{13}C NMR (CDCl$_3$): 195.2 (C=O), 214.3 (CO) [120] IR (methylcyclohexane): 1972, 2015 (both CO), 1625 (C=O) [20]
*81	FpCON(CH$_3$)$_2$	orange crystals, m.p. 111 to 112° [4] ^{1}H NMR (CHCl$_3$): 2.98 (s, CH$_3$), 4.88 (s, C$_5$H$_5$) [4] ^{57}Fe-γ (78 K): δ = 0.312(5), Δ = 1.85(1) [72] IR: 1933, 2020 (both CO), 1530, 1545 (both C=O), 2910, 3060 (both CH), other bands 832 to 1475 in KBr [4], 1960, 2015 (both CO), 1612 (C=O) in CH$_2$Cl$_2$ [20]
82	FpCONHC$_2$H$_5$ III (62%) [20]	yellow, m.p. 97° (dec.) [20] IR (methylcyclohexane): 1968, 2015 (both CO), 1622 (C=O) [20] for the reaction with COS, see "Further information" on No. 80 [85]
*83	FpCON(C$_2$H$_5$)$_2$	orange crystals, m.p. 75 to 76.5° [4] ^{1}H NMR (CHCl$_3$): 1.09 (t, CH$_3$, J = 7), 3.43 (q, CH$_2$, J = 7), 4.85 (s, C$_5$H$_5$) [4] ^{57}Fe-γ (78 K): δ = 0.021 (^{57}Co/Cr), Δ = 0.173 [6] IR: 1942, 1964, 2004, 2017 (all CO), 1534 (C=O) in halocarbon oil mull, 2850, 2925, 3025 (all CH), other bands 828 to 1450 in KBr [4]

References on pp. 47/51

Table 2 (continued)

No.	compound method of preparation (yield)	properties and remarks
84	FpCONHC$_3$H$_7$-i III (70%) [20]	yellow, m.p. 78 to 80° (dec.) [20] IR (methylcyclohexane): 1965, 2015 (both CO), 1620 (C=O) [20]
85	FpCONHC$_4$H$_9$-n III [20]	yellow solid [20] IR (methylcyclohexane): 1968, 2019 (both CO), 1622 (C=O) [20]
*86	FpCONHC$_4$H$_9$-t	yellow, m.p. 95 to 97° [41] ^{1}H NMR (CD$_3$COCD$_3$): 1.26 (s, C(CH$_3$)$_3$), 4.96 (s, C$_5$H$_5$) [41] IR (C$_6$H$_{14}$): 1959, 2022 (both CO), 1633 (C=O) [41] mass spectrum: [M]$^+$ [41]
*87	FpCONHC$_6$H$_{11}$-cyclo III [40]	yellow crystals, m.p. 116 to 117° [40] ^{13}C NMR (CDCl$_3$): 193.0 (C=O), 214.4 (CO) [120] IR (C$_7$H$_{16}$): 1971, 2032 (both CO), 1635 (C=O) [40]
88	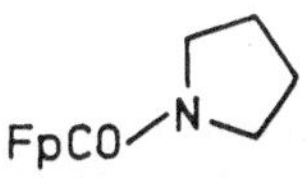 FpCO–N⟨pyrrolidine⟩ III (40%) [20]	yellow, m.p. 68° (dec.) [20] IR (C$_7$H$_{16}$): 1967, 2023 (both CO), 1566 (C=O) [20]
89	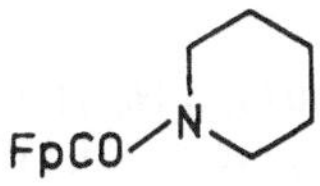 FpCO–N⟨piperidine⟩ III (45%) [20]	yellow, m.p. 70° (dec.) [20] ^{13}C NMR (CDCl$_3$): 194.7 (C=O), 214.1 (CO) [120] IR (C$_6$H$_{12}$): 1968, 2020 (both CO), 1560 (C=O) [20]
*90	FpCONHNH$_2$	bright yellow [26] IR (CH$_2$Cl$_2$): 1967, 2023 (both CO), 1588 (C=O) [26]
*91	FpCONHNHCH$_3$	IR (CH$_2$Cl$_2$): 1969, 2025 (both CO), 1592 (C=O) [26]
*92	FpCON(CH$_3$)NHCH$_3$	IR (CH$_2$Cl$_2$?): 1957, 2037 (both CO), 1600 (C=O) [26]

supplement

No.	compound method of preparation (yield)	properties and remarks
*93	FpCOCH=C(CH$_3$)$_2$ I (80%) [150]	yellow needles, m.p. 62 to 63° [150]
*94	FpCOCH$_2$C(CH$_3$)$_2$NHCH$_2$C$_6$H$_5$ VI (62%) [150]	m.p. 68 to 70° [150] ^{1}H NMR (CDCl$_3$): 1.13 (s, 6H, CH$_3$), 4.83 (s, C$_5$H$_5$) [150] IR (KBr): 1965, 2020 (both CO), 1631 (C=O), 3320 (NH) [150]
95	FpCOCH$_2$CH(CH$_3$)NHCH$_2$C$_6$H$_5$ VI (78%) [150]	oil [150] ^{1}H NMR (CDCl$_3$): 1.10 (d, CH$_3$, J=5.7), 4.82 (s, C$_5$H$_5$) [150] IR (neat): 1963, 2019 (both CO), 1645 (C=O), 3320 (NH) [150]

References on pp. 47/51

Table 2 (continued)

No.	compound method of prepara- tion (yield)	properties and remarks
96	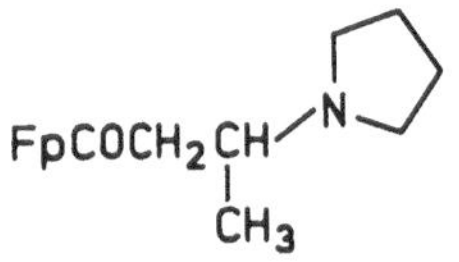 VI (51%) [150]	oil [150] ^{1}H NMR (CDCl$_3$): 1.06 (d, CH$_3$, J = 6.2), 4.85 (s, C$_5$H$_5$) [150] IR (neat): 1975, 2022 (both CO), 1647 (C=O) [150]
97	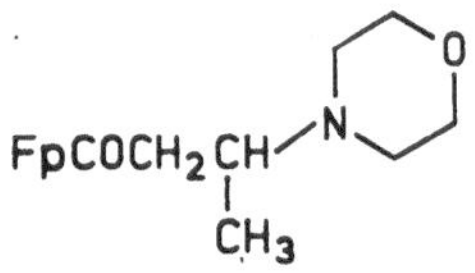 VI (57%) [150]	m.p. 69 to 70° [150] ^{1}H NMR (CDCl$_3$): 0.98 (d, CH$_3$, J = 6.3), 4.87 (s, C$_5$H$_5$) [150] IR (KBr): 1945, 2015 (both CO), 1659 (C=O) [150]
98	FpCOCH$_2$CH(CH$_3$)SC$_6$H$_5$ VI (95%) [150]	oil [150] ^{1}H NMR (CDCl$_3$): 1.14 (d, CH$_3$, J = 6.6), 4.82 (s, C$_5$H$_5$) [150] IR (neat): 1950, 2020 (both CO), 1630 (C=O) [150]
99	FpCOCH$_2$CH(CH$_3$)SCH$_2$CH$_2$OH VI (66%) [150]	oil [150] ^{1}H NMR (CDCl$_3$): 1.21 (d, CH$_3$, J = 6.5), 4.86 (s, C$_5$H$_5$) [150] IR (neat): 1950, 2015 (both CO), 1625 (C=O), 3440 (OH) [150]
100	FpCOCH$_2$CH(CH$_3$)NHCH$_2$CH=CH$_2$ VI (89%) [150]	oil [150] ^{1}H NMR (CDCl$_3$): 0.95 (d, CH$_3$, J = 5.4), 4.80 (s, C$_5$H$_5$) [150] IR (neat): 1955, 2020 (both CO), 1640 (C=O), 3310 (NH) [150]
101	FpCOCH$_2$CH(CH$_3$)NHC$_4$H$_9$-n VI (98%) [150]	oil [150] ^{1}H NMR (CDCl$_3$): 0.96 (d, CH$_3$, J = 5.7), 4.83 (s, C$_5$H$_5$) [150] IR (neat): 1955, 2020 (both CO), 1640 (C=O), 3315 (NH) [150]
102	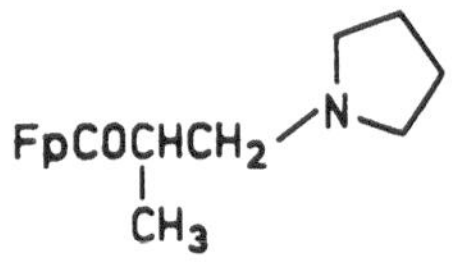 VI (50%) [150]	oil [150] ^{1}H NMR (CDCl$_3$): 0.98 (d, CH$_3$, J = 6.9), 4.89 (s, C$_5$H$_5$) [150] IR (neat): 1955, 2026 (both CO), 1639 (C=O) [150]
103	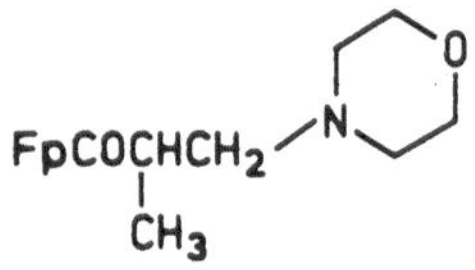 VI (75%) [150]	oil [150] ^{1}H NMR (CDCl$_3$): 0.94 (d, CH$_3$, J = 6.9), 4.90 (s, C$_5$H$_5$) [150] IR (neat): 1956, 2015 (both CO), 1644 (C=O) [150]
104	FpCOCH(CH$_3$)CH$_2$SC$_6$H$_5$ VI (90%) [150]	oil [150] ^{1}H NMR (CDCl$_3$): 1.11 (d, CH$_3$, J = 6.6), 4.84 (s, C$_5$H$_5$) [150] IR (neat): 1955, 2020 (both CO), 1630 (C=O) [150]

References on pp. 47/51

Table 2 (continued)

No.	compound method of prepara- tion (yield)	properties and remarks
105	FpCOCH(CH₃)CH₂SCH₂CH₂OH VI (78%) [150]	oil [150] ¹H NMR (CDCl₃): 1.06 (d, CH₃, J = 6.9), 4.89 (s, C₅H₅) [150] IR (neat): 1950, 2015 (both CO), 1625 (C=O), 3420 (OH) [150]
*106	FpCOC(CH₃)=P(C₂H₅)₃ II [155]	—

No.	compound method of prepara- tion (yield)	properties and remarks

*107	
	Fp−C=O
	I [158]
		m.p. 91 to 92° [158]
		¹H NMR (CDCl₃): 1.30 to 2.0 (m, 12H, CH₂), 2.26 (br s, 1H,
		  CH), 4.84 (s, C₅H₅) [158]
		¹³C {¹H} NMR (CDCl₃): 19.7, 28.5, 31.2, 32.1, 34.8, 42.6 (all
		  CH₂), 35.3 (CH), 71.6 (>C<), 86.4 (C₅H₅), 215.2, 215.4 (both
		  CO), 262.1 (C=O) [158]
		mass spectrum: [M]⁺, [M−CO]⁺, [M−2 CO]⁺, [M−3 CO]⁺
		  [158]

108	FpCOC₆H₄CH₃-4
	I (70%) [161]
		yellow crystals, m.p. 87 to 88° (from petroleum ether at low
		  temperature) [161]
		¹H NMR (CS₂): 2.31 (s, CH₃), 4.77 (s, C₅H₅), 7.01, 7.27 (both:
		  complex, 2H, C₆H₄) [161]
		IR (mulls): 1945, 1980 sh, 2020 (all CO), 1595 (C=O), 3115,
		  3140 (both CH), further bands given [161]
		photodecarbonylation (UV) proceeds smoothly to give
		  FpC₆H₄CH₃-4 [161]

109	FpCOC₆H₄COOCH₃-4
	I (70%) [161]
		yellow crystals, m.p. 129 to 131° (from petroleum ether, b.p.
		  100 to 120°)
		¹H NMR (CDCl₃): 3.91 (s, CH₃), 4.93 (s, C₅H₅), 7.41, 8.10 (both:
		  complex, 2H, C₆H₄) [161]
		IR (mulls): 1945, 1990 sh, 2010 (all CO), 1269 (C−O), 1590
		  (FeC=O), 1720 (C=O of COOC), 3010, 3105, 3120 (all CH),
		  further bands given [161]
		gives with HCl/H₂O/CH₃COCH₃ No. 71 [161]

110	FpCOC₆H₄Br-2
	I (70%) [161]
		yellow crystals, m.p. 82 to 83° [161]
		less soluble in concentrated HCl than No. 70 [161]
		¹H NMR (CS₂): 4.83 (s, C₅H₅), multiplet centered at 7.0 (4H,
		  C₆H₄) [161]
		IR (mulls): 1965, 2010 (both CO), 1604 (C=O), 3100, 3115,
		  3140 (all CH), further bands given [161]
		photodecarbonylation (UV) proceeds smoothly to give
		  FpC₆H₄Br-2 [161]

* Further information:

C₅H₅Fe(CO)₂CHO (Table 2, No. 1) is formed as an intermediate in the reaction of [C₅H₅Fe-
(CO)₃]PF₆ with NaBH₄ in acetone at 77 K and has been identified by its ¹H NMR resonances

at $-80\,°C$. Raising the temperature converts it into FpH (-50 to $-20\,°C$) and then into Fp_2. Attempts to isolate the formyl complex have been unsuccessful [136].

$C_5H_5Fe(CO)_2COCH_3$ (Table **2**, No. **2**) is isolated from the reaction residue (Method I) by extraction with CH_2Cl_2, filtration through Al_2O_3, and after solvent removal, sublimation at 70 to $90\,°C/0.3$ Torr [4]. The reaction mixture (Method I) may also be worked up by pouring it into ice–water and subsequent sublimation at $30\,°C/1$ Torr of the solid formed on standing [2, 3] or by chromatography on Al_2O_3 with petroleum ether [151]. $FpCOCH_3$ is obtained in an almost quantitative yield by addition of CH_3COCl to a suspension of $[PR_4^1][Fp]$ ($R^1 = CH_3$, C_2H_5, $n-C_4H_9$) in benzene. After 10 to 30 min the salts $[PR_4^1]Cl$ are filtered off and the solution is evaporated to dryness. The remaining product is purified by crystallization from pentane at $-78\,°C$ [101]. The use of $[N(C_4H_9-n)_4][Fp]$ (in place of Na[Fp], Method I), prepared from $FpSi(CH_3)_3$ and $[N(C_4H_9)_4-n]F$ in THF at -77 K, was described, giving a yield of 47% (determined by ^1H NMR spectroscopy) [122]. The compound has also been prepared by the reaction of Fp^-, formed in situ by reduction of Fp_2 with the sodium adduct of naphthalene in THF, with CH_3COCl, evaporation of the solution under vacuum, chromatography of the residue on SiO_2, and elution with $CHCl_3/O(C_3H_7-i)_2$ (1:1). After evaporation of the solvent, sublimation at $80\,°C/10^{-3}$ Torr affords the product in a 56% yield [94]. $FpCOCH_3$ is formed along with the main product Fp_2 in an 11% yield by the reaction of Na[Fp] with $ClCH_2COCl$. Extraction of the crude product with petroleum ether followed by chromatography on Florisil gives an orange-red powder containing traces of Fp_2 which could not be removed by recrystallization. The pure product is obtained by sublimation at $80\,°C/10^{-2}$ Torr [82]. Refluxing a mixture of $FpCH_3$ and Fp^+ (1:1 mole ratio) in $ClCH_2CH_2Cl$ for 1 h and subsequent workup by chromatography gives a 41% yield of $FpCOCH_3$ [112]. $FpCOCH_3$ is also obtained by carbonylation of $FpCH_3$ in ether at $125\,°C$ and 210 atm (3000 $lb/in.^2$) of CO for 2 h [2, 3] or in tetradecane at $97\,°C$ and 325 atm of CO [76]; see also [1, 68]. After solvent removal under reduced pressure the residue is sublimed between 25 and $40\,°C/1$ Torr to give a mixture of the starting material and $FpCOCH_3$. The pure product is obtained by dissolving the sublimate in a minimum of $HCON(CH_3)_2$ and addition of small amounts of water. Since a clean separation was not possible by this method, the yield could not be determined accurately [2, 3]. Large amounts of Fp_2 are also reported to be formed in this reaction [76]. No CO uptake could be detected by stirring $FpCH_3$ in toluene under CO (377 Torr) for 25 h [95]. However, rapid CO insertion in $FpCH_3$ is induced by addition of $AlBr_3$ (1:1 mole ratio) to the complex in toluene at $0\,°C$ and subsequent reaction of the adduct formed, $C_5H_5Fe(CO)(COAlBr_3)CH_3$ (not isolable), with CO at $20\,°C/\sim400$ Torr. To prepare $FpCOCH_3$ from the adduct $C_5H_5Fe-(CO)_2(COAlBr_3)CH_3$ now formed (neither isolable), toluene is removed from the filtered mixture at reduced pressure and the remaining oil is redissolved in toluene and hydrolyzed at $0\,°C$. After solvent removal $FpCOCH_3$ is purified by sublimation for a 50% yield. Attempts to carry out the reaction with a catalytic amount of Lewis acid were unsuccessful [91, 95]. The reaction of $FpCOOCH_3$ (No. 77) with CH_3MgBr (1:2 mole ratio) in ether at room temperature for 16 h and subsequent hydrolysis with saturated aqueous NH_4Cl also leads to formation of $FpCOCH_3$. Solvent removal at reduced pressure of the ether layer, extraction of the residue with pentane, and cooling the filtered solution at $-78\,°C$ for 4 h afford the compound, which is purified by sublimation at 55 to $65\,°C/0.1$ Torr, 78% yield [12]; see also [9]. This method has also been employed with $FpCON(CH_3)_2$ and CH_3MgBr, giving $FpCOCH_3$ in a 41% yield [12]. The reaction of $[Fp=C(OCH_3)CH_3]BF_4$ with CH_3MgI, affording predominantly $FpCOCH_3$ was briefly mentioned [114, 145]. In addition to $FpC(OCH_3)=CH_2$ (55% yield) and $FpC(OCH_3)_2CH_3$ (39% yield), the compound is formed in a 6% yield by addition of $NaOCH_3$ in THF to a solution of $[Fp=C(OCH_3)CH_3]BF_4$ in CH_2Cl_2 at $-78\,°C$ [115]. Treatment of $C_5H_5Fe-(CO)(P(C_6H_5)_3)COCH_3$ with $[N(P(C_6H_5)_3)_2][C_5H_5V(CO)_3H]$ ($\sim1:1$ mole ratio) in THF under a CO atmosphere (600 Torr) induces exchange of $P(C_6H_5)_3$ by CO. After 2d at $25\,°C$, 50%

of $C_5H_5Fe(CO)(P(C_6H_5)_3)COCH_3$ had been converted into $FpCOCH_3$ [102]. A mixture of products including No. 2 is obtained by the reaction of $FpCH_3$ with $[C(C_6H_5)_3]BF_4$ in CH_2Cl_2 at -10 to $+10$ °C for 100 h [127].

A comparison of the Mössbauer isomer shifts and $v(CO)$ data of $FpCOCH_3$ and $FpSO_2CH_3$ indicates a similar electron density at the iron nucleus for both compounds (comparable isomer shift) but a greater σ-donor ability and thus a smaller π-bonding ability for the RCO (lower $v(CO)$ frequency) than for the RSO_2 group [52]. A linear correlation between the partial isomer shift of the C_5H_5 ligand and the chemical shift of the C_5H_5 protons in a series of FpR compounds including $FpCOCH_3$ was cited [6].

Integrated IR intensities of the $v(CO)$ vibrations in $CHCl_3$ have been determined for $FpCOCH_3$ and related compounds and used to calculate FeCO group dipole derivatives. In all cases the dipole moment derivatives for the symmetric stretching motion are smaller (for $FpCOCH_3$: $\mu'_{FeCO}(sym) = 7.76$ D) than those for the corresponding asymmetric stretching vibration ($\mu'_{FeCO}(asym) = 9.58$ D). This shows that the C_5H_5 ligand primarily acts as a donor ligand in these complexes. Variations in the derivatives as a function of the angle between MCO (M=Fe, Sn, Mn) groups were considered in detail [47].

The IR spectrum of isolated $FpCOCH_3$ molecules (at high dilution in CH_4, Ar, N_2, and CO matrices) at 12 K exhibits multiply split A' and A'' terminal CO bands attributable to matrix effects and a much larger splitting ($\sim$36 cm^{-1}) in the acetyl CO band, presumably arising from conformational isomerism about the metal–acetyl bond. The observed relative intensity of the two terminal CO bands in an N_2 matrix has been used to calculate an OC–Fe–CO angle of $99 \pm 1°$ [107].

The IR spectrum of $FpCOCH_3$ in a polyvinyl chloride film matrix at 12 K shows two terminal bands at 1956 and 2016 cm^{-1} and two C=O bands at 1626 and 1649 cm^{-1} [137].

The low $v(C=O)$ frequency at 1655 cm^{-1} compared to the carbonyl absorption in ketones is suggested to result from resonance contributions (VII) which lead to metal–carbon double bonding and reduction of the CO bond order [4].

VII

$FpCOCH_3$ is soluble in C_6H_6 and CH_2Cl_2 but only sparingly soluble in saturated hydrocarbons such as pentane or hexane [4].

The compound is fairly stable in the solid state but somewhat air-sensitive in solution [4]. Attempts to decompose $FpCOCH_3$ in refluxing $CHCl_3$ (18 h) failed [45].

A sufficient number of metastable ions in the mass spectrum indicated the following two main sequences of fragmentation of the molecular ion which occurred with very low abundance in the spectrum [23]:

References on pp. 47/51

The observation of the ions $[(C_5H_5)_2Fe]^+$ and $[C_5H_4CH_3FeC_5H_5]^+$ suggests thermal decomposition (200 to 250 °C inlet temperature) into ferrocene and methylferrocene. This assumption is supported by the mass spectrum run at 70 °C in which the ion $[(C_5H_5)_2Fe]^+$ was weaker by a factor of 11. The appearance of a stronger molecular ion (factor 4.5) and $[C_5H_5Fe(CO)_3]^+$ fragment (2.5) at 70 °C is attributed to the fact that in the mass spectrum taken at 200 °C some thermal decarbonylation of $FpCOCH_3$ to give $FpCH_3$ occurred prior to ionization. All fragments and their relative intensities (at inlet temperatures of 70 and 200 °C) and the metastable ions are listed [23].

$FpCOCH_3$ in polystyrene is rapidly cleaved by UV irradiation. Afterwards, the IR spectrum corresponds to that of $FpCH_3$ [162]. Wavelength-selective photolysis of $FpCOCH_3$ at high dilution in frozen gas matrices such as Ar, CH_4, N_2, and CO at 12 K produces $FpCH_3$ via the coordinatively unsaturated intermediate $C_5H_5Fe(CO)COCH_3$ as is assumed on the basis of IR measurements (cf. p. 9 ff.). The formation of the latter species was discussed in relation to the mechanisms of thermal and photochemical decarbonylation reactions of $FpCOCH_3$ [107]. For a discussion of the IR data (CH_4 matrix) in relation to the bonding of the acetyl ligand (σ-acyl or η^2-acyl) in $C_5H_5Fe(CO)COCH_3$, see [125]. According to the observed $\nu(CO)$ frequencies, photolysis in a ^{13}CO-doped (5%) Ar matrix leads to formation of C_5H_5Fe-$(^{12}CO)_2CH_3$, rather than $C_5H_5Fe(^{12}CO)(^{13}CO)COCH_3$ or $C_5H_5Fe(^{12}CO)(^{13}CO)CH_3$ [107].

IR spectroscopic experiments using polyvinyl chloride film matrices at 12 to 200 K have shown that the photoinduced decarbonylation of $FpCOCH_3$ is thermally reversible, that is, the intermediate $C_5H_5Fe(CO)COCH_3$ formed at 12 K by irradiation of the film can react with CO at 200 K to regenerate $FpCOCH_3$ [137].

The polarographic reduction of $FpCOCH_3$ in $CH_3OCH_2CH_2OCH_3/0.1$ M $[N(C_4H_9-n)_4]ClO_4$ at $E_{1/2} = -2.5$ V (vs. 10^{-3} M $AgClO_4/Ag$) takes place in a one-electron step. Attempted reoxidation of the electrochemically generated species back to the initial compound failed [11].

The compound is reduced by treatment with a tenfold excess of $LiAlD_4$ in toluene within 30 min to give in addition to Fp_2 and $Li[Fp]$ the primary products $C_2H_3D_3$, $C_3H_3D_3$, and $C_3H_3D_5$ (3:0.8:1 ratio), C_2HD_3, C_2D_4, and $C_2H_2D_2$ (2:1:1 ratio) and other hydrocarbons ranging from C_1 to C_4. Reduction of $Fp^{13}COCH_3$ with $LiAlH_4$ produces ethane, propene, and propane (each containing one ^{13}C), C_2H_4 and $H_2^{13}C=CH_2$ (2:1 ratio), and some products with no ^{13}C. In both of these examples extensive mixing of labels has occurred in the ethylene product, contrary to the results for the primary derivatives. The reaction of $FpCOCH_3$ with $LiAlH_4$ under ^{13}CO leads to no incorporation of ^{13}C into the C_2 and C_3 hydrocarbons. From these results and the reaction behavior of Fp_2 and FpR^1 ($R^1=CH_3$, C_2H_5) under similar conditions, it is suggested that chain extension occurs by a CO insertion reaction, probably by CO insertion into an alkylidene–metal bond [105]; see also [92].

Addition of $BH_3 \cdot THF$ to $FpCOCH_3$ causes rapid reduction (2 min at 20 °C) of the acetyl group giving FpC_2H_5. A possible mechanism was discussed. When, however, B_2H_6 in C_6H_6 is used in place of $BH_3 \cdot THF$, no reduction is observed even after several hours at room temperature [61].

Treatment of a tetradecane solution of $FpCOCH_3$ with 313 atm of H_2 at 155 °C results in removal of the acetyl group, yielding Fp_2. The expected formation of FpH and CH_3CHO did not take place, as indicated by the absence of IR bands in the range 1600 to 1780 cm^{-1}. Cooling the solution to 31 °C while maintaining the H_2 pressure gives $\nu(CO)$ frequencies which can be assigned to $FpCH_3$ [76].

References on pp. 47/51

Controlled-potential oxidation of FpCOCH$_3$ in CH$_3$CN/C$_2$H$_5$OH (10%)/0.1 M [N(C$_2$H$_5$)$_4$]ClO$_4$ at 1.3 V (vs. SCE) requires 2.90±0.05 electrons per Fe, suggesting a further oxidation of the reputed cation-radical [FpCOCH$_3$]$^{+\cdot}$. In accord with this assumption, CH$_3$COOC$_2$H$_5$ (70%) has been found in the resultant solution by gas chromatography. Extensive degradation of the iron moiety has been qualitatively noted by the absence of CO absorptions in the IR spectrum of the residue [88]. For the cyclic voltammogram, see p. 9.

Oxidation of FpCOCH$_3$ in CH$_3$CN with [NH$_4$]$_2$[Ce(NO$_3$)$_6$] or anhydrous Cu(SO$_3$CF$_3$)$_2$ at −78 °C gives a green solution, the color of which fades rapidly. The short lifetime ($\tau_{1/2}$ = 0.44 s at 20 °C) of the species responsible for the color suggests the presence of an intermediate cation radical like [FpCOCH$_3$]$^{+\cdot}$, because this species shows two visible absorptions at 580 (log ε = 2.2) and 445 nm (log ε = 3.0) in CH$_3$CN at 20 °C. Its decomposition is first-order for three half-lives in CH$_3$CN at 20 °C [87]. For the oxidation with AgPF$_6$, see p. 9.

FpCOCH$_3$ reacts with HCl to form a cream-colored powder that dissociates very readily and could not be kept pure outside an HCl atmosphere. On the basis of IR data, Formula VIII is proposed for the protonated complex. The shift of the C$_5$H$_5$ proton resonance to lower fields in the ^{1}H NMR spectrum measured in CF$_3$COOH (δ = 5.24 ppm) indicates an increased positive charge on the metal in the protonated species [15].

$$\left[\begin{array}{c} \text{CO} \\ | \\ \bigodot\!\!-\!\text{Fe}\!=\!\text{C}\!\!<\!\!\begin{array}{l}\text{OH}\\\text{CH}_3\end{array} \\ | \\ \text{CO} \end{array}\right]^+$$

VIII

Dissolution of the compound in liquid SO$_2$ at −40 °C led to no SO$_2$ insertion after 12 h [10].

Addition of HgCl$_2$ to a solution of FpCOCH$_3$ in C$_6$H$_6$ and heating at 50 °C for 1 h afford FpHgCl and CH$_3$COCl [17].

FpCOCH$_3$ reacts with P$_4$S$_{10}$ or B$_2$S$_3$ (1:2 mole ratio) in dry ether for 4 days to give C$_5$H$_5$Fe(CO)$_2$SC(S)CH$_3$, in which the dithioacetate ligand is S-monodentate, and C$_5$H$_5$Fe-(CO)S$_2$CCH$_3$, in which the dithio ligand is bidentate [62, 69].

[C$_5$H$_5$Fe(P(OC$_6$H$_5$)$_3$)$_2$]$_2$ and a small quantity of C$_5$H$_5$Fe(CO)(P(OC$_6$H$_5$)$_3$)CH$_3$ are formed on irradiation of FpCOCH$_3$ and P(OC$_6$H$_5$)$_3$ in benzene at 80 °C for 8 h [17]. Photolysis in the presence of the ^{2}D ligands P(CH$_3$)$_3$, CNR^1 (R^1 = CH(C$_6$H$_5$)CH$_3$, C$_6$H$_{11}$-cyclo; 1:1 mole ratio) in toluene at −70 °C produced C$_5$H$_5$Fe(CO)(^{2}D)CH$_3$ and traces of C$_5$H$_5$Fe(CO)(^{2}D)COCH$_3$ [97]. Photolysis in the presence of P(C$_6$H$_5$)$_3$ is described on p. 10.

The reactions of FpCOCH$_3$ with the ylides (CH$_3$)$_3$P=CH$_2$ (in ether at 0 °C/1 h), C$_2$H$_5$(CH$_3$)$_2$P=CH$_2$ (in THF at 20 °C/1 h), and (C$_2$H$_5$)$_3$P=CHCH$_3$ or (n-C$_4$H$_9$)$_3$P=CH$_2$ (in C$_6$H$_6$ at 20 °C/3 h) in a 1:2 mole ratio yield the corresponding acetyl-substituted ylides R$_3^1$P=CR^{2}COCH$_3$ (R^1 = CH$_3$, C$_2$H$_5$, n-C$_4$H$_9$; R^2 = H, CH$_3$) and the phosphonium metalates [PR$_4^1$][Fp] which precipitate. A mechanism involving transfer of the acetyl group from iron to the phosphorus ylide was discussed [101].

The reactions of FpCOCH$_3$ with transition metal carbonyl Lewis acids such as [FpL]PF$_6$ (L = isobutene, THF) in refluxing CH$_2$Cl$_2$ for 1 to 6 h, with C$_5$H$_5$M(CO)$_3$FPF$_5$ (M = Mo, W) in CH$_2$Cl$_2$ at −20 °C for 0.5 h [112], or with C$_5$H$_5$Mo(CO)$_3$FSbF$_5$ in CH$_2$Cl$_2$ at −25 °C for 1.5 h

References on pp. 47/51

[128] give the bimetallic μ–acetyl adducts IX, which may be formulated as carboxonium salts. In acetone or by treatment with $[N(C_4H_9-n)_4]I$ in CH_2Cl_2, these adducts are reconverted into the starting $FpCOCH_3$ [112].

IX   M = Fe (n = 0)
     M = Mo, W (n = 1)

The carbene complex $[Fp=C(OCH_3)CH_3]^+$ is obtained by treatment of $FpCOCH_3$ with $CH_3O-SO_2CF_3$ [98] or with $[O(CH_3)_3]BF_4$ [114, 145] in CH_2Cl_2. Alkylation of $FpCOCH_3$ with $[HC(OR^1)_2]PF_6$ ($R^1 = CH_3$, C_2H_5), prepared in situ from $HC(OR^1)_3$ and $[C(C_6H_5)_3]PF_6$ in CH_2Cl_2 (20 °C), quantitatively gave the alkoxycarbene salts $[FpC(OR^1)CH_3]PF_6$ after 18 h [149].

Attempts to alkylate $FpCOCH_3$ with $[Fp=CH_2]PF_6$ or $[C_5H_5(CO)_3Mo=CH_2]PF_6$ in CH_2Cl_2 at −78 to +20 °C failed. In both reactions nearly 80% of the $FpCOCH_3$ was recovered, with the remaining $FpCOCH_3$ largely found as its bimetallic μ-(η^1-C,O)-acetyl compounds $[FpC(CH_3)OFp]^+$ and $[FpC(CH_3)OMo(CO)_3C_5H_5]^+$ [148].

For the reaction with $LiN(Si(CH_3)_3)_2$ to give $FpC(OLi)=CH_2$, which further reacts with a series of alkylating agents, see Preparation Method V. The formation of $FpC(OLi)=CH_2$ from $FpCOCH_3$ and $LiN(C_3H_7-i)_2$ in THF was briefly mentioned [130].

Low–temperature reaction of $FpCOCH_3$ with $(CF_3SO_2)_2O$ followed by addition of CH_3SH or C_6H_5SH produces the carbene complex $[Fp=C(SR^1)CH_3]CF_3SO_3$ ($R^1 = CH_3$, C_6H_5) [153].

$C_5H_5Fe(CO)_2{}^{13}COCH_3$ (Table 2, No. **2a**). Extraction of the reaction residue (Method I) with pentane and cooling the concentrated extracts to −78 °C afford the crude product which is further purified by repeated vacuum sublimation at 40 °C onto a dry ice–cooled probe [48, 58]. Using $CH_3{}^{13}COCl$ (90% labeled) as starting material for Method I gives an 88% labeled product (determined mass spectroscopically) [99].

The mass spectrum of $Fp^{13}COCH_3$ (28% labeled) shows a similar fragmentation pattern as the unlabeled complex, that is, stepwise loss of CO and competitive loss of CH_3. The extent of labeling in the corresponding fragments indicates that decarbonylation in the mass spectrometer occurs with elimination of a terminal CO group and the relative abundances of various peaks indicate preferential loss of terminal carbonyls as compared to the CH_3 group. A dependence of the fragmentation pattern on the inlet temperature and ionization energy has been observed. Thus at 70 eV and 260 °C the major pathway proceeds by loss of terminal CO groups so that the amount of labeling in all CO–containing ions is the same as that in the starting compound. In contrast, at 20 eV and 330 °C the presumably longer lifetime of ions excited by less energetic electrons permits more facile loss of the CH_3 group with formation of $[C_5H_5Fe(CO)_2{}^{13}CO]^+$ and scrambling of the labeled CO in the subsequent fragments. All fragments and their relative intensities at 70 eV/260 °C and 20 eV/330 °C were given [58].

For the reduction of the compound with $LiAlH_4$, see "Further information" on $FpCOCH_3$ (No. 2).

UV irradiation of $Fp^{13}COCH_3$ with an excess of $P(C_6H_5)_2N(CH_3)R^1$ ($R^1=CH(C_6H_5)CH_3$) in toluene at $-70\,°C$ gives $C_5H_5Fe(^{13}CO)(P(C_6H_5)_2N(CH_3)R^1)CH_3$ (71% yield) and C_5H_5Fe-$(CO)(P(C_6H_5)_2N(CH_3)R^1)(^{13}COCH_3)$ (7% yield), both compounds as diastereomers [99]. For the corresponding reaction with $P(C_6H_5)_3$, see on p. 10.

$C_5H_5Fe(CO)_2COCF_3$ (Table 2, No. 3) is formed along with appreciable amounts of Fp_2 by treatment of Na[Fp] with $(CF_3CO)_2O$ (1:1 mole ratio) in THF at $-78\,°C$ and stirring the mixture at room temperature for 16 h. After solvent removal under reduced pressure, the product is separated by distillation of the residue at $110\,°C/0.02$ Torr and purified by crystallization from pentane followed by sublimation at $60\,°C/0.1$ Torr, 23% yield [28]. Workup of the above reaction residue as described for $FpCOC_2F_5$ (see No. 16) gives only a 9% yield [8].

$C_5H_5Fe(CO)_2COCH_2OCH_3$ (Table 2, No. 4) is extracted from the reaction residue (Method I) with ether. After the extracts have been washed with water and dried, the solvent is removed, leaving the product as an oil [55].

Treatment of the compound with $HPF_6 \cdot O(C_2H_5)_2$ in CH_2Cl_2 for 5 min affords a yellow crystalline precipitate of $[FpCH_2=C(OH)OCH_3]PF_6$ [55]. Electrophilic activation (not specified) of the alkoxyacetyl ligand to give $[Fp=C(OR^1)CH_2OCH_3]^+$ followed by reduction with $Li[BH(C_2H_5)_3]$ leads to formation of $FpCH(OR^1)CH_2OCH_3$, and ultimately of $FpCH_2CHO$ [138].

Isomerization in the presence of acid to give $FpCH_2COOCH_3$ was mentioned [147].

$C_5H_5Fe(CO)_2COCHS(O)(CH_3)_2$ (Table 2, No. 6). All ^1H NMR signals appear as sharp singlets at room temperature. Temperature-dependent ^1H NMR studies show no significant broadening or splitting of the signals down to $-90\,°C$, indicating the presence of only one isomer. For steric reasons and charge distribution, the isomer with the (E)–configuration must be the thermodynamically preferred one [64].

Comparison of the ^{57}Fe Mössbauer data with those of the carbamoyl compounds $FpCONR_2^1$ ($R^1=CH_3$, C_2H_5) reveals a similar electronic and geometric environment of the iron atom for all complexes studied [64].

The IR band at $1148\;cm^{-1}$ in Nujol or $1175\;cm^{-1}$ in CH_2Cl_2 is tentatively assigned to a $\nu(SO)$ vibration [64].

The NMR and IR data support the formulation of the compound as an ylide [64].

The diamagnetic compound is readily soluble in polar organic solvents such as THF, acetone, pyridine, CH_3CN, and CH_2Cl_2, moderately soluble in ether and toluene, and insoluble in saturated hydrocarbons. It is stable in air for a short period, however, it becomes orange–red to red–violet after several hours [64].

The mass spectrum does not show the molecular ion $[M]^+$, however the fragments $[M-CH_3]^+$, $[M-CO]^+$, $[M-(CH_3)_2S=O]^+$, and $[M-CHS(O)(CH_3)_2]^+$ were observed [64].

Addition of excess HCl in ether to $FpCOCHS(O)(CH_3)_2$ in THF at room temperature gives a yellow precipitate consisting of a mixture of $[C_5H_5Fe(CO)_3]Cl$ and $[(CH_3)_3SO]Cl$ [64].

$C_5H_5Fe(CO)_2C^1OC^2H=P(CH_3)_3$ (Table 2, No. 7) is also obtained from $[C_5H_5Fe(CO)_3]BF_4$ and $(CH_3)_3P=CH_2$ [84] or by treatment of $FpCOC(CH_3)=P(C_2H_5)_3$ (No. 106) with $(CH_3)_3P=CH_2$ (1:1.1 mole ratio) in benzene at room temperature for 1 h, removal of the solvent and $(C_2H_5)_3P=CHCH_3$ in vacuum and recrystallization of the residue from benzene/pentane (1:2), 98% yield [109].

References on pp. 47/51

The compound crystallizes in the monoclinic space group $P2_1/a-C_{2h}^5$ (No. 14) with $a =$ 10.721(3), $b = 10.293$(3), $c = 12.329$(3) Å, $\beta = 93.84$(2)°; $Z = 4$, $d_c = 1.438$ g/cm³. The molecular structure is shown in **Fig. 1**. The almost planar bonding system comprising Fe, C(1)-O, C(2), P as well as the elongation of 0.05 Å of the P–C(2) distance and the shortening of ~0.08 Å of the P–CH₃ bonds indicate a participation of the phosphonium enolate form (X) in the ground state of the complex, while the Fe–C(1) distance (2.039(1) Å) reveals that electron transfer from the metal to the acyl function is unimportant [84].

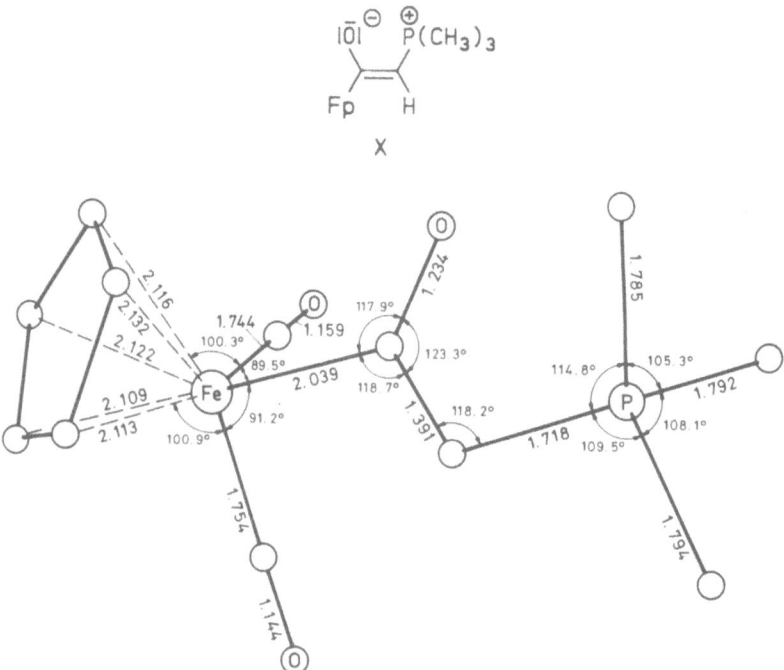

Fig. 1. Molecular structure of $C_5H_5Fe(CO)_2COCH=P(CH_3)_3$ [84].

The compound reacts with R^1Cl ($R^1 = H$, $Si(CH_3)_3$) [84] or R^1I ($R^1 = CH_3$) in benzene [84, 109] to give the salts $[FpCOCH(R^1)P(CH_3)_3]X$ ($X = Cl$, I). For the corresponding reaction with CH_3OSO_2F, see p. 12. Heating or UV irradiation of the complex in the presence of $P(CH_3)_3$ does not lead to CO exchange, nor does the acyl group react with $(CH_3)_3P=CH_2$, both indicating strong interaction between the ylidic carbanion and the acyl carbon C(1). This is supported by the relatively short C(1)–C(2) distance (1.391(1) Å) [84], cf. Formula X [155].

$C_5H_5Fe(CO)_2COCH=P(CH_3)_2C_2H_5$ (Table 2, No. 9) has been obtained by treatment of $FpCOC(CH_3)=P(CH_3)_3$ (No. 17) with a catalytic amount of $(CH_3)_3P=CH_2$ in benzene at room temperature for 10 h. Addition of pentane to the concentrated reaction mixture affords the product in a 97% yield [109].

$C_5H_5Fe(CO)_2COCH_2C_6H_5$ (Table 2, No. 12) is isolated from the reaction residue (Method I) by sublimation. The sublimate is further purified by chromatography on Florisil with light petroleum/ether (9:1) as eluent. The compound has also been obtained by hydrolysis of $FpC\equiv CC_6H_5$ in THF with 5% aqueous HCl for 4 h, extraction of the mixture with ether, solvent removal, and workup of the residue as above, 51% yield [54]. It also forms by the reaction of FpC≡CC$_6$H$_5$ with HBF$_4$·O(CH$_3$)$_2$ in anhydrous CH$_3$OH, 30% yield [74] and by exposure of [FpC(OC$_2$H$_5$)CH$_2$C$_6$H$_5$]BF$_4$ to H$_2$O [104].

The mass spectrum shows ions at m/e 296, 268, 240, and 212, together with strong lines at 205 and 91 corresponding to [C$_5$H$_5$Fe(CO)$_3$]$^+$ and [C$_7$H$_7$]$^+$, respectively [54].

Oxidation of the compound with [NH$_4$]$_2$[Ce(NO$_3$)$_6$] in R^{1}OH (R^1 = CH$_3$, C$_2$H$_5$, i-C$_3$H$_7$) gives C$_6$H$_5$CH$_2$COR^1, C$_6$H$_5$CH$_2$COOR^1, C$_6$H$_5$CHO, and C$_6$H$_5$CH$_2$CH$_2$C$_6$H$_5$. Based on the product distribution resulting from varying reaction parameters, a reaction mechanism is proposed involving [FpCOCH$_2$C$_6$H$_5$]$^{+·}$ as the intermediate [123]; see also [87] and "Further information" on FpCOCH$_3$ (No. 2). The same products, however in another proportion, are formed on oxidation of FpCOCH$_2$C$_6$H$_5$ with AgNO$_3$ [123]. For other oxidation procedures, see p. 9.

C$_5$H$_5$Fe(CO)$_2$COCH(C$_6$H$_5$)$_2$ (Table 2, No. **13**) is isolated from the reaction residue (Method I) by extraction with hexane/CH$_2$Cl$_2$ (1:1) and cooling of the extracts to −20 °C for 48 h [65].

C$_5$H$_5$Fe(CO)$_2$COC(C$_6$H$_5$)$_3$ (Table 2, No. **14**). Workup of the reaction residue (Method I) is carried out as described [4] for FpCOCH$_3$ [48]; see No. 2.

C$_5$H$_5$Fe(CO)$_2$COC$_2$H$_5$ (Table 2, No. **15**). The reaction residue (Method I) is worked up by chromatography on Al$_2$O$_3$ with petroleum ether [151] or by extraction with light petroleum (b.p. 40 to 60 °C), filtration and evaporation of the solvent at 40 °C/0.1 Torr. Chromatography of the resulting oil on Al$_2$O$_3$ in light petroleum/benzene indicated that no further purification was necessary after extraction of the residue [5]. The compound has also been obtained by hydrolysis of FpC≡CCH$_3$ or FpCH$_2$C≡CH in ethanol with 10% ethanolic HCl, extraction with ether and chromatography of the extracts on Florisil with ether/petroleum ether (1:1) as eluent. However, the main product from the reaction of FpCH$_2$C≡CH was FpCH$_2$COCH$_3$ [24]. A 1:2 mixture of FpCOC$_2$H$_5$ and FpCH$_2$COCH$_3$ forms by protonation of FpCH=C=CH$_2$ with HPF$_6$ etherate at −20 °C, giving first [FpHC≡CCH$_3$]PF$_6$ which then rapidly reacts with water [51]. Treatment of FpC(OCOCH$_3$)=CHCH$_3$ with ethanolic KOH also results in formation of FpCOC$_2$H$_5$ [24].

The very slightly volatile compound is stable in air and soluble in common organic solvents [5].

For the reaction with LiN(Si(CH$_3$)$_3$)$_2$ and then with CH$_3$I, see Preparation Method V.

C$_5$H$_5$Fe(CO)$_2$COC$_2$F$_5$ (Table 2, No. **16**) is obtained from the reaction residue (Method I) by extraction with CH$_2$Cl$_2$, solvent removal from the extracts (filtered through Al$_2$O$_3$) and chromatography of a benzene solution of the residue (containing Fp$_2$, FpCOC$_2$F$_5$, and tarry by-products) on Al$_2$O$_3$. After solvent removal of the benzene eluate at 30 Torr, the residue is extracted with pentane (Fp$_2$ is nearly pentane-insoluble). On cooling of the extracts to −78 °C FpCOC$_2$F$_5$ precipitates. It is further purified by sublimation at 50 to 80 °C/0.1 to 0.5 Torr [8].

C$_5$H$_5$Fe(CO)$_2$COC(CH$_3$)=P(CH$_3$)$_3$ (Table 2, No. 17) has been obtained by addition of (CH$_3$)$_3$P=CH$_2$ in ether to a suspension of [FpCOCH(CH$_3$)=P(CH$_3$)$_3$]I (1:1 mole ratio) in the same solvent at room temperature, filtration of [P(CH$_3$)$_4$]I formed after 45 min, and concentration of the mixture. Addition of pentane gives the product in a 90% yield [109].

The compound is thermally stable in the pure state; it does not decompose even on refluxing in benzene. It remains unchanged on storing at room temperature in daylight for several weeks. However, the compound rearranges irreversibly to the thermodynamically favored isomer FpCOCH=P(CH$_3$)$_2$C$_2$H$_5$ (No. 9) on treatment with a catalytic amount of (CH$_3$)$_3$P=CH$_2$ in benzene at room temperature for 10 h. A mechanism for the rearrangement is estab-

                                            References on pp. 47/51

lished involving the intermolecular transfer of the $C_5H_5Fe(CO)_3$ group between the tautomeric ylides $(CH_3)_3P=CHCH_3$ and $C_2H_5(CH_3)_2P=CH_2$ [109].

$C_5H_5Fe(CO)_2COCH_2CH_2C_5H_5$ (Table 2, Nos. **19** and **20**) have been obtained as a 1:1 mixture by carbonylation of $FpCH_2CH_2(C_5H_5$-cyclo), formed from spiro[2.4]hepta-4,6-diene and Na[Fp] in THF and subsequent hydrolysis, in benzene at 110 °C/15 bar CO pressure in an autoclave for 6 h, followed by chromatography of the concentrated mixture on Al_2O_3 with hexane/ether (1:1) as eluent, 72% yield [118].

The mass spectrum of the mixture shows fragments (relative intensities) at m/e = 298 (2), 270 (16), 242 (5), 214 (100), 121 (96), and 56 (70) [118].

$C_5H_5Fe(CO)_2COCH(CH_3)C_6H_5$ (Table 2, No. **23**) is isolated from the reaction residue (Method I) by extraction with pentane, filtration of the extracts through Al_2O_3, and chromatography of the concentrated and cooled (-78 °C) filtrate on Al_2O_3. The product precipitates on concentration and cooling to -78 °C of the pentane(?) eluate. The optically active compound has been prepared similarly (Method I) by using $(-)_{546}$-$C_6H_5CH(CH_3)COCl$ [33].

$C_5H_5Fe(CO)_2COCH(CH_3)C_6H_5$ gradually turns brown over a period of several weeks even when stored under N_2 in a freezer. The optically active compound decomposes after several days. Solutions in organic solvents decompose within hours [33]. The complex exhibits a negative Cotton effect curve [22].

The optically active species melts at a lower temperature (oil at room temperature) than its respective inactive form (crystals at room temperature). The latter is therefore most probably a racemic compound [33].

$C_5H_5Fe(CO)_2COC_3H_7$-n (Table 2, No. **25**) also forms by addition of Fp^+ to a CO-blanketed solution of FpC_3H_7-n in CH_2Cl_2 at room temperature [131].

$C_5H_5Fe(CO)_2COC_3H_7$-i (Table 2, No. **26**). Workup of the reaction residue (Method I) is performed by extraction with petroleum ether, solvent removal from the filtered extracts under reduced pressure and chromatography of the residue on SiO_2 with benzene as eluent. The golden brown liquid remaining after removing the benzene is crystallized at -20 °C from pentane [65]. Workup by chromatography on Al_2O_3 with petroleum ether is also possible [151].

$C_5H_5Fe(CO)_2COC_3F_7$-n (Table 2, No. **27**). The reaction residue (Method I) is worked up as described for $FpCOC_2F_5$ (No. 16). The crystals which separate on cooling the pentane extracts to -78 °C are isolated by removing the supernatant liquid with a syringe. They melt on drying in a stream of N_2 while warming to room temperature, giving an orange oil which has not been further purified [8].

The observation of the molecular ion and of the fragment $[C_5H_5Fe(CO)_3]^+$ in the mass spectrum reveals that decarbonylation prior to ionization does not take place (in contrast to $FpCOC_6H_5$) in the mass spectrometer (inlet temperature 200 to 230 °C, 70 eV). The molecular ion undergoes fragmentation first by loss of the acyl CO group, then of the two terminal CO groups in one step. It also easily loses the C_3F_7 ligand to give the ion $[C_5H_5Fe(CO)_3]^+$ which then fragments into Fe^+ by stepwise elimination of the three CO groups followed by cleavage of the C_5H_5 ring. The $[C_5H_5Fe(CO)_nC_3F_7]^+$ ion can also lose one F atom, forming the corresponding $[C_5H_5Fe(CO)_nC_3F_6]^+$ fragments (n = 0, 1, and 2). Most of the fragmentation steps are supported by the presence of metastable ions. All fragments, their relative intensities, and the metastable transitions were reported [27].

$C_5H_5Fe(CO)_2CO(CH_2)_3Cl$ (Table 2, No. **28**) is extracted from the oily reaction residue (Method I) with hexane. The filtered extracts are cooled to -78 °C, and after removal of the supernatant liquid the residue is recrystallized twice at -78 °C from hexane [56].

Treatment of the compound with $AgPF_6$ in acetone for 1 h affords the cyclic complex XI. No reaction takes place with I^- [56].

$$\left[C_5H_5(CO)_2Fe \diagup\!\!\!\!\diagdown_O \overset{\oplus}{\diagdown}\!\!\!\!\diagup \right]^+ [PF_6]^-$$

XI.

$C_5H_5Fe(CO)_2COCH_2CH(CH_3)C_6H_5$ (Table 2, No. **29**). Workup of the reaction residue (Method I) is carried out by dissolution in $CHCl_3$, filtration through Al_2O_3, evaporation of the filtrate, and recrystallization of the residue from pentane/benzene (5:2). The $(-)_D$-compound has been prepared similarly using the optically pure $C_6H_5CH(CH_3)CH_2COCl$ [57].

$C_5H_5Fe(CO)_2COC_4H_9$-t (Table 2, No. **32**). The reaction residue (Method I) is mixed with warm hexane and the product obtained after filtration and solvent removal is recrystallized from hexane/benzene (10:1) [60].

$C_5H_5Fe(CO)_2COCH_2CH_2CH(CH_3)_2$ (Table 2, No. **34**). The reaction according to Method IV has also been carried out at 0 °C/9 h (85% yield, one crystalline by-product of unknown structure) and at 25 °C/1 h (75%) [131].

$C_5H_5Fe(CO)_2COCH=CH_2$ (Table 2, No. **38**). The reaction residue (Method I) is worked up as described for $FpCOC_3F_7$-n (No. 27) [8] or for $FpCOCH=C(CH_3)_2$ (No. 93) [150].

The ^1H NMR spectrum of the vinyl group in the complex has been analyzed as an AMX system [8].

The compound reacts with $Fe_2(CO)_9$ in benzene at 40 °C (2 h) to give $(FpCOCH=CH_2)Fe(CO)_4$ (35% yield); see Formula XII on p. 38 [37].

(E)-$C_5H_5Fe(CO)_2COCH=CHC_6H_5$ (Table 2, No. **39**). The reaction residue (Method I) is extracted with $CHCl_3$ and after solvent removal from the extracts (filtered through Al_2O_3) the product is chromatographed on Al_2O_3 with benzene as eluent. Evaporation of the filtered eluate at 30 Torr gives orange crystals which are purified by crystallization from CH_2Cl_2/hexane [8]. Only the (E)-complex is obtained by treatment of Na[Fp] with either (Z)- or (E)-$C_6H_5CH=CHCOCl$, the reaction times are 17 h for the (E)-chloride and 1 h for the (Z)-chloride [90]. Method I has also been carried out with K[Fp], prepared from Fp_2 and $K[BH(C_4H_9$-s$)_3]$ in THF, and $C_6H_5CH=CHCOCl$ at -78 °C (0.5 h). After warming to room temperature the solvent is removed and the residue extracted with hexane. Concentration and storage at 0 °C for 12 h affords the product in 72% yield [80, 160]. Also the (E)-isomer has been obtained by heating XIII in benzene at 60 °C for 7 h or by UV irradiation in THF for 5 h, solvent removal under vacuum from the filtered solution and thin-layer chromatography of the residue on Al_2O_3 in hexane/$CHCl_3$ (1:1), 40 and 20% yield, respectively [36, 37].

The assignment of the geometry of (E)-$FpCOCH=CHC_6H_5$ is based on the coupling constant of the vinyl protons. The signals in the NMR spectrum are well-resolved, consistent with interconversion of the conformers which is rapid on the NMR time scale [90]. The three $\nu(CO)$ bands observed in the IR spectra (in C_6H_{14}) of (E)-$FpCOCH=CHC_6H_5$ are believed to result from conformational isomerism about the metal-acyl bond. However, the resolution of the bands due to the conformers is not complete, with only the high-frequency band showing splitting. The other carbonyl-stretching and acyl-stretching bands are slightly broadened singlets [90].

References on pp. 47/51

The most distinctive feature of the rather complex mass spectrum, which shows only a very weak molecular ion, is the presence of fragments with C_6H_5 as well as C_8H_7 substituents, that is, $[C_5H_5FeC_6H_5]^+$ and the substituted ferrocene ions $[C_5H_5FeC_5H_4C_6H_5]^+$ and $[C_5H_5FeC_5H_4COC_8H_7]^+$. All fragments, their relative intensities, and the metastable ions were reported [23].

The polarographic reduction of (E)-FpCOCH=CHC$_6$H$_5$ in $CH_3OCH_2CH_2OCH_3$/0.1 M [N(C$_4$H$_9$-n)$_4$]ClO$_4$ takes place in a two-electron step giving an ill-defined half-wave potential at $E_{1/2} = -(2.4$ to $2.7)$ V (vs. 10^{-3} M AgClO$_4$/Ag). Attempts to reoxidize the reduced solution to generate the starting material have been unsuccessful [11].

The compound reacts with $Fe_2(CO)_9$ in benzene at 40 °C (2 h) to give (FpCOCH=CHC$_6$H$_5$)Fe(CO)$_4$ (XIII, 50% yield) and small amounts of (FpCOCH=CHC$_6$H$_5$)Fe(CO)$_3$ (XIV). Traces of these two products, Fp$_2$ and FpCH=CHC$_6$H$_5$ are obtained in addition to complex XV (8%), when the reaction is carried out at 60 °C for 7 h. The (E)-structure of XIII and XIV is tentatively assigned [36, 37].

XII: R^1 = H
XIII: R^1 = C$_6$H$_5$ XIV   XV

Treatment of (E)-FpCOCH=CHC$_6$H$_5$ with trimethylamine oxide (1.1 equivalent) gives a 32% yield of (E)-FpCH=CHC$_6$H$_5$ [86].

C$_5$H$_5$Fe(CO)$_2$COCH=CHCH$_3$ (Table 2, No. **41**). The reaction residue (Method I) is worked up as described for (E)-FpCOCH=CHC$_6$H$_5$ (No. 39). However, the product is further purified by vacuum sublimation onto a dry-ice/acetone cooled probe. Only the (E)-complex is obtained by treatment of Na[Fp] with either (Z)- or (E)-CH$_3$CH=CHCOCl, the reaction times are 17 h for the (E)-chloride and 1 h for the (Z)-chloride [90].

The assignment of the geometry of (E)-FpCOCH=CHCH$_3$ is based on the coupling constant of the vinyl protons. The signals in the NMR spectrum are well-resolved, consistent with rapid interconversion of the conformers on the NMR time scale. The three ν(CO) bands observed in the IR spectrum (in C$_6$H$_{14}$) of (E)-FpCOCH=CHCH$_3$ are believed to result from conformational isomerism about the metal-acyl bond. However, the resolution of the bands due to the conformers is not complete, only the high-frequency band shows splitting. The other carbonyl-stretching and acyl-stretching bands are slightly broadened [90].

An (E)/(Z) mixture (75% yield) results from preparation according to Method I and workup as described for FpCOCH=C(CH$_3$)$_2$ (No. 93). The geometry of CH$_3$CH=CHCOCl used was not specified [150].

C$_5$H$_5$Fe(CO)$_2$COC(CH$_3$)=CH$_2$ (Table 2, No. **42**) is isolated from the oily reaction residue (Method I) by washing with CH_2Cl_2 after addition of water, filtration of the extracts through Al$_2$O$_3$ and evaporation of the solvent under vacuum. The crude product is distilled at 88 to 92 °C/10^{-2} Torr [145]. Another workup of the reaction residue was described in [150]. For details, see "Further information" on FpCOCH=C(CH$_3$)$_2$ (No. 93).

The compound is quite air-stable, it may also be purified by chromatography on SiO$_2$ in the air [113].

References on pp. 47/51

$C_5H_5Fe(CO)_2CO(CH_2)_nCH=CH_2$ (n = 1 to 3) and (E)-$C_5H_5Fe(CO)_2COCH_2CH_2CH=CHCH_3$ (Table 2, Nos. **43**, **44**, **47**, and **48**). Compounds No. 43, 44, and 47 prepared according to Method I are purified by chromatography on Al_2O_3. No. 48 ((E)-structure?) has been obtained in 15% yield by heating $FpCH_2CH_2CH=CHCH_3$ in CH_3CN at 90 °C for three days and purified as aforementioned. Under similar conditions $FpCH_2CH_2CH=CH_2$ gives compound No. 44 [35].

$C_5H_5Fe(CO)_2COC_3H_5$ (Table 2, No. **50**) is extracted from the reaction residue (Method I) with CH_2Cl_2. Chromatography on Florisil with light petroleum (b.p. 40 to 60 °C)/benzene (9:1) as eluent gives an oil which is further purified by vacuum distillation (bath temperature 140 °C/0.01 Torr) [25]. The reaction mixture (Method I) is transferred to a flask containing Al_2O_3 followed by removal of the solvent in vacuum, chromatography of the products (Fp_2 and title compound coated on the Al_2O_3) on Al_2O_3 with hexane/CH_2Cl_2 (1:1) as eluent, and solvent removal from the eluate [110].

The compound remains liquid on a −78 °C probe. It slowly decomposes in air and light at room temperature, but is stable under N_2. Little decomposition occurs during distillation. The complex is soluble in hydrocarbons and other common organic solvents [25].

Data from the literature concerning the decarbonylation by UV irradiation are contradictory. During the "UV irradiation of a light petroleum solution for 24 h, or of the complex alone for 16 h" no FpC_3H_5-cyclo was formed [25], whereas the decarbonylation "with a 450 Watt Hg lamp in a pyrex well" at 10 °C gave yields of 75% (70 min in CD_3COCD_3), 52% (2 h in C_6D_6), or 16% (105 min in petroleum ether) [110]. Attempts to decarbonylate the compound either by heating in light petroleum at 120 °C for 24 h in a sealed tube [25], or by treatment with $Rh(P(C_6H_5)_3)_3Cl$ were not successful [50].

The mass spectrum shows stepwise loss of the terminal carbonyl and acyl carbonyl groups to give $[FeC_3H_5]^+$ which undergoes further fragmentation [25].

Reaction with $[O(CH_3)_3]BF_4$ in CH_2Cl_2 at 25 °C for 12 h gives $[Fp=C(OCH_3)(C_3H_5$-cyclo)]$^+$, which, on quenching directly in a cold (−78 °C) stirred solution of $NaBH_4$ in CH_3ONa/CH_3OH, affords $FpCH(OCH_3)C_3H_5$-cyclo [140].

$C_5H_5Fe(CO)_2COC_3H_2(CH_3)_2OCH_3$ (Table 2, No. **51**). For preparation, $C_3H_2(CH_3)_2(OCH_3)$-COOH is combined with $SOCl_2$ in ether/dimethylformamide for 8 h and (without isolation of the chloride?) converted to No. 51 by K[Fp] in THF at −78 to +20 °C (6 h) [144].

Photolysis in benzene results in formation of the ring-expanded carbene complex XVI. The proposed mechanism involves migration of alkyl from carbon to 16-electron coordinately-unsaturated iron, thus generating the carbene complex XVII which subsequently ring expands to give XVI [143].

XVI   XVII

$C_5H_5Fe(CO)_2COC_3H_3(C_2H_5)_2$ (Table **2**, No. **52**). The reaction mixture (Method I) is worked up in a manner similar to that described for $FpCOC_3H_5$–cyclo (No. 50) [110].

$C_5H_5Fe(CO)_2COC_3H_3(C_6H_5)_2$ (Table **2**, No. **53**). Similar workup of the reaction mixture (Method I) as described for compound No. 50. The compound also forms when, in Method I, 1–chloro–2,3–diphenylcyclopropane–1–carbonylchloride is used in place of 2,3–diphenylcyclopropane–1–carbonylchloride, 15% yield [110].

$C_5H_5Fe(CO)_2COC_3H(C_6H_5)_2$ (Table **2**, No. **54**). The reaction residue (Method I) is worked up as described for compound No. 50, except that the product is eluted with ether and further purified by recrystallization from CH_2Cl_2/hexane at $-30\,°C$ [116].

The compound is thermally stable. It was recovered unchanged from refluxing hexane [81], THF, and benzene solutions [116].

UV irradiation in hexane leads to decomposition affording Fp_2 as the only metal–containing product [81].

The mass spectrum shows the molecular ion $[M]^+$ and the fragments $[M-CO]^+$, $[M-2CO]^+$, and $[C_3H(C_6H_5)_2]^+$ [116].

$C_5H_5Fe(CO)_2COC_4H_7$ (Table **2**, No. **55**). The reaction residue (Method I) consists of a red oil, which is purified by distillation under reduced pressure [50].

The presence of two (ketonic) $\nu(C=O)$ bands in the IR spectrum suggests two isomeric forms of the complex. Attempts to separate these isomers have not been made [50].

On extended exposure to air, the compound begins to decompose. It is freely miscible with organic solvents [50].

$C_5H_5Fe(CO)_2COC_6H_{10}CH_3$ (Table **2**, No. **56**) is extracted from the reaction residue (Method I) with petroleum ether (b.p. 30 to 60 °C) and the extracts are chromatographed on Al_2O_3 with petroleum ether as eluent. Solvent removal in vacuum affords the product [86].

$C_5H_5Fe(CO)_2COC_7H_{11}$ (Table **2**, No. **57**). Similar workup of the reaction mixture (Method I) as described for compound No. 50. The complex also forms when, in Method I, 7–chloro(or bromo)–bicyclo[4.1.0]heptan–7–carbonylchloride is used in place of the acyl chloride unsubstituted in position 7; 17 and 3% yields are obtained from Method I and this latter procedure, respectively [110].

$C_5H_5FeCOC_7H_{11}$ (Table **2**, Nos. **58**, endo–compound; and **61**, exo–compound) have also been prepared according to Method I, although by addition of K[Fp] (in place of Na[Fp]) to endo– and exo–2–norbornanecarbonylchloride (6:5 mole ratio) in THF at 0 °C and further reaction at room temperature for 10 h. After solvent removal under reduced pressure the residue is washed with pentane and the pentane suspension is filtered through Celite and concentrated. Chromatography on Al_2O_3 at 10 °C using ether/pentane (1:50) as eluent gives the products after solvent removal in 69 and 73% yields, respectively. Lower yields of the endo–complex are obtained when the preparation according to Method IV is carried out under Ar at 0 °C/15 h (45%) or under N_2 at $-20\,°C/24$ h (28%). Similarly, the exo–compound is formed in only 17% yield, when the corresponding reaction is performed under N_2 or Ar at 25 °C/7 h [146].

Both compounds are oxidatively and thermally stable [146].

$C_5H_5Fe(CO)_2COC_7H_{10}D$ (Table **2**, No. **59**, endo–compound) has been prepared from K[Fp] and the corresponding R'COCl (see No. 58) in 76% yield [146].

References on pp. 47/51

The $^{13}C\{^1H\}$ NMR spectrum is identical to that of No. 58 except for the C-2 signal ($\delta =$ 76.4 ppm), which is split and shifted slightly upfield by deuterium [146].

$C_5H_5Fe(CO)_2COC_7H_9D_2$ (Table 2, No. 60, endo-compound) has been obtained from compound No. 62 which is combined with 10% Pd on charcoal in ethanol and stirred under a D_2 atmosphere for 2 h. After solvent removal from the filtered mixture the residue is chromatographed on Al_2O_3 at 10 °C with ether/pentane (1:50) as eluent to give a 73% yield of a yellow solid. The $^{13}C\{^1H\}$ NMR spectrum is consistent with the structure given in Table 2. All resonances coincide with those of No. 58 except for the signals at $\delta = 24.4$ and 29.3 ppm, which are of reduced intensity and show splitting and slight upfield shifts characteristic of deuteration at C-5 and C-6 [146].

$C_5H_5Fe(CO)_2COC_7H_9$ (Table 2, No. 62, endo-compound) has been prepared in the same manner from K[Fp] and C_7H_9COCl as described for Nos. 58 and 61, 84% yield [146].

$C_5H_5Fe(CO)_2COC_7H_{11}$ (Table 2, No. 63) is alkylated by treatment with $[O(C_2H_5)_3]BF_4$ in CH_2Cl_2 at 25 °C to give the ethoxycarbene salt $[FpC(OC_2H_5)C_7H_{11}]BF_4$ [132].

$C_5H_5Fe(CO)_2COC_7H_{10}OCH_3$ (Table 2, No. 64). The reaction mixture (Method I) is worked up as described [110] for compound No. 50. An analytically pure sample is obtained by rechromatographing the oil on Al_2O_3 with hexane/CH_2Cl_2 (4:1) as eluent [142].

$C_5H_5Fe(CO)_2COC_{10}H_{15}$-1 (Table 2, No. 65). The reaction residue (Method I) is extracted with hot pentane, the solvent removed, and the product obtained after recrystallization from hexane [60].

The compound is stable in air for at least three years. It is sublimable in vacuum (80 °C/10^{-3} Torr) without decomposition, although at slightly higher temperatures minor decarbonylation does occur to afford $FpC_{10}H_{15}$-1. Decarbonylation also proceeds slowly with prolonged refluxing of the complex in heptane, whereas UV irradiation at room temperature in saturated hydrocarbon solvents gives moderate yields of $FpC_{10}H_{15}$-1 together with substantial amounts of Fp_2 [60].

$C_5H_5Fe(CO)_2COC_{10}H_{15}$-2 (Table 2, No. 66) has been prepared by bubbling CO through a solution of $FpC_{10}H_{15}$-2 in boiling heptane for 2 h, followed by solvent removal and sublimation of the residue at 75 °C/0.1 Torr, 97% yield [60].

$C_5H_5Fe(CO)_2COC_6H_5$ (Table 2, No. 67) is extracted from the reaction residue (Method I) with $CHCl_3$. After filtration through Al_2O_3 and solvent removal at reduced pressure the remaining crystals are washed with pentane and sublimed at 90 °C/0.2 Torr [8]. The preparation according to Method I has also been carried out with K[Fp], formed from $K[BH(C_4H_9\text{-s})_3]$ and Fp_2 in THF. After 0.5 h at −78 °C the mixture is warmed to room temperature and worked up as above. The oil remaining after solvent removal from the filtrate is crystallized by dissolution in $CHCl_3$, addition of hexane, and cooling to −78 °C, 67% yield [80]. The use of $[N(C_4H_9\text{-n})_4][Fp]$ (in place of Na[Fp], Method I), prepared from $[N(C_4H_9\text{-n})_4]F$ and $FpSi(CH_3)_3$ in THF at −77 K, gave a yield of 51% (determined by ^1H NMR spectroscopy) [122]. Also, the reaction of Fp^- (formed in situ by reduction of Fp_2 with the sodium adduct of naphthalene in THF) with C_6H_5COCl in THF, evaporation of the solution under vacuum, chromatography of the residue on SiO_2, and elution with $CHCl_3/(i\text{-}C_3H_7)_2O$ (1:1) gives the compound. It is further purified by sublimation at 100 °C/10^{-3} Torr, 52% yield [94]. $FpCOC_6H_5$ forms in 3.1% yield by the reaction of $FpCOOCH_3$ (No. 77) with C_6H_5MgBr in ether at room temperature for 1 h and hydrolysis with saturated aqueous NH_4Cl solution. Evaporation of the ether layer gives a yellow-brown liquid which is chromatographed on Al_2O_3 with ether as eluent. After solvent removal the residue is dissolved in pentane, filtered, and cooled overnight at −78 °C [12]; see also [9]. The reaction of FpMgBr with C_6H_5COCl to

References on pp. 47/51

give $FpCOC_6H_5$ was reported [67]. The compound has been obtained along with Fp_2 by addition of C_6H_5Li to $[C_5H_5Fe(CO)_3][B(C_6H_5)_4]$ in THF at 0 °C, solvent removal after 15 min, and sublimation of the residue at 80 °C/0.5 Torr. IR and NMR data indicate the sublimate consists of $FpCOC_6H_5$ (~80%) and Fp_2 (~20%) [45]. Carbonylation of FpC_6H_5 under conditions described for $FpCOCH_3$ (see No. 2) represents another method to prepare $FpCOC_6H_5$ [2, 3].

The mass spectrum exhibits neither the molecular ion [23, 42] nor $[C_5H_5Fe(CO)_3]^+$, a characteristic fragment of acyl derivatives of the type FpCOR′. Instead, the highest mass ion corresponds to $[C_5H_5Fe(CO)_2C_6H_5]^+$, indicating facile decarbonylation within the mass spectrometer. Thus, the remainder of the spectrum is actually that of FpC_6H_5. All fragments and their relative intensities were reported [23].

The anionic formyl complex obtained from $FpCOC_6H_5$ and $Li[BH(C_2H_5)_3]$ (see VI on p. 12), may also be generated using the chiral borohydride XVIII. The latter and similar reaction mixtures reduce acetophenone at −78 °C/2 h to 1-phenylethanol, of which the specific rotation has been determined. The results together with a control done in the absence of a metal complex have been used to interpret the mechanism of hydride transfer reactions of several anionic metal formyl complexes [77].

XVIII

Oxidatively induced nucleophilic cleavage of the Fe–acyl bond with rapid formation of $C_6H_5COOCH_3$ takes place on treatment of $FpCOC_6H_5$ with LiCl and Ce^{4+} in CH_3OH [49].

The compound reacts with P_4S_{10} or B_2S_3 in a similar way to that described for $FpCOCH_3$ (see No. 2) [62, 69].

UV irradiation of $FpCOC_6H_5$ and $(C_6H_5)_2PCH_2CH_2P(C_6H_5)_2$ in benzene for 2.5 h affords $C_5H_5Fe(CO)[(C_6H_5)_2PCH_2CH_2P(C_6H_5)_2]COC_6H_5$ and after 18 h $C_5H_5Fe[(C_6H_5)_2PCH_2CH_2-P(C_6H_5)_2]COC_6H_5$ with the phosphine coordinated as a bidentate ligand [67]. For the analogous reaction with $P(C_6H_5)_3$, see p. 10.

Attempts to insert SO_2 by dissolution of $FpCOC_6H_5$ in liquid SO_2 at −40 °C for some hours failed [10].

Quantitative alkylation to the alkoxycarbene salt, $[Fp=C(OCH_3)C_6H_5]PF_6$, occurs when the compound is treated for 18 h with $[HC(OCH_3)_2]PF_6$, prepared in situ from $HC(OCH_3)_3$ and $[C(C_6H_5)_3]PF_6$ in CH_2Cl_2 at 20 °C [149].

$C_5H_5Fe(CO)_2COC_6H_4F$-4 (Table 2, No. **68**) is isolated from the reaction residue (Method I) by extraction with CH_2Cl_2/hexane and cooling the filtered and concentrated solution to −78 °C [32].

References on pp. 47/51

$C_5H_5Fe(CO)_2COC_6F_5$ (Table 2, No. **69**) has been prepared by addition of a slurry of $[C_5H_5Fe(CO)_3]PF_6$ in THF to a solution of C_6F_5Li (1:2 mole ratio) in ether at $-78\,°C$, stirring the mixture for 1 h and then at 0 °C for an additional 2 h. After filtration and evaporation, the residue is dissolved in benzene and chromatographed on Al_2O_3 using 10% benzene in hexane as eluent. $FpCOC_6F_5$ is obtained in 18% yield. Other products isolated are FpC_6F_5 (13%) and a very small amount of exo-$(1-C_6F_5)C_5H_5Fe(CO)_3$ [18]; see also [14].

$C_5H_5Fe(CO)_2COC_6H_4OCH_3$-4 (Table 2, No. **70**) is isolated from the reaction residue (Method I) by extracting first with $CHCl_3$, then concentrated HCl. The HCl solution is filtered and No. 70 crystallizes upon addition of H_2O [161]. The compound is soluble in concentrated HCl, and crystallizes on dilution with H_2O. The protonation is believed to occur at the FpC=O oxygen with formation of $[Fp=C(OH)C_6H_4OCH_3-4]^+$ [161].

Photodecarbonylation (UV) proceeds smoothly to give $FpC_6H_4OCH_3$-4 [161].

$C_5H_5Fe(CO)_2COC_6H_4COOH$-4 (Table 2, No. **71**) is prepared by boiling No. 109 under reflux for 16 h in a concentrated HCl/acetone/H_2O mixture (5:7:5). The yellow crystals formed (mixture of Nos. 71 and 109) are chromatographed on silica gel. With C_6H_6/petroleum ether (1:1) No. 109 is eluted, with $CHCl_3$ No. 71, yield 40% (from $C_6H_5CH_3$) [161].

Treatment of a suspension of $(C_6H_6Mo(\pi-C_3H_5)Cl)_2$ in ethanol with $FpCOC_6H_4COOH$-4/ KOH in H_2O at 60 °C for 30 min affords $FpCOC_6H_4COOMo(\pi-C_3H_5)C_6H_6$ [53].

$C_5H_5Fe(CO)_2COR'$ ($R' = C_4H_3O$, $C_4H_2O(CH_3)$, C_8H_5O, C_4H_3S) (Table 2, Nos. **72 to 75**). The reaction residue (Method I) is worked up by chromatography on Al_2O_3 plates with petroleum ether/$CHCl_3$ (1:1) as eluent [63].

Compound No. 75, 47% enriched with ^{13}C at the acyl group, has also been prepared according to Method I [117].

The mass spectra of both labeled and unlabeled compound No. 75 show the same fragmentation pattern. The difference appears mainly in the ratio of the intensities of the parallel fragments as can be seen from the reported data [117].

UV irradiation of No. 75 and $(C_6H_5)_3GeC\equiv CC_6H_5$ (1:1 mole ratio) in benzene for 4 h gives, in addition to the decarbonylated complex, FpC_4H_3S (16% yield), a 51% yield of the cyclization product XIX (structure confirmed by X-ray analysis), which was the only structural isomer of four possible ones which could be isolated [129], cf. similar reaction with $C_6H_5C\equiv CC_6H_5$, p. 11.

XIX

Replacement of one CO group takes place on photolysis of compound No. 75 and the ^2D ligand $(C_6H_5)_2PC\equiv CC_6H_5$ (5:6 mole ratio) in benzene for 5 h, giving $C_5H_5Fe(CO)[(C_6H_5)_2PC\equiv CC_6H_5]COC_4H_3S$, whereas under similar conditions, but in the presence of $CH_3OOCC\equiv CCOOCH_3$ (1:1 mole ratio), the unstable product $C_5H_5Fe(CO)(CH_3OOCC\equiv CCOOCH_3)COC_4H_3S$ is obtained in which the acetylenic ligand is π-bonded to iron [129].

$C_5H_5Fe(CO)_2COOH$ (Table 2, No. **76**). The yellow precipitate formed on treatment of a cold aqueous solution of $[C_5H_5Fe(CO)_3]BF_4$ with an equimolar amount of KOH is assumed to be FpCOOH. However, the material rapidly decomposes to yield Fp_2 with evolution of

References on pp. 47/51

H_2 and CO_2. Apparently, FpCOOH decarbonylates to give FpH, which can be identified in a benzene extract of the reaction mixture, and which is known to decompose into hydrogen and Fp_2 [83]. In connection with other metallocarboxylic acids, the compound is mentioned in [157]. Protonating [FpCO$_2$]Na gives [$C_5H_5Fe(CO)_3$]$^+$, purportedly via No. 76 [163].

$C_5H_5Fe(CO)_2COOCH_3$ (Table 2, No. **77**) has been prepared by refluxing FpCON(CH$_3$)$_2$ in methanol for 1 h followed by cooling to room temperature, solvent removal under reduced pressure and extraction of the residue with pentane. The crystals, which precipitate from the filtered extracts upon cooling to −78 °C overnight, are purified by sublimation at 60 to 90 °C/0.1 Torr onto a cooled (−78 °C) probe, 49% yield [12]; see also [9]. The compound has also been obtained by treatment of [FpCOO]$_2$Mg (prepared in situ from Fp$_2$Mg and CO_2 in THF at −90 to +25 °C) with CH$_3$SO$_3$CF$_3$ (1:2 mole ratio). The yield, determined by quantitative IR measurements, was 71%. Isolation of the product by extraction with pentane and crystallization reduces the yield due to the interference of polymerized THF, induced by CH$_3$SO$_3$CF$_3$ [139]; see also [126]. The compound also forms by stirring a mixture of [$C_5H_5Fe(CO)_3$]PF$_6$ in absolute methanol in the presence of a tenfold excess of NaOCH$_3$ at room temperature for 4 h. Solvent removal, dissolution of the residue in pentane, and cooling the filtered solution to −78 °C overnight affords the product which has been identified by its IR spectrum [20].

A correlation of the CO stretching force constants of a series of transition–metal thiocarbonyls with those of their carbonyl analogues was pointed out. It revealed equal force constants for pairs where the CS and CO ligands have the same electronic and symmetry properties, for example, in $C_5H_5Fe(CO)_2COOCH_3$ and $C_5H_5Fe(CO)(CS)COOCH_3$ [70].

The volatile compound decomposes slowly into Fp$_2$ upon storage in a closed container over a period of weeks even in a freezer at −10 °C. It is readily soluble, even in nonpolar solvents [12].

Treatment of FpCOOCH$_3$ with excess gaseous HCl in ether or benzene for several minutes affords a pale yellow hygroscopic precipitate, [$C_5H_5Fe(CO)_3$][HCl$_2$] [9, 12]. The reaction mechanism may be similar to the acid-catalyzed hydrolysis of organic esters [20]. The reaction with CH$_3$SO$_3$F or CH$_3$SO$_3$CF$_3$ causes cleavage of the OCH$_3$ group and formation of [$C_5H_5Fe(CO)_3$]$^+$ as the only isolable organometallic product [108]. Similarly, protonation of FpCOOCH$_3$ (formed in situ from [FpCOO]$_2$Mg and CH$_3$SO$_3$CF$_3$) with HBF$_4$ · O(C$_2$H$_5$)$_2$ gives [$C_5H_5Fe(CO)_3$]BF$_4$ [139].

The compound reacts with R′MgBr (1:2 mole ratio) in ether at room temperature in 1 h (R′ = C$_6$H$_5$) or 16 h (R′ = CH$_3$) to give FpCOR′ after hydrolysis with saturated aqueous NH$_4$Cl [12]; see also [9].

$C_5H_5Fe(CO)_2COOC_6H_{11}$ (Table 2, No. **78**) has been prepared by treatment of the carbene complex XX with HBF$_4$ · O(C$_5$H$_5$)$_2$ in the presence of excess CO followed by demethylation of the cation formed (XXI) with N(C$_2$H$_5$)$_3$, 83% yield [71].

XX   XXI

The compound reacts with $HBF_4 \cdot O(C_2H_5)_2$ to give $[C_5H_5Fe(CO)_3]BF_4$ and cyclohexanol [71].

$C_5H_5Fe(CO)_2CONH_2$ (Table 2, No. 79) is readily soluble in C_6H_6, CH_2Cl_2, and liquid NH_3. It is stable under N_2 for weeks and can be isolated even from refluxing benzene without decomposition [44]. In contrast, $FpCONH_2$, prepared from $[C_5H_5Fe(CO)_3]PF_6$ and gaseous NH_3, is extremely unstable [20].

Warming $FpCONH_2$ (formed from $[C_5H_5Fe(CO)_3]PF_6$ and liquid NH_3 at $-50\,°C$) to $+10\,°C$ for 10 min gives FpH and NH_4NCO [89].

The mass spectrum shows a fragmentation pattern similar to that described for $FpCONHCH_3$ [20] and $FpCOR'$ ($R' = CH_3$, C_6H_5) [23]. There are also peaks of low intensity resulting from primary cleavage of the NH_2 group. The following ions and their relative intensities were reported: $[C_5H_5Fe(CO)_2CONH_2]^+$ (13), $[C_5H_5Fe(CO)_2NH_2]^+$ (92), $[C_5H_5Fe(CO)NH_2]^+$ (96), $[C_5H_5FeNH_2]^+$ (100), and $[C_5H_5Fe]^+$ (100) [44].

$C_5H_5Fe(CO)_2CONHCH_3$ (Table 2, No. 80). The mass spectrum shows the molecular ion and fragments corresponding to the loss of CO and $NHCH_3$ [20]. In an attempt to decarbonylate $FpCONHCH_3$ to the unknown $FpNHCH_3$, a solution of the compound in THF was irradiated for 15 h. Only Fp_2 and unidentified materials were formed [20].

The compound reacts with $COCl_2$ in CH_2Cl_2 in the presence of $N(C_2H_5)_3$ to give $[FpC{\equiv}NCH_3]Cl$. However, the reaction mixture must contain enough base so that all HCl formed can be neutralized. Otherwise, $FpCONHCH_3$ reacts with HCl to yield $[C_5H_5Fe(CO)_3]^+$ and CH_3NH_2 (reverse formation reaction) [59].

$FpSC(S)NHCH_3$ and a small amount of $C_5H_5Fe(CO)SC(S)NHCH_3$ (<5%) are the products of the reaction with CS_2 at room temperature for 24 h. Higher temperature (boiling CS_2) or longer reaction time slightly increases the yield of the latter bidentate species. If COS is bubbled into a toluene solution of $FpCONHR^1$ ($R^1 = CH_3$, C_2H_5) only $FpSC(O)NHR^1$ is formed after three days. No bidentate complexes have been detected even at higher temperature or longer reaction time. The mechanism of the (di)thiocarbamate formation is discussed on the basis of a nucleophilic attack of the carbamoyl N atom on the carbon of CS_2 or COS, respectively [85].

It has been found that acyl chlorides react with $FpCONHCH_3$ to give the amides (e.g. $C_6H_5CONHCH_3$ from C_6H_5COCl) and FpCl, thus indicating the aminating ability of the carbamoyl derivatives [96].

The reaction with $HgCl_2$ proceeds in a similar way to that described for $FpCONHC_6H_{11}$-cyclo (see No. 87) [34].

$C_5H_5Fe(CO)_2CON(CH_3)_2$ and $C_5H_5Fe(CO)_2CON(C_2H_5)_2$ (Table 2, Nos. 81 and 83) have been prepared by treatment of Na[Fp] with R_2^1NCOCl ($R^1 = CH_3$ or C_2H_5; 1:1 mole ratio) in THF at room temperature for 16 h, solvent removal at 30 Torr, extraction of the residue with CH_2Cl_2, and filtration of the extracts through Al_2O_3. The crude products obtained after evaporation of the solvent can be purified either by recrystallization from CH_2Cl_2/hexane or sublimation at $80\,°C/0.1$ Torr; the yields are 42 and 59%, respectively. Recrystallized products, however, decompose completely at room temperature within one day in a closed container, giving Fp_2 and developing pressure. Sublimed samples can be stored for several weeks in sealed vials under N_2 before decomposition is appreciable [4]. Destructive hydrolysis of $FpCON(CH_3)_2$ by moisture may be prevented by direct sublimation of the residue obtained from evaporation of the CH_2Cl_2 extracts at $90\,°C/0.1$ Torr without any intermediate washing with pentane or similar solvents [12].

References on pp. 47/51

The sensitivity of the dialkylcarbamoyl derivatives toward water may account for the difficulty in obtaining reproducible IR spectra in KBr. The $v(C=O)$ band at 1534 cm^{-1} in the IR spectrum of No. 83 is discussed in terms of bonding character [4].

Compounds No. 81 and 83 are volatile at 80 °C/0.1 Torr. They are soluble in C_6H_6 and CH_2Cl_2, but only sparingly soluble in saturated aliphatic hydrocarbons such as pentane and hexane [4].

$FpCON(CH_3)_2$ reacts with CH_3OH under reflux for 1 h to afford $FpCOOCH_3$ [12]; see also [9]. With C_2H_5OH a viscous orange-brown liquid, probably $FpCOOC_2H_5$, is obtained but has not been investigated in detail [12].

$C_5H_5Fe(CO)_2CONHC_4H_9$-t (Table 2, No. **86**) has been obtained by insertion of $(CH_3)_3CNCO$ into the Fe–H bond of FpH, which is generated in situ by the reaction of Na[Fp] with $(CH_3)_3CCl$ in THF at −78 °C. After addition of $(CH_3)_3CNCO$ the mixture is warmed to room temperature and stirred for 12 h. Evaporation to dryness, extraction with warm pentane, and cooling to −20 °C affords $(C_5H_5)_2Fe_2(CO)_3[CNC(CH_3)_3]$ as a precipitate. The extracts are then evaporated and the residue is sublimed under high vacuum at 70 °C onto a water-cooled probe, followed by resublimation at 30 °C over a period of 3 days to give the product [41].

$C_5H_5Fe(CO)_2CONHC_6H_{11}$ (Table 2, No. **87**) shows no reaction with CH_3I or $n-C_3F_7I$ at room temperature for 18 h [40].

Addition of $HgCl_2$ to an ether solution of the compound at −78 °C and warming the mixture to near reflux for 1 h gives FpHgCl, $[C_5H_5Fe(CO)_3]Cl$, $[cyclo-C_6H_{11}NH_3]Cl$, and cyclo-$C_6H_{11}NCO$ [40]; see also [34].

$C_5H_5Fe(CO)_2CONHNH_2$ (Table 2, No. **90**) is formed as intermediate in the reaction of NH_2NH_2 with a suspension of $[C_5H_5Fe(CO)_3]PF_6$ (2:1 mole ratio) in C_6H_6 at 7 °C during 3 to 5 min. The voluminous bright yellow precipitate consists of a mixture of presumably $[NH_2NH_3]PF_6$ and $FpCONHNH_2$, the latter of which has been identified by its carbonyl frequencies in the IR spectrum. Also, an impure form of the compound is obtained by carrying out the reaction in CH_2Cl_2 and precipitation of the products with dropwise addition of pentane [26].

The compound is not stable. It loses NH_3 to form FpNCO. In $CHCl_3$ at room temperature 70% of the complex is converted into the isocyanate compound within 1 h, whereas in the solid state the reaction is much slower with about 10% conversion in the same time [26].

$C_5H_5Fe(CO)_2CONHNHCH_3$ (Table 2, No. **91**) is formed as intermediate from $[C_5H_5Fe-(CO)_3]PF_6$ and CH_3NHNH_2 in CH_2Cl_2 within 3 min (cf. No. 90) and has been identified by its IR spectrum. The final product of this reaction is also FpNCO [26].

$C_5H_5Fe(CO)_2CON(CH_3)NHCH_3$ (Table 2, No. **92**) forms in a manner similar to Nos. 90 and 91 from $[C_5H_5Fe(CO)_3]PF_6$ and $CH_3NHNHCH_3$. The compound is too unstable to be isolated, but its IR spectrum shows the characteristic carbonyl absorptions. Unlike the other carbazoyl complexes, this one does not react further to give FpNCO. The over-all mechanism for the reaction of $[C_5H_5Fe(CO)_3]^+$ with hydrazines was discussed [26].

$C_5H_5Fe(CO)_2COCH=C(CH_3)_2$ (Table 2, No. **93**). Workup of the reaction mixture (Method I) is performed by extraction with $CHCl_3$, filtration through Al_2O_3, solvent removal, and flash chromatography of the residue on SiO_2 [150]; see also [156].

$C_5H_5Fe(CO)_2COCH_2C(CH_3)_2NHCH_2C_6H_5$ (Table 2, No. **94**). Treatment with Br_2 (3 equivalents) in CH_2Cl_2 at −78 °C and after 1 h with $N(C_2H_5)_3$ (6 equivalents) followed by gradually warming to ambient temperature affords the β-lactam XXII [150]; see also [156].

$C_5H_5Fe(CO)_2COC(CH_3)=P(C_2H_5)_3$ (Table **2,** No. **106**) reacts with $(CH_3)_3P=CH_2$ in benzene at room temperature for 1 h to give $FpCOCH=P(CH_3)_3$ (No. 7) and $(C_2H_5)_3P=CHCH_3$ (transfer of the FpCO group) [109]. Treatment with CH_3I in benzene at 25 °C for 30 h leads to C–alkylation and formation of $[FpCOC(CH_3)_2P(C_2H_5)_3]I$ [155].

$C_5H_5Fe(CO)_2COC_8H_{13}$ (Table **2,** No. **107**) reacts with $[O(C_2H_5)_3]BF_4$ in CH_2Cl_2 at 25 °C to give XXIII [158].

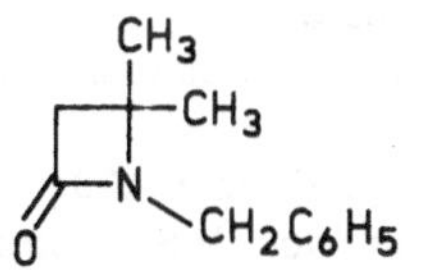

XXII 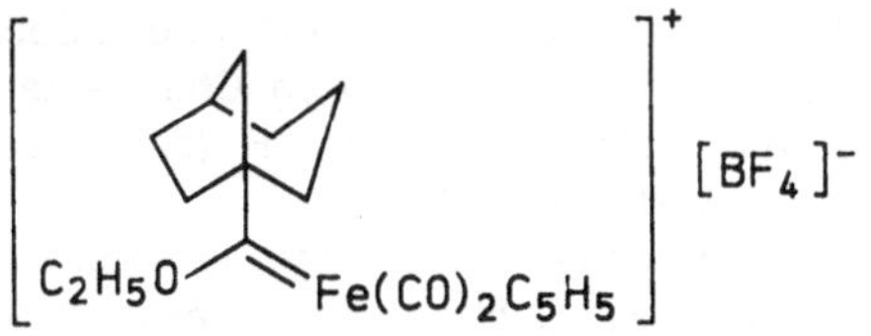   XXIII

References:

[1] Coffield, T.H.; Kozikowski, J.; Closson, R.D. (5th Intern. Conf. Coord. Chem., London 1959, Spec. Publ. No. 13, p. 126).
[2] Closson, R.D.; Coffield, T.H.; Ethyl Corp. (U.S. 3029266 [1959/62]).
[3] Closson, R.D.; Coffield, T.H.; Ethyl Corp. (U.S. 3159660 [1962/64]).
[4] King, R.B. (J. Am. Chem. Soc. **85** [1963] 1918/22).
[5] McCleverty, J.A.; Wilkinson, G. (J. Chem. Soc. **1963** 4096/9).
[6] Herber, R.H.; King, R.B.; Wertheim, G.K. (Inorg. Chem. 3 [1964] 101/7).
[7] Jolly, P.W.; Stone, F.G.A. (unpublished observations from Treichel, P.M.; Stone, F.G.A.; Advan. Organometal. Chem. 1 [1964] 143/220, 179/80, 189).
[8] King, R.B.; Bisnette, M.B. (J. Organometal. Chem. 2 [1964] 15/37).
[9] King, R.B.; Bisnette, M.B.; Fronzaglia, A. (J. Organometal. Chem. 4 [1965] 256/7).
[10] Bibler, J.B.; Wojcicki, A. (J. Am. Chem. Soc. **88** [1966] 4862/70).

[11] Dessy, R.E.; Stary, F.E.; King, R.B.; Waldrop, M. (J. Am. Chem. Soc. **88** [1966] 471/6).
[12] King, R.B.; Bisnette, M.B.; Fronzaglia, A. (J. Organometal. Chem. 5 [1966] 341/56).
[13] Nesmeyanov, A.N.; Chapovskii, Yu.A.; Lokshin, B.V.; Polovyanyuk, I.V.; Makarova, L.G. (Dokl. Akad. Nauk SSSR 166 [1966] 1125/8; Dokl. Chem. Proc. Acad. Sci. USSR **166/171** [1966] 213/6).
[14] Treichel, P.M.; Shubkin, R.L. (J. Organometal. Chem. 5 [1966] 488/90).
[15] Green, M.L.H.; Hurley, C.R. (J. Organometal. Chem. 10 [1967] 188/90).
[16] Nesmeyanov, A.N.; Chapovskii, Yu.A.; Polovyanyuk, I.V.; Makarova, L.G. (J. Organometal. Chem. 7 [1967] 329/37).
[17] Nesmeyanov, A.N.; Chapovskii, Yu.A. (Izv. Akad. Nauk SSSR Ser. Khim **1967** 2075/7; Bull. Acad. Sci. USSR Div. Chem. Sci. **1967** 1988/90).
[18] Treichel, P.M.; Shubkin, R.L. (Inorg. Chem. 6 [1967] 1328/34).
[19] King, R.B. (Inorg. Chim. Acta 2 [1968] 454/8).
[20] Busetto, L.; Angelici, R.J. (Inorg. Chim. Acta 2 [1968] 391/4).

[21] Alexander, J.J.; Wojcicki, A. (J. Organometal. Chem. 15 [1968] P23/P24).
[22] Wojcicki, A.; Alexander, J.J.; Graziani, M.; Thomasson, J.E.; Hartman, F.A. (New Aspects Chem. Metal. Carbonyls Deriv. Proc. 1st Intern. Symp., Venice 1968, Abstr. C6, pp. 1/6).
[23] King, R.B. (J. Am. Chem. Soc. **90** [1968] 1417/29).
[24] Jolly, P.W.; Pettit, R. (J. Organometal. Chem. 12 [1968] 491/5).
[25] Bruce, M.I.; Iqbal, M.Z.; Stone, F.G.A. (J. Organometal. Chem. 20 [1969] 161/8).

[26] Angelici, R.J.; Busetto, L. (J. Am. Chem. Soc. **91** [1969] 3197/200).

[27] King, R.B. (Appl. Spectrosc. **23** [1969] 137/47).

[28] King, R.B.; Kapoor, R.N.; Pannell, K.H. (J. Organometal. Chem. **20** [1969] 187/93).

[29] Whitesides, G.M.; Boschetto, D.J. (J. Am. Chem. Soc. **91** [1969] 4313/4).

[30] King, R.B.; Epstein, L.M.; Gowling, E.W. (J. Inorg. Nucl. Chem. **32** [1970] 441/5).

[31] Susuki, T.; Tsuji, J. (J. Org. Chem. **35** [1970] 2982/6).

[32] Stewart, R.P.; Treichel, P.M. (J. Am. Chem. Soc. **92** [1970] 2710/8).

[33] Alexander, J.J.; Wojcicki, A. (Inorg. Chim. Acta **5** [1971] 655/8).

[34] Angelici, R.J.; Jetz, W.; Green, C.R. (5th Intern. Conf. Organometal. Chem., Moscow 1971, Vol. 1, Abstr. No. 139).

[35] Mérour, J.Y.; Charrier, C.; Roustan, J.L.; Benaim, J. (Compt. Rend. C **273** [1971] 285/8).

[36] Nesmeyanov, A.N.; Rybin, L.V.; Rybinskaya, M.I.; Kaganovich, V.S.; Petrovskii, P.V. (Izv. Akad. Nauk SSSR Ser. Khim. **1971** 2733/8; Bull. Acad. Sci. USSR Div. Chem. Sci. **1971** 2592/6).

[37] Nesmeyanov, A.N.; Rybinskaya, M.I.; Rybin, L.V.; Kaganovich, V.S.; Petrovskii, P.V. (J. Organometal. Chem. **31** [1971] 257/67).

[38] Farnell, L.F.; Randall, E.W.; Rosenberg, E. (Chem. Commun. **1971** 1078/9).

[39] Hyano, J.K. (Diss. Univ. Alberta 1971).

[40] Jetz, W.; Angelici, R.J. (Inorg. Chem. **11** [1972] 1960/2).

[41] Jetz, W.; Angelici, R.J. (J. Organometal. Chem. **35** [1972] C37/C39).

[42] Müller, J. (Angew. Chem. **84** [1972] 725/37; Angew. Chem. Intern. Ed. Engl. **11** [1972] 653/62).

[43] Nesmeyanov, A.N.; Leshcheva, I.F.; Polovyanyuk, I.V.; Ustynyuk, Yu.A.; Makarova, L.G. (J. Organometal. Chem. **37** [1972] 159/65).

[44] Ellermann, J.; Behrens, H.; Krohberger, H. (J. Organometal. Chem. **46** [1972] 119/38).

[45] Darensbourg, M.Y. (J. Organometal. Chem. **38** [1972] 133/8).

[46] Gansow, O.A.; Schexnayder, D.A.; Kimura, B.Y. (J. Am. Chem. Soc. **94** [1972] 3406/8).

[47] Darensburg, D.J. (Inorg. Chem. **11** [1972] 1606/9).

[48] Alexander, J.J.; Wojcicki, A. (Inorg. Chem. **12** [1973] 74/6).

[49] Anderson, S.N.; Fong, C.W.; Johnson, M.D. (J. Chem. Soc. Chem. Commun. **1973** 163).

[50] Iqbal, M.Z. (Pakistan J. Sci. Res. **25** [1973] 81/5).

[51] Raghu, S.; Rosenblum, M. (J. Am. Chem. Soc. **95** [1973] 3060/2).

[52] Ingletto, G.; Tondello, E.; Di Sipio, L.; Carturan, G.; Graziani, M. (J. Organometal. Chem. **56** [1973] 335/7).

[53] Green, M.L.H.; Mitchard, L.C.; Silverthorn, W.E. (J. Chem. Soc. Dalton Trans. **1973** 1403/8).

[54] Abu Salah, O.M.; Bruce, M.I. (J. Chem. Soc. Dalton Trans. **1974** 2302/4).

[55] Cutler, A.; Raghu, S.; Rosenblum, M. (J. Organometal. Chem. **77** [1974] 381/91).

[56] Game, C.H.; Green, M.; Moss, J.R.; Stone, G.A. (J. Chem. Soc. Dalton Trans. **1974** 351/7).

[57] Reich-Rohrwig, P.; Wojcicki, A. (Inorg. Chem. **13** [1974] 2457/64).

[58] Alexander, J.J. (J. Am. Chem. Soc. **97** [1975] 1729/32).

[59] Fehlhammer, W.P.; Mayr, A. (Angew. Chem. **87** [1975] 776/7; Angew. Chem. Intern. Ed. Engl. **14** [1975] 757/8).

[60] Moorhouse, S.; Wilkinson, G. (J. Organometal. Chem. **105** [1976] 349/55).

[61] van Doorn, J.A.; Masters, C.; Volger, H.C. (J. Organometal. Chem. **105** [1976] 245/54).

[62] Busetto, L.; Palazzi, A. (Chim. Ind. [Milan] **58** [1976] 804).

[63] Nesmeyanov, A.N.; Kolobova, N.E.; Goncharenko, L.V.; Anisimov, K.N. (Izv. Akad. Nauk SSSR Ser. Khim. **1976** 153/9; Bull. Acad. Sci. USSR Div. Chem. Sci. **1976** 142/6).

[64] Weber, L. (J. Organometal. Chem. **122** [1976] 69/75).

[65] Lukehart, C.M.; Zeile, J.V. (J. Am. Chem. Soc. **99** [1977] 4368/72).

[66] Labinger, J.A.; Madhavan, S. (J. Organometal. Chem. **134** [1977] 381/9).

[67] Felkin, H.; Meunier, B.; Pascard, C.; Prange, T. (J. Organometal. Chem. **135** [1977] 361/72).

[68] King, R.B.; Iqbal, M.Z.; Frazier, C.C.; King, A.D. (8th Intern. Conf. Organometal. Chem., Kyoto, Japan, 1977, Abstr. No. 5 B 01).

[69] Busetto, L.; Palazzi, A.; Foresti Serantoni, E.; Riva Di Sanseverino, L. (J. Organometal. Chem. **129** [1977] C 55/C 58).

[70] Andrews, M.A. (Inorg. Chem. **16** [1977] 496/9).

[71] Klemarczyk, P.; Price, T.; Priester, W.; Rosenblum, M. (J. Organometal. Chem. **139** [1977] C 25/C 28).

[72] Pebler, J.; Petz, W. (Z. Naturforsch. **32 b** [1977] 1431/4).

[73] Knoll, L. (J. Organometal. Chem. **152** [1978] 311/3).

[74] Davison, A.; Solar, J.P. (J. Organometal. Chem. **155** [1978] C 8/C 12).

[75] Gladysz, J.A.; Selover, J.C. (Tetrahedron Letters **1978** 319/22).

[76] King, R.B.; King, A.D.; Iqbal, M.Z.; Frazier, C.C. (J. Am. Chem. Soc. **100** [1978] 1687/94).

[77] Gladysz, J.A.; Merrifield, J.H. (Inorg. Chim. Acta **30** [1978] L 317/L 318).

[78] Kuhlmann, E.J.; Alexander, J.J. (J. Organometal. Chem. **174** [1979] 81/7).

[79] Kuhlmann, E.J.; Alexander, J.J. (Inorg. Chim. Acta **34** [1979] L 193/L 195).

[80] Gladysz, J.A.; Williams, G.M.; Tam, W.; Johnson, D.L.; Parker, D.W.; Selover, J.C. (Inorg. Chem. **18** [1979] 553/8).

[81] Chidsey, C.E.; Donaldson, W.A.; Hughes, R.P.; Sherwin, P.F. (J. Am. Chem. Soc. **101** [1979] 233/5).

[82] Dilgassa, M.; Curtis, M.D. (J. Organometal. Chem. **172** [1979] 177/84).

[83] Grice, N.; Kao, S.C.; Pettit, R. (J. Am. Chem. Soc. **101** [1979] 1627/8).

[84] Blau, H.; Malisch, W.; Voran, S.; Blank, K.; Krüger, C. (J. Organometal. Chem. **202** [1980] C 33/C 38).

[85] Busetto, L.; Palazzi, A.; Foliadis, V. (Inorg. Chim. Acta **40** [1980] 147/52).

[86] Laycock, D.E.; Hartgerink, J.; Baird, M.C. (J. Org. Chem. **45** [1980] 291/9).

[87] Magnuson, R.H.; Zulu, S.; T'sai, W.M.; Giering, W.P. (J. Am. Chem. Soc. **102** [1980] 6887/8).

[88] Klingler, R.J.; Kochi, J.K. (J. Organometal. Chem. **202** [1980] 49/63).

[89] Jungbauer, A.; Behrens, H. (J. Organometal. Chem. **186** [1980] 361/70).

[90] Quinn, S.; Shaver, A. (Inorg. Chim. Acta **38** [1980] 243/5).

[91] Butts, S.B. (Diss. Northwestern Univ. 1980, pp. 1/150; Diss. Abstr. Intern. B **41** [1980] 2170).

[92] Wong, A.; Atwood, J.D. (J. Organometal. Chem. **199** [1980] C 9/C 12).

[93] Stimson, R.E.; Shriver, D.F. (Inorg. Chem. **19** [1980] 1141/5).

[94] Gompper, R.; Bartmann, E. (Liebigs Ann. Chem. **1980** 229/40).

[95] Butts, S.B.; Strauss, S.H.; Holt, E.M.; Stimson, R.E.; Alcock, N.A.; Shriver, D.F. (J. Am. Chem. Soc. **102** [1980] 5093/100).

[96] Busetto, L.; Palazzi, A. (13th Congr. Nazl. Chim. Inorg. Atti, Camerino, Italy, 1980, pp. 122/4).

[97] Brunner, H.; Vogt, H. (J. Organometal. Chem. **210** [1981] 223/36).

[98] Brookhart, M.; Tucker, J.R.; Husk, G.R. (J. Am. Chem. Soc. **103** [1981] 979/81).

[99] Brunner, H.; Vogt, H. (Chem. Ber. **114** [1981] 2186/207).

[100] Lukehart, C.M.; Srinivasan, K. (J. Am. Chem. Soc. **103** [1981] 4166/70).

[101] Malisch, W.; Blau, H.; Haaf, F.J. (Chem. Ber. **114** [1981] 2956/70).

[102] Jones, W.D.; Huggins, J.M.; Bergman, R.G. (J. Am. Chem. Soc. **103** [1981] 4415/23).

[103] Eilbracht, P.; Henkes, E. (Chem. Ber. **114** [1981] 1559/61).

[104] Bates, D.J.; Rosenblum, M.; Samuels, S.B. (J. Organometal. Chem. **209** [1981] C55/C59).

[105] Wong, A.; Atwood, J.D. (J. Organometal. Chem. **210** [1981] 395/401).

[106] Hudson, A.; Lappert, M.F.; Lednor, P.W.; MacQuitty, J.J.; Nicholson, B.K. (J. Chem. Soc. Dalton Trans. **1981** 2159/63).

[107] Fettes, D.J.; Narayanaswamy, R.; Rest, A.J. (J. Chem. Soc. Dalton Trans. **1981** 2311/6).

[108] McCormick, F.B.; Angelici, R.J. (Inorg. Chem. **20** [1981] 1111/7).

[109] Blau, H.; Malisch, W. (J. Organometal. Chem. **235** [1982] C1/C6).

[110] Manganiello, F.J.; Christensen, L.W.; Jones, W.M. (J. Organometal. Chem. **235** [1982] 327/34).

[111] Moss, J.R. (J. Organometal. Chem. **231** [1982] 229/35).

[112] LaCroce, S.T.; Cutler, A.R. (J. Am. Chem. Soc. **104** [1982] 2312/4).

[113] Kremer, K.A.M.; Kuo, G.-H.; O'Connor, E.J.; Helquist, P.; Kerber, R.C. (J. Am. Chem. Soc. **104** [1982] 6119/21).

[114] Casey, C.P.; Miles, W.H.; Tukada, H.; O'Connor, J.M. (J. Am. Chem. Soc. **104** [1982] 3761/2).

[115] Casey, C.P.; Tukada, H.; Miles, W.H. (Organometallics **1** [1982] 1083/4).

[116] DeSimone, D.M.; Desrosiers, P.J.; Hughes, R.P. (J. Am. Chem. Soc. **104** [1982] 4842/6).

[117] Goncharenko, L.V.; Kolobova, N.E.; Petrovskii, P.V. (Izv. Akad. Nauk SSSR Ser. Khim. **1982** 422/7; Bull. Acad. Sci. USSR Div. Chem. Sci. **1982** 382/5).

[118] Eilbracht, P.; Totzauer, W. (Chem. Ber. **115** [1982] 1669/81).

[119] Stimson, R.E. (Diss. Northwestern Univ. 1981, pp. 1/167; Diss. Abstr. Intern. B **42** [1982] 3682).

[120] Angelici, R.J.; Formanek, T. (Inorg. Chim. Acta **76** [1983] L9/L11).

[121] Lukehart, C.M.; Srinivasan, K. (Organometallics **2** [1983] 1640/2).

[122] Marten, D.F.; Wilburn, S.M. (J. Organometal. Chem. **251** [1983] 71/8).

[123] Ts'ai, W.-M. (Diss. Univ. Boston 1983, pp. 1/244; Diss. Abstr. Intern. B **44** [1983] 1462).

[124] Zulu, S.J. (Diss. Univ. Boston 1983, pp. 1/273; Diss. Abstr. Intern. B **44** [1983] 493).

[125] Hitam, R.B.; Narayanaswamy, R.; Rest, A.J. (J. Chem. Soc. Dalton Trans. **1983** 615/8).

[126] Cutler, A.R.; Forschner, T.C.; Menard, K.P. (Abstr. Papers 186th Natl. Meeting Am. Chem. Soc., Washington, D.C., 1983, INOR 245).

[127] Sünkel, K.; Nagel, U.; Beck, W. (J. Organometal. Chem. **251** [1983] 227/43).

[128] Sünkel, K.; Schloter, K.; Beck, W.; Ackermann, K.; Schubert, U. (J. Organometal. Chem. **241** [1983] 333/42).

[129] Goncharenko, L.V.; Kolobova, N.E.; Kuz'mina, L.G.; Struchkov, Yu.T. (Izv. Akad. Nauk SSSR Ser. Khim. **1983** 1162/70; Bull. Acad. Sci. USSR Div. Chem. Sci. **1983** 1049/56).

[130] Liebeskind, L.S.; Welker, M.E. (Organometallics **2** [1983] 195/7).

[131] Bly, R.S.; Silverman, G.S.; Mahmun Hossain, M.; Bly, R.K. (Organometallics **3** [1984] 642/4).

[132] Bly, R.S.; Silverman, G.S. (Organometallics **3** [1984] 1765/7).

[133] Kuo, G.-H.; Helquist, P.; Kerber, R.C. (Organometallics **3** [1984] 806/8).

[134] Chang, B.-H.; Coil, P.C.; Brown, M.J.; Barnett, K.W. (J. Organometal. Chem. **270** [1984] C23/C25).

[135] Butler, I.R.; Lindsell, W.E.; Thomas, M.J.K. (J. Organometal. Chem. **262** [1984] 59/68).

[136] Brown, D.A.; Glass, W.K.; Ubeid, M.T. (Inorg. Chim. Acta **89** [1984] L47/L48).

[137] Hooker, R.H.; Rest, A.J.; Whitwell, I. (J. Organometal. Chem. **266** [1984] C27/C30).

[138] Bodnar, T.W. (Diss. Wesleyan Univ. 1984, pp. 1/461; Diss. Abstr. Intern. B **45** [1984] 1775).

[139] Forschner, T.; Menard, K.; Cutler, A. (J. Chem. Soc. **1984** 121/2).

[140] Brookhart, M.; Studabaker, W.B.; Husk, G.R. (Organometallics **4** [1985] 943/4).

[141] Brinkman, K.; Helquist, P. (Tetrahedron Letters **26** [1985] 2845/8).

[142] Manganiello, F.J.; Oon, S.M.; Radcliffe, M.D.; Jones, W.M. (Organometallics **4** [1985] 1069/72).

[143] Lisko, J.R.; Jones, W.M. (Organometallics **4** [1985] 944/6).

[144] Lisko, J.R.; Jones, W.M. (Organometallics **4** [1985] 612/4).

[145] Casey, C.P.; Miles, W.H.; Tukada, H. (J. Am. Chem. Soc. **107** [1985] 2924/31).

[146] Bly, R.S.; Silverman, G.S.; Bly, R.K. (Organometallics **4** [1985] 374/83).

[147] Crawford, E.J.; Lambert, C.; Menard, K.P.; Cutler, A.R. (J. Am. Chem. Soc. **107** [1985] 3130/9).

[148] Bodnar, T.W.; Cutler, A.R. (Organometallics **4** [1985] 1558/65).

[149] Bodnar, T.W.; Cutler, A.R. (Syn. React. Inorg. Metal-Org. Chem. **15** [1985] 31/42).

[150] Ojima, I.; Kwon, H.B. (Chem. Letters **1985** 1327/30).

[151] Brookhart, M.; Tucker, J.R.; Husk, G.R. (J. Am. Chem. Soc. **105** [1983] 258/64).

[152] Bly, R.S.; Bly, R.K.; Silverman, G.S. (Abstr. Papers 186th Natl. Meeting Am. Chem. Soc., Washington, D.C., 1983, ORGN 33).

[153] Kuo, G.-H. (Diss. State Univ. New York 1985, pp. 1/176; Diss. Abstr. Intern. B **46** [1985] 1925).

[154] Lukehart, C.M. (U.S. 4089881 [1978]).

[155] Malisch, W.; Blau, H.; Blank, K.; Krüger, C.; Liu, L.K. (J. Organometal. Chem. **296** [1985] C32/C36).

[156] Ojima, I.; Kwon, H.-B. (12th Intern. Conf. Organometal. Chem., Vienna 1985, Abstr. 115).

[157] Gibson, D.H.; Ong, T.-S.; Owens, K.; Mandal, S.; Sattich, W.E.; Franco, J. (12th Intern. Conf. Organometal. Chem., Vienna 1985, Abstr. 467).

[158] Bly, R.S.; Hossain, M.M.; Lebioda, L. (J. Am. Chem. Soc. **107** [1985] 5549/50).

[159] Kolobova, N.E.; Goncharenko, L.V. (Khim. Geterotsikl. Soedin. **1979** 1461/5; Chem. Heterocycl. Compounds [USSR] **15** [1979] 1173/6).

[160] Gladysz, J.A.; Williams, G.M.; Tam, W.; Johnson, D.L. (J. Organometal. Chem. **140** [1977] C1/C6).

[161] Duncan, J.D. (Diss. Balliol College, Oxford 1969, pp. 1/157).

[162] Shaver, A.; Butler, I.S.; Eisenberg, A.; Gao, J.P.; Xu, Z.H.; Bertwin, F.; Uhm, H.; Klein, D. (Appl. Organometallic Chem. **1** [1987] 383/92).

[163] Giuseppetti, M.E.; Cutler, A.R. (Organometallics **6** [1987] 970/3).

1.5.2.3.16.1.8   Iminoacyl Compounds of the Type $C_5H_5Fe(CO)_2C(=NR')R''$

The compounds FpC(=NR')R'' listed in Table 3 have been prepared by the following methods (C_6H_{11} = cyclohexyl):

Method I: Treatment of Na[Fp] with $C_6H_5(Cl)C=NR'$ in THF at $-78\,°C$ followed by warming to room temperature and stirring for 1.25 h. After solvent removal in vacuum the residue is taken up in benzene and chromatographed on $Al_2O_3 \cdot 6\,H_2O$ with C_6H_6/THF (1:1) as eluent. The products (R'' = C_6H_5) are recrystallized from toluene/hexane at $-20\,°C$ [5]; see also [4].

Method II: Reaction of $C_5H_5Fe(CO)(C=NR')R''$ under 30 atm of CO in benzene (R' = C_6H_{11}, R'' = C_6H_4Cl-4 or $CH_2C_6H_4Cl$-4; R' = $CH_2C_6H_5$, R'' = $CH_2C_6H_4Cl$-4) or in THF (R' =

C_6H_{11}, $R'' = CH_2C_6H_5$) at 25 °C for 100 to 144 h in an autoclave. After solvent removal the residue is worked up by chromatography on Al_2O_3 with benzene as eluent [2].

The preparation according to Method I has also been carried out at 60 °C. Whereas $FpC(=NC_6H_5)C_6H_5$ was formed exclusively in this case (yield not given), $FpC(=NC_3H_7-i)C_6H_5$ was obtained in only 10% yield along with $[FpC(C_6H_5)NHC_3H_7-i]PF_6$ [6]. Studies show that the yields of the products prepared according to Method II decrease with shorter reaction times and increase with the dielectric constant of the solvent employed ($C_6H_6 < THF < CH_3CN$). Thus $FpC(=NC_6H_{11})CH_2C_6H_5$ is obtained in 83% yield when the reaction is performed in CH_3CN for 140 h (cf. 77% in THF/144 h) [2]. The relative reaction rates have been found to decrease in the order $C_6H_5CH_2NC > C_6H_{11}NC \gg (CH_3)_3CNC$ (no detectable iminoacyl complex formation) for the isocyanide and $4\text{-}ClC_6H_4CH_2 > C_6H_5CH_2 > 4\text{-}ClC_6H_4$ for the aralkyl or aryl group [2].

The mass spectra of $FpC(=NC_6H_{11})C_6H_4Cl-4$, $FpC(=NC_6H_{11})CH_2C_6H_5$, and $FpC(=NC_6H_{11})$-$CH_2C_6H_4Cl-4$ do not exhibit the molecular ion peak. In the spectrum of $FpC(=NC_6H_{11})CH_2C_6H_5$, loss of CO from $[C_5H_5Fe(CO)C(=NC_6H_{11})CH_2C_6H_5]^+$ gives $[C_5H_5FeC(=NC_6H_{11})CH_2C_6H_5]^+$, which appears to break down via two pathways. In the first pathway, successive loss of C_6H_{11} and CN forms $[C_5H_5FeCH_2C_6H_5]^+$, which further eliminates the $C_6H_5CH_2$ and C_5H_5 groups to give Fe^+. In the second, loss of $C_6H_5CH_2$ affords the ion $[C_5H_5FeCNC_6H_{11}]^+$. A similar fragmentation pattern is observed for $FpC(=NC_6H_{11})CH_2C_6H_4Cl-4$, whereas in the spectrum of $FpC(=NC_6H_{11})C_6H_4Cl-4$ the second path leading to formation of $[C_5H_5FeCNC_6H_{11}]^+$ from $[C_5H_5FeC(=NC_6H_{11})C_6H_4Cl-4]^+$ is not detectable [2]. The mass spectrum of $FpC(=NC_6H_5)C_6H_5$ shows the molecular ion and fragments resulting from successive loss of CO groups [4].

Photolytic decarbonylation of $FpC(=NR')C_6H_5$ proceeds, in the case of $R' = i\text{-}C_3H_7$ (in THF for 1 h), by means of phenyl migration from the iminoacyl ligand to the Fe atom with formation of an isocyanide ligand to give $C_5H_5Fe(CO)(CNR')C_6H_5$ [5]. The same type of product is obtained upon irradiation of $FpC(=NC_6H_{11})CH_2C_6H_5$ (THF, 25 °C, 21 h), a reaction that is reversible [2]; cf. Method II. However, when $R' = C_6H_5$ or $C_6H_4CH_3-4$ the Fe,N-heterocycle I is formed (hexane, 75 min) [5]. Attempts to thermally decarbonylate $FpC(=NC_6H_{11})CH_2C_6H_5$ failed [2].

R^1 = H or CH_3

I

The reaction of $FpC(=NC_6H_{11})R''$ ($R'' = CH_2C_6H_5$ or $CH_2C_6H_4Cl-4$) with 42% HBF_4 in benzene at room temperature for 2 h proceeds with protonation at the imino nitrogen atom to give the carbene complex $[FpC(NHC_6H_{11})R'']BF_4$ in nearly quantitative yield. Treatment of this compound with $NaOCH_3$ or C_6H_5Li regenerates the parent complex [3].

Remarks Concerning Table 3: The IR absorptions of $\nu(CO)$ in the range 1940 to 2020 cm^{-1} are strong bands, whereas the absorption between 1580 and 1610 cm^{-1} resulting from the $\nu(C=N)$ vibration is of medium to weak intensity.

References on p. 54

Table 3
Iminoacyl Compounds of the Type $C_5H_5Fe(CO)_2C(=NR')R''$.
Further information on compounds with numbers preceded by an asterisk is given at the
end of the table.
For abbreviations and dimensions, see p. X.

No.	compound method of preparation (yield)	properties and remarks
1	$FpC(=NCH_3)C_6H_5$ I (38%) [5]	m.p. 87 to 89° [5] ^{1}H NMR (CD_3COCD_3): 3.51 (s), 5.16 (s), 7.13 (m) [5] IR: 1945, 2001 (both CO), 1596 (C=N) [5]
*2	$FpC(=NCH_3)C_6F_5$	yellow crystals, m.p. 108 to 109° [1] ^{1}H NMR $(CDCl_3)$: 3.58 (s, CH_3), 5.00 (s, C_5H_5) [1] IR (CCl_4): 1980, 2020 (both CO); bands at 560, 595, 630, 705, 960, 980, 1100, 1130, 1300, 1390, 1490, 1520, 2860, 2920, 2960 [1]
3	$FpC(=NC_3H_7-i)C_6H_5$ I (32%) [5]	m.p. 109 to 111° [5] ^{1}H NMR (CD_3COCD_3): 1.12 (d, 6H), 3.66 (sept, 1H, J = 6), 5.12 (s), 7.07 (m) [5] IR (C_6H_{14}): 1960, 2010 (both CO), 1590 (C=N) [5]
4	$FpC(=NC_6H_{11}-cyclo)C_6H_4Cl-4$ II (27%) [2]	m.p. 87 to 89° (dec.) [2] ^{1}H NMR $(CDCl_3)$: 0.8 to 2.2 (br, 10H, CH_2), 2.9 to 3.5 (br, =NCH<), 4.88 (C_5H_5), 7.01 (q, C_6H_4, J_{AB}=8.4) [2] IR (KBr): 1953, 2007 (both CO), 1598 (C=N) [2]
5	$FpC(=NC_6H_5)C_6H_5$ I (60%) [5]	m.p. 106° (dec.) [5] ^{1}H NMR (CD_3COCD_3): 4.90 (s), 7.1 (m) [5] IR (THF): 1950, 2010 (both CO), 1586 (C=N) [5]
6	$FpC(=NC_6H_4CH_3-4)C_6H_5$ I (38%) [5]	m.p. 125 to 130° [5] ^{1}H NMR (CD_3COCD_3): 2.21 (s), 4.85 (s), 6.56 (m), 7.12 (m) [5] IR (THF): 1940, 2000 (both CO), 1590 (C=N) [5]
*7	$FpC(=NC_6H_{11}-cyclo)CH_2C_6H_5$ II (77%) [2]	m.p. 94 to 96° (dec.) [2] ^{1}H NMR $(CDCl_3)$: 1.0 to 2.15 (br, 10H, CH_2), 2.9 to 3.4 (br, =NCH<), 4.0 (s, $FeCCH_2$), 4.49 (s, C_5H_5), 7.0 to 7.6 (br, C_6H_5) [2] IR (KBr): 1932, 1993 (both CO), 1609 (C=N) [2]
8	$FpC(=NC_6H_{11}-cyclo)CH_2C_6H_4Cl-4$ II (63%) [2]	m.p. 114 to 115° (dec.) [2] ^{1}H NMR $(CDCl_3)$: 1.1 to 2.1 (br, 10H, CH_2), 2.9 to 3.4 (br, =NCH<), 3.98 (s, $FeCCH_2$), 4.58 (C_5H_5), 7.26 (s, C_6H_4) [2] IR (KBr): 1946, 2000 (both CO), 1607 (C=N) [2]
9	$FpC(=NCH_2C_6H_5)CH_2C_6H_4Cl-4$ II (67%) [2]	m.p. 105 to 107° (dec.) [2] ^{1}H NMR $(CDCl_3)$: 4.08 (s, $FeCCH_2$), 4.62 (C_5H_5), 4.69 (s, NCH_2), 7.0 to 7.5 (complex, C_6H_4) [2] IR (KBr): 1942, 1997 (both CO), 1604 (C=N) [2]

                          References on p. 54

Further information:

$C_5H_5Fe(CO)_2C(=NCH_3)C_6F_5$ (Table 3, No. 2) is formed in 1.1% yield together with six other products by addition of $[FpCNCH_3]PF_6$ to C_6F_5Li in THF at $-78\,°C$ and stirring the mixture at $0\,°C$ overnight. After solvent removal the residue is worked up by repeated chromatography on Al_2O_3. The product is crystallized from ether/pentane [1].

$C_5H_5Fe(CO)_2C(=NC_6H_{11})CH_2C_6H_5$ (Table 3, No. 7). Attempts to replace CO by $P(C_6H_5)_3$ or cyclo-$C_6H_{11}NC$ in benzene or THF at reflux (8 h) have been unsuccessful. Only the starting material could be recovered [2].

References:

[1] Treichel, P.M.; Stenson, J.P. (Inorg. Chem. **8** [1969] 2563/7).
[2] Yamamoto, Y.; Yamazaki, H. (Inorg. Chem. **13** [1974] 2145/50).
[3] Yamamoto, Y.; Yamazaki, H. (Bull. Chem. Soc. Japan **48** [1975] 3691/7).
[4] Adams, R.D.; Chodosh, D.F.; Golembeski, N.M. (J. Organometal. Chem. **139** [1977] C39/C43).
[5] Adams, R.D.; Chodosh, D.F.; Golembeski, N.M.; Weissman, E.D. (J. Organometal. Chem. **172** [1979] 251/67).
[6] Brunner, H.; Kerkien, G.; Wachter, J. (J. Organometal. Chem. **295** [1982] 295/300).

1.5.2.3.16.1.9 Thioxo Compounds of the Type $C_5H_5Fe(CO)_2C(S)R'$

The methods used to prepare the compounds FpC(S)R' in Table 4 are described below. Although the reactions according to Method I and II must be carried out under an N_2 atmosphere, the workup may be done in air [7, 10].

Method I: Addition of CS_2 to a solution of Na[Fp] in THF and after 20 to 30 s further reaction with CH_3I [7] or $C_6H_5CH_2Br$ [9]. After ~5 min (longer reaction times reduce the yields) the mixture containing $FpC(S)SCH_3$ is worked up by decantation, evaporation to dryness, extraction of the residue with warm pentane, concentration of the filtered extracts, and cooling to $-20\,°C$ [7]. $FpC(S)SCH_2C_6H_5$ is isolated from the reaction residue by extraction with ether, filtration through Celite, evaporation, and repeated fractional crystallization of the resulting oil from hexane and/or ether at $-20\,°C$ [9].

Similarly, the Si- and Sn-containing compounds FpC(S)R' have been prepared by addition of CS_2 to a THF solution of $K_xNa_{1-x}[Fp]$ at $-60\,°C$ (R' = $SSi(CH_3)_3$ and $SSn(CH_3)_3$) or of K[Fp] at $-20\,°C$ (R' = $SSn(C_6H_5)_3$) followed by treatment with neat $(CH_3)_3SiCl$ or with R^1_3SnCl ($R^1 = CH_3$, C_6H_5) in THF for 20 to 30 min, respectively. After warming to $0\,°C$ and evaporation of the solvent, the residue is extracted with petroleum ether and the filtered and concentrated extracts are cooled to $-40\,°C$ [13]. In the case of $FpC(S)SSn(C_6H_5)_3$, workup of the reaction residue is performed by extraction with CH_2Cl_2, filtration through SiO_2, addition of heptane to the filtrate, and concentration [6].

Method II: Treatment of $[FpCS]CF_3SO_3$ with $NaOC_6H_5$ (prepared from phenol and NaH) [8] or with $NaSeC_6H_5$ (from $C_6H_5Se_2C_6H_5$ and Na/Hg) [10] in THF for 1 h and 10 min, respectively, followed by evaporation of the solvent, and extraction of the residue with CS_2. Starting with $NaOC_6H_5$, solvent removal from the filtered extracts gives $FpC(S)OC_6H_5$ as an oil which could not be crystallized. Attempted purification by chromatography on Florisil led to decomposition [8]. In the case of $FpC(S)SeC_6H_5$ evaporation of the extracts affords a solid which is recrystallized from hexane by cooling to $-20\,°C$ [10].

References on pp. 58/9

FpC(S)OCH$_3$ has been prepared by a similar procedure, i.e., [FpCS]PF$_6$ in anhydrous CH$_3$OH is stirred at room temperature for 20 min in the presence of an excess of Na$_2$CO$_3$ and the reaction residue is worked up by dissolution in ether, filtration, solvent removal, and vacuum sublimation at 40 °C. Similarly FpC(S)OC$_2$H$_5$ has been obtained from [FpCS]$^+$ and C$_2$H$_5$O$^-$ in C$_2$H$_5$OH [4, 5].

Method III: Treatment of [FpC(SR^1)SFp]X, prepared from FpC(S)SFp and R^{1}X, with [N(C$_4$H$_9$-n)$_4$]I in boiling acetone for 4 h, where R^1=CH$_3$, X=I; R^1=CH$_3$, X=CF$_3$SO$_3$; R^1=C$_2$H$_5$, X=CF$_3$SO$_3$; R^1=C$_6$H$_5$CH$_2$, X=Br; and R^1=CH$_2$=CHCH$_2$, X=Br. The reaction mixture containing FpI and FpC(S)SR^1 is evaporated and chromatographed on Al$_2$O$_3$ with light petroleum/CH$_2$Cl$_2$ as eluent. The dithioesters are crystallized from n-pentane at −20 °C [15].

The dithioester complexes Fp(S)SR^1 (R^1=CH$_3$, CH$_2$C$_6$H$_5$) are stable towards air in the solid state [7, 9], but they decompose slowly when exposed to light [9]. The thermal instability of FpC(S)SCH$_3$ was mentioned [6]. The compounds No. 10 and 11 containing the groups SSi(CH$_3$)$_3$ and SSn(CH$_3$)$_3$ are thermally labile and very sensitive to moisture. Decomposition, presumably involving the intermediate FpC(S)SH, gives finally FpC(S)SFp [13]. FpC(S)SSn-(C$_6$H$_5$)$_3$ is significantly more thermally stable than FpC(S)SCH$_3$ [6].

Dry HCl bubbled through a benzene solution of FpC(S)OCH$_3$ [2, 3], (also in ether [5]), FpC(S)OC$_2$H$_5$ [1], or FpC(S)SCH$_3$ [3, 7] leads to formation of [FpCS]$^+$. Also the compounds FpC(S)NHR^1 (R^1=CH$_3$, C$_2$H$_5$) undergo rapid reaction with HCl [5]; see also [4].

The compounds FpC(S)OR^1 (R^1=CH$_3$ [10], C$_6$H$_5$ [8]), FpC(S)SR^2 (R^2=CH$_3$, CH$_2$C$_6$H$_5$ [9]), and FpC(S)SeC$_6$H$_5$ [10] react with CH$_3$SO$_3$F in CH$_2$Cl$_2$ for 0.5 to 1.5 h to give the cationic carbene complexes [Fp=C(SCH$_3$)OR^1]SO$_3$F, [Fp=C(SCH$_3$)SR^2]SO$_3$F, or [Fp=C(SCH$_3$)SeC$_6$H$_5$]-SO$_3$F, which are isolated as PF$_6^-$ salts. The same type of products has been obtained by alkylation of FpC(S)OCH$_3$ and FpC(S)SCH$_3$ with CH$_3$SO$_3$CF$_3$ [10] and of FpC(S)SCH$_3$ with C$_2$H$_5$SO$_3$F [9].

Remarks Concerning Table 4: The compounds listed in Table 4 generally show two strong CO bands in the range 1965 to 2040 cm^{-1} and one medium to strong C=S band in the range 1150 to 1250 cm^{-1}.

Table 4
Thioxo Compounds of the Type C$_5$H$_5$Fe(CO)$_2$C(S)R′.
Further information on compounds with numbers preceded by an asterisk is given at the end of the table.
For abbreviations and dimensions, see p. X.

No. compound method of preparation (yield)	properties and remarks
*1 FpC(S)OCH$_3$ II (70%) [4, 5]	orange-red, m.p. 75 to 76° [4, 5] ^{1}H NMR (CS$_2$): 4.03 (CH$_3$), 4.84 (C$_5$H$_5$) [5] IR: 1993, 2031 (both CO), 1193 (C=S) in C$_6$H$_{12}$ [5], 1993, 2024 (both CO), 1193 (C=S) in C$_6$H$_{14}$ [2]
*2 FpC(S)OC$_2$H$_5$ II [4, 5]	orange-brown [1] IR: 1989, 2031 (both CO), 1250 (C=S) in C$_5$H$_{12}$ [1], 1980, 2031 (both CO), 1172 (C=S) in CS$_2$ [4, 5]

References on pp. 58/9

Table 4 (continued)

No.	compound method of preparation (yield)	properties and remarks
*3	FpC(S)OC$_6$H$_5$ II [8]	not obtained pure [8] dark yellow oil [8] IR (C$_6$H$_{14}$): 1996, 2042 (both CO) [8, 10]
*4	FpC(S)SH	red-brown, m.p. 72 to 74° (dec.), stable at $-30°$ under N$_2$ [14] ^{1}H NMR (CCl$_4$): 5.00 (s, C$_5$H$_5$), 6.85 (s, broad, SH) [14] ^{13}C NMR ((CD$_3$)$_2$CO): 88.2 (C$_5$H$_5$, J(C,H) = 189), 213.1 (CO), 293.3 (CSSH) [14] IR (CH$_2$Cl$_2$): 1986, 2030 (both CO), 1045 (C=S), 2485 (SH) [14] IR (KBr): 1953, 2015 (both CO), 448, 500 (Fe–C), 560, 587, 620 (δ-FeCO), 719 (C–S), 902 (δ-CSH), 1040 (C=S), 2340 (broad, SH) [14] mass spectrum: [M]$^+$ (2), [M−CO]$^+$ (77), [M−2CO]$^+$ (94), [C$_5$H$_5$Fe(CS)]$^+$ (5), [C$_5$H$_5$Fe(SH)]$^+$ (43), [C$_5$H$_5$FeH]$^+$ (60), [C$_5$H$_5$Fe]$^+$ (100) [14]
*5	FpC(S)SCH$_3$ I (62%) [7] III (72%) [15]	brown crystals, m.p. 72° [7], 70 to 71° [15], 70° [12] ^{1}H NMR: 2.60 (s, CH$_3$), 4.92 (s, C$_5$H$_5$) in CCl$_4$ [7], 2.64 (s, CH$_3$), 4.94 (s, C$_5$H$_5$) in CDCl$_3$ [12, 15] IR: 1981, 2033 in CH$_2$Cl$_2$ [15], 1982, 2031 (impure substance) in THF [6], 1988, 2035 [7], 1990, 2040 (all CO) [12] in C$_6$H$_{14}$ mass spectrum: [M]$^+$ [12]
6	FpC(S)SC$_2$H$_5$ III (70%) [15]	yellow-orange, m.p. 55 to 58° [15] ^{1}H NMR (CDCl$_3$): 1.21 (t, CH$_3$), 3.23 (q, CH$_2$), 4.87 (s, C$_5$H$_5$) [15] IR (CH$_2$Cl$_2$): 1983, 2032 (both CO) [15]
7	FpC(S)SCH$_2$CH=CH$_2$ III (48%) [15]	yellow-orange, m.p. 45 to 48° [15] ^{1}H NMR: 3.96 (d, SCH$_2$), 4.88 (s, C$_5$H$_5$), 5.18 (m, =CH$_2$), 5.86 (m, CH) in CDCl$_3$, 3.96 (d, SCH$_2$, J=6.9), 4.03? (s, C$_5$H$_5$), 4.88 (d, =CH$_2$, 1H, J=9.8?), 5.02 (d, =CH$_2$, 1H, J=17), 5.79 (ddt, =CH–, J=6.7?, 6.9, 17.0) in C$_6$D$_6$ [15] IR (CH$_2$Cl$_2$): 1984, 2030 (both CO) [15]
8	FpC(S)SCH$_2$C$_6$H$_5$ I (37%) [9] III (65%) [15]	dark orange crystals, m.p. 73 to 75° [9], 71 to 73° [15] ^{1}H NMR (CDCl$_3$): 4.61 (s, CH$_2$), 4.97 (s, C$_5$H$_5$), 7.39 (s, C$_6$H$_5$) [9], 4.54 (s, CH$_2$), 4.91 (s, C$_5$H$_5$), 7.13 (m, C$_6$H$_5$) [15] IR (CH$_2$Cl$_2$): 1982, 2031 [9], 1982, 2032 (all CO) [15]
9	FpC(S)SC$_6$H$_5$ II (not isolated) [10]	IR (THF): 1979, 2028 (both CO) [10]

References on pp. 58/9

Table 4 (continued)

No.	compound method of preparation (yield)	properties and remarks
10	FpC(S)SSi(CH$_3$)$_3$ I (72%) [13]	red–brown very hygroscopic crystals, m.p. 45° (dec.) [13] IR: 1978, 2028 (both CO) in CH$_2$Cl$_2$, 690 (ν_{as}, SiC$_3$), 743 (ϱ, SiC$_3$), 708, 1010 (both CS), 1235 (δ, SiC$_3$), 2890, 2950 (both CH$_3$) in KBr [13] mass spectrum: [M]$^+$, [M−CO]$^+$, [M−2CO]$^+$ [13]
11	FpC(S)SSn(CH$_3$)$_3$ I (88%) [13]	brown needles, m.p. 75° (dec.) [13] IR: 1978, 2020 (both CO) in CH$_2$Cl$_2$, 770 (ϱ, SnC$_3$), 728, 998 (both CS), 1174, 1183 (both δ, SnC$_3$), 2910, 2980 (both CH$_3$) in KBr [13] mass spectrum: [M]$^+$, [M−CH$_3$]$^+$, [M−CO]$^+$, [M−2CO]$^+$ [13]
*12	FpC(S)SSn(C$_6$H$_5$)$_3$ I (85%) [6]	golden brown crystals, dec. >50° [6] IR: 1989, 2035 (both CO) in THF, 967, 997, 999, 1006, 1022, 1073 in Nujol, the bands at 1006 and either 997 or 999 are attributed to ν(CS) [6]
13	FpC(S)SeC$_6$H$_5$ II (56%) [10]	large red crystals, m.p. 76 to 79° [10] ^{1}H NMR (CS$_2$): 4.91 (s, C$_5$H$_5$), 7.27 (s, C$_6$H$_5$) [10] IR (CH$_2$Cl$_2$): 1983, 2032 (both CO) [10] k(CO) = 16.28 [10]
*14	FpC(S)TeC$_6$H$_5$	—
*15	FpC(S)NHCH$_3$	IR (C$_6$H$_{14}$): 1965, 2007 (both CO), 1158 (C=S) [4, 5]
*16	FpC(S)NHC$_2$H$_5$	—
17	FpC(S)SRu(CO)$_2$C$_5$H$_5$	see Section 1.5.2.3.16.1.10
18	FpC(S)SRe(CO)$_5$	see Section 1.5.2.3.16.1.10

Further information:

C$_5$H$_5$Fe(CO)$_2$C(S)OCH$_3$ (Table **4**, No. **1**). Performing the reaction of [FpCS]$^+$ with CH$_3$O$^-$ (Method II) in non-dried CH$_3$OH yields a mixture of No. 1 and C$_5$H$_5$Fe(CO)(CS)C(O)OCH$_3$ [5]. FpC(S)OCH$_3$ has also been obtained by stirring a mixture of Na[Fp] and ClC(S)OCH$_3$ in THF for 12 h under N$_2$ followed by solvent removal at room temperature/20 Torr, extraction of the residue with benzene, filtration through Celite, and evaporation of the solvent. The residue is further purified by dissolution in CH$_2$Cl$_2$, addition of hexane, and cooling in a dry ice/acetone bath, 43% yield [2]; see also [3].

C$_5$H$_5$Fe(CO)$_2$C(S)OC$_2$H$_5$ (Table **4**, No. **2**) is obtained as an impure product from Na[Fp] and ClC(S)OC$_2$H$_5$ [1]; cf. "Further information" on No. 1.

C$_5$H$_5$Fe(CO)$_2$C(S)OC$_6$H$_5$ (Table **4**, No. **3**) decomposes upon storage in air for 4 to 5 days. It reacts with Fp$^-$ in THF to give (C$_5$H$_5$)$_2$Fe$_2$(CO)$_3$CS and Fp$_2$ [8].

C$_5$H$_5$Fe(CO)$_2$C(S)SH (Table **4**, No. **4**) is prepared in ether by adding to K$_x$Na$_{1-x}$[Fp] at −78 °C first CS$_2$ and after 3 min HCl in ether (molar ratio 1:1.3:1). The evaporation residue is extracted with ether, concentrated, and after addition of n-pentane cooled to $-40\,°C$. The microcrystalline FpC(S)SH (yield 65%) separates and is dried in high vacuum. It is easily soluble in THF, CCl_4, CH_2Cl_2, and benzene. The solutions are not stable [14].

In humid air No. 4 decomposes slowly to give FpC(S)SFp. The protonation in CH_2Cl_2 gives $[FpCS]^+$ [14].

In solid FpC(S)SH hydrogen bonding is expected, for the SH band in KBr ($2340\ cm^{-1}$) and in Nujol ($2330\ cm^{-1}$) is at much longer wavelength than in CH_2Cl_2 ($2485\ cm^{-1}$) [14].

$C_5H_5Fe(CO)_2C(S)SCH_3$ (Table **4**, No. **5**). Method I carried out for a longer reaction time has led to the formation of an oily product from which no pure compound could be isolated. The existence of FpC(S)SCH$_3$ is based only on the IR spectrum. However, the reported ν(CO) frequencies at 2062 and $2088\ cm^{-1}$ [2, 3] deviate markedly from those given by other authors [6, 7, 12]. FpC(S)SCH$_3$ has been obtained in a 46% yield by bubbling CH_3NH_2 through a suspension of $[FpC(SCH_3)SRe(CO)_5]PF_6$ in ether for 1 h, evaporation of the filtered solution to dryness, extraction of the residue with pentane, and solvent removal [12]. The compound also forms by the reaction of $[FpC(SCH_3)SR\dot{e}(CO)_5]PF_6$ with a fivefold excess of LiBr in acetone at room temperature for 3 h and subsequent workup as aforementioned, 44% yield [12]. The complexes FpC(SCH$_3$)SM(CO)$_5$ (M = Cr, Mo, W) and FpC(SCH$_3$)SMn(CO)$_2$(C$_5$H$_4$R^1) with R^1 = H or CH$_3$ (for their formation, see reactions below) completely decompose in solution at room temperature within 1 d to give FpC(S)SCH$_3$ and M(CO)$_6$ or (C$_5$H$_4$R^1)Mn(CO)$_3$, respectively [13]. The formation of FpC(S)SCH$_3$ from $[FpCS]^+$ and CH$_3$S$^-$ was briefly mentioned [11].

FpC(S)SCH$_3$ is soluble in polar and nonpolar organic solvents [7]. Addition of FpC(S)SCH$_3$ to solutions of M(CO)$_5\cdot$THF (M = Cr, Mo, W) or (C$_5$H$_4$R^1)Mn(CO)$_2\cdot$THF (R^1 = H, CH$_3$), prepared by irradiation of M(CO)$_6$ or (C$_5$H$_4$R^1)Mn(CO)$_3$ in THF, respectively, and stirring the mixture at room temperature for 1 to 2 h in the dark affords the complexes FpC(SCH$_3$)SM(CO)$_5$ and FpC(SCH$_3$)SMn(CO)$_2$(C$_5$H$_4$R^1). Also, treatment of FpC(S)SCH$_3$ with "Mo(CO)$_4$(η^4-C$_7$H$_8$)" gives FpC(SCH$_3$)SMo(CO)$_5$. However, from the reaction with Co$_2$(CO)$_8$ in ether for 1 h, only CH$_3$SCCo$_3$(CO)$_9$ could be isolated. The expected formation of FpCCo$_3$(CO)$_9$ did not take place [13].

$C_5H_5Fe(CO)_2C(S)SSn(C_6H_5)_3$ (Table **4**, No. **12**). Due to its thermal instability it was difficult to obtain an analytically pure product by Method I [6].

$C_5H_5Fe(CO)_2C(S)TeC_6H_5$ (Table **4**, No. **14**). Reaction of $[FpCS]^+$ with $C_6H_5Te^-$ according to Method II proceeds with gas evolution and formation of an unstable, intractable green tar. Based on IR spectra of the reaction solution it is assumed that the unstable green $C_5H_5Fe(CO)(CS)TeC_6H_5$ has formed along with a very small amount of FpC(S)TeC$_6$H$_5$ [10].

$C_5H_5Fe(CO)_2C(S)NHCH_3$ and **$C_5H_5Fe(CO)_2C(S)NHC_2H_5$** (Table **4**, Nos. **15** and **16**). FpC(S)NHCH$_3$ is formed by treating a suspension of $[FpCS]PF_6$ in N$_2$-saturated ether with CH$_3$NH$_2$ for 20 min, filtration, and evaporation of the solvent under reduced pressure. Attempts to isolate the compound from the remaining brown oil failed. Thus, it was characterized only by its IR spectrum [4, 5]. FpC(S)NHC$_2$H$_5$ has been formed by an analogous procedure [5].

The compounds are rather unstable [5].

References:

[1] Busetto, L.; Angelici, R.J. (J. Am. Chem. Soc. **90** [1968] 3283/4).
[2] Busetto, L.; Belluco, U.; Angelici, R.J. (J. Organometal. Chem. **18** [1969] 213/5).

[3]   Busetto, L.; Belluco, U.; Angelici, R.J. (4th Intern. Conf. Organometal. Chem. Bristol 1969, Abstr. U 3).

[4]   Busetto, L.; Graziani, M.; Belluco, U. (Ger. Offen. 2 116 226 [1970/71]).

[5]   Busetto, L.; Graziani, M.; Belluco, U. (Inorg. Chem. **10** [1971] 78/80).

[6]   Ellis, J.E.; Fennell, R.W.; Flom, E.A. (Inorg. Chem. **15** [1976] 2031/6).

[7]   Dombek, B.D.; Angelici, R.J. (Inorg. Syn. **17** [1977] 100/3).

[8]   Quick, M.H.; Angelici, R.J. (J. Organometal. Chem. **160** [1978] 231/9).

[9]   McCormick, F.B.; Angelici, R.J. (Inorg. Chem. **18** [1979] 1231/5).

[10]   McCormick, F.B.; Angelici, R.J. (Inorg. Chem. **20** [1981] 1111/7).

[11]   McCormick, F.B.; Matachek, J.R.; Angelici, R.J. (unpublished observations from [10]).

[12]   Busetto, L.; Palazzi, A.; Monari, M. (J. Chem. Soc. Dalton Trans. **1982** 1631/4).

[13]   Stolzenberg, H.; Fehlhammer, W.P.; Monari, M.; Zanotti, V.; Busetto, L. (J. Organometal. Chem. **272** [1984] 73/80).

[14]   Mayr, A.; Stolzenberg, H.; Fehlhammer, W.P. (J. Organometal. Chem. **338** [1988] 223/6).

[15]   Busetto, L.; Bordoni, S.; Zanotti, V. (J. Organometal. Chem. **339** [1988] 125/31).

1.5.2.3.16.1.10 $C_5H_5Fe(CO)_2C(S)SRu(CO)_2C_5H_5$ and $C_5H_5Fe(CO)_2C(S)SRe(CO)_5$

Both compounds have been prepared by the reaction of [Fp]$^-$ with CS_2 in THF, followed by addition of $C_5H_5Ru(CO)_2I$ at $-60\,°C$ or $Re(CO)_5Br$ at $-78\,°C$, respectively. After 30 to 40 min the mixture is warmed to room temperature, the solvent removed, and the residue extracted with CH_2Cl_2 (Ru complex) or light petroleum, b.p. 40 to 70 °C (Re complex). The former complex is precipitated from the filtered (through Florisil) and concentrated extracts by addition of hexane and cooling to $-40\,°C$. It is further purified from traces of Fp_2 and $FpRu(CO)_2C_5H_5$ by repeated recrystallization from CH_2Cl_2/hexane, and crystallizes with 1 mole CH_2Cl_2, 77% yield [5]. For the latter complex, evaporation of the light petroleum extracts affords $FpC(S)SRe(CO)_5$ contaminated with small amounts of the chelate complex I [1].

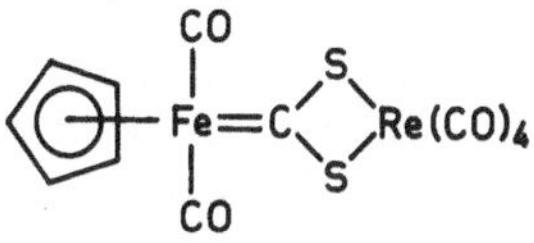

I

The compounds decompose rapidly in solution [5]. They are readily alkylated at the thione-S atom by $CF_3SO_3R^1$ ($R^1=CH_3$, C_2H_5) in CH_2Cl_2 at 0 °C to room temperature for 10 to 20 min to give the dithiocarbene salts $[Fp=C(SR^1)SRu(CO)_2C_5H_5]SO_3CF_3$ or $[Fp=C(SR^1)SRe(CO)_5]SO_3CF_3$, respectively [5]; see also [1, 2].

$C_5H_5Fe(CO)_2C(S)SRu(CO)_2C_5H_5 \cdot CH_2Cl_2$. The orange compound melts at 85 °C (dec.) [5].

^1H NMR spectrum ($CDCl_3$): $\delta=4.88$ (s, C_5H_5), 5.32 (s, CH_2Cl_2), 5.40 (s, C_5H_5) ppm [5].

IR spectrum: 1975 (sh), 1987 (vs), 2023 (vs), 2035 (vs), all CO, in CH_2Cl_2; 745 (m), 1010 (s) cm^{-1}, both C=S, in KBr [5].

$C_5H_5Fe(CO)_2C(S)SRe(CO)_5$. ^1H NMR spectrum ($CDCl_3$): $\delta=4.88$ (s, C_5H_5) ppm [1, 3].

IR spectrum: 1995 (s), 2030 (s), 2070 (w), 2130 (m) in C_6H_{14} [1], 1970 (s), 2019 (vs), 2052 (w), 2125 (m), all CO, in $CHCl_3$ [3]; 920 (m), 998 (s) cm^{-1}, both C=S, in Nujol [1, 3].

The compound reacts rapidly with solutions of $M(CO)_5 \cdot THF$ (M = Cr, Mo, W), prepared by irradiation of $M(CO)_6$ in THF, at room temperature for 30 min to give II. Similarly, the reaction proceeds with $C_5H_5Mn(CO)_2 \cdot THF$, prepared by irradiation of $C_5H_5Mn(CO)_3$ in THF, to yield III [3].

Treatment of the compound with $[FpCS]^+$ in CH_2Cl_2 at room temperature for 0.5 h results in formation of the cyclic complex IV [4].

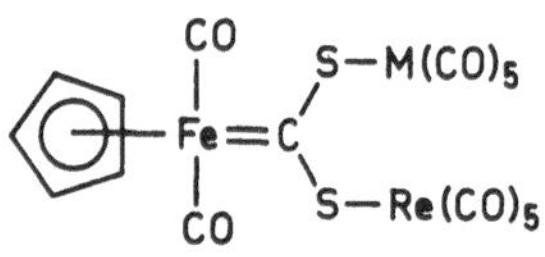

M = Cr, Mo, W

II   III 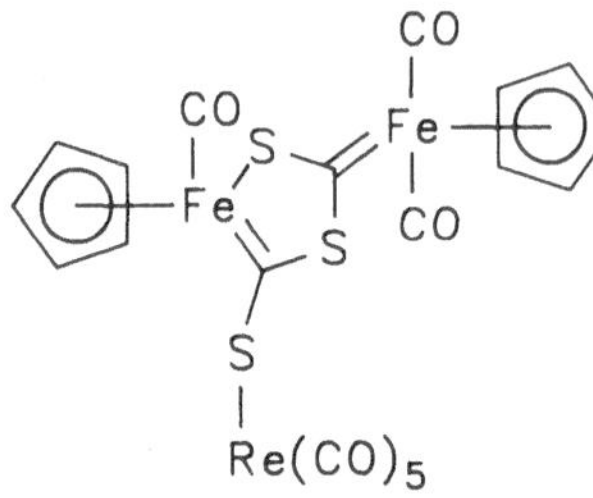   IV

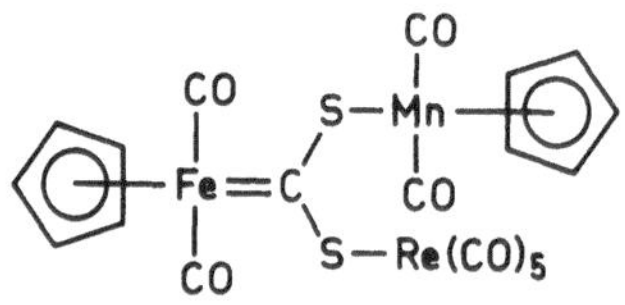

References:

[1] Busetto, L.; Palazzi, A.; Monari, M. (J. Chem. Soc. Dalton Trans. **1982** 1631/4).

[2] Busetto, L.; Palazzi, A.; Monari, M. (J. Organometal. Chem. **228** [1982] C19/C20).

[3] Busetto, L.; Monari, M.; Palazzi, A.; Albano, V.G.; Demartin, F. (J. Chem. Soc. **1983** 1849/55).

[4] Albano, V.G.; Braga, D.; Busetto, L.; Monari, M.; Zanotti, V. (J. Chem. Soc. Chem. Commun. **1984** 1257/8).

[5] Stolzenberg, H.; Fehlhammer, W.P.; Monari, M.; Zanotti, V.; Busetto, L. (J. Organometal. Chem. **272** [1984] 73/80).

1.5.2.3.16.1.11 $C_5H_5Fe(CO)_2R$ Compounds with R = Alkenyl

The compounds in Table 5, pp. 84/113, can be subdivided into four main groups:

(1) FpR where Fe is directly bonded to a C=C bond (Nos. 1 to 91) including R ligands containing a second conjugated (Nos. 76 to 87) or isolated (Nos. 9 and 89 to 91) C=C bond, or containing a conjugated C=N (No. 70) or C≡C bond (No. 88),

(2) FpR where R is an allyl-type ligand containing the structural element $FeC^1C^2=C^3$ (Nos. 92 to 146). C^3 can be part of a cyclohexanone ring (Nos. 93 and 94) or can be substituted by 1,3-dioxolan-2-yl (Nos. 92, 113 to 116). R can also contain a second C=C bond (Nos. 144, 145, and 146),

(3) FpR with a structure $FeC_nC=C$, where n is 1 to 5 (Nos. 147 to 165, 177, 178), No. 177 also containing a 1,3-dioxolan-2-yl group,

(4) FpR, where R is or contains an allenyl group (Nos. 166 to 175), or where R is C=C=O (No. 176). In Nos. 166 to 171 and 174 Fe is directly bonded to C=C=C, whereas Nos. 172, 173, and 175 have an FeCH$_2$CH$_2$C=C=C structure.

The methods used to prepare the compounds in Table 5 are described below.

Method I: Reaction of Na[Fp] with RX (X = F, Cl, Br [2, 6, 10, 13, 16, 22, 30, 42, 44, 45, 47, 67, 68, 78, 80, 86, 90, 94, 99, 105, 127, 136, 148], $OSO_2C_6H_5$ [78, 86], or $OSO_2C_6H_4CH_3$-4 [45, 99]) in THF usually for several hours at room temperature [6, 42, 44, 47, 78, 86, 90, 136, 148, 154], at 0 °C to room temperature [13, 80, 86, 94, 114] followed by warming to 60 °C [2], at −70 °C to room temperature [16, 127] and further at 50 to 60 °C [105, 106], or at −78 °C to room temperature

[10, 22, 45, 67, 86, 99, 166]. After solvent removal the residue is generally worked up by chromatography on Al_2O_3. Other workup procedures used, are described under "Further information".

Method II: The reactions of Na[Fp] in THF with $HC{\equiv}CCHR^1X$ ($R^1 = H$ or CH_3, $X = Br$, both at 0 °C/2 h [36]; $R^1 = H$, $X = OSO_2C_6H_5$ [69]), with $CH_2(OCH_3)C{\equiv}CCH_2OSO_2C_6H_5$ at 0 °C/1 h [118], with $CF_3C{\equiv}CCF_3$ at -78 °C/2 h and subsequent warming to room temperature [23] or with $CH_2(OH)C{\equiv}CCH_2Cl$ at -78 °C, then at -25 °C/2 h [164] give the allene complexes No. 166 to 170. After solvent removal the residue is worked up by extraction with petroleum ether and chromatography on Al_2O_3 with ether/hexane (3:7) (No. 170) [118] or on Florisil with petroleum ether (b.p. 40 to 60 °C) (No. 168) [23]. In the case of No. 169 evaporation of the solvent is followed by filtration through Celite with CH_2Cl_2/ether and recrystallization from ether/hexane [164]. Compounds No. 166 and 167 are purified by distillation [36].

Method III: Addition of (E)–$R^1COCH{=}CHI$ ($R^1 = C_6H_5$, $4\text{-}CH_3C_6H_4$, $4\text{-}BrC_6H_4$) in benzene to a solution of Fp_2 in the same solvent. After refluxing for 2.5 h the solvent is removed in vacuum and the residue is chromatographed on Al_2O_3. Elution with benzene gives the compounds (E)–$FpCH{=}CHCOR^1$, which are further purified by crystallization from petroleum ether at -70 °C [61], see also [70].

Similarly, the (Z)- and (E)-isomers of $FpCH{=}CHCOOR^2$ ($R^2 = H$, CH_3, C_2H_5) are obtained from Fp_2 and (Z)- or (E)–$R^2OCOCH{=}CHI$, respectively, in refluxing benzene for ~6 h and workup of the reaction mixture by solvent removal in vacuum and chromatography of the residue on SiO_2 with petroleum ether and mixtures of petroleum ether/$CHCl_3$ as eluents. The isomers of the acid $FpCH{=}CHCOOH$ are recrystallized from water [107].

Method IV a: UV irradiation of $FpE(CH_3)_3$ ($E = Si$, Ge, Sn) in the presence of excess $CF_3C{\equiv}CR^1$ ($R^1 = H$, CF_3) in hexane at 25 °C for 24 h (No. 49), 48 h (No. 48), or 84 h (No. 61), or at 76 °C for 42 h (Nos. 59 and 60) in a sealed tube affords the insertion products $FpC(CF_3){=}C(R^1)E(CH_3)_3$. After removal of the reaction volatiles and extraction of the residue with hexane, the extracts are chromatographed on SiO_2 with a hexane/benzene mixture as eluent. The yields of $FpC(CF_3){=}CHE(CH_3)_3$ (Nos. 48 and 49) increase (from 6 to 11% for $E = Si$, from 7 to 20% for $E = Ge$), when the reactions are carried out at 76 °C for 48 h. However, impurities are present due to the extensive polymerization of the propyne [40], see also [31].

Method IV b: The insertion products (Z)–$FpC(CF_3){=}C(CF_3)SR^1$ ($R^1 = CF_3$, C_6F_5) have been obtained by condensation of $CF_3C{\equiv}CCF_3$ and pentane onto $FpSR^1$ at 77 K, warming to 350 K for several hours, removal of the volatiles in vacuum, extraction of the residue with CH_2Cl_2 and centrifugation after addition of hexane. Solvent removal from the decanted solutions gives the products, which are purified by repeated recrystallization [87].

Method IV c: $NCC{\equiv}CCN$ undergoes an insertion reaction with $FpCH_2CH{=}CHR^1$. Solutions of the reactants in CH_2Cl_2 are mixed at -78 °C, then (after 2 to 4 h when $R^1 = H$, after 30 min when $R^1 = CH_3$) slowly warmed to room temperature. Workup of the mixture to obtain No. 89 is carried out after 2 h by filtration through Al_2O_3, evaporation of the solvent and chromatography of the residual oil on Al_2O_3 with benzene as eluent. After solvent removal and dissolution of the oily residue in benzene, the product is precipitated by addition of pentane. Compound

References on pp. 141/7

No. 90 is isolated from the reaction mixture by dilution with pentane and chromatography of the filtered and concentrated solution on Al_2O_3 with pentane/ CH_2Cl_2 (1:1) as eluent. The product is recrystallized from pentane/benzene (2:1) [129].

Method V: Treatment of $[Fp(CH_2=CHCH_2CH_2CHR^1Fp)]BF_4$, obtained from $FpCH_2CH=CH_2$ and $[Fp(CH_2=CHR^1)]BF_4$ in CH_3NO_2, with a solution of NaI in acetone at room temperature for 15 min to 2 h and workup after solvent removal by extraction of the residue with CH_2Cl_2 and chromatography of the filtered and concentrated extracts on Al_2O_3, elution with Skellysolve B ($R^1=H$) or 1% CH_3OH/CH_2Cl_2 ($R^1=CHO$) affords $FpCHR^1CH_2CH_2CH=CH_2$. Similarly, the compounds $FpCH(COOR^2)CH(CH_3)CH_2CH=CH_2$ ($R^2=CH_3$, C_2H_5) result from the reactions of $FpCH_2CH=CH_2$ with (E)-$[Fp(CH_3CH=CHCOOR^2)]X$ ($X=BF_4$, PF_6) in CH_3NO_2 ($R^2=C_2H_5$) or CH_2Cl_2/CH_3NO_2 (6:1) ($R^2=CH_3$) and subsequent treatment with NaI. Mixtures of $FpCHR^3CH_2CH_2CH=CH_2$ and $FpCH_2CHR^3CH_2CH=CH_2$ ($R^3=CH_3$ or C_6H_5) are the products from the corresponding reactions of $FpCH_2CH=CH_2$ with $[Fp(CH_2=CHR^3)]BF_4$ and NaI [155].

The same demetalation procedure (i.e., treatment with NaI) has also been used to convert the dinuclear complexes $[Fp(CH_2=CR^4CH=CHFp)]^+$ ($R^4=H$, CH_3, C_2H_5, $CH(OCH_3)_2$), prepared by protonation of $FpCH=CHCH=CHFp$ ($R^4=H$) or from $FpCH_2C\equiv CCH_2Fp$ and R^4OSO_2F or $[CH(OCH_3)_2]^+$ in CH_2Cl_2, to the respective mononuclear compounds $FpCH=CHCR^4=CH_2$ [119].

Method VI: Deprotonation of the salts $[Fp(CH_2=CHCH_2R^1)]X$ ($X=BF_4$, PF_6) in CH_2Cl_2 [86] or CH_3NO_2 [114] by treatment with $N(C_2H_5)_3$ ($\sim$1:1 mole ratio) at room temperature (Nos. 115 and 125) or at 0 °C (all other compounds) for 15 to 30 min causes formation of the allyl complexes $FpCH_2CH=CHR^1$ predominantly as (E)-isomers [86]. Exceptions are $FpCH_2CH=CHR^1$ ($R^1=OCH_3$, Br [86], Cl, OC_6H_5 [114]) where the (Z)-isomers are the exclusive products (see "General Remarks" on pp. 63/ 4). After solvent removal the residue is extracted with ether and the concentrated extracts are chromatographed on Al_2O_3 with CH_2Cl_2/ether (1:1) as eluent [86]. A similar procedure, but with $[Fp(CH_2=CHCH_2CH_2R^2)]^+$ ($R^2=CH_3$, C_6H_5) as starting material, gives a mixture of (Z)- and (E)-$FpCH_2CH=CHCH_2R^2$ [120]. (E)-$FpCH_2CH=CHCH(OCH_3)_2$ is isolated from the reaction residue by extraction with CH_2Cl_2/ether (1:9), filtration through Celite and concentration to dryness [86]. $FpCH_2C(OCH_3)=CH_2$, formed by addition of $N(C_2H_5)_3$ to $[Fp(CH_2=C(OCH_3)CH_3)]BF_4$ in CH_2Cl_2 at -78 °C, remains after addition of ether to the reaction mixture and removal of the solvent from the filtrate [175].

For the preparation of $FpCH_2CR^3=CHX$, the amine $N(C_3H_7-i)_2C_2H_5$ is used as deprotonating agent either in CH_3NO_2 ($X=Cl$) [114] or CH_2Cl_2 ($X=Br$) [86] at 0 °C and the reaction residue is worked up by washing with ether/petroleum ether (1:1) ($R^3=H$, $X=Cl$) or by extraction with ether ($R^3=H$, $X=Br$) or ether/ petroleum ether (1:1) ($R^3=CH_3$, $X=Br$) and concentration of the extracts. In the case of $FpCH_2CH=CHBr$, the resulting oil may be crystallized by dissolution in petroleum ether and slow evaporation with a stream of N_2 [86]. $FpCH=C=CH_2$ resulting from deprotonation of $[Fp(CH_2=C=CH_2)]BF_4$ with $N(C_6H_{11}-cyclo)_2C_2H_5$ in CH_2Cl_2 at 0 °C (30 min) remains after evaporation of the solvent from the filtered (through Al_2O_3) reaction mixture [134].

The preparation of compounds No. 93 and 94 from the corresponding salts and $N(C_2H_5)_3$ in CH_2Cl_2 at 25 °C is briefly mentioned in [161].

Method VII: Treatment of the allene salts [Fp(CH$_2$=C=CHR^1)]BF$_4$ (R^1 = CH$_3$, C$_6$H$_5$) with various nucleophiles (Nu) at room temperature leads to addition reactions giving
vinyl complexes of the type FpC(CH$_2$Nu)=CHR^1. The reactions have been performed with NaBH$_4$ (Nu = H) in THF for 3.5 h (R^1 = C$_6$H$_5$) and 20 h (R^1 = CH$_3$),
with aqueous NaOH in acetone followed by addition of pentane, with NaOCH$_3$
in CH$_3$OH for 15 min, with NH(C$_2$H$_5$)$_2$ in pentane for 1 h, and with NaI in acetone
for 10 min. After evaporation of the solvent the residue is extracted with pentane (Nu = H, N(C$_2$H$_5$)$_2$), CH$_2$Cl$_2$ (Nu = OH, I), or CH$_2$Cl$_2$/pentane (1:4) (Nu =
OCH$_3$) and the extracts are chromatographed on Al$_2$O$_3$ with the same solvents
as eluents [91], see also [57]. Different products are obtained from the reactions
of [Fp(CH$_2$=C=CHR^1)]BF$_4$ (R^1 = H, CH$_3$, C$_6$H$_5$) with 0.1 N NaOH in dry acetone
and workup by chromatography on Al$_2$O$_3$. [Fp(CH$_2$=C=CH$_2$)]BF$_4$ and the
corresponding (Z)-3-methyl- and (Z)-3-phenylallene complexes afford
FpC(CH$_2$OH)=CHR^1, whereas the (E)-3-methyl- and (E)-3-phenylallene salts
give FpC(=CH$_2$)CH(OH)R^1 due to steric effects [134].

FpC(CH$_2$OCOCH$_3$)=CH$_2$, prepared from [Fp(CH$_2$=C=CH$_2$)]BF$_4$ and
[N(CH$_3$)$_3$CH$_2$C$_6$H$_5$]OCOCH$_3$ in CH$_2$Cl$_2$ (30 min), is isolated after addition of ether
and evaporation of the solvent from the filtrate [134]. The addition of NaN$_3$
or C$_6$H$_5$NH$_2$ to [Fp(CH$_2$=C=CHR^1)]BF$_4$ afforded the complexes FpC(CH$_2$Nu)=CHR^1
(R^1 = CH$_3$, Nu = N$_3$; R^1 = C$_6$H$_5$, Nu = NHC$_6$H$_5$) as impure oils. Attempts to purify
these rather unstable materials were unsuccessful [91].

(Z)-[Fp(CH$_2$=C=CHCH$_3$)]BF$_4$ in anhydrous CH$_3$OH is allowed to react with
Na$_2$CO$_3$ at 0 °C for 30 min for the preparation of (Z)-FpC(CH$_2$OCH$_3$)=CHCH$_3$
(No. 65). The reaction mixture is filtered, stripped, and the residue is extracted
with hexane. Removal of the solvent gives the product [104].

General Remarks

Method I: In general, allyl chlorides constitute the best reactants. In preparing FpCH$_2$CH=
CH$_2$, the yields are as high as 91% and lower with either the bromides (25%) or tosylates.
Those substrates that in principle could give either primary or secondary (η^1-allyl)Fp com-
plexes are found to yield only the former. Thus, metalation of CH$_3$CH=CHCH$_2$Cl or CH$_2$=
CHCHClCH$_3$ is reported to give (Z)- and (E)-FpCH$_2$CH=CHCH$_3$, while metalation of C$_6$H$_5$CH=
CHCH$_2$Cl or CH$_2$=CHCHClC$_6$H$_5$ yields only (E)-FpCH$_2$CH=CHC$_6$H$_5$ [99], see also [42]. On the
basis of the products from Na[Fp] and CH$_2$=CHCD$_2$OSO$_2$C$_6$H$_4$CH$_3$-4, i.e., FpCH$_2$CH=CD$_2$ (95%
yield) and FpCD$_2$CH=CH$_2$ (5%), an S$_N$2' mechanism is supposed. Distillation (40 °C/10^{-3}
Torr during 45 min), warming in CDCl$_3$ (37 °C for 15 min) or storing at room temperature
(4 h) of FpCH$_2$CH=CD$_2$ leads to a total isomerization [42]. The reactions of Na[Fp] with (Z)-
or (E)-ClCH=CHCONH$_2$ are stereospecific, giving the respective isomers FpCH=CHCONH$_2$
[166]. Similarly the (E)-isomers are the sole products from the reactions of Na[Fp] with
(E)-ClCH=CHCOR^1 [48].

Method III: The reaction of Fp$_2$ with (E)-C$_2$H$_5$OCOCH=CHI at different temperatures (40
to 80 °C range) results in the formation of equivalent amounts of (E)-FpCH=CHCOOC$_2$H$_5$
and FpI, whereas in the case of the (Z)-isomer the (Z)-FpCH=CHCOOC$_2$H$_5$/FpI ratio depends
on the reaction temperature leading to increasing relative amounts of FpI at lower temperatures. The difference in the behavior of the two isomers is discussed [107].

Method VI: The stereochemistry of the allyliron compounds derived by deprotonation
of the complex cations is explained in terms of preferred base abstraction of an allylic
proton trans to the Fe-olefin bond [86], see also [88]. Model considerations indicate that

References on pp. 141/7

the CH_3O oxygen atom in $[Fp(CH_2=CHCH_2OCH_3)]^+$ may readily approach within bonding distance one of the CO groups, while the longer C–Br bond in $[Fp(CH_2=CHCH_2Br)]^+$ allows the halogen to interact with both CO ligands. Thus, deprotonation through these conformations must give rise to the (Z)–allyl complexes $FpCH_2CH=CHR^1$ [86].

It is apparent from the ^1H NMR spectra that $FpC(CH_3)=CHCH_3$, $FpC(CH_3)=CHC_6H_5$, and $FpC(CH_2N(C_2H_5)_2)=CHC_6H_5$ were isolated as mixtures of geometric isomers. It has not been possible to ascertain the stereochemistry of any of these complexes, since in general the coupling constant $^4J(CH_2,CH)$ is approximately the same ($\sim 1.5 \pm 0.3$ Hz) for both (Z)– and (E)–isomers of $Fp(CH_2R^1)C=CHR^2$ ($R^1 = H$, $N(C_2H_5)_2$; $R^2 = CH_3$, C_6H_5) [91].

Progressing from $R^1COCH=CHCl$ ($R^1 = CH_3$, C_6H_5, $4-CH_3C_6H_4$) to the corresponding keto-vinyl iron derivatives $FpCH=CHCOR^1$, all ^1H NMR signals undergo a small shift towards weaker field strength. Also the ketonic CO stretching frequencies are somewhat lowered relative to those of the starting ketones. Both shifts seem being associated with increased contribution of bipolar structures, see [16].

An analysis of the ^1H NMR spectra of $FpCH=CHCR^1=CH_2$ with $R^1 = H$, CH_3, C_2H_5, or $CH(OCH_3)_2$ excludes the formulation as $FpC(=CH_2)CR^1=CH_2$, also possible from the formation reaction. The coupling constants of the vicinal hydrogens of the butadiene moieties are 15 to 20 Hz and prove an (E)–structure [119].

The vinyl complexes $FpCR^1=CHR^1$ ($R^1 = CH_3$, C_2H_5, C_6H_5) are yellow, air–sensitive oils that thermally decompose at room temperature in two days [149].

The compounds (E)–$FpCH=CHCOR^1$ ($R^1 = CH_3$, C_6H_5, $4-CH_3C_6H_4$) are yellow solids, reasonably stable to storage in the dark. They gradually darken on exposure to light and are particularly unstable in solution [16].

The complexes $FpCH=CHCOOR^1$ ($R^1 = H$, CH_3, C_2H_5) are stable in the solid state, but they decompose slowly in solution [107].

In general the allyl iron compounds $FpCR^1R^2CR^3=CR^4R^5$ are amber, air– and heat–sensitive liquids, which are best characterized as their $(NC)_2C=C(CN)_2$ or SO_2 adducts [99].

The cyclic voltammograms of $FpCH_2CH=CH_2$, $FpCH_2CH=CHC_6H_5$, $FpCH_2CH=CHCH_3$, $FpCH_2CH=C(CH_3)_2$, and $FpCH=C=CH_2$ are reported in [185].

$FpCH=CH_2$ reacts with $HPF_6 \cdot O(C_2H_5)_2$ in CH_2Cl_2 at $-25\,°C$ to give the carbonium complex $[FpCH_2CHCH(CH_3)Fp]^+$ (structure I) [167], see also [168]. However, protonation with CF_3COOH at $-10\,°C$ affords a 5:1 mixture of I and $[Fp(CH_2=CH_2)]^+$ [167].

I

Treatment of $FpC(CH_2OH)=CHR^1$ ($R^1 = CH_3$, C_6H_5) with a catalytic amount of HBF_4 in acetone converts the complexes to the corresponding aldehydes $FpCH(CHO)CH_2R^1$, whereas brief exposure of $FpC(CH_2OH)=CH_2$ to one equivalent of HBF_4 in aqueous acetone at room temperature gives a 1:2 mixture of $FpCH_2COCH_3$ and $FpCH(CHO)CH_3$. The same two products are obtained as 10:1 mixture from $FpC(CH_2OCOCH_3)=CH_2$ and aqueous CH_3COOH/CH_3COOK after 30 min at room temperature [134]. However, treatment of $FpC(CH_2OH)=C=CH_2$ with HBF_4 in THF yields the cation $[Fp(CH_2=CHCOCH_3)]^+$ [204]. Also, $FpC(CH_2OCH_3)=C=CH_2$

is reported to give $[Fp(CH_2=CHCOCH_3)]^+$ when allowed to react with 48% aqueous HBF_4 in ether at 0 °C [188].

The reaction of (E)–$FpCH=CHC(CH_3)_2OH$ with HBF_4 in ether at −23 °C [193] or −15 °C [201] leads to precipitation of the carbene salt $[Fp=CHCH=C(CH_3)_2]BF_4$. The same type of products, i.e., $[Fp=CHCH=C(CH_3)_2]X$ and $[Fp=CHCH=CHCH_2CH_3]X$ (X = CF_3SO_3), result from $FpCH=CHC(CH_3)=CH_2$ and (E)–$FpCH=CHCH=CHCH_3$, respectively, and CF_3SO_3H in CH_2Cl_2 (−20 to −60 °C) [190], see also [198].

Protonation of $FpCH_2CR^1=CHR^2$ in light petroleum (b.p. 30 to 40 °C) with dry HCl (R^1 = H, R^2 = H or CH_3) [6], see also [1], or with mineral acids ($R^1 = R^2$ = H) [7], in CH_2Cl_2 with $HPF_6 \cdot O(C_2H_5)_2$ at 0 °C (R^1 = H, $R^2 = C_6H_5$), in CH_2Cl_2 with HPF_6 in acetic anhydride (R^1 = H, R^2 = COC_6H_5) [86], or with 48% aqueous HBF_4 in acetic anhydride ($R^1 = CH_3$, R^2 = H) [41] gives the cationic complexes $[Fp(CH_2=CR^1CH_2R^2)]^+$. Studies of the products prepared from $FpCH_2CH=CHR^2$ (R^2 = H, CH_3) and DCl indicate that C-3 is protonated [6, 7]. The deuteration of $FpCH_2CH=CHC_6H_5$ with CF_3COOD is found to be non-stereospecific, affording equal amounts of the diastereomerically deuterated cations [86]. While the protonation of $FpCH_2CH=CHR^2$ (R^2 = H, CH_3) does not appear to be reversible [1], treatment of II (R^3 = H, CH_3; Nos. 93 and 94) in CH_2Cl_2 with $HBF_4 \cdot O(C_2H_5)_2$ at −78 °C for 0.5 h gives the cationic olefin complex III, which is the starting material for the preparation of the compounds No. 93 and 94 according to Method VI [161].

R^3 = H or CH_3

The reactions of $FpCR^1=CH_2$ (R^1 = H, CH_3) in CH_2Cl_2 with ether solutions of HBF_4 (2 equivalents) at −78 °C followed by addition of alkenes and after warming to 25 °C by quenching the mixtures with saturated aqueous $NaHCO_3$ (in order to minimize acid-catalyzed rearrangement of the products) give 1-R^1-1-CH_3 cyclopropanes. Thus, dimethylcyclopropanes are obtained from n-dec-1-ene, isobutene, $C_6H_5CH=CH_2$, and $C_6H_5C(CH_3)=CH_2$ [174], see also [176], while 2-methylpent-1-ene, (Z)-n-dec-5-ene, 4-methylcyclohexene, cis-cyclooctene, and cis-cyclododecene failed to give detectable amounts of cyclopropanes upon reaction with $FpC(CH_3)=CH_2$. With respect to stereochemical behavior, the reactions of $FpCH=CH_2$ with cis-cyclooctene and $C_6H_5C(CH_3)=CH_2$ proceed with syn-selectivity. Cyclopropanes in detectable quantities are not produced when $FpC(C_6H_5)=CH_2$ is treated under the conditions described above. Instead, the styrene complex $[Fp(CH_2=CHC_6H_5)]BF_4$ is obtained, presumably by rearrangement of the intermediately formed carbene complex $[Fp=C(C_6H_5)-CH_3]BF_4$ [174]. Treatment of $FpC(CH_3)=CH_2$ with HBF_4 in ether at −23 °C [176, 201] or −30 °C [174], see also [177, 198], or with CF_3SO_3H in ether at −78 °C [201] in the absence of alkenes

IV   V

causes immediate formation of the isopropylidene complex [Fp=C(CH$_3$)$_2$]X (X = BF$_4$, CF$_3$SO$_3$), whereas the reaction of (E)-FpCH=CHC(CH$_3$)$_2$OH with [C(C$_6$H$_5$)$_3$]PF$_6$ and styrene in CH$_2$Cl$_2$ at −15 °C gives a 1:3 mixture of IV (cis-compound) and V (trans-compound) [201].

On the basis of all these results the cyclopropanation reactions are assumed to proceed via the carbene derivatives [Fp=C(R^1)CH$_3$]X. Intramolecular cyclopropanation with formation of norcarane takes place by treatment of FpCH=CH(CH$_2$)$_3$CH=CH$_2$ in CH$_2$Cl$_2$ with HBF$_4$ at −50 °C for 1 h and then with aqueous NaHCO$_3$ [174], see also [198, 199], or of FpCH(SC$_6$H$_5$)(CH$_2$)$_4$CH=CH$_2$ with [O(CH$_3$)$_3$]BF$_4$ in CH$_2$Cl$_2$ [199].

UV irradiation of FpCH$_2$CH=CHR^1 in an evacuated vessel for 6 h (R^1=H) or 24 h (R^1=CH$_3$) leads to CO evolution and formation of π-allyl complexes [6]. Examinations of the reaction products resulting from photolysis of FpCH$_2$CR^2=CH$_2$ (R^2=H, CH$_3$) in petroleum ether at 10 °C revealed that mixtures of the isomers VI and VII are formed. VI thermally isomerizes to the more stable system VII with a moderate rate at ambient temperatures. This is thought to be the reason for the failure of earlier workers [6] to detect both isomers [114]. The possibility that both isomers are formed, but that one decomposes or interconverts during isolation is considered by [84], who assigned structure VII to the only isomer isolated, based on magnetic anisotropies and shielding of protons calculated and observed for the analoguous indenyl complexes (η^5-C$_9$H$_7$)M(CO)$_n$(η^3-C$_3$H$_5$) (M = Mo, W, n = 2; M = Fe, Ru, n = 1). Under the same conditions as above, photolysis of (Z)-FpCH$_2$CH=CHR^1 (R^1=Cl, OCH$_3$, OC$_6$H$_5$) proceeds with retention of stereochemistry, yielding a single isomer shown by ^{1}H NMR and X-ray analysis (R^1=OC$_6$H$_5$) to have the structure IX. However, when the irradiation is carried out at −45 °C, both possible isomers VIII and IX are observed (^{1}H NMR). The unstable VIII rapidly isomerizes above 0 °C to give IX, which convert to XI on heating in benzene. Correspondingly, photolysis of (E)-FpCH$_2$CH=CH(OCH$_3$) at 10 °C gives the complex X, which isomerizes to give XI (R^1=OCH$_3$) at a moderate rate at room temperature [114].

Irradiation of FpCH$_2$CH$_2$CH=CH$_2$ in cyclohexane for 4 h [45] or in pentane for 20 h results in formation of a 35:65 mixture of IX (R^1=CH$_3$) and XI (R^1=CH$_3$) [47, 148]. XII is obtained on irradiating FpC[=C(CF$_3$)$_2$]C(CF$_3$)=CF$_2$ in pentane for 3 h [106, 108], see also [93].

$FpCH_2C(CH_3)_2CH=CH_2$ converts during photolysis in cyclohexane for 15 min into XIIIa [45], whereas XIIIb is believed to result from the irradiation of $FpC(CN)=C(CN)CH_2CH=CH_2$ in benzene for 2 h [129].

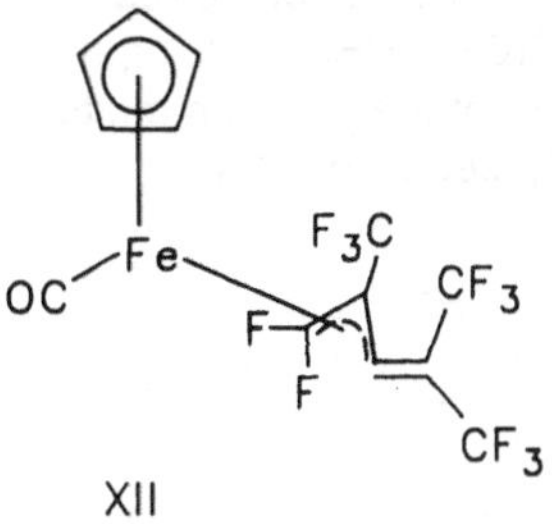

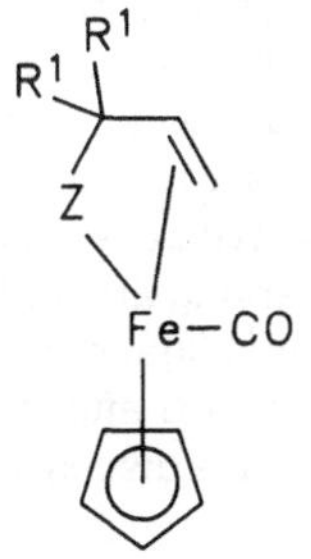

a: $R^1=CH_3$; $Z=CH_2$
b: $R^1=H$; $Z=$ $C(CN)=C(CN)$

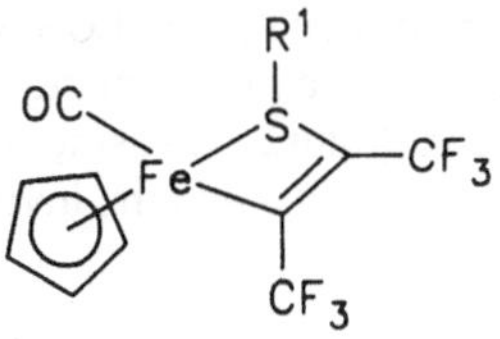

XII   XIII   XIV

Another type of product (XIV) forms by the photochemical decarbonylation of (Z)-$FpC(CF_3)=C(CF_3)SR^1$ ($R^1=CF_3$, C_6F_5; Nos. 57 and 58) in pentane. The compound with $R^1=CF_3$ also gives some $FpSCF_3$ in addition to XIV [87]. The compounds $FpCH=CHCOR^2$ ($R^2=CH_3$, C_6H_5) remain unchanged on irradiation in absolute benzene or THF under argon for 9 to 10 h [35].

Similar products as under UV irradiation are obtained by refluxing $FpC[=C(CF_3)_2]C(CF_3)=CF_2$ in decalin for 10 h [105, 106], see also [93, 108], or by heating (Z)-$FpC(CF_3)=C(CF_3)SCF_3$ in hexane at 60 °C [87], whereas attempts to prepare the π-allyl complexes by heating $FpCH_2CH=CHR^1$ ($R^1=H$, CH_3) in the pure state or in dilute solution in an inert solvent have been unsuccessful, giving only Fp_2 and unidentified hydrocarbons [6]. Heating $FpCH_2CH_2CH=CHR^2$ ($R^2=H$, CH_3) in CH_3CN at 90 °C for 3 d leads to formation of the acyl complexes $FpCOCH_2CH_2CH=CHR^2$ [47], while heating of $Fp(CH_2)_3CH=CH_2$ in CH_3NO_2 affords only 10% 2-methylcyclopentanone and decomposition products [148].

Initially, structure XV was assigned to the products formed by heating the allene complexes $FpCH_2CH_2CH=C=CH_2$, $FpCH_2CH_2CH=C=CHCH_3$, and $FpCH_2CH_2C(CH_3)=C=CH_2$ in THF at 60 °C for 3 d [44]. However, based on ^1H NMR and IR data the same authors in later publications propose structure XVI for the products obtained under similar conditions (50 °C, 48 h [62], 60 °C, 24 h [148]). They could further isolate a compound of structure XVII (assumed to be the precursor of XVI, $R^1=R^2=H$), when the reaction of $FpCH_2CH_2CH=C=CH_2$ was carried out at 30 to 35 °C for 2 h [62, 148]. The ring closure is viewed as a cis-migration, suggesting a concerted pathway in the reactions where an η^2-allenic ligand is converted into an η^3-allyl ligand [148].

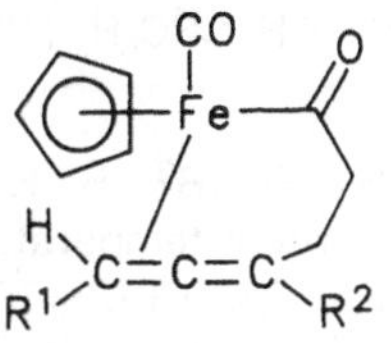

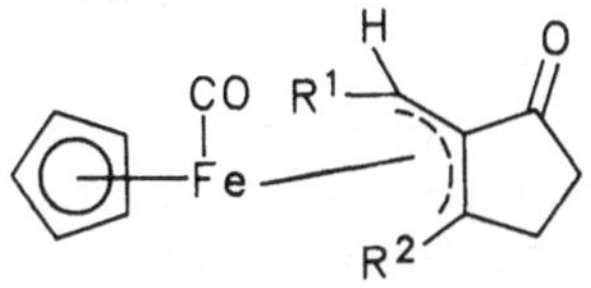

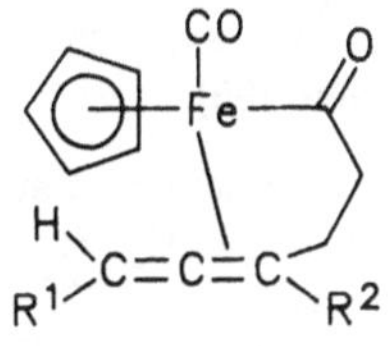

XV   XVI   XVII

$R^1,R^2=H$, CH_3

References on pp. 141/7

The compounds $FpCH_2CR^1=CR^2R^3$ undergo regiospecific coupling reactions through the terminal carbon of the alkenyl ligand when oxidized with $AgPF_6$ in CH_2Cl_2 or CH_2ClCH_2Cl at -78 to $+24\,°C$. The relative amounts of the resulting products XVIII and XIX, mono- and dinuclear hexadiene complexes, are dependent on the nature of the substituents R^2 and R^3 at the allyl ligand. Thus, $FpCH_2CH=CH_2$ affords primarily complex XVIII and considerably lesser amounts of XIX. $FpCH_2C(CH_3)=CH_2$ gives XVIII as the only coupling product, whereas $(E)-FpCH_2CH=CHC_6H_5$ yields XIX and only a trace of a substance thought to be XVIII. $FpCH_2CH=C(CH_3)_2$ produces both XVIII and XIX. With the exception of $(E)-FpCH_2CH=CHC_6H_5$, the oxidation of $FpCH_2CR^1=CR^2R^3$ also gives varying amounts of XX. A variant of the reaction with $AgPF_6$ consists of the transfer of the methallyl ligand of $FpCH_2C(CH_3)=CH_2$ to 1,3-diphenylbenzo[c]furan, causing formation of the adduct XXI after treatment of the reaction mixture with NaI. Mechanisms involving cation radicals are discussed [130], see also [141].

XVIII   XIX

XX

$R^1=R^2=R^3=H$
$R^1=CH_3,\ R^2=R^3=H$
$R^1=H,\ R^2=R^3=CH_3$
$R^1=R^3=H,\ R^2=C_6H_5$

XXI

The reactions of $FpCH_2CR^1=CH_2$ ($R^1=H$, CH_3) with $HgCl_2$ (slight deficiency) in THF at $25\,°C$ rapidly afford 1:1 adducts, which are formulated as the η^2-olefin iron complexes $[Fp(CH_2=CR^1CH_2)(HgCl)]Cl$. In contrast, $FpCH_2CH=C(CH_3)_2$ and $(E)-FpCH_2CH=CHC_6H_5$ react more slowly with cleavage of the Fe–C bond to give FpHgCl (almost quantitative yield) after 1 or 2 h. In the first case, $(CH_3)_2C=CHCH_2Cl$ was isolated. Both elimination and 1:1 adduct formation are observed for $(E)-FpCH_2CH=CHCH_3$ after 3 h, the products are FpHgCl (65%), $FpCH_2CH=CHCH_3 \cdot HgCl_2$ (6%), and FpCl (trace). Possible mechanisms for both cleavage and adduct formation are discussed [95, 100, 121]. The rates of the cleavage reactions of $FpCH_2CH=C(CH_3)_2$ and $(E)-FpCH_2CH=CHC_6H_5$ have been monitored by IR spectroscopy in the $\nu(CO)$ region at $25\,°C$ under pseudo-first-order conditions. The cleavage is found to exhibit a second-order dependence on $HgCl_2$ and to obey a third-order rate expression overall: $-d[FpCH_2CH=C(CH_3)_2]/dt = k_3[FpCH_2CH=C(CH_3)_2][HgCl_2]^2$ with $k_3=6.6\times10^{-3}\ M^{-2}\cdot s^{-1}$. An analogous expression with $k_3=1.3\times10^{-3}\ M^{-2}\cdot s^{-1}$ is valid for $(E)-FpCH_2CH=CHC_6H_5$ [121], see also [95].

The reactions between $FpCH_2CR^1=CR^2R^3$ ($R^1=R^2=R^3=H$; $R^1=CH_3$, $R^2=R^3=H$; $R^1=R^2=H$, $R^3=CH_3$; $R^1=H$, $R^2=R^3=CH_3$) and anhydrous $SnCl_2$ in THF at room temperature for 3 h (when $R^1=R^2=R^3=H$) or 24 h (all others) give the insertion products $FpSnCl_2CH_2CR^1=CR^2R^3$. In addition, $FpSnCl_3$ is formed in the reactions of $FpCH_2C(CH_3)=CH_2$ and an $(E)/(Z)$-mixture of $FpCH_2CH=CHCH_3$. When treated with $SnCl_2$ in CH_3OH or with excess $SnCl_2$ in THF, $FpCH_2C(CH_3)=CH_2$ produces primarily $FpSnCl_3$. Possible reaction mechanisms are discussed [78]. These results differ in some points from those obtained from more extended studies on analogous reactions. Thus, $FpCH_2CR^1=CH_2$ ($R^1 = H$, CH_3) reacts with $SnCl_2$, $SnBr_2$, or $GeCl_2 \cdot$ dioxane (1:1 mole ratio) in THF at room temperature to give the insertion products $FpEX_2CH_2CR^1=CH_2$ (E = Sn, Ge; X = halogen) in essentially quantitative yields. The reactions are very rapid and complete within 10 min for $GeCl_2 \cdot$ dioxane and within 20 min for SnX_2. In the corresponding reactions of an (E)/(Z)-mixture of $FpCH_2CH=CHCH_3$ the rearranged product $FpEX_2CH(CH_3)CH=CH_2$ is formed first, and then converts to the unrearranged form at a rate proportional to the excess of EX_2. For equimolar amounts of $FpCH_2CH=CHCH_3$ and $GeCl_2 \cdot$ dioxane, the isomerization is about 50% complete after 1 h, but for a ratio of 1:2.5, isomerization is 75% complete within 3 min. In a reaction mixture with 40% deficiency of $GeCl_2 \cdot$ dioxane the rearranged product is stable indefinitely. In the presence of excess EX_2 the insertion derivatives $FpEX_2CH_2CR^1=CH_2$ slowly (5 to 24 d) react to form $FpEX_3$. In CH_3OH the reactions of $FpCH_2CR^1=CH_2$ and an (E)/(Z)-mixture of $FpCH_2CH=CHCH_3$ with $SnCl_2$ and $SnBr_2$ are more complex, an unknown species and substantial amounts of $FpEX_3$ are also produced. The insertion reactions are completely inhibited by the addition of a small amount (~ 5 mol%) of the radical scavenger 1,1-diphenylpicrylhydrazyl. A mechanistic scheme for these processes is discussed [131].

$FpSnCl_3$ results as sole product from the reaction of $FpCH=C=CH_2$ with $SnCl_2$ in THF after 24 h. Treatment of $FpCH_2CH=CH_2$ with SnI_2 in benzene affords FpI. $PbCl_2$ in CH_3OH failed to react with $FpCH_2CR^1=CR^2R^3$ [78].

The vinyl derivatives $FpCH=CHR^1$ ($R^1 = H$, C_6H_5 [48, 49], $COCH_3$, COC_6H_5 [21, 29, 37, 38], $CON(CH_2)_5$ [166]) produce on heating with $Fe_2(CO)_9$ in benzene at 40°C for 2 h the σ, π-binuclear complexes XXII, in which the two iron atoms are connected by an intermetallic bond and a bridging CO group.

XXII

Addition of excess $FpCH=CH_2$ to a CH_2Cl_2 solution of $[Fp=CH_2]PF_6$, prepared by halide abstraction from $FpCH_2Cl$ with $AgPF_6$ in CH_2Cl_2 at −80°C, and subsequent warming to 20°C leads to formation of a mixture of $[(FpCH_2)_2CH]PF_6$ (57%), $[Fp(CH_2=CH_2)]PF_6$ (8%), and $[C_5H_5Fe(CO)_3]PF_6$ (8%). A mechanism is proposed, according to which the dinuclear salt is formed by electrophilic attack of $[Fp=CH_2]^+$ at the vinylic β-carbon, followed by isomerization of the resulting intermediate $FpCH^+CH_2CH_2Fp$. Similar treatment of $FpCH_2CH=CH_2$ gives $[Fp(CH_2=CHCH_2CH_2Fp)]PF_6$ (41%), $[Fp(CH_2=CHCH_3)]PF_6$ (18%), $[Fp(CH_2=CH_2)]PF_6$ (16%), and $[C_5H_5Fe(CO)_3]PF_6$ (19%) [202].

Dinuclear condensation products $[Fp(CH_2=CHCH_2CHR^1CHR^2Fp)]^+$ result from the reaction of $FpCH_2CH=CH_2$ with $[Fp(R^1CH=CHR^2)]BF_4$ in CH_3NO_2 at room temperature for 4 to 5 h [155]. For $R^1=R^2=H$, see also [71]. In the cases, where $R^1=CH_3$ or C_6H_5 and $R^2=H$, mixtures of two regioisomeric adducts ($R^1=H$, $R^2=CH_3$ and $R^1=CH_3$, $R^2=H$, 1:1 mixture; $R^1=H$, $R^2=C_6H_5$ and $R^1=C_6H_5$, $R^2=H$, 1:3 mixture) are obtained. The proportion of the latter isomeric adducts is not markedly affected by solvent. It remains 1:3 in CH_2Cl_2 as in CH_3NO_2 and is 2:3 in CH_3COCH_3. However, an attempt to perform the condensation reaction in dimethylformamide led to the recovery of 60% of unreacted $FpCH_2CH=CH_2$ even after heating at 50 °C for 5 h. No condensation product could be detected. Presumably the more basic solvent impedes the reaction by solvation of the cationic complex [155], see also [81].

The reactions of $FpCH_2CH=CH_2$ with $[Fp(CH_2=CHCHO)]BF_4$ or $(E)-[Fp(CH(COOC_2H_5)=CHCH_3)]BF_4$ in CH_3NO_2 at 0 °C for 3 h afford the adducts $[Fp(CH_2=CHCH_2CH_2CH(Fp)CHO)]^+$ and $[Fp(CH_2=CHCH_2CH(CH_3)CH(Fp)COOC_2H_5)]^+$, respectively, resulting from conjugate addition. The diastereoselectivity of both reactions prefers configurations in which the olefin-olefin interactions are maximized, this is discussed. The analogous reaction of $FpCH_2CH=CH_2$ with $[Fp(CH(COOCH_3)=CHCH_3)]PF_6$ to give $[Fp(CH_2=CHCH_2CH(CH_3)CH(Fp)COOCH_3)]^+$ has been carried out in CH_2Cl_2/CH_3NO_2 (6:1) at 0 °C (2 h), then at room temperature (30 min) [155]. The reaction of $FpCH_2CH=CH_2$ with $[Fp(CH_2=CHCOCH_3)]^+$ in CH_2Cl_2 at 0 °C to afford $[Fp(CH_2=CHCH_2CH_2CH(Fp)COCH_3)]^+$ is briefly mentioned in [192]. $FpCH_2CH=CH_2$ also condenses rapidly with $[Fp(CH_2=CHOCH_3)]^+$ at room temperature to give $[Fp(CH_2=CHCH_2CH(OCH_3)CH_2Fp)]^+$, whereas no reaction occurs with $[Fp(CH_2=CHN(CH_3)_2)]^+$ on prolonged standing or even on warming to 40 °C. Kinetic studies, carried out on the BF_4^- salts in CD_3NO_2 and monitored by NMR spectrometry, show that the half-life time for the reaction of $FpCH_2CH=CH_2$ with $[Fp(CH_2=CH_2)]^+$ at 10 °C is 82 min and with $[Fp(CH_2=CHOCH_3)]^+$, same concentration of reactants, but at −3 °C, is 3 min. Rate plots exhibit good second-order behavior throughout. The activation energy for the reaction with $[Fp(CH_2=CH_2)]^+$ has been calculated to be 19.4 kcal/mol [169].

In contrast to the above addition reactions, treatment of $FpCH_2C(CH_3)=CH_2$ with $[Fp(CH_2=C(CH_3)_2)]BF_4$ in CH_2ClCH_2Cl at 55 °C and further at 70 °C for 12 min affords the exchange product $[Fp(CH_2=C(CH_3)CH_2Fp)]BF_4$ [171].

The reaction of $FpCH_2CH=CH_2$ with $[Fp(CH_2=CHCH=CH_2)]BF_4$ in CH_3NO_2 at room temperature for 3 h gives a 1:1 mixture of the dinuclear complexes XXIII and XXIV [155], see also [71, 81].

Condensation of $FpCH_2CH=CH_2$ with $[Fp(C_5H_8)]BF_4$ failed (C_5H_8 = cyclopentene). Even under forcing conditions of refluxing the reactants in CH_2Cl_2 for several hours, the only product obtained was the dinuclear complex XXV [155].

XXIII   XXIV

XXV

References on pp. 141/7

Both carbenium ions, $[\eta^1$-benzocyclobutenylidene-Fp$]^+$ and $[C(C_6H_5)_3]^+$, add readily to the allyl complexes FpCH$_2$CR^1=CR^{2}R^3 to give the addition products XXVI. Reaction conditions are reported for FpCH$_2$CH=CH$_2$. It is added to $[\eta^1$-benzocyclobutenylidene-Fp$]^+$ [109] or $[C(C_6H_5)_3]^+$ [154] in CH$_2$Cl$_2$, and the mixture is refrigerated at $-18\,°C$ for 1 h or allowed to reach room temperature, respectively. In contrast, FpCH$_2$CH=C(CH$_3$)$_2$ undergoes in both cases exclusive hydride abstraction to form the isoprene complex [Fp(CH$_2$=CHC(CH$_3$)=CH$_2$)]$^+$ [73] and FpCH$_2$CH$_2$CH=CH$_2$ reacts with $[C(C_6H_5)_3]BF_4$ in CH$_2$Cl$_2$ at $0\,°C$ to room temperature to give [Fp(CH$_2$=CHCH=CH$_2$)]BF$_4$ [154]. According to [152] treatment of (Z)-FpCH$_2$CH=CHCH$_3$ with $[C(C_6H_5)_3]BF_4$ in CHCl$_3$ causes protonation to give [Fp(CH$_2$=CHCH$_2$CH$_3$)]BF$_4$. Isotope experiments revealed that the proton originates not from the solvent, but from traces of water. It hydrolyzes $[C(C_6H_5)_3]BF_4$ to HBF$_4$, which then protonates (Z)-FpCH$_2$CH=CHCH$_3$ to the η^2-bonded but-1-ene complex. Similarly (E)-FpCH=CHC$_6$H$_5$ reacts with $[C(C_6H_5)_3]BF_4$ in CH$_2$Cl$_2$ at $0\,°C$ to room temperature, producing [Fp(CH$_2$=CHC$_6$H$_5$)]BF$_4$ [154]. FpCH$_2$CH=CH$_2$ has also been combined with $[\eta^2$-1,2-benzocyclobutadiene-Fp$]^+$ in CH$_2$Cl$_2$ at $24\,°C$ for 15 min, yielding the cationic binuclear complex XXVII [109].

$R^1=R^2=R^3=H$

$R^1=CH_3,\ R^2=R^3=H$

$R^1=H,\ R^2=CH_3,\ R^3=H$

XXVI

XXVII

The reaction of FpCH$_2$CH=CH$_2$ with tropyliumiron tricarbonyl in CH$_2$ClCH$_2$Cl at $55\,°C$ for 3 h proceeds via the intermediate XXVIII (detected at lower temperatures by ^{1}H NMR) to give the dinuclear hydroazulene complex XXIX as a mixture of C$_2$ epimers. Similarly, FpCH$_2$CH=CHR^1 (R$^1\neq$H) affords, at room temperature after 10 min, compound XXX as a mixture of trans-1,2-disubstituted complexes. The corresponding reaction of FpCH=C=CH$_2$ in CH$_3$NO$_2$ at room temperature for 15 min results in formation of complex XXXI [101].

Addition of FpCH$_2$CH=CHR^1 (R^1 = H, CH$_3$) or FpCH=C=CH$_2$ in CH$_2$Cl$_2$ to the trimethylsilanyl-oxy- or di-n-butylboranyloxytropylium salts XXXII, prepared from tropeneiron tricarbonyl and (CH$_3$)$_3$SiOSO$_2$CF$_3$ or (n-C$_4$H$_9$)$_2$BOSO$_2$CF$_3$ in CH$_2$Cl$_2$, at $-78\,°C$ (2 h), followed by warming to room temperature, refluxing for 2 h and quenching the reaction with absolute ethanol and solid K$_2$CO$_3$ at room temperature for 12 h, affords the ketohydroazulene complexes XXXIII and XXXIV, respectively, as single regioisomers [209].

The alk-2-enyl complexes FpCH$_2$CR^1=CR^{2}R^3 react rapidly with liquid SO$_2$ at $-10\,°C$ or below via the zwitterionic intermediates XXXV, characterized by IR and ^{1}H NMR spectros-

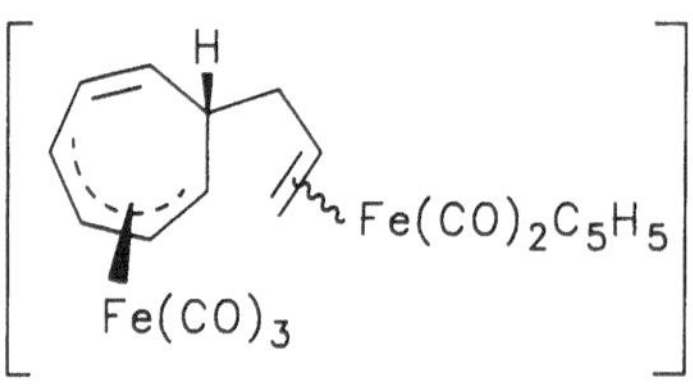

XXVIII

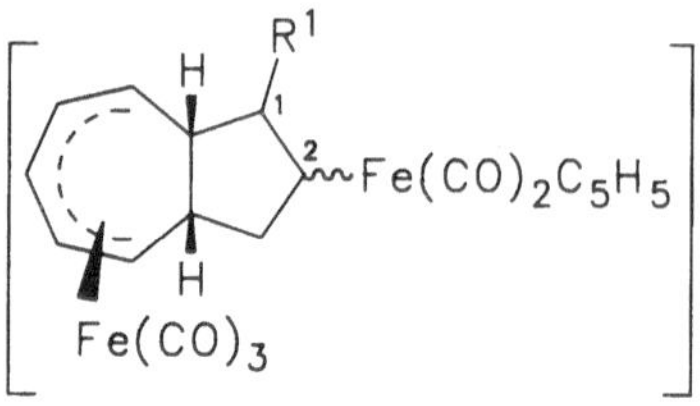

XXIX: R^1=H

XXX: R^1= [structure] or [structure]

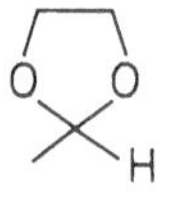

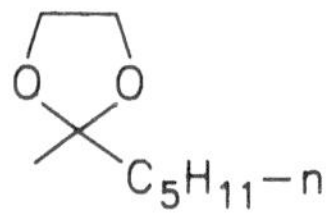

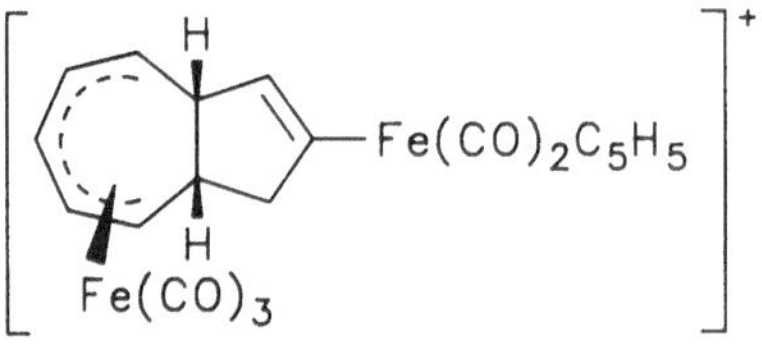

XXXI

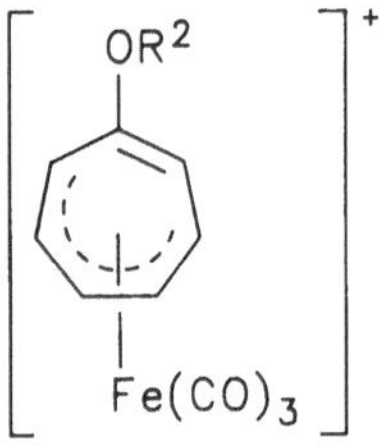

R^2=Si(CH$_3$)$_3$ or
B(C$_4$H$_9$-n)$_2$

XXXII   XXXIII 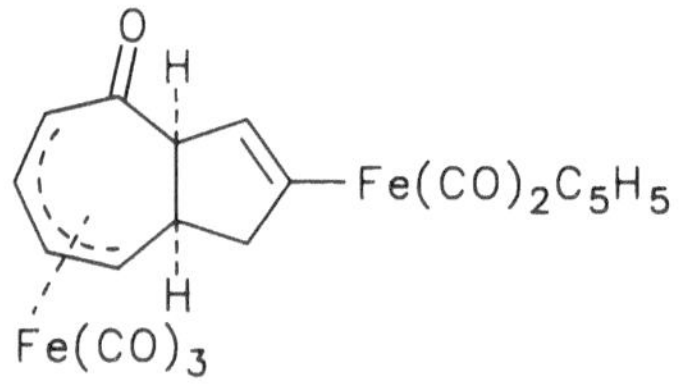   XXXIV

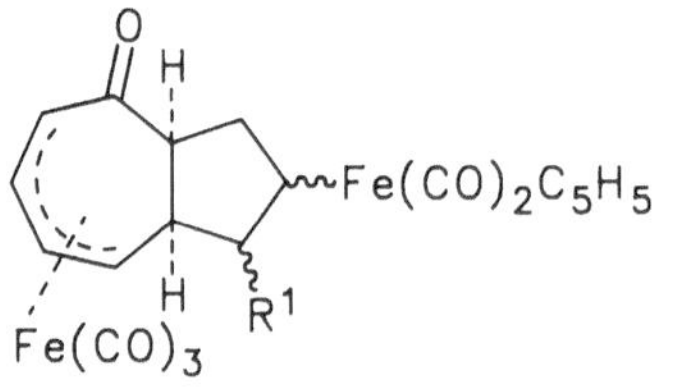

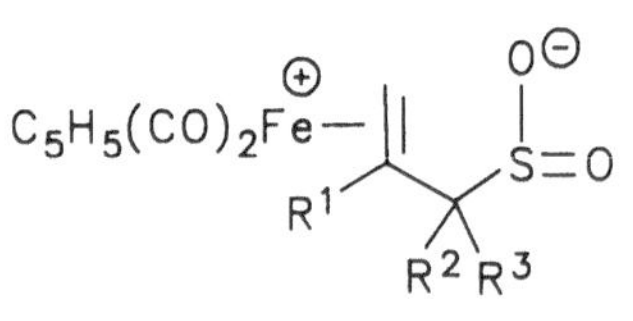

R^1=R^2=R^3=H
R^1=R^2=H, R^3=CH$_3$
R^1=H, R^2=R^3=CH$_3$
R^1=CH$_3$, R^2=R^3=H

XXXV

copy, and in the case where R^1=R^2=R^3=H, also by conductivity measurements and alkylation with [O(CH$_3$)$_3$]BF$_4$ (for this reaction see also [86]) and [C(C$_6$H$_5$)$_3$]Cl to give [Fp(CH$_2$=CHCH$_2$SO$_2$R^4)]$^+$ (R^4=CH$_3$, C(C$_6$H$_5$)$_3$). Attempts to isolate these zwitterions by rapid removal of the excess SO$_2$, as well as storage in liquid SO$_2$, lead to rearrangement to the respective S-sulfinato complexes FpSO$_2$CH$_2$CR^1=CR^{2}R^3 and/or FpSO$_2$CR^{2}R^{3}CR^1=CH$_2$. The rates increase with the increasing extent of alkyl substitution in the olefin fragment [74, 132]. For an (E)/(Z)-mixture of $FpCH_2CH=CHCH_3$ the reaction with SO_2 at $-40\,°C$ [51] or $-65\,°C$ [207] is so fast that it has not been possible to measure the pseudo-first-order rate constant by low-temperature IR spectroscopy. The formation of the rearranged S-sulfinato isomers $FpSO_2CR^2R^3\ CR^1=CH_2$ is promoted by low temperatures of neat SO_2, and nonpolar (hexane) over polar (CH_3CN) organic solvents. Moreover, the ratio of the rearranged to the unrearranged isomer decreases with the nature of the substituents R^1, R^2, and R^3 in the order $R^1=R^2=H$, $R^3=CH_3>R^1=H$, $R^2=R^3=CH_3\geqq R^1=R^2=H$, $R^3=C_6H_5$. Thus, from an (E)/(Z)-mixture of $FpCH_2CH=CHCH_3$ in refluxing SO_2 only the rearranged $FpSO_2CH(CH_3)CH=CH_2$ has been obtained, while $FpCH_2CH=C(CH_3)_2$ and (E)-$FpCH_2CH=CHC_6H_5$ afford mixtures of both isomers [136], see also [43]. It has further been found that the formation of (E)-$FpSO_2CH_2CH=CHC_6H_5$ proceeds with retention of the (E)-configuration of the starting material [136]. The rearrangement of the zwitterion resulting from $FpCH_2CH=CH_2$ and SO_2 occurs concomitantly with the formation of an unstable cycloaddition product XXXVI, which undergoes dimerization to XXXVII and/or trimerization upon storage in solution, finally yielding $[Fp(C_3H_5SO_2)_n]$ (n>3). Studies show that at lower concentrations the zwitterion appears to rearrange preferentially to $FpSO_2CH_2CH=CH_2$, whereas at higher concentrations its rearrangement to XXXVI seems to be favored [132]. Possible mechanisms of these reactions are presented and discussed [132, 136]. The formation of $FpSO_2CH_2CH=CH_2$ from $FpCH_2CH=CH_2$ and SO_2 is also briefly mentioned in [55].

XXXVI   XXXVII

$FpCH_2CH=CHCl$ reacts only slowly and with considerable decomposition with neat SO_2 at reflux or with an SO_2-saturated hexane/ether (1:1) solution at 25 to 30 °C for 11 h to yield a small amount of what appears to be the corresponding S-sulfinate from its IR spectrum [136].

Cycloadducts of structures XXXVIII and XXXIX result from the reactions of $FpC(=CH_2)CH=CH_2$ or $FpCH=CHC(CH_3)=CH_2$, respectively, with SO_2 at $-78\,°C$ to $+24\,°C$. However, XXXIX could not be obtained pure [200].

XXXVIII   XXXIX

No detectable amount of an SO_2-containing product forms on holding $FpCH=CH_2$, $FpCH=C=CH_2$, or $FpCH=C=CHCH_3$ in liquid SO_2 at reflux for 12 to 48 h. Also no evidence of reaction other than decomposition to intractable solids could be obtained by carrying out the reaction of $FpCH=C=CH_2$ with liquid SO_2 in a pressure vessel at 25 to 30 °C for 24 to 48 h or with

SO_2 in $CHCl_3$ at 25 °C for 8 h. The lack of reactivity is ascribed to the relatively strong Fe–C bonds in these compounds resulting from the participation of the metal in π-bonding with the unsaturated σ-bonded hydrocarbon ligand [52], see also [50].

The reaction of $FpCH_2CH=CH_2$ with CH_3SO_2Cl in CH_2Cl_2 and further with $N(C_2H_5)_3$ at 0 °C for 1 h gives the cycloadduct XXXVII [99]. The same product results from $FpCH_2CH=CH_2$ and $CH_2=SO_2$ [83].

[3+2]cycloaddition with formation of the pyrrolid-2-one complex XL with $R^1 = ClSO_2$ and $R^2 = H$ ($R^3 = R^4 = CH_3$; $R^3 = H$, $R^4 = C_6H_5$) takes place on treatment of $FpCH_2CH=CR^3R^4$ in CH_2Cl_2 with $ClSO_2NCO$ in benzene at −70 to −45 °C for 20 min, whereas the similar reaction of $FpCH_2C(CH_3)=CH_2$ with $ClSO_2NCO$ at −40 °C for 15 min gives the insertion product $FpN(SO_2Cl)COCH_2C(CH_3)=CH_2$. Further reaction of the mixture resulting from $FpCH_2CH=CHR^3$ ($R^3 = CH_3$, C_6H_5) at 10 °C for 10 to 20 min affords XL with $R^1 = R^3 = H$ ($R^4 = CH_3$ or C_6H_5) and FpCl. Reaction mechanisms are discussed [63, 64, 72]. Instantaneous formation of the cycloadduct XL ($R^1 = ClSO_2$, $R^2 = R^3 = R^4 = H$) from $FpCH_2CH=CH_2$ and $ClSO_2NCO$ in CH_2Cl_2 occurs at −80 °C [99], see also [65].

$$C_5H_5(CO)_2Fe \overset{R^3 \quad R^4}{\underset{R^2}{\diagup}} \overset{O}{\diagdown} N-R^1$$

XL

$R^1 = H$, $ClSO_2$, CH_3OSO_2, $4\text{-}CH_3C_6H_4SO_2$, $2,5\text{-}Cl_2C_6H_3$
$R^2 = H$, CH_3
$R^3 = H$, CH_3, C_6H_5
$R^4 = H$, CH_3, OCH_3, C_6H_5

The reactions of $FpCH_2CR^2=CR^3R^4$ in CH_2Cl_2 with other isocyanates such as 4-$CH_3C_6H_4SO_2NCO$ (for $R^2 = R^3 = H$, $R^4 = H$, OCH_3 (see also [86]), C_6H_5 (see also [88]); $R^2 = CH_3$, $R^3 = R^4 = H$; $R^2 = H$, $R^3 = R^4 = CH_3$) [99], see also [65], or CH_3OSO_2NCO ($R^2 = R^3 = R^4 = H$) [65] at room temperature for 30 min or $2,5\text{-}Cl_2C_6H_3NCO$ ($R^2 = R^3 = R^4 = H$) at room temperature for several days [65] or by refluxing for 75 h [99] proceed in a similar way, giving XL ($R^1 = 4\text{-}CH_3C_6H_4SO_2$, CH_3OSO_2, or $2,5\text{-}Cl_2C_6H_3$, respectively). The reaction fails with ethyl or phenylisocyanate [99]. The pyrrolinone XLI has been obtained in a rapid reaction of $FpCH=C=CH_2$ with $4\text{-}CH_3C_6H_4SO_2NCO$ at room temperature [69], and XLII results from $FpCH_2C(CH_2Cl)=CH_2$ [141]. Kinetic studies on the reactions of $FpCH_2CH=CR^3R^4$ ($R^3 = R^4 = H$; $R^3 = H$, $R^4 = CH_3$ or C_6H_5; $R^3 = R^4 = CH_3$) with $4\text{-}CH_3C_6H_4SO_2NCO$ in CH_2Cl_2 at 25 °C revealed that the cycloaddition is first-order in each reactant and second-order overall. The bimolecular rate constants ($k_2 \cdot 10^2$ in $M^{-1} \cdot s^{-1}$) decrease in the order $FpCH_2CH=CHCH_3$ (11) > $FpCH_2CH=C(CH_3)_2$ (7.2) > $FpCH_2CH=CH_2$ (0.26) $\gtrsim FpCH_2CH=CHC_6H_5$ (0.25), indicating that both steric and electronic effects are contributing. Additionally, the less hindered (Z)-isomer of $FpCH_2CH=CHCH_3$ undergoes cycloaddition twice as rapidly as the more encumbered (E)-isomer. It has further been found that $FpCH_2CH=CHR^3$ and the structurally related $FpCH_2C\equiv CR^3$ undergo cycloaddition with $4\text{-}CH_3C_6H_4SO_2NCO$ at comparable rates. On the basis of the data obtained it has not been possible to distinguish between a two-step dipolar mechanism or a concerted one [163], see also [138].

References on pp. 141/7

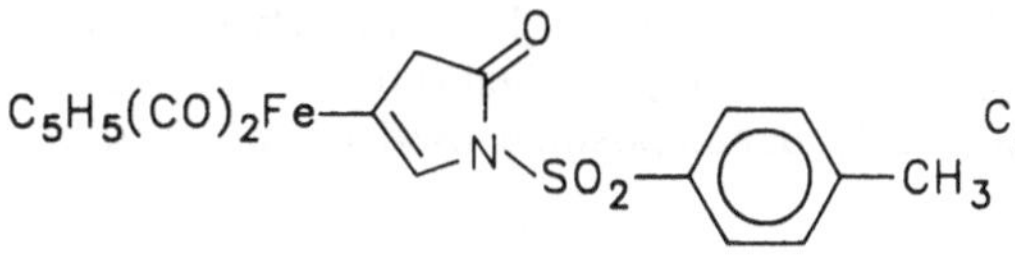

XLI

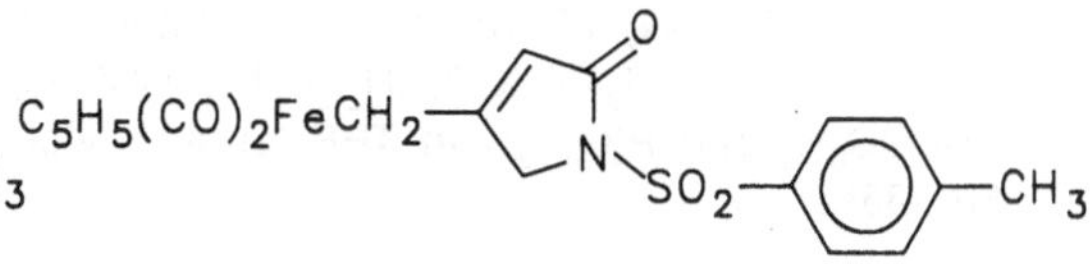

XLII

The reaction of $FpC(=CH_2)CH=CH_2$ with $4\text{-}CH_3C_6H_4SO_2NCO$ in CH_2Cl_2 at 22 °C causes immediate formation of the imino lactone XLIII. A stepwise cycloaddition of the isocyanate via dipolar intermediates is assumed [200], see also [141].

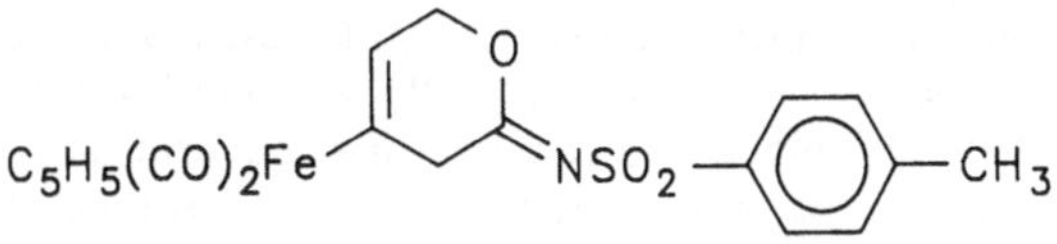

XLIII

Treatment of $FpCH_2CH=CH_2$ with CCl_3CONCO in CH_2Cl_2 at room temperature for 1 h affords exclusively $(E)\text{-}FpCH_2CH=CHCONHCOCCl_3$, whereas the corresponding reaction of $FpCH_2CH=C(CH_3)_2$ at −78 °C to room temperature for 2 h gives the lactam XL ($R^1 = CCl_3CO$, $R^2 = H$, $R^3 = R^4 = CH_3$) isolated after brief treatment with aqueous NaOH as the lactam XL ($R^1 = R^2 = H$, $R^3 = R^4 = CH_3$) [99].

$CH_3OC(O)N=SO_2 \cdot THF$, formed by treatment of methyl N-(chlorosulfonyl)urethane with NaH in THF, has also been found to enter into a [3+2]cycloaddition reaction with $FpCH_2CH=CH_2$ in THF at −10 °C to room temperature (2 h) yielding the γ-sultam XLIV [65, 99].

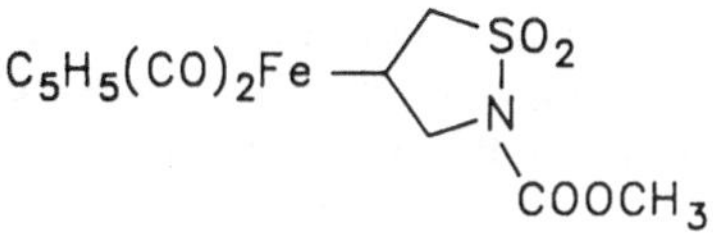

XLIV

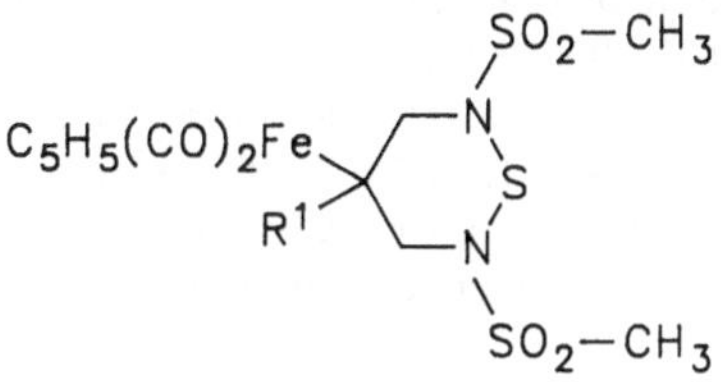

XLV

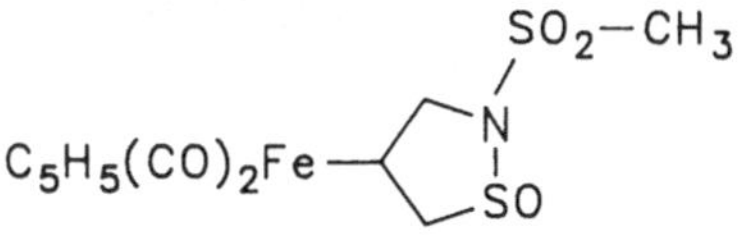

XLVI

[3+3]cycloaddition to give complex XLV is effected by the reaction of $FpCH_2CR^1=CH_2$ ($R^1 = H$ or CH_3) with one equivalent of $S(NS(O)_2CH_3)_2$ in CH_2Cl_2 at 25 °C for 30 min. A possible reaction mechanism is proposed. In contrast, reaction of $FpCH_2CH=CH_2$ with $CH_3S(O)_2NSO$ under comparable conditions yields the [3+2]cycloadduct XLVI [180].

The reactions of $Fp(CH_2)_nCH=CH_2$ (n=2 to 4) with $P(C_6H_5)_3$ in CH_3CN or THF at 50 to 55 °C for 4 to 5 d [47, 148], see also [44], or of $Fp(CH_2)_2CH=C=CHCH_3$ with $P(C_4H_9\text{-}n)_3$ in

References on pp. 141/7

THF at 60 °C [44] afford the acyl complexes $C_5H_5Fe(CO)(P(C_6H_5)_3)CO(CH_2)_nCH=CH_2$ and $C_5H_5Fe(CO)(P(C_4H_9-n)_3)CO(CH_2)_2CH=C=CHCH_3$, respectively. Refluxing (E)-FpCH= $CHCOC_6H_5$ and $P(C_6H_5)_3$ in THF for 42 h yields only decomposition products along with Fp_2 [35].

One CO group is substituted when a mixture of $FpCH_2CH=CH_2$ and $P(OCH_3)_3$ in benzene is either allowed to stand for 2 h or is UV-irradiated in benzene [189] or in petroleum ether for 0.5 h [157] in the presence of a trace of Fp_2. Under the latter conditions the same type of product, $C_5H_5FeCO(^2D)CH_2CH=CH_2$, also results from $FpCH_2CH=CH_2$ and the ^2D ligands $P(OC_6H_5)_3$ [157, 189], $P(CH_3)_3$, $P(CH_3)_2C_6H_5$, or $P(C_6H_5)_2CH_3$ [189]. A radical chain mechanism is proposed, which is initiated by the radical $C_5H_5Fe(CO)_2\cdot$ (produced by photochemical cleavage of Fp_2), which undergoes rapid CO substitution, followed by transfer of the $CH_2CH=CH_2$ ligand from $FpCH_2CH=CH_2$ to $C_5H_5Fe(CO)(^2D)\cdot$. The alkyl-transfer step could be evidenced by crossover experiments [189], see also [147]. In contrast, a mixture of $FpCH_2CH=CH_2$ and $P(C_6H_5)_3$ showed no reaction after several days in the dark, and on irradiation in the presence of Fp_2, only slow formation of $C_5H_5Fe(CO)(\eta^3-CH_2CHCH_2)$ and a small amount of $(C_5H_5)_2Fe_2(CO)_3(P(C_6H_5)_3)$ were observed. It is assumed that $P(C_6H_5)_3$ substitutes $C_5H_5Fe(CO)_2\cdot$ too slowly to keep the radical chain path going [189]. UV irradiation of $P(C_6H_5)_3$ with $FpC(CH_3)=CH_2$ in toluene/hexane (8:92) for 75 min [176, 201], or with FpCH= $CHOR^1$ in THF for 6 h ($R^1 = CH_3$, C_6H_5 [35]) or 20 h ($R^1 = C_6H_4CH_3-4$ [140]), or with (E)-FpCH= $CHCONH_2$ or with (E)-FpCH=CHCON(CH_2)_5$ in THF for 20 h [166] has been found to give the corresponding CO substitution products, whereas from the analogous reaction of (E)- $FpCH=CHCOCH_3$ with $P(OC_6H_5)_3$ only the starting material could be recovered [35]. Also, photolysis of $FpCH_2CR^2=CH_2$ ($R^2 = H$, CH_3 [157], or OCH_3 [187]) and the cyclic phosphite XLVII in petroleum ether for 15 min [157] or in pentane for 30 to 40 min [187] in the presence of a catalytic amount of Fp_2 gives the monosubstitution product $C_5H_5Fe(CO)(P(OCH_2)_3CCH_3)$- $CH_2CR^2=CH_2$. However, similar treatment of (E)-$FpCH_2CH=CHCH_3$ affords only the dinuclear complex $(C_5H_5)_2Fe_2(CO)_3(P(OCH_2)_3CCH_3)$ [157].

Elimination of both CO groups and formation of complex XLVIII occurs on photolysis of $FpC(=C(CF_3)_2)C(CF_3)=CF_2$ at 10 to 15 °C, either in the presence of $P(C_6H_5)_3$ in pentane for 10 h [108, 128], see also [105, 106, 117], or of $P(OC_6H_5)_3$ in benzene for 16 h, or of benzothiazole in benzene for 26 h [128], see also [93]. As indicated by IR measurements, these reactions proceed via an η^3-allylidene complex, in which the remaining CO group is slowly replaced by the respective ^2D ligand [128].

$^2D = P(C_6H_5)_3, P(OC_6H_5)_3,$

$C_5H_5Fe(^2D)$

XLVII   XLVIII

Condensation of excess F_3CCOCF_3 onto the complexes $FpCH_2CR^1=CR^2 R^3$ in pentane ($R^1 = R^2 = H$, $R^3 = CH_3$; $R^1 = H$, $R^2 = R^3 = CH_3$) or CH_2Cl_2 (all others) and stirring the solutions for 0.5 to 2.5 h under a dry ice condenser before warming to room temperature causes formation of the cycloaddition products IL. In a similar reaction $FpCH_2C(CH_3)=CH_2$ affords the cycloadduct along with the insertion product $FpOC(CF_3)_2CH_2C(=CH_2)CH_2C(CF_3)_2OH$. No reaction has been observed between the alkenyl complexes and neat F_3CCOCF_3. The reactions are compared with other related [3+2]cycloadditions involving transition metal–carbon σ-bonded compounds [92].

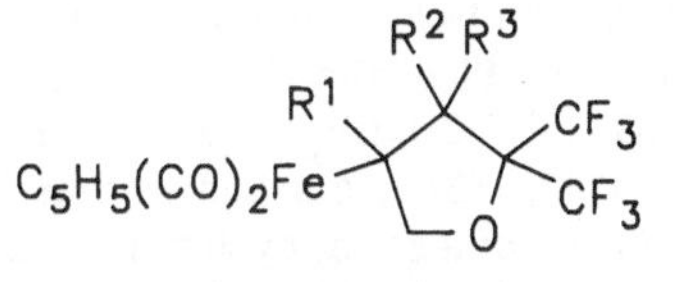

$$C_5H_5(CO)_2Fe \overset{R^1}{\underset{O}{\longleftarrow}} \overset{R^2\ R^3}{\underset{}{\bigtriangleup}} \overset{CF_3}{\underset{CF_3}{}}$$

IL

R^1=H, R^2=R^3=CH$_3$
R^1=R^2=H, R^3=H, CH$_3$, C$_6$H$_5$, or Cl
R^1=CH$_3$, R^2=R^3=H

The compounds FpCH$_2$CH=CR^{1}R^2 (R^1=R^2=H; R^1=H, R^2=CH$_3$) react with NCC≡CCN in CH$_2$Cl$_2$ at −78 °C to room temperature during several hours to give the insertion products FpC(CN)=C(CN)CR^{1}R^{2}CH=CH$_2$ with a rearranged allyl ligand. In a similar reaction FpCH$_2$CH= C(CH$_3$)$_2$ affords a mixture of two compounds that could not be separated, but of which the more abundant product is postulated on the basis of IR and ^{1}H NMR data to be FpC(CN)= C(CN)C(CH$_3$)$_2$CH=CH$_2$. A reaction mechanism is proposed [129], see also [112].

The reactions of FpCH$_2$CR^1=CR^{2}R^3 with excess F$_3$CC≡CCF$_3$ in CHCl=CCl$_2$ at 60 to 65 °C for 1.5 to 12 h proceed with insertion of the acetylene, giving the monocarbonyls L with an η^1,η^2-alkadienyl ligand containing an unrearranged allyl moiety. Possible reaction mechanisms are discussed. Because of instability, the product resulting from FpCH$_2$CH=C(CH$_3$)$_2$ could only be identified by a ν(CO) absorption at 1990 cm^{-1}. Besides the main product L, the reactions afford a number of other complexes in low yields, e.g. (E)-FpC(CF$_3$)=CHCF$_3$ and the norbornadienyl complex LI [129], see also [112]. Insertion, with formation of L (R^1=R^2=R^3=H) or of the analogous product LII, takes place on irradiation of FpCH$_2$CH=CH$_2$ in the presence of excess F$_3$CC≡CCF$_3$ or F$_3$CC≡CH in hexane at room temperature for 24 h [113], see also [115], or for 15 d in a sealed tube [158], respectively. Other products resulting from the latter reaction are ferrocene and C$_5$H$_5$Fe(CO)(η^3-C$_3$H$_5$) [158].

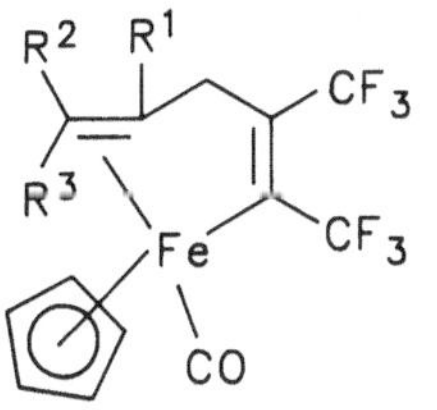

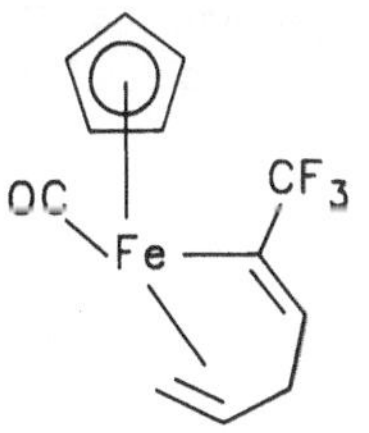

L

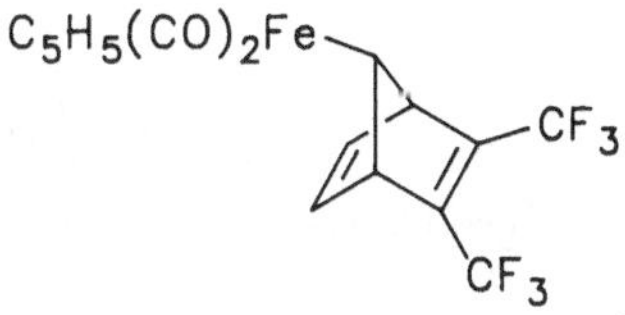

R^1=R^2=H, R^3=CH$_3$
R^1=CH$_3$, R^2=R^3=H
R^1=H, R^2=R^3=CH$_3$

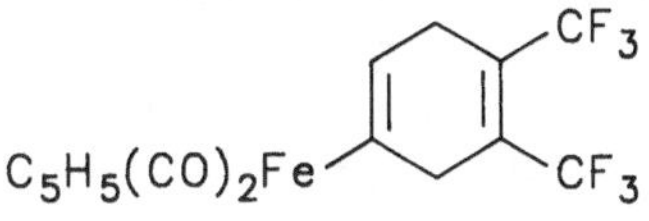

LI

LII

LIII

FpC(=CH$_2$)CH=CH$_2$ reacts with F$_3$CC≡CCF$_3$ in CH$_2$Cl$_2$ when sealed in a tube at −78 °C and then allowed to stand at 24 °C for 3 h to give the cycloadduct LIII [200].

The reaction of FpCH$_2$CH=CH$_2$ with two equivalents of CH$_3$OOCC≡CCOOCH$_3$ in dimethyl-formamide at room temperature for 90 h affords the cyclic compound LIV (R^1=H) (42% yield), FpC(COOCH$_3$)=C(COOCH$_3$)CH$_2$CH=CH$_2$ (9%), and FpCH$_2$CH=CHC(COOCH$_3$)=CHCOOCH$_3$ (3%) [172], see also [142, 143]. It is assumed that the collapse of the intermediate dipolar zwitterion LV follows three routes: (a) cyclization, (b) insertion, and (c) H–transfer [172]. Under similar conditions FpCH$_2$C(OCH$_3$)=CH$_2$ generates the H–transfer product FpCH$_2$C(OCH$_3$)=CHC(COOCH$_3$)=CHCOOCH$_3$ (63%) [173], see also [142, 143]. (Z)-FpCH$_2$CH=CHOCH$_3$ gives only the cyclic product LIV (R^1=OCH$_3$, 77%), when it is combined with four equivalents of CH$_3$OOCC≡CCOOCH$_3$ in dimethylformamide at room temperature for 48 h, followed by addition of another two equivalents and stirring further for 24 h. It is suggested that the absence of linear products might be due to the presence of the 3–methoxy group which could disfavour abstraction of hydrogen from C–3 in the intermediate zwitterion [178], see also [143]. For the reaction of FpC(=CH$_2$)CH=CH$_2$ with CH$_3$OOCC≡CCOOCH$_3$ see under No. 78. Only starting material has been recovered from the reaction of FpCH$_2$CH=CH$_2$ with four equivalents of CH≡CCOOCH$_3$ in dimethylformamide at room temperature for 80 h [172].

LIV   LV

R^1=H or OCH$_3$

[3+2]cycloaddition products LVI are obtained from FpCH$_2$CR^1=CR^{2}R^3 in CH$_2$Cl$_2$ [46, 53, 86, 99, 139], CHCl=CHCl [46], C$_6$H$_6$ [80, 86], CH$_3$CN [80] or THF [53, 80] and (NC)$_2$C=C(CN)$_2$ in THF [86, 99, 139], see also [142], or in the aforementioned solvents at room temperature generally during several minutes. The corresponding reactions of FpCH$_2$CH=CR^{2}R^3 (R^2=H, R^3=CH$_2$C$_6$H$_5$ or C$_2$H$_5$; R^2=CH$_3$, R^3=C$_2$H$_5$) have been carried out in CH$_2$Cl$_2$/THF at 0 °C to room temperature [120]. Possible reaction mechanisms are discussed [46, 53, 80, 99], see also [54, 66, 88]. Treatment of (Z)-FpCH$_2$CH=CHBr with (NC)$_2$C=C(CN)$_2$ in CH$_2$Cl$_2$/THF at 0 °C gives FpBr as the only isolable organometallic substance [86].

LVI

R^1=R^2=H, R^3=H, CH$_3$, CH$_2$C$_6$H$_5$, C$_2$H$_5$, OCH$_3$, SO$_2$CH$_3$, CH(OCH$_3$)$_2$, COCH$_3$, COCH(CH$_3$)$_2$,

COOCH$_3$, C$_6$H$_5$, , , or

R^1=H, R^2=R^3=CH$_3$
R^1=CH$_3$ or OCH$_3$, R^2=R^3=H
R^1=H, R^2=CH$_3$, R^3=C$_2$H$_5$

References on pp. 141/7

The cycloadduct LVII results from the reaction of FpCH=C=CH$_2$ with (NC)$_2$C=C(CN)$_2$ at room temperature for 15 min [69] and LVIII is obtained from FpCH$_2$C(CH$_2$Cl)=CH$_2$ [141]. The [4+2]cycloadducts LIX and LX are formed by treatment of FpC(=CH$_2$)CH=CH$_2$ or (E)-FpCH=CHCR^2=CH$_2$ (R^2=H or CH$_3$) with (NC)$_2$C=C(CN)$_2$ in CH$_2$Cl$_2$ at 22 °C for 0.5 h [200]. The reaction of FpC(=CH$_2$)CH=CH$_2$ was too fast to be monitored by ^{1}H NMR spectroscopy. However, competition experiments showed that FpC(=CH$_2$)CH=CH$_2$ is, by at least a factor of three, more reactive toward (NC)$_2$C=C(CN)$_2$ than the electron-rich 2-trimethoxysiloxybuta-1,3-diene [200], see also [141].

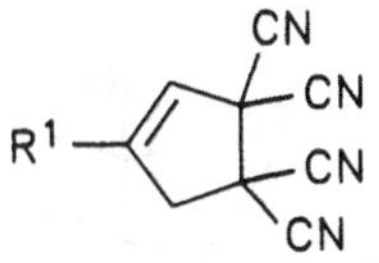

LVII: R^1=C$_5$H$_5$(CO)$_2$Fe

LVIII: R^1=C$_5$H$_5$(CO)$_2$FeCH$_2$

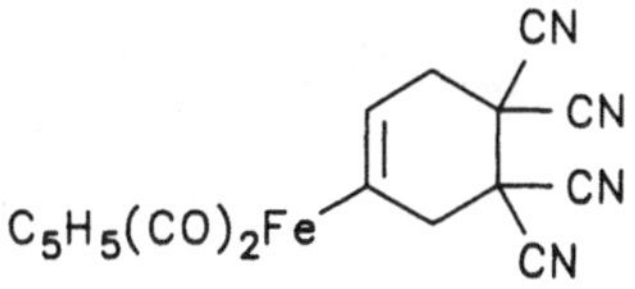

LIX

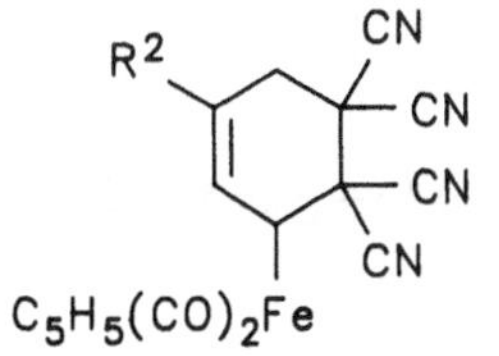

LX

R^2=H or CH$_3$

Mixtures of diastereomers result from the reactions of FpCH$_2$C(R^1)=CR$_2^2$ (R^1=R^2=H; R^1=CH$_3$, R^2=H; R^1=H, R^2=CH$_3$) or FpCH=C=CHR^3 (R^3=H or CH$_3$) with (Z)- or (E)-CF$_3$(NC)C=C(CN)CF$_3$ in benzene or CH$_3$CN at room temperature for various lengths of time to give the cycloadducts LXI and LXII. A two-step mechanism involving a dipolar intermediate is proposed [208]. Adducts of the type LXIII (R^4=H or CH$_3$) are generated in the reaction of FpCH$_2$CH=CH$_2$ or FpCH$_2$C(CH$_3$)=CH$_2$ with 2-ClC$_6$H$_4$CH=C(CN)$_2$ in CH$_2$Cl$_2$ at room temperature for 3 h or in refluxing CH$_2$Cl$_2$ for 2 h, respectively [99]. Similarly, compound LXIV has been obtained from FpCH=C=CH$_2$ [69] and complex LXV from FpC(=CH$_2$)CH=CH$_2$ in refluxing CH$_2$ClCH$_2$Cl after 2 h [200].

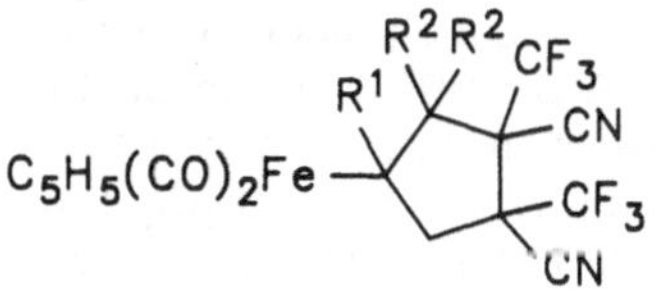

LXI

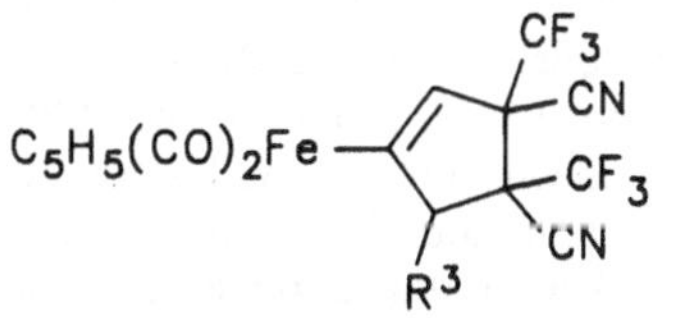

LXII

R^1, R^2, R^3=H, CH$_3$

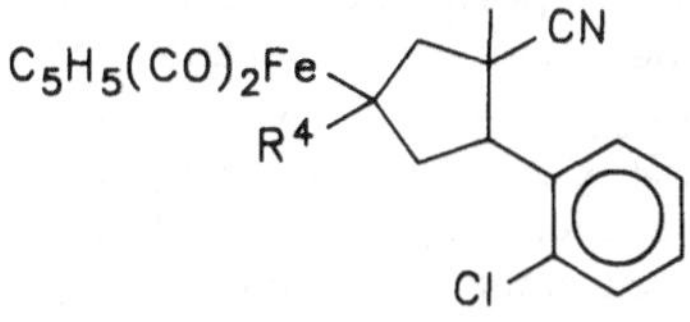

R^4=H or CH$_3$

LXIII

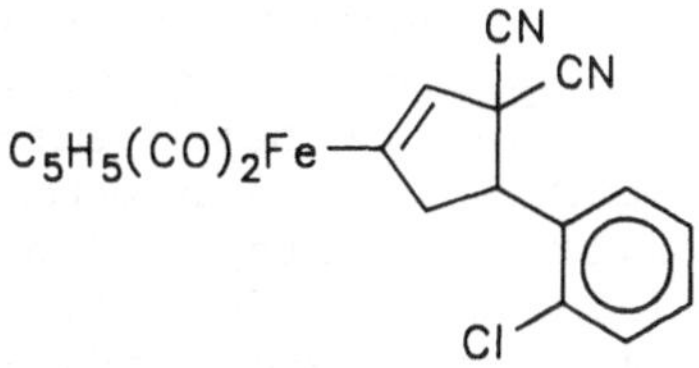

LXIV

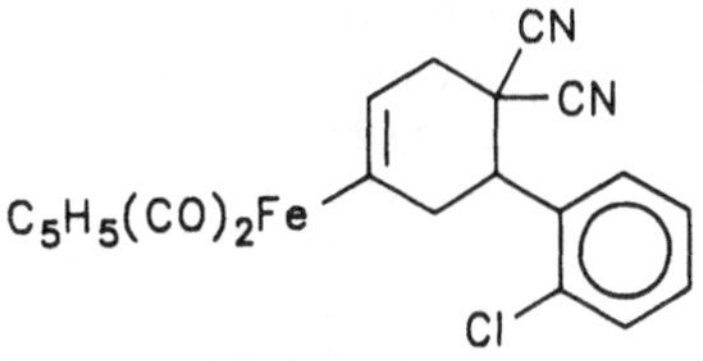

LXV

References on pp. 141/7

The cyclic adducts LXVII (mixture of diastereomers) are formed by the reaction of $FpCH_2CH=CH_2$ with $R^1R^2C=CHR^3$ (LXVIa to LXVIg [172]; LXVIb, d [143]; LXVIg [99, 175]), generally in CH_2Cl_2 at room temperature for 1 to 20 h. An exception is the reaction of $FpCH_2CH=CH_2$ with $(CH_3OOC)_2C=CHCOOCH_3$ which does not proceed in CH_2Cl_2 or in THF, but gives a good yield of LXVIIc in dimethylformamide after 90 h [172], see also [143]. Similarly, the reaction of $FpCH_2CH=CHCH_3$ ((E)/(Z)-mixture) with $(C_2H_5OOC)_2C=CH_2$ in ether overnight gives LXVIIIb [175]. (Z)-$FpCH_2CH=CH(OCH_3)$ and $R^1R^2C=CHR^3$ (LXVIa to f [178], see also [143]; LXVIh [194]) in CH_2Cl_2 for 3 h (LXVIa to d) or dimethylformamide for 90 h (LXVIe and f) at room temperature give the cyclopentyl derivatives LXIX.

	R^1	R^2	R^3
a:	COOC_2H_5	CN	COOC_2H_5
b:	COOC_2H_5	COOC_2H_5	H
c:	COOCH_3	COOCH_3	COOCH_3
d:	CN	CN	COOC_2H_5
e:	COOC_2H_5	CN	CN
f:	CN	COOC_2H_5	CN
g:	COOCH_3	COOCH_3	H
h:	COOCH_3	CN	COOCH_3

In contrast, H transfer leading to formation of the linear compounds $FpCH_2C(OCH_3)= CH(CHR^3CR^1R^2)_nH$ (n=1 from LXVIc, n≥2 from LXVIb, f, g) becomes the predominant process in the analogous reactions of $FpCH_2C(OCH_3)=CH_2$ and $R^1R^2C=CHR^3$ (LXVI) in CH_2Cl_2 (LXVIc, f), dimethylformamide (LXVIb), or benzene (LXVIg) (reaction time 2.5 h for LXVIf, 20 h for LXVIb, c, g), whereas with the olefins LXVId (20 h) and LXVIe (1 h) in CH_2Cl_2 only the cyclic products LXX are obtained. Steric factors are thought to be mainly responsible for the different behavior. The linear adducts are formed exclusively by treatment of $FpCH_2C(OCH_3)=CH_2$ with $(R^4OOC)_2C=C(COOR^4)_2$ ($R^4 = CH_3$, C_2H_5) in dimethylformamide at room temperature after 70 or 48 h, respectively. However, no reaction has been observed between $FpCH_2CH=CH_2$ or (Z)-$FpCH_2CH=CH(OCH_3)$ and $(CH_3OOC)_2C=C(COOCH_3)_2$. Due to hydrolysis, $FpCH_2COCH_3$ is isolated as a side-product in all the reactions of $FpCH_2C(OCH_3)= CH_2$. It was the only product obtained from the reaction of $FpCH_2C(OCH_3)=CH_2$ with $(CH_3OOC)_2C=CH_2$ (LXVIg) in dimethylformamide, THF, or ether, whereas in benzene the polymeric H-transfer adduct $FpCH_2C(OCH_3)=CH[CH_2CH(COOCH_3)_2]_nH$ with n≥2 was formed along with $FpCH_2COCH_3$ [173], see also [143]. The keto ester $FpCH_2COCH_2CH_2CH(COOC_2H_5)_2$, contaminated with some $FpCH_2COCH_3$, was isolated from the reaction of $FpCH_2C(OCH_3)=CH_2$ with $(C_2H_5OOC)_2C=CH_2$ in ether after 4 h, resulting from hydrolysis of the first-formed, highly sensitive enol ether $FpCH_2C(OCH_3)=CHCH_2CH(COOC_2H_5)_2$ during workup [175]. It has been found that solvent polarity influences the reaction pathway, because the linear complex $FpCH_2C(OCH_3)=CHCH(COOCH_3)CH(COOCH_3)_2$ was the only product formed in the reaction of $FpCH_2C(OCH_3)=CH_2$ with the olefin LXVIc in THF and benzene. However, a mixture of the latter and the cyclic derivative LXX was obtained in dimethylformamide and CH_2Cl_2. The reaction was faster in the more polar dimethylformamide and CH_2Cl_2 than in THF and benzene [173], see also [143].

A mixture of the isomeric [4+2]cycloadducts LXXI and LXXII (1:1, identified by ^{1}H NMR and IR) and Fp_2, which could not be separated, has been obtained by refluxing $FpC(=CH_2)CH=CH_2$ with an excess of $CH_3OOCCH=CH_2$ in CH_2ClCH_2Cl in the presence of a small amount of hydroquinone. After 120 h 70% of the starting material was recovered unchanged [200].

LXXI: $R^1=COOCH_3$, $R^2=H$
LXXII: $R^1=H$, $R^2=COOCH_3$
LXXIII: $R^1=R^2=COOCH_3$

The cycloadducts LXXIII and LXXIV result from the reactions of $FpC(=CH_2)CH=CH_2$ with (E)-$CH_3OOCCH=CHCOOCH_3$ in refluxing CH_2Cl_2 (96 h) or $C_2H_5OOCN=NCOOC_2H_5$ in CH_2Cl_2 at 24 °C (2 h), respectively [200].

$FpCH_2CH=CH_2$ functions as a moderately reactive 1,3-dipole in [3+2]cycloaddition reactions with Lewis acid activated cyclohex-2-enone (C_6H_8O) or with 2-carbethoxycycloalk-2-enones. Thus, treatment with $[Fp(C_6H_8O)]BF_4$ in refluxing CH_2Cl_2 or addition of $FpCH_2CH=CH_2$ to a CH_2Cl_2 solution of C_6H_8O containing freshly sublimed $AlBr_3$ at −78 °C and further reaction at 0 °C for 1 h afford the cycloadduct LXXV as a mixture of two stereoisomers in 10 and 45% yields, respectively. The corresponding reaction of $FpCD_2CH=CD_2$ with 4,4-D_2-cyclohex-2-enone yields LXXVI, the ^{1}H NMR spectrum of which is consistent with the assignment of a cis-hydrindane structure to the two stereoisomers. $FpCH=C=CH_2$ reacts with cyclohex-2-enone in the presence of $AlBr_3$ at −78 to 0 °C over a period of 2 h to form the cycloadduct LXXVII. Possible reaction mechanisms are discussed [175].

LXXV   LXXVI   LXXVII

Further cycloadducts result from the reaction of $FpCH_2CH=CH_2$ with the keto esters LXXVIII, LXXIX, and LXXXI in refluxing CH_2Cl_2 for 24 h. The products LXXXII are obtained in 50 (n=2), 63 (n=3), and 9% (n=4) yields, respectively. These reactions are very sensitive to steric effects. Thus, under similar conditions $FpCH_2CH=CH_2$ fails to react with the keto ester LXXX or with the cyanoacrylic ester LXXXIII, and the starting material has been recovered unchanged from the attempted reaction of (E)-$FpCH_2CH=CHCH(OCH_3)_2$ or (E)-$FpCH_2CH=CHC_3H_4O_2CH_3$ (No. 114) with the keto ester LXXVIII. Also, $FpCH_2C(OCH_3)=CH_2$ failed to give any cycloadduct with cyclohex-2-enone either alone or in the presence of $AlBr_3$ or with the keto ester LXXIX, whereas from $FpCH_2C(CH_3)=CH_2$ and LXXVIII in refluxing CH_2Cl_2 overnight, the adduct LXXXIV is formed [175].

References on pp. 141/7

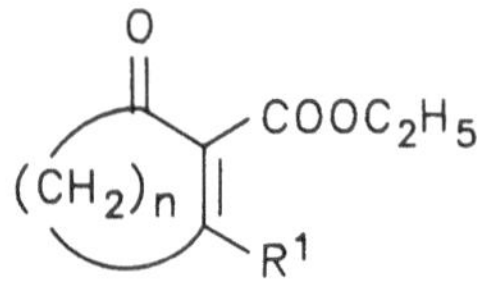

LXXVIII: $R^1 = H$, n = 2
LXXIX: $R^1 = H$, n = 3
LXXX: $R^1 = CH_3$, n = 3
LXXXI: $R^1 = H$, n = 4

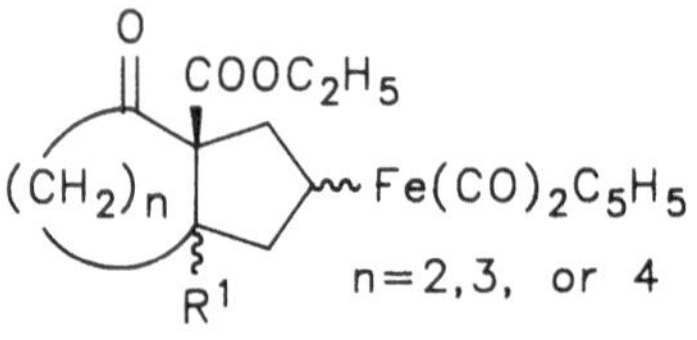

LXXXII

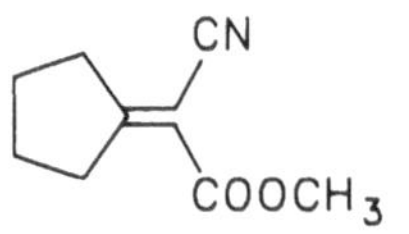

LXXXIII 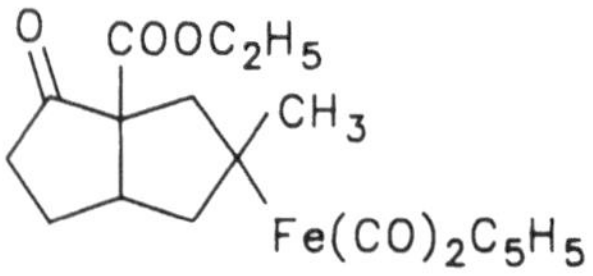   LXXXIV

The cycloadduct LXXXV is formed in the reaction of $FpC(=CH_2)CH=CH_2$ with p-benzoquin-one in refluxing CH_2ClCH_2Cl for 1.5 h, whereas similar treatment of $(E)-FpCH=CHCR^1=CH_2$ ($R^1 = H$ or CH_3) results in formation of a material which could not be identified [200]. Also, no identifiable products could be obtained from $FpCH_2CH=CH_2$ and either p-benzoquinone or tetrachloro-p-benzoquinone [46]. However, $FpCH_2CH=CH_2$ reacts rapidly with 2,3-di-chloro-5,6-dicyanobenzoquinone to give the adduct LXXXVI [46, 99].

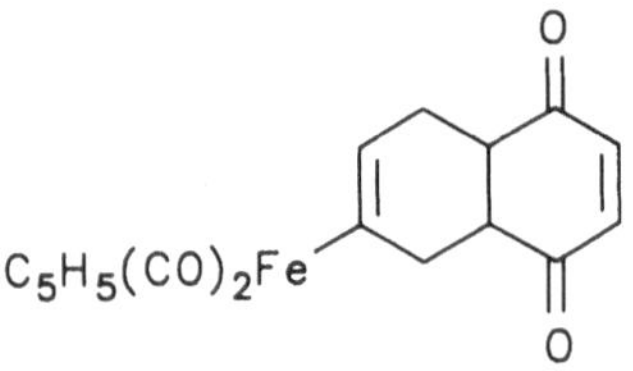

LXXXV 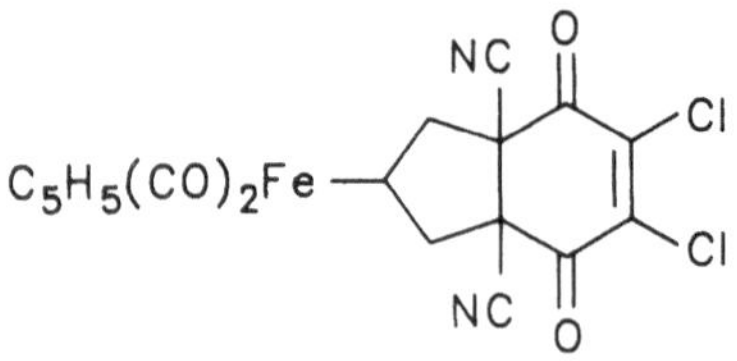   LXXXVI

Quinazarinquinone, which has two potential dienophilic sites ($>C=C<$ and $-CH=CH-$), produces the cycloadduct LXXXVII with $FpC(=CH_2)CH=CH_2$ via the internal double bond ($>C=C<$) only [200].

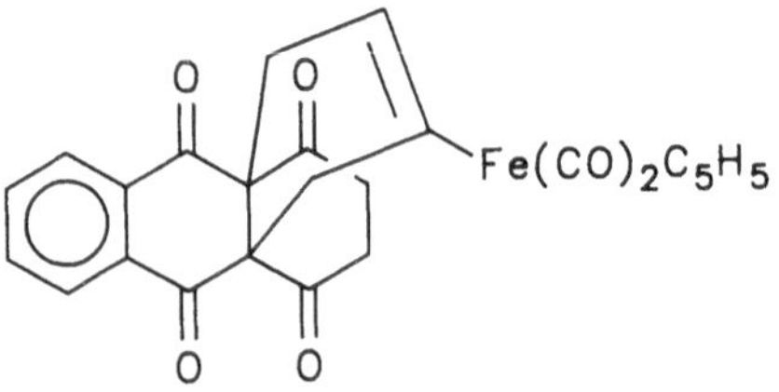

LXXXVII

Maleic anhydride reacts with $FpC(=CH_2)CH=CH_2$ and with $(E)-FpCH=CHC(CH_3)=CH_2$ in refluxing CH_2ClCH_2Cl for 1.5 h and 15 min, respectively, to give the cycloadducts LXXXVIII and LXXXIX. However, product LXXXVIII could not be obtained pure [200].

References on pp. 141/7

LXXXVIII   LXXXIX

The reactions of $FpCH_2CR^1=CR^2R^3$ with organic halides R^4X ($R^4=CCl_3$, $CH_2COOC_2H_5$, $CH(CH_3)COOCH_3$, $CH_2C_6H_5$, or CH_2COCH_3; $X=Cl$, Br, or I) afford $CH_2=CR^1CR^2R^3R^4$. Generally substitution occurs at the γ-position of the allyl ligand in $FpCH_2CR^1=CR^2R^3$. Thus, $FpCH_2CH=CHCH_3$ reacts with $BrCH_2COOC_2H_5$ to give $CH_2=CHCH(CH_3)CH_2COOC_2H_5$. The yields are high when $R^2=R^3=H$ or $R^2=H$, $R^3=CH_3$, but α-substitution and cyclopropanation become competitive when $R^2=R^3=CH_3$. The yields are generally higher, when an excess of the Fp complex is employed. Increasing the reaction temperature causes no marked change in yields; above 60 °C extensive decomposition occurs. Also, no solvent effect on the rate of reaction or product distribution is observed. The order of reactivity of the organic halides is $I>Br>Cl$. With $C_6H_5CCl_3$, the major product is 1,2-diphenyl-1,1,2,2-tetrachloroethane (40%); 1,2-dichloro-1-phenyl-but-3-ene is not observed. Galvinoxyl inhibits the rate of reaction between $FpCH_2CH=CH_2$ and CCl_4. The addition of Fp_2 increases the rate slightly, but does not alter the yields. Photolysis increases the rate. Thus, the reaction of $FpCH_2CH=CH_2$ with $BrCH_2COOC_2H_5$ requires one week at 24 °C, but only 5 h under UV irradiation. Addition of $(C_6H_5C(=O)O)_2$ and N-bromosuccinimide both alter the kinetics of the reaction of $FpCH_2CH=CH_2$ with CCl_4 and increase overall reaction time. The reaction between $FpCH_2CH=CH_2$ and $D(+)-BrCH(CH_3)COOCH_3$ affords a racemized substitution product and racemized starting material $BrCH(CH_3)COOCH_3$ [181], see also [165].

Kinetic studies of the reaction between $FpCH_2CH=CH_2$ and CCl_4 in CH_3CN revealed a first-order reaction in $FpCH_2CH=CH_2$ and two-thirds in CCl_4. When nitrosodurene was added, the ESR spectrum showed the presence of trapped allyl radicals and product analysis showed the existence of hexachloroethane. These results suggest a radical mechanism involving homolytic cleavage of $FpCH_2CH=CH_2$ into Fp and allyl radicals, followed by abstraction of a chlorine atom from CCl_4 by the Fp radical producing a CCl_3 radical. This then displaces the Fp radical from $FpCH_2CH=CH_2$ via an S_H2' reaction [181], see also [165]. An S_H2' mechanism has also been proposed for the reactions of $FpCH_2C(CH_3)=CH_2$ or $FpCH_2CH=CHCH_3$ with CCl_4 in the presence of Fp_2 to give $CH_2=C(CH_3)CH_2CCl_3$ and $CH_2=CHCH(CH_3)CCl_3$, respectively [141].

The polarographic reduction in CH_3CN (0.1 N $[N(C_2H_5)_4]ClO_4$) takes place in an irreversible two-electron step, giving the anions $[Fp]^-$ and R^-, the half-wave potential is $E_{1/2}=-1.96$ V for $FpCH_2CH=CH_2$ (No. 95), referred to SCE. A good linear correlation between the half-wave potentials of FpR compounds and the pK_a values of the corresponding RH has been found [205, 206].

Remarks Concerning Table 5 on pp. 84/113

IR bands resulting from the terminal CO groups (two strong bands in the range 1920 to 2060 cm^{-1}) are presented first under "IR", followed by further bands given in the respective publication.

Table 5

$C_5H_5Fe(CO)_2R$ Compounds with R = Alkenyl.

Further information on compounds with numbers preceded by an asterisk is given at the end of the table on pp. 113/40.

For abbreviations and dimensions, see p. X.

No.	compound method of preparation (yield)	properties and remarks

structures of the type FeC=C

No.	compound	properties and remarks
*1	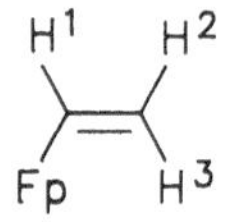 (H¹, H², H³ on vinyl; Fp)	brown-yellow oil, m.p. −5 to −6° [9], orange liquid, solidified in a freezer at −15° [8] ^{1}H NMR: 4.71 (s, C_5H_5), 5.23 (d, H-3), 5.70 (d, H-2), 6.82 (dd, H-1, J = 16.5, 8.5) in $CDCl_3$ [174], 4.81 (C_5H_5), 5.42 (H-3), 5.90 (H-2), 7.14 (H-1), J(H-1,3) = 17.4, J(H-1,2) = 9.0, J(H-2,3) = 1.4 in $CDCl_3$ [48], 4.75 (C_5H_5), 5.32 (H-3), 5.82 (H-2), 6.93 (H-1), J(H-1,3) = 17, J(H-1,2) = 8, J(H-2,3) not observed in CS_2 [8] IR: 1971, 2022 (both CO) in C_6H_{12}, 1560 (C=C) in KBr [48], 1950, 2020 (both CO), 1550 (C=C), 2850 sh, 2920, 3010, 3080 (all CH) as liquid film, bands from 828 to 1465 [8] k = 15.91, k′ = 0.56 [25]
*2	(E)-FpCH=CHC$_6$H$_5$	light yellow crystals, m.p. 46 to 47° (dec., from pentane) [48, 49], m.p. 44 to 45° (from CH_2Cl_2/C_6H_{14}) [159] ^{1}H NMR: 4.70 (s, C_5H_5), 7.04 (q, CH=CH, J(H,H) = 16), 7.00 to 7.35 (m, C_6H_5) [159], 4.83 (s, C_5H_5), 6.60 (d, FeCH=, J(H,H) = 16), 7.2 (m, C_6H_5), 7.70 (d, =CH–aryl) in $CDCl_3$ [154], 4.83 (C_5H_5), 7.66 (FeCH=), 7.08 to 7.48 (C_6H_5), 6.63 (=CH–aryl, J(H,H) = 16) in $CDCl_3$ [48, 49] IR: 1979, 2029 (both CO) in C_6H_{12}, 1556 (C=C) in KBr [48,49], 1977, 2022 in petroleum ether [154], 1974, 2021 (all CO) in C_6H_{14} [159]
*3	FpC(C_6H_5)=CH$_2$	yellow-brown oil [191] ^{1}H NMR ($CDCl_3$): 4.73 (s, C_5H_5), 5.30, 5.60 (both: s, 1 H, =CH$_2$), 7.10 (m, C_6H_5) [174]
*4	FpC(C_6H_5)=CHC$_6$H$_5$	yellow oil, air-sensitive [149] ^{13}C NMR (CH_2Cl_2): 86.19 (C_5H_5), 124.29 to 136.71 (C_6H_5), 141.90 (=CH), 149.60 (FeC), 215.54 (CO) [149] IR (C_6H_{14}): 1982, 2025 (both CO) [149]

References on pp. 141/7

Table 5 (continued)

No.	compound method of preparation (yield)	properties and remarks

*5

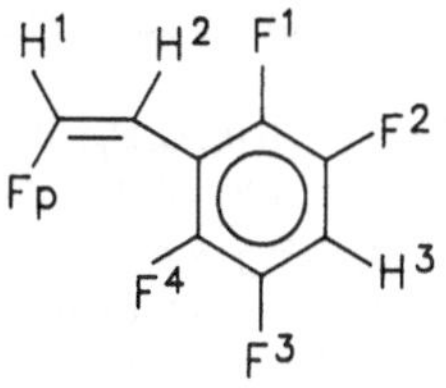

yellow, m.p. 59 to 61° [17]
^{1}H NMR: ~6.85 (complex, H-3, and one half
   of an AB quartet from H-1, H-2), 7.51 (d,
   J = 10.5, other half of the AB quartet) [17]
^{19}F NMR: 138.4 (F-1,4), 138.4 (F-2,3) in THF
   at 35° [20], 140.8 [17]
IR (C_6H_{12}): 1973, 2024 (both CO), 829 (C_5H_5),
   708, 1624, 1642 (all C_6-ring bands), 930,
   985, 1167, 1249 (all CF), 1490, 1498 sh (both
   C_6F_4), other bands 1115, 1303, 1385 [17]

*6

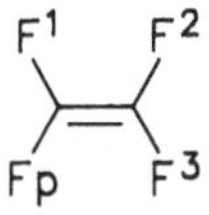

I (20%) [10], (5%) [13]

bright yellow-orange crystals [10, 13], m.p.
   27° [10], ~18° [11]
^{19}F NMR (CS_2): 89.5 (F-2), 139.6 (F-3), 147.3
   (F-1), J(F-1,3) = 122.4, J(F-2,3) = 107.1,
   J(F-1,2) = 46.8 [10]
IR: 2000, 2050 in C_6H_{12} [10], 2006, 2056 (all
   CO), 1707 (C=C) [11], other bands 827, 960,
   1040, 1235, 1710 in CS_2 [10]

*7 FpC(OCH$_3$)=CH$_2$

yellow oil [179]
^{1}H NMR (CDCl$_3$): 3.52 (s, OCH$_3$), 4.02 (d, 1 H,
   =CH$_2$, J = 1.5), 4.59 (d, 1 H, =CH$_2$), 4.80 (s,
   C$_5$H$_5$) [176, 179]
^{13}C NMR (C$_6$D$_6$): 56.3 (OCH$_3$) 85.6 (C$_5$H$_5$), 97.3
   (=CH$_2$), 176.6 (FeC), 216.2 (CO) [179]
IR (CH$_2$Cl$_2$): 1969, 2018 (both CO) [179]

*8 FpC(OLi)=CH$_2$

—

*9 (E)-FpCH=CH(CH$_2$)$_3$CH=CH$_2$

—

*10 FpCH=C(CN)$_2$
    I (32%) [67]

yellow, m.p. 115 to 116° [58, 67, 75, 85]
^{1}H NMR (CDCl$_3$): 5.08 (s, C$_5$H$_5$), 10.27 (s,
   =CH) [67]
^{13}C NMR (CH$_2$Cl$_2$ or CS$_2$, referred to CS$_2$):
   −20.0 (=CH−), −18.4 (CO), 76.5, 78.8 (both
   CN), 93.2 (=C<), 105.5 (C$_5$H$_5$) [76]
IR: 2002, 2053 (both CO), 2223, 2234 (both
   CN) in CH$_2$Cl$_2$ [58, 67], see also [76], 1452
   (C=C), 3111, 3132 (both CH) in KBr [67]

*11 (E)-FpCH=CHCH$_3$

oil, isolated slightly impure [159]
^{1}H NMR (CDCl$_3$): 1.75 (dd, CH$_3$, J(CH$_3$, CH) =
   1.5 and 6), 4.77 (s, C$_5$H$_5$), 5.85 (m, CH=CH,
   J(H, H) = 15) [159]
IR (C$_6$H$_{14}$): 1966, 2015 (both CO) [159]

References on pp. 141/7

Table 5 (continued)

No.	compound method of preparation (yield)	properties and remarks
*12	$FpC(CH_3){=}CH_2$	m.p. 28 to 31° [176, 201], rather air-stable [174] ^{1}H NMR: 2.12 (m, CH_3), 4.79 (s, C_5H_5), 4.95 (m, 1 H, $=CH_2$), 5.58 (m, 1 H, $=CH_2$) [174], 2.09 (s, CH_3), 4.73 (s, C_5H_5), 4.89 (s, 1 H, $=CH_2$), 5.53 (br s, 1 H, $=CH_2$) [176] in $CDCl_3$, 2.18 (s, CH_3), 4.09 (s, C_5H_5), 5.19 (s, 1 H, $=CH_2$), 5.87 (q, 1 H, $=CH_2$, J = 1.3) in C_6D_6 [201] ^{13}C NMR ($CDCl_3$): 39.11 (CH_3), 85.58 (C_5H_5), 124.89 ($=CH_2$), 153.67 (FeC), 216.43 (CO) [174] ^{13}C {^{1}H} NMR (C_6D_6): 39.2 (CH_3), 85.5 (C_5H_5), 125.4 ($=CH_2$), 152.3 (FeC), 216.8 (CO) [176, 201] IR ($CHCl_3$): 1961, 2005 (both CO), 1581 (C=C) [176, 201]
*13	$FpC(CH_3){=}CHC_6H_5$ (E)/(Z)-mixture VII (42%) [91]	yellow-orange needles, m.p. 76 to 77° [91] ^{1}H NMR ($CDCl_3$): isomer A: 2.30 (d, CH_3), 4.82 (s, C_5H_5) 7.25 (q, =CH), 7.29 (s, C_6H_5), $^4J(CH_3, CH) = 1.6$; isomer B: 2.36 (d, CH_3), 4.67 (s, C_5H_5), 7.25 (q, =CH), 7.29 (s, C_6H_5), $^4J(CH_3, CH) = 1.6$; A:B ratio ~2:1 [91], see also [57] IR: 1967, 2020 in C_5H_{12} [57], 1960, 2016 (all CO) in $CHCl_3$ [91] mass spectrum (150°): $[M]^+$ (7), $[M-CO]^+$ (18), $[M-2\,CO]^+$ (93), $[C_5H_5Fe]^+$ (100) [91]
*14	(Z)-$FpCH{=}CHCF_3$	yellow-orange oil [17, 40] ^{1}H NMR: 4.8 (s, C_5H_5), 6.54 ("2 overlapping q's, one half of an AB pattern", J(CF_3, H) = 8.5), 7.91 (d, HC=CH, J(H,H) = 11.5, other half of AB pattern) [17] ^{19}F NMR: 58.2 (d, J = 8.5) [17] IR: 1984, 1996 sh, 2035, 2040 sh (all CO) in C_6H_{12}, 1590 (C=C), 835, 3130 (both C_5H_5), 1115, 1195, 1280, 1340 (all CF_3), other bands 955, 1005, 1017, 1065, 1425, 2950 in CCl_4 [17], the complete IR spectrum (neat, 500 to 3140 range) is given in [40]
*15	$FpC(CF_3){=}CH_2$	mixture with 30% (Z)-$FpCH{=}CHCF_3$ [26, 27] yellow oil [26, 27] ^{1}H NMR (CCl_4): 5.5 (s), 6.25 (q, J = 1.9), also represented graphically [26]

Table 5 (continued)

No.	compound method of preparation (yield)	properties and remarks

<table>
<tr>
<td></td>
<td></td>
<td>^{19}F NMR: 60.5 (d, J (H, CF$_3$) = 1.9) [26]
IR (CCl$_4$ or neat): 1983, 2033 (both CO),
   1581 sh (C=C), other bands 670 to 1432 [26]</td>
</tr>
</table>

*16

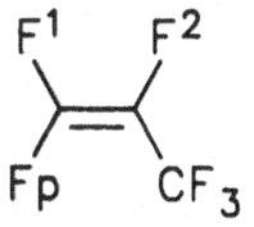

I (4.6%) [2], (56%) [10]

orange crystals, m.p. 69.5 to 71° [2, 3, 10]
^{19}F NMR (THF): 66 (q, CF$_3$), 86 (F-1), 166 (F-2)
  (d's of q's), relative intensities 3 : 1 : 1 [2, 4],
  J (F-1, 2) = 131, J (CF$_3$, F-2) = 13,
  J (CF$_3$, F-1) = 22 [4]
IR: 2004, 2050 (both CO), 1973 (^{13}CO?) in
  Cl$_2$C=CCl$_2$ [4], 1995, 2040 (both CO) [2], 678
  (CF), 812 (C–C), 844 (C$_5$H$_5$), 1055, 1320
  (both ν_{sym} C–F of CF$_3$), 1124, 1185 (both
  νC–F of =CF), 1645 (C=C), other bands at
  1015, 1264 in CS$_2$ [4], see also [2]

*17   FpCF=CClCF$_3$
    I (4%) [30]

mixture of 7% (E)-isomer (?) and 93%
  (Z)-isomer [30]
^{1}H NMR: 5.04 (s, C$_5$H$_5$) [30]
^{19}F NMR: 58.0 (d, CF$_3$, ^{4}J (CF$_3$, CF) = 8.7) for
  the (E)-isomer, 19.4 (CF), 60.1 (CF$_3$),
  ^{4}J (CF$_3$, CF) = 23.6 for the (Z)-isomer [30]
IR (CHCl$_3$): 1564 (C=C) [30]

*18   FpC(CH$_2$I)=CHC$_6$H$_5$
    VII (small amount) [91]

brown tar [91]
IR (CHCl$_3$): 1965, 2018 (both CO) [91]

*19   FpC(CH$_2$OH)=CH$_2$
    VII (35%) [134]

yellow needles, m.p. 58 to 59° [134]
^{1}H NMR (CS$_2$): 1.55 (br s, OH), 4.0 (br s,
  CH$_2$O), 4.78 (s, C$_5$H$_5$), 5.0 (br s, 1 H,
  =CH$_2$), 5.75 (t, 1 H, =CH$_2$, J = 1.5) [134]
IR (CH$_2$Cl$_2$): 1960, 2020 (both CO), 3600 (OH)
  [134]

*20   FpC(CH$_2$OH)=CHC$_6$H$_5$
    VII (13%) [91]

cream yellow, m.p. 101 to 102° [91]
^{1}H NMR (CDCl$_3$): 4.29 (br s, CH$_2$), 4.61 (s,
  C$_5$H$_5$), 7.18 (s, C$_6$H$_5$), 7.48 (br s, =CH), sig-
  nal for OH not observed [91]
IR (CHCl$_3$): 1965, 2018 (both CO) [91]
mass spectrum (100°): [M − CO]$^+$ (14),
  [M − CO − H$_2$O]$^+$ (4), [M − 2 CO]$^+$ (39),
  [M − 2 CO − H$_2$O]$^+$ (50) [91]

21   (Z)-FpC(CH$_2$OH)=CHC$_6$H$_5$
    VII (25%) [134]

m.p. 109 to 110° [134]
^{1}H NMR (CD$_3$COCD$_3$): 3.8 (t, OH, ^{3}J = 6), 4.28
  (d, CH$_2$), 4.88 (s, C$_5$H$_5$), 7.29 (m, C$_6$H$_5$), 7.66
  (br s, =CH) [134]
IR (CH$_2$Cl$_2$): 1960, 2030 (both CO), 3590 (OH) [134]

References on pp. 141/7

Table 5 (continued)

No.	compound method of preparation (yield)	properties and remarks
22	FpC(=CH$_2$)CH(OH)C$_6$H$_5$ VII (16%) [134]	^{1}H NMR (CS$_2$): 1.65 (d, OH, ^{3}J = 4), 4.4 (s, C$_5$H$_5$), 5.1 (s, =CH$_2$), 5.94 (s, CH), 7.1 to 7.4 (m, C$_6$H$_5$) [134] IR (CH$_2$Cl$_2$): 1960, 2030 (both CO), 3590 (OH) [134]
23	FpC(CH$_2$OCH$_3$)=CH$_2$	obtained from FpC(CH$_2$OH)=CH$_2$ by reaction with Li(C$_4$H$_9$-t) at $-20°$ and subsequent methylation with [O(CH$_3$)$_3$]BF$_4$ [122], see also No. 19
24	(Z)–FpC(CH$_2$OCH$_3$)=CHC$_6$H$_5$ VII (16%) [91]	no exact proof for the (Z)–configuration orange oil [57, 91] ^{1}H NMR (CDCl$_3$): 3.41 (s, OCH$_3$), 4.09 (d, CH$_2$), 4.66 (s, C$_5$H$_5$), 6.97 (s, =CH), 7.29 (s, C$_6$H$_5$), ^{4}J(CH$_2$, CH) = 1.4 [91], see also [57] IR: 1967, 2018 in C$_5$H$_{12}$ [57], 1963, 2017 (all CO) in CHCl$_3$ [91] mass spectrum (60°): [M]$^+$ (0.5), [M−CO]$^+$ (3), [M−2 CO]$^+$ (8) [91]
*25	FpC(CH$_2$OCOCH$_3$)=CH$_2$ VII (83%) [134]	orange oil (crude product) [134] ^{1}H NMR (CS$_2$): 1.95 (s, CH$_3$), 4.4 (t, CCH$_2$O, ^{4}J = 1.5), 4.82 (s, C$_5$H$_5$), 5.0 (t, 1 H, =CH$_2$), 5.65 (t, 1 H, =CH$_2$) [134] IR (CH$_2$Cl$_2$): 1960, 2030 (both CO), 1725 (C=O) [134]
*26	(Z)–FpC(CH$_2$OCOCH$_3$)=CHC$_6$H$_5$	no exact proof for the (Z)–configuration yellow oil [91] ^{1}H NMR (CDCl$_3$): 2.10 (s, CH$_3$), 4.72 (s, C$_5$H$_5$), 4.77 (d, CH$_2$), 7.27 (s, C$_6$H$_5$), 7.47 (m, =CH) [91] IR (CHCl$_3$): 1966, 2020 (both CO) [91] mass spectrum (60°): [M]$^+$ (0.2), [M−CO]$^+$ (0.4), [M−2 CO]$^+$ (6) [91]
*27	FpC(=CHC$_6$H$_5$)CH$_2$N(C$_2$H$_5$)$_2$ (E)/(Z)–mixture VII (49%) [91]	orange oil [57, 91] ^{1}H NMR (CDCl$_3$): isomer A: 1.02 (t, CH$_3$), 2.58 (q, CCH$_2$N, ^{3}J(CH$_3$, CH$_2$) = 7.0), 3.19 (d, =CCH$_2$N,^{4}J(CH$_2$, CH) = 1.4), 4.53 (s, C$_5$H$_5$), 7.27 (m, C$_6$H$_5$; separate =CH signal not ob- served, presumably covered by the C$_6$H$_5$ resonance) [57, 91] isomer B: 0.87 (t, CH$_3$), 2.39 (q, CCH$_2$N, ^{3}J(CH$_3$, CH$_2$) = 7.0), 3.12 (d, =CCH$_2$N, ^{4}J(CH$_2$, CH) = 1.4), 4.79 (s, C$_5$H$_5$), 7.27 (m,

Table 5 (continued)

No.	compound method of preparation (yield)	properties and remarks
		C_6H_5; separate =CH signal also not observed; A:B ratio ~7:3 [91], see also [57] IR: 1965, 2016 in C_5H_{12} [57], 1959, 2014 (all CO) in $CHCl_3$ [91] mass spectrum (150°): $[M]^+$ (2), $[M-CO]^+$ (5), $[M-2 CO]^+$ (11), $[M-2 CO-C_2H_5]^+$ (13) [91]
28	$FpC(CH_2NHC_6H_5)=CHC_6H_5$ VII [91]	obtained only as impure oil, rather unstable [91] IR ($CHCl_3$): 1962, 2016 [91] mass spectrum (150°): $[M]^+$ (4), $[M-CO]^+$ (4), $[M-2 CO]^+$ (27) [91]
*29	$FpC(N(CH_3)_2)=C(CH_3)_2$ I [90]	brown liquid, air-sensitive [79, 89, 90] ^{1}H NMR ($CDCl_3$): 1.65, 1.80 (s's, $=C(CH_3)_2$), 2.21 (s, $N(CH_3)_2$), 4.69 (s, C_5H_5) [79, 90] IR (C_6H_{14}): 1949, 2000 (both CO) [79, 90]
30	$FpC(NC_5H_{10})=C(CH_3)_2$ ($NC_5H_{10}=$ piperid-1-yl) I [89]	yellow-brown liquid [89]
*31	I (45%) [16] III (95%) [61]	yellow, m.p. 83° (dec.) [70], 80° (dec.) [16] ^{1}H NMR: 4.87 (C_5H_5), 7.16 (H-2), 7.32 (C_6H_5, H-3,4), 7.75 (C_6H_5, H-2(?)), 9.21 (H-1), J(H-1, H-2)=16.3 in $CDCl_3$ [35, 48, 49], same in CCl_4 [16] IR: 1980, 2030 (both CO) in C_6H_{12} [48], 1638 (C=O), 1512 (C=C) [29, 35, 48, 49], 1960, 2025 (both CO), 1645 (C=O), 3100, 3115 (both CH of C_5H_5), other bands 765 to 1600 [16] in KBr
*32	I (50%) [16] III (75%) [61]	yellow, m.p. 110° (dec.) [16] ^{1}H NMR: 2.40 (CH_3), 4.95 (C_5H_5), 7.42 (H-2), 9.67 (H-1), J(H-1,2)=16.5 in $CDCl_3$ [140], 2.32 (CH_3), 4.89 (C_5H_5), 7.08 (C_6H_4, H-4,5), 7.15 (H-2), 7.67 (C_6H_4, H-3,6), 9.20 (H-1, J(H-1,2)=16.3) in CCl_4 [16] IR: 1981, 2030 (both CO) in $CHCl_3$, 1645 (C=O) [140], 1950, 2000 (both CO), 1625 (C=O), 3095 (CH of C_5H_5), other bands 790 to 1600 [16] in KBr

Table 5 (continued)

No.	compound method of preparation (yield)	properties and remarks
*33	(E)-FpCH^1=CH^{2}COC$_6$H$_4$Br-4 III (60%) [61]	dark red, fine crystals, m.p. 137 to 140° (dec.) [61] ^{1}H NMR (CH$_3$COCH$_3$): 5.20 (s, C$_5$H$_5$), 7.28, 9.57 (CH=CH, AB quartet, J(H-1,2) = 16.5) [61] IR (KBr): 1980, 2030 (both CO), 1645 (C=O), 800 to 860 (δ CH of C$_5$H$_5$), 3050 (ν CH of C$_5$H$_5$), 980 ((E)-CH=CH) [61]
34	(E)-FpCH=CHC(=NC$_6$H$_5$)C$_6$H$_4$CH$_3$-4	see No. 32 on p. 119
35	(Z)-FpCH^1=CH^{2}COOH III (68%) [107]	yellow crystals, m.p. 123 to 127° [107] ^{1}H NMR (CDCl$_3$): 5.01 (C$_5$H$_5$), 6.68 (H-2), 9.04 (H-1), J(H-1,2) = 11.0 [107] IR: 1982, 2030 (both CO) in CHCl$_3$, 1530 (C=C), 1670 (C=O) in KBr [107]
36	(E)-FpCH^1=CH^{2}COOH III (68%) [107]	yellow crystals, m.p. 137 to 139° [107] ^{1}H NMR (CDCl$_3$): 5.00 (C$_5$H$_5$), 6.03 (H-2), 9.10 (H-1), J(H-1,2) = 16.5 [107] IR: 1978, 2030 (both CO) in CHCl$_3$, 1540 (C=C), 1660 (C=O) in KBr [107]
37	(Z)-FpCH^1=CH^{2}COOCH$_3$ III (78%) [107]	yellow crystals, m.p. 116 to 119° [107] ^{1}H NMR (CDCl$_3$): 3.61 (CH$_3$), 4.82 (C$_5$H$_5$), 6.65 (H-2), 8.52 (H-1), J(H-1,2) = 10.5 [107] IR: 1972, 2032 (both CO) in CHCl$_3$, 1540 (C=C), 1695 (C=O) in KBr [107]
38	(E)-FpCH^1=CH^{2}COOCH$_3$ III (91%) [107]	yellow crystals, m.p. 59 to 61° [107] ^{1}H NMR (CDCl$_3$): 3.57 (CH$_3$), 4.88 (C$_5$H$_5$), 6.08 (H-2), 8.80 (H-1), J(H-1,2) = 16.5 [107] IR: 1982, 2030 (both CO) in CHCl$_3$, 1555 (C=C), 1705 (C=O) in KBr [107]
*39	(Z)-FpC(OCOCH$_3$)=CHCH$_3$	amber oil [28] ^{1}H NMR (CS$_2$): 1.41 (d, =CCH$_3$, ^{3}J = 6.5), 1.97 (s, CH$_3$CO), 4.88 (q, =CH), 4.89 (s, C$_5$H$_5$) [28] IR: CH$_3$COO bands at 1226 and 1731 in C$_6$H$_{12}$ and CS$_2$ [28]
40	(Z)-FpCH^1=CH^{2}COOC$_2$H$_5$ III (82%) [107]	yellow crystals, m.p. 69 to 72° [107] ^{1}H NMR (CDCl$_3$): 1.24 (CH$_3$), 4.04 (CH$_2$), 4.81 (C$_5$H$_5$), 6.59 (H-2), 8.49 (H-1), J(H-1,2) = 10.5 [107] IR: 1980, 2040 (both CO) in CHCl$_3$, 1545 (C=C), 1693 (C=O) in KBr [107]

Table 5 (continued)

No.	compound method of preparation (yield)	properties and remarks
41	(E)-FpCH^1=CH^{2}COOC$_2$H$_5$ III (93%) [107]	yellow crystals, m.p. 55 to 56° (from petro- leum ether) [107] ^{1}H NMR (CDCl$_3$): 1.25 (CH$_3$), 4.0 (CH$_2$), 4.89 (C$_5$H$_5$), 5.98 (H-2), 8.76 (H-1), J(H-1,2) = 16.5 [107] IR: 1978, 2030 (both CO) in CHCl$_3$, 1552 (C=C), 1695 (C=O) in KBr [107]
*42	(E)-FpCF=C(CF$_3$)COOC$_2$H$_5$ I (6%) [127]	yellow crystals, m.p. 68 to 69° (from petro- leum ether) [127] ^{19}F NMR (THF): −8.1 (q, =CF), 55.9 (d, CF$_3$), J(CF$_3$, F) = 21.1 [127] IR: 2005, 2050 (both CO) in C$_6$H$_{12}$, 1550, 1570 (both C=C), 1720 (C=O) in KBr [127]
*43	(Z)-FpCF=C(CF$_3$)COOC$_2$H$_5$ I (6%) [127]	yellow crystals, m.p. 96 to 97° (from petro- leum ether/benzene = 1:1) [127] ^{19}F NMR (THF): −7.8 (q, =CF), 53.3 (d, CF$_3$), J(CF$_3$, F) = 10.6 [127] IR: 2010, 2055 (both CO) in C$_6$H$_{12}$, 1580 (C=C), 1720 (C=O) in KBr [127]
*44	(Z)-FpCH^1=CH^{2}CONH$_2$ I(64%) [166]	yellow, m.p. 135 to 137° (dec., from toluene), air-stable [166] ^{1}H NMR(CD$_3$SOCD$_3$): 5.06 (C$_5$H$_5$), 6.53 (H-2), 8.46 (H-1), J(H-1,2) = 10.0 [166] ^{13}C NMR (CD$_3$SOCD$_3$): 86.14 (C$_5$H$_5$), 134.29 (C-2), 150.76 (C-1), 168.5 (CONH$_2$), 214.96 (CO) [166] IR: 1980, 2030 (both CO) in CHCl$_3$, 1535 (C=C), 1665 (C=O), 1620 (NH) in KBr [166]
*45	(E)-FpCH^1=CH^{2}CONH$_2$ I (70%) [166]	yellow, m.p. 147 to 149° (dec., from toluene), air-stable [166] ^{1}H NMR (CD$_3$SOCD$_3$): 5.27 (C$_5$H$_5$), 6.24 (H-2), 8.77 (H-1), J(H-1,2) = 16.0 [166] ^{13}C NMR (CD$_3$SOCD$_3$): 86.67 (C$_5$H$_5$), 138.0 (C-2), 158.85 (C-1), 164.31 (CONH$_2$), 215.64 (CO) [166] IR:1970, 2025 (both CO) in CHCl$_3$, 1550 (C=C), 1650 (C=O), 1600 (NH) in KBr [166]
*46	(E)-FpCH=CHCON(CH$_2$)$_5$ I (52%) [166]	yellow, m.p. 132 to 134° (dec., from petro- leum ether), air-stable [166] IR: 1965, 2025 (both CO) in CHCl$_3$, 1540 (C=C), 1615 (C=O) in KBr [166]

       References on pp. 141/7

Table 5 (continued)

No.	compound method of preparation (yield)	properties and remarks
*47	(Z)–FpC^1(CF$_3$)=C^{2}HSCH$_3$	yellow, m.p. 61° [145] ^{1}H NMR: 2.40 (s, CH$_3$), 5.00 (s, C$_5$H$_5$), 7.12 (q, CH) [145] ^{13}C NMR (CDCl$_3$): 17.2 (dq, SCH$_3$, ^{1}J(H,C) = 139.6 and ^{3}J(H,C) = 4.9), 85.6 (m, C$_5$H$_5$, ^{1}J(H,C) = 181.5 and J(H,C) = 6.8), 123.3 (q, C-1, ^{2}J(F,C) = 30.3), 127.7 (dq, CF$_3$, ^{1}J(F,C) = 272.5, ^{3}J(H,C) = 6.9), 144.9 (q, C-2, ^{3}J(F,C) = 11.0, ^{1}J(H,C) = 168.0 and · J(H,C) = 5.8), 212.5 (CO) [145] ^{19}F NMR: 56.2 (d, CF$_3$, ^{4}J(F,H) = 2.1) [145] IR: 1995, 2040 (both CO), 1550 (C=C) in Nujol, 1985, 2030 (both CO) in CHCl$_3$ [145]
48	(Z)–FpC(CF$_3$)=CHSi(CH$_3$)$_3$ IVa (6%, 11%) [40]	yellow oil [40] ^{1}H NMR (CHCl$_3$): 0.2 (s, CH$_3$), 4.9 (s, C$_5$H$_5$), 7.0 (q, =CH, ^{4}J(H,F) = 2) [40] ^{19}F NMR (C$_6$H$_{12}$): 60.1 (d, ^{4}J(H,F) = 2) [40] IR (neat): 1980, 2020 (both CO), 1090, 1125, 1210 (all CF), 1550 (C=C), other bands 560 to 3130 [40]
49	(Z)–FpC(CF$_3$)=CHGe(CH$_3$)$_3$ IVa (7%, 20%) [40]	yellow oil [40] ^{1}H NMR (CHCl$_3$): 0.3 (s, CH$_3$), 4.8 (s, C$_5$H$_5$), 7.2 (q, =CH, ^{4}J(H,F) = 2) [40] ^{19}F NMR (C$_6$H$_{12}$): 59.4 (d, ^{4}J(H,F) = 2) [40] IR (neat): 1980, 2020 (both CO), 1090, 1125, 1205 (all CF), 1555 (C=C), other bands 550 to 3130 [40]
*50	FpC(C^{1}H$_3$)=CHC^{2}H$_3$ VII (10%) [91]	yellow oil [91], air-sensitive [149], mixture of two isomers (~1:1) [91]; no information about the geometry in [149] ^{1}H NMR (CDCl$_3$): 1.70 (dq, CH$_3$), 1.99, 2.14 (2 m, CCH$_3$), 4.73, 4.79 (2 s, C$_5$H$_5$), 5.45, 6.03 (2 m, =CH), assignment precluded by the 1:1 ratio of isomers, ^{5}J(CH$_3$, CH$_3$) = 1.3, ^{3}J(CH$_3$, CH) = 6.7 [91] ^{13}C NMR (CH$_2$Cl$_2$): 20.63 (q, C-2, J = 123.1), 40.03 (q, C-1, J = 124.5), 85.66 (d, C$_5$H$_5$, J = 176.7), 130.02 (d, =CH, J = 145.3), 138.46 (s, FeC), 216.53 (s, CO) [149] IR: 1955, 2013 in CHCl$_3$ [91], 1952, 2002 (all CO) in C$_6$H$_{14}$ [149] mass spectrum (150°): [M]$^+$ (10), [M−CO]$^+$ (33), [M−2 CO]$^+$ (88) [91]

Table 5 (continued)

No.	compound method of preparation (yield)	properties and remarks
*51	FpC(C$_2$H$_5$)=CH$_2$	oil [120] ^{1}H NMR (CS$_2$): 1.02 (t, CH$_3$, J=7), 2.3 (m, CCH$_2$C), 4.73 (s, C$_5$H$_5$), 4.90 (t, 1 H, =CH$_2$, J=1.0), 5.55 (t, 1 H, =CH$_2$, J=1.5) [120] IR (neat): 1957, 2028 (both CO) [120]
*52	(Z)-FpC(CF$_3$)=CHCH$_3$	inseparable mixture with 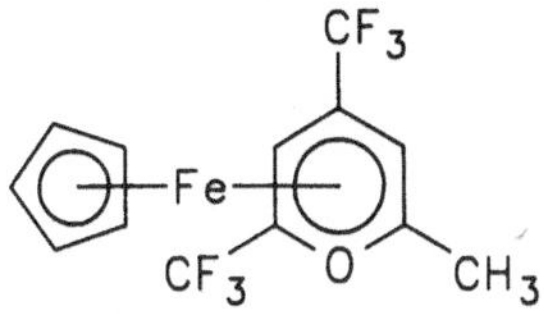 (ratio 1:7.5) [158] ^{1}H NMR(CDCl$_3$): 1.9 (d, CH$_3$, J(CH,CH$_3$)=7), 4.9 (s, C$_5$H$_5$), 6.9 (m, CH, J(CH,CF$_3$)=2.5) [158] ^{19}F NMR (CD$_3$COCD$_3$): 56.9 (q, CF$_3$, J(CF$_3$, CH$_3$)=2.5, J(CF$_3$, CH)=2.5) [158] IR (C$_6$H$_{14}$): 1984, 2032 (both CO) [158] mass spectrum: [M]$^+$, [M−CO]$^+$, [M−2 CO]$^+$ [158]
*53	(Z)-FpC(CF$_3$)=CHCH$_2$C$_6$H$_5$	inseparable mixture with 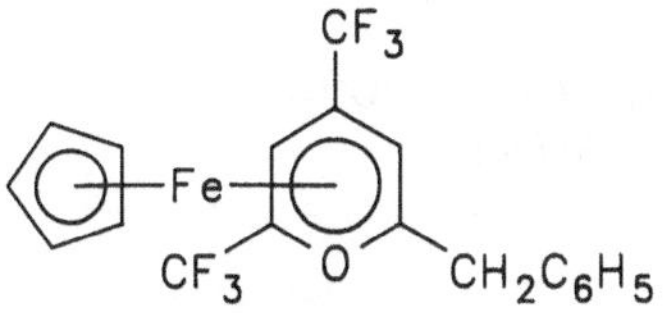 (ratio 1:8) [158] ^{1}H NMR (CDCl$_3$): 3.6 (m, CH$_2$, J(CH$_2$, CF$_3$)=2.5, J(CH,CH$_2$)=7), 4.9 (s, C$_5$H$_5$), 6.9 (tq, CH, J(CH,CH$_2$)=7, J(CH,CF$_3$)=2.5), 7.3 (br m, C$_6$H$_5$) [158] ^{19}F NMR (CD$_3$COCD$_3$): 56.9 (q, CF$_3$, J(CF$_3$, CH$_2$)=2.5, J(CF$_3$, CH)=2.5) [158] IR (C$_6$H$_{14}$): 1986, 2035 (both CO) [158] mass spectrum: [M]$^+$, [M−CO]$^+$, [M−2 CO]$^+$ [158]
*54	(E)-FpC(CF$_3$)=CHCF$_3$	yellow, m.p. 53 to 54° [129] ^{1}H NMR (CDCl$_3$): 4.97 (s, C$_5$H$_5$), 6.09 (q, CH,^{3}J(H,CF$_3$)=9.5) [129] ^{19}F NMR: 54.89 (q, FeCCF$_3$, ^{5}J(F,F)=12.8), 58.73 (q, d, =CHCF$_3$, ^{3}J(H,CF$_3$)=9.2) [129] IR (C$_5$H$_{12}$): 1999, 2044 [129]

References on pp. 141/7

Table 5 (continued)

No.	compound method of preparation (yield)	properties and remarks
*55	(Z)–FpC(CF$_3$)=CHCF$_3$	yellow powder, m.p. 37.5 to 39° [40] IR (mull): 1960 sh, 1990, 2040 (all CO), 1120, 1220, 1290 (all CF), 1610 (C=C), other bands 545 to 3150 [40]
*56	(E)–FpC(C^{1}F$_3$)=CFC^{2}F$_3$	yellow crystals, m.p. 75°, volatile [59] ^{19}F NMR (CH$_2$Cl$_2$): 47.8 (d, C^{1}F$_3$), 63.3 (d, C^{2}F$_3$), 71.8 (br, =CF–, ^{3}J(F,F) ~6, ^{4}J(F,F) = 27.3) [59] IR: 2012, 2054 (both CO) in C$_6$H$_{12}$, 1054, 1113 sh, 1134, 1148, 1194, 1228, 1320 (all CF), 1630 (C=C), 3150 (CH), other bands 654 to 1432 in KBr [59]
*57	(Z)–FpC(CF$_3$)=C(CF$_3$)SCF$_3$ IVb [87]	yellow crystals, air-stable [87] ^{19}F NMR (CH$_2$Cl$_2$): 42.7 (SCF$_3$) [87]
*58	(Z)–FpC(CF$_3$)=C(CF$_3$)SC$_6$F$_5$ IVb [87]	yellow crystals, air-stable [87]
59	FpC(CF$_3$)=C(CF$_3$)Si(CH$_3$)$_3$ IVa (8%) [40]	yellow solid [40] ^{19}F NMR (C$_6$H$_{12}$?): 50.0, 51.7 (2 broad signals of equal intensity) [40] IR (Nujol): 1960 sh, 1990, 2030 (all CO), 1105, 1130, 1160, 1210, 1250 (all CF), 1515 (C=C), other bands 550 to 3150 [40]
60	FpC(CF$_3$)=C(CF$_3$)Ge(CH$_3$)$_3$ IVa (15%) [40]	^{19}F NMR (C$_6$H$_{12}$?): 50.0, 50.5 (2 broad signals of equal intensity) [40] IR (Nujol): 1960 sh, 1990, 2030 (all CO), 1100, 1130, 1165, 1210, 1240 (CF), 1530 (C=C), other bands 550 to 3150 [40]
*61	FpC(CF$_3$)=C(CF$_3$)Sn(CH$_3$)$_3$ IVa (15%) [40]	yellow crystals, m.p. 86 to 88° [15, 40] ^{19}F NMR (C$_6$H$_{12}$?): 49.4, 50.4 [40] IR (Nujol or C$_6$H$_{12}$): 1940 sh, 1975, 2020 (all CO), 1010, 1110, 1120, 1160, 1210, 1240 (all CF), 1535 (C=C), 2920 (ν_{sym}CH), 2980 (ν_{as}CH), other bands 510 to 3140 [40], see also [15]
*62	(E)–FpC^{1}H=C^{2}HC(CH$_3$)$_2$OH	red oil [193] ^{1}H NMR (C$_6$D$_6$): 1.31 (s, OH, exchange with D$_2$O), 1.34 (s, 6H, CH$_3$), 4.05 (s, C$_5$H$_5$), 6.00 (d, =C^{2}H–, J = 15.8), 6.70 (d, =C^{1}H–) [193, 201] ^{13}C{^{1}H} NMR (C$_6$D$_6$): 30.7 (CH$_3$), 72.8 (COH), 85.2 (C$_5$H$_5$), 121.9 (C-2), 153.1 (C-1), 216.5 (CO) [193, 201]

Table 5 (continued)

No.	compound method of preparation (yield)	properties and remarks
		IR (CH$_2$Cl$_2$): 1958, 2006 (both CO) [193, 201], other bands 3585 [193], 3685 [201] mass spectrum: [M]$^+$ [193, 201]
*63	(Z)-FpC(CH$_2$OH)=CHCH$_3$ VII (30%) [134]	according to [91] determination of the stereochemistry was not possible m.p. 91.5 to 93.5° [129] ^{1}H NMR: 1.18 (br s, OH), 1.74 (d, CH$_3$, ^{3}J(CH$_3$, CH)=7), 4.0 (br s, CH$_2$), 4.8 (s, C$_5$H$_5$), 6.22 (q, =CH–) in CS$_2$ [134], 1.83 (d, CH$_3$, ^{3}J(CH$_3$, CH)=7), 4.20 (br s, CH$_2$), 4.91 (s, C$_5$H$_5$), 6.37 (q, =CH–), OH signal not observed, in CDCl$_3$ [91] IR: 1955, 2020 (both CO), 3590 (OH) in CH$_2$Cl$_2$ [134], 1962, 2017 (both CO) in CHCl$_3$ [91]
*64	FpC(=CH$_2$)CH(OH)CH$_3$ VII (10%) [134]	^{1}H NMR (CS$_2$): 1.75 (d, CH$_3$, J=7), 4.1 (br s, OH), 4.15 (br s, 2 H, CH$_2$(?)), 4.8 (s, C$_5$H$_5$), 5.47 (q, 1 H, =CH(?), J=7) [134] IR (CH$_2$Cl$_2$): 1950, 2020 (both CO), 3590 (OH) [134]
*65	(Z)-FpC(CH$_2$OCH$_3$)=CHCH$_3$ VII (92%) [104]	amber liquid [104] ^{1}H NMR (CS$_2$): 1.73 (d m, CCH$_3$, ^{3}J(CH$_3$, CH)=6.5), 3.18 (s, OCH$_3$), 3.78 (m, CH$_2$O), 4.78 (s, C$_5$H$_5$), 6.18 (q, =CH–) [104] IR (neat): 1949, 2010 (both CO) [104]
66	FpC(CH$_2$N$_3$)=CHCH$_3$ VII [91]	obtained only as impure oil, rather unstable [91] IR (CHCl$_3$): 1965, 2019 [91] mass spectrum (150°): [M]$^+$(0.1), [M–CO]$^+$ or [M–N$_2$]$^+$ (3), [M–2 CO]$^+$ or [M–CO–N$_2$]$^+$ (18) [91]
*67	FpC(=CHCH$_3$)CH$_2$N(C$_2$H$_5$)$_2$ VII (43%) [91]	geometry not given [57, 91] orange oil [57, 91] ^{1}H NMR (CDCl$_3$): 0.99 (t, CH$_3$), 2.47 (q, CH$_2$), both from N(C$_2$H$_5$)$_2$, 1.77 (d, =CCH$_3$), 3.02 (br s, =CCH$_2$N), 4.82 (s, C$_5$H$_5$), 6.23 (q, =CH–), ^{3}J(CH$_3$, =CH–)=6.5, ^{3}J(CH$_3$, CH$_2$)=7.5 [91], see also [57] IR: 1964, 2015 in C$_5$H$_{12}$ [57], 1953, 2011 (all CO) in CHCl$_3$ [91] mass spectrum (150°): [M]$^+$ (3), [M–CO]$^+$ (21), [M–2 CO]$^+$ (37) [91]

   References on pp. 141/7

Table 5 (continued)

No.	compound method of preparation (yield)	properties and remarks
*68	(E)-FpCH[1]=CH[2]COCH$_3$ I (60%) [16], (54%) [193, 201]	yellow, m.p. 83° (dec.) [16] ^{1}H NMR: 2.12 (CH$_3$), 4.92 (C$_5$H$_5$), 6.45 (H-2), 9.32 (H-1), J(H-1,2) = 16.5 in CDCl$_3$ [140], 1.98 (CH$_3$), 4.86 (C$_5$H$_5$), 6.36 (H-2), 8.77 (H-1), J(H-1,2) = 16.4 in CDCl$_3$ [48], see also [29, 35], same in CCl$_4$ [16] IR: 1980, 2031 in C$_6$H$_{12}$ [29, 35, 48] or CHCl$_3$ [29], 1975, 2027 [140], 1530 (C=C), 1641 (C=O) in CHCl$_3$ [29] ·IR (KBr): 1975, 2030 (both CO) [16], 1516, 1534 (both C=C) [29], 1534 (C=C) [35, 48], 1635 [16], 1641 [29, 35, 48], 1643 [140] (all C=O), 3090, 3105 (both CH of C$_5$H$_5$), other bands 800 to 1600 [16]
*69	FpC(=CH$_2$)COCH$_3$	oil [133] ^{1}H NMR (CS$_2$?): 2.20 (s, CH$_3$), 4.90 (s, C$_5$H$_5$), 5.75 (s, 1 H, =CH$_2$), 6.17 (s, 1H, =CH$_2$) [98, 133] IR (neat): 1960, 2008 (both CO), 1654 (C=O) [98, 133]
*70	FpCH=CHC(CH$_3$)=NNHC$_6$H$_3$(NO$_2$)$_2$-2,4	probably the (E)-compound brick red, m.p. 133° (dec., from CH$_3$OH) [39] IR (C$_6$H$_{12}$): 1980, 2029 (both CO) [39]
*71	FpC(COOCH$_3$)=CHCOOCH$_3$	—
*72	CH$_3$CH$_2$$\overset{1}{}C\overset{2}{=}$CHCH$_2CH_3$ ($\overset{3}{}$ $\overset{4}{}$) Fp	yellow oil, air-sensitive [149] ^{1}H NMR (CS$_2$): 0.98 (m, 6H, CH$_3$), 2.18 (m, 4H, CH$_2$), 4.76 (s, C$_5$H$_5$), 5.82 (t, =C[3]H-, [3]J = 6.1) [149] ^{13}C NMR (CH$_2$Cl$_2$): 14.88, 15.78 (both CH$_3$), 28.65 (C-4), 45.32 (C-1), 86.76 (C$_5$H$_5$), 137.31 (C-3), 142.90 (C-2), 216.94 (CO) [149] IR (CH$_2$Cl$_2$): 1950, 2004 (both CO) [149]
*73	(E)-FpC(COOCH$_3$)=CHCH$_2$C(CH$_3$)$_2$OH	^{1}H NMR (CS$_2$): 1.09 (s, C(CH$_3$)$_2$), 2.58 (br s, OH), value not given (d, 2H, [3]J = 7.5), 3.62 (s, OCH$_3$), 4.87 (s, C$_5$H$_5$), 5.47 (t, =CH-) [150] IR (CHCl$_3$): 1685 (C=O) [150]
*74	(Z)-FpC(=CHCH$_3$)CH$_2$CH(COOCH$_3$)$_2$	m.p. 54.5 to 55.5° [120] ^{1}H NMR (CS$_2$): 1.65 (dt, CCH$_3$, J = 7.1), 2.77 (br d, CH$_2$, [3]J = 7), 3.60 (s, 6H, OCH$_3$), 3.81 (t, >CH-), 4.86 (s, C$_5$H$_5$), 5.86 (q, =CH-, [3]J = 7) [120] IR (neat): 1955, 2010 (both CO) [120]

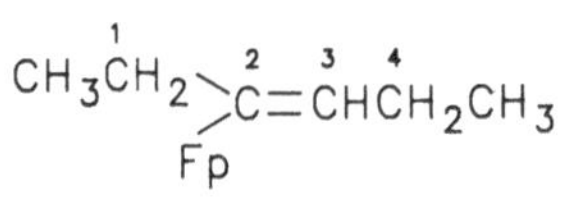

References on pp. 141/7

Table 5 (continued)

No.	compound method of preparation (yield)	properties and remarks

*75   (E)–FpC(=CHCH$_3$)CH$_2$CH(COOCH$_3$)$_2$ — ^{1}H NMR (CS$_2$): 1.62 (d, CCH$_3$, J=7), 2.75 (br d, CH$_2$, ^{3}J=7), 3.60 (s, 6H, OCH$_3$), 3.8 (t, >CH–), 4.78 (s, C$_5$H$_5$), 5.34 (q, =CH–, ^{3}J=7) [120]
IR (neat): 1955, 2010 (both CO) [120]

structures of the type Fe(C=C)$_2$, FeC=CC=N, and FeC=CC≡C

76   (E)–FpCH^1=CH^{2}CH=CH$_2$
      V [119]

^{1}H NMR: 6.52 (H-1), J(H-1,2)=15 to 20 [119]

*77   (E)–FpCH^1=CH^{2}C(CH$_3$)=CH$_2$
      V [119]

^{1}H NMR: 1.80 (m, CH$_3$), 4.57 (br s, =CH$_2$), 4.84 (s, C$_5$H$_5$), 6.42 (d, 1H, J=16.1), 7.02 (d, 1H, J=16.1) in CDCl$_3$ [190], 6.92 (H-1), J(H-1,2)=15 to 20 [119]
^{13}C NMR (CD$_2$Cl$_2$): 19.1, 86.4, 108.7, 134.0, 144.5, 147.8, 216.4 [190]
IR (Nujol): 1964, 1995 (both CO) [190]

*78

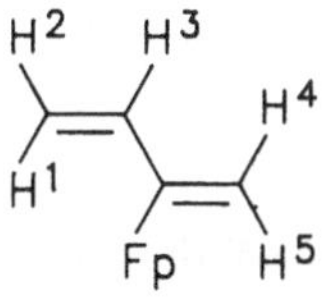

deep orange oil [102]
^{1}H NMR (CS$_2$): 4.76 (s, C$_5$H$_5$), 4.76 (m, H-1), 4.92 (m, H-2, J(H-1,2)=2.4), 5.20, 5.78 (m's, H-4 and H-5), 6.50 (m, H-3, J(H-1,3)=15.0, J(H-2,3)=9.0) [102]
IR (neat): 1949, 2003 (both CO) [102]
mass spectrum: [M]$^+$ [102]

*79

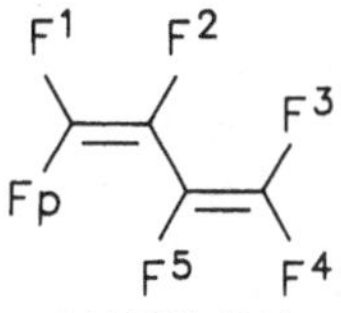

I (16%) [22]

yellow crystals, m.p. 32 to 34° [22]
^{19}F NMR (ether): 55.9 (F-2), 75.9 (F-1), 108.9 (F-4), 120.2 (F-3), 156.4 (F-5), J(F-1,2)=41, J(F-1,5)=7, J(F-1,3)=3, J(F-1,4)−3, J(F-2,5)−14, J(F-2,3)−13, J(F-2,4)=4, J(F-3,5)=114, J(F-4,5)=25, J(F-3,4)=79 [22]
IR (C$_6$H$_{12}$): 1993, 2040 (both CO), 1673 (FeC= C), 1763 (F^{5}C=C) [14, 22], other bands 633 to 1800 [22]

*80

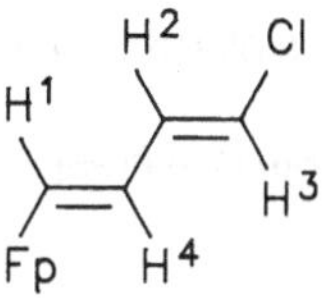

^{1}H NMR (CS$_2$): 4.73 (s, C$_5$H$_5$), 5.65 (d, H-1, J(H-1,4)=12.8), 6.19 (m, H-2 and H-4), 7.02 (d, H-3, J(H-2,3)=14.3) [96, 110]
IR (neat): 1947, 2003 (both CO) [96, 110]
mass spectrum: [M]$^+$ [110]

*81   FpC(=CH$_2$)CBr=CH$_2$

only about 90% pure [98]
^{1}H NMR: 4.90 (s, C$_5$H$_5$), 5.19 (br s, 2H, =CH$_2$), 5.41 (s, 1H, =CH$_2$), 5.92 (s, 1H, =CH$_2$) [98]
IR (neat): 1965, 2010 (both CO) [98]

References on pp. 141/7

Table 5 (continued)

No.	compound method of preparation (yield)	properties and remarks
*82		gold, m.p. 61 to 63° [111] ^{1}H NMR (CS$_2$): 3.51 (s, OCH$_3$), 4.83 (s, C$_5$H$_5$), 5.12 (d, H–3), 6.32 (m, H–4), 6.88 (m, H–2, J(H–2,4) = 9.8, J(H–2,3) = 13.5), 7.75 (d, H–1, J(H–1,4) = 14.3) [111], see also [82] IR (Nujol): 1938, 2000 (both CO) [82, 111]
83	(E)-FpCH^1=CH^{2}C(=CH$_2$)CH(OCH$_3$)$_2$ V [119]	^{1}H NMR: J(H–1,2) = 15 to 20 [119]
84	FpC(=CH$_2$)C(OC$_2$H$_5$)=CH$_2$	obtained from [CH$_2$=C(Fp)C(=OC$_2$H$_5$)CH$_2$Fp]$^+$ and CH$_3$O$^-$, no further details given [133]
*85	(E,E)-FpCH=CHCH=CHCH$_3$	^{1}H NMR (CDCl$_3$): 1.71 (d, CH$_3$, J = 6.3), 4.83 (s, C$_5$H$_5$), 5.29 (m, 1H), 6.06 (dd, 1H, J = 14, 9.9), 6.31 (dd, 1H, J = 15.3, 9.9), 6.84 (d, 1H, J = 15.3) [190] ^{13}C NMR (CD$_2$Cl$_2$): 18.0, 86.3, 119.8, 134.7, 136.1, 146.2, 216.4 [190] IR (Nujol): 1964, 1997 (both CO) [190]
86	(E)-FpCH^1=CH^{2}C(C$_2$H$_5$)=CH$_2$ V [119]	^{1}H NMR: 7.28 (FeCH), J(H–1,2) = 15 to 20 [119]
*87	 I (22%) [105, 106], (7%) [94]	yellow crystals, m.p. 55 to 56° (from C$_5$H$_{12}$) [68, 94, 105, 106] ^{1}H NMR (THF): 5.02 (s, C$_5$H$_5$) [68, 105, 106] ^{19}F NMR (THF): 53.3 (CF$_3^4$), ~55.1 (CF$_3^5$), ~55.1 (CF$_3^3$), 80.5 (F–1), 88.8 (F–2), signals of CF$_3^3$ and CF$_3^5$ are overlapped, J(CF$_3^4$, CF$_3^5$) = 10, J(CF$_3^3$, F–2) = 14, J(CF$_3^3$, F–1) = 9.5, J(F–1,2) = 26, J(CF$_3^5$, F–1) = 4.8 [105, 106] IR: 1995, 2050 (both CO), 1000, 1100, 1420 (all C$_5$H$_5$), 1558, 1728 (both C=C), 3120 (C$_5$H$_5$) in KBr or C$_6$H$_{12}$ [68, 105, 106] R (neat): 1558, 1728 (both C=C) [105, 106]
*88	(Z)-FpC(=CHC≡CCH$_3$)CH$_2$N(C$_2$H$_5$)$_2$	amber oil [195] ^{1}H NMR (CDCl$_3$): 2 (CH$_3$), 3.05 (CH$_2$), 4.8 (C$_5$H$_5$), 6.3 (C=CH) [195] IR (CHCl$_3$): 1955, 2010 (both CO) [195]

structures of the type FeC=CC$_n$C=C (n = 1 or 3)

*89	FpC(CN)=C(CN)CH$_2$CH=CH$_2$ IV c (30%) [129]	yellow-brown, m.p. 97 to 99° [129] ^{1}H NMR (CDCl$_3$): 3.35 (d, –CH$_2$–, ^{3}J(H,H) = 6.5), 5.04 (s, C$_5$H$_5$), 5.17 to 5.43 (m, =CH$_2$), 5.43 to 6.03 (m, –CH=) [129]

Table 5 (continued)

No.	compound method of preparation (yield)	properties and remarks
		IR: 1996, 2042 (both CO) in CH_2Cl_2, 1532, 1643 (both C=C), 2184, 2209 (both CN) in KBr [129]
*90	$FpC(CN)=C(CN)CH(CH_3)CH=CH_2$ IVc [129]	yellow-brown, m.p. 119.5 to 120° [129] ^{1}H NMR ($CDCl_3$): 1.27 (d, CH_3), 3.85 (dq, >CH-, $^3J(H,H) \sim {}^3J(H,CH_3) \sim 6.5$), 5.03 (s, C_5H_5), 4.9 to 6.1 (m, $CH=CH_2$) [129] IR: 1999, 2046 (both CO) in CH_2Cl_2, 1529, 1637 (both C=C), 2180, 2212 (both CN) in KBr [129]
*91	$FpC(COOCH_3)=C(COOCH_3)CH_2CH=CH_2$	amber oil [172] ^{1}H NMR ($CDCl_3$): 3.09 (dd, $-CH_2-$, J=6 and 1.5), 3.76 (s, 6H, $COOCH_3$), 4.77 to 5.18 (m, $=CH_2$), 4.97 (s, C_5H_5), 5.46 to 6.15 (m, $-CH=$) [172] IR (CCl_4): 1980, 2022 (both CO), 1700 (C=O), 1633 (C=C) [172] mass spectrum: $[M-2\,CO]^+$ (100), 180 (26), 152 (56), 122 (50), 121 (55), 69 (33), 65 (27), 56 (26), no parent peak [172]

structures of the type FeCC=C

92	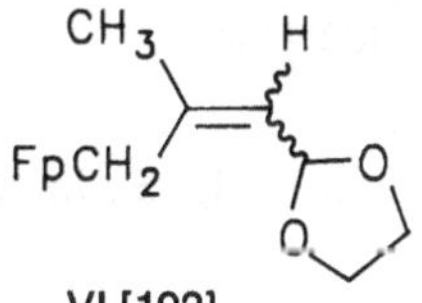 VI [192]	for preparation, see No. 177 on p. 139
*93	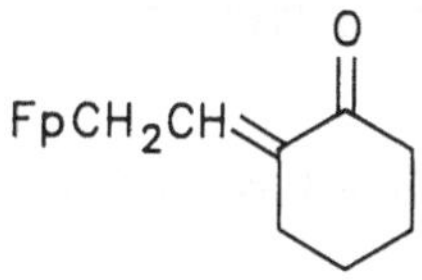 VI [161]	^{1}H NMR: 7.15 (t, $-CH=$, J=9) [161]
94	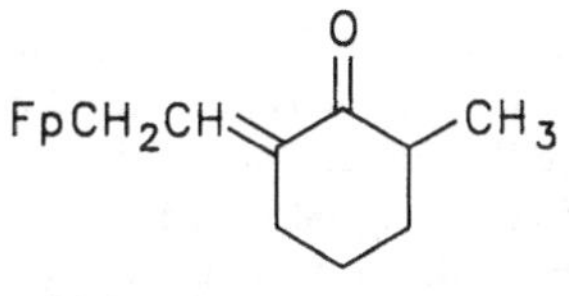 VI [161]	^{1}H NMR ($CDCl_3$): 7.15 (t, CH=, J=9) [161]

References on pp. 141/7

Table 5 (continued)

No.	compound method of preparation (yield)	properties and remarks

*95

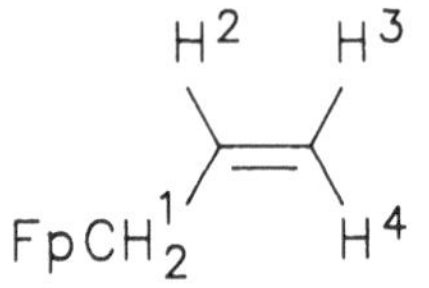

FpCH$_2^1$ ... structure with H^2, H^3, H^4

I (50%) [154], (34%) [6], (91%) [99], (77%) [114]

amber oil [6], yellow oil [32], decomposes at ~65° [6]

^{1}H NMR: 2.07 (q, CH$_2$, J(H–1,2) = 8.4, J(H–1,3 or H–1,4) = 1), 4.55 (s, C$_5$H$_5$), 4.69 (complex, H–3 and H–4), 6.16 (complex, J(H–1,2) = 8.2, J(H–2,3) = 9.2, J(H–2,4) = 17.1) as pure liquid [6], 2.05 (d, CH$_2$, J = 8), 4.54 (s, C$_5$H$_5$), 4.37 to 4.96 (m, =CH$_2$), 5.94 (ddt, CH=, J = 16.0, 10.0, 8.0) in CS$_2$ [99], 2.10 (d, CH$_2$, J = 8), 4.59 (s, C$_5$H$_5$), 4.32 to 4.98 (m, =CH$_2$), 5.95 (ddt, CH=, J = 16, 10, 8) in CDCl$_3$ [154]

^{13}C NMR: 5.11 (FeC) in CD$_3$NO$_2$ [182], −23.6 (CO), 20.6 (C–2), 57.9 (C–3), 160.1 (C–1) referred to CS$_2$ [56]

IR: 1961, 2014 [131], 1950, 2010 [97] in C$_6$H$_{12}$, 1955, 2005 in petroleum ether [154], 1952, 2000 in CH$_2$Cl$_2$ [99]

IR (neat): 1948, 2010 (all CO) [6], see also [18, 32], 982 (δ CH$_2$ out of plane), 998, 1013, 1112, 1358, 1419 (all C$_5$H$_5$), 1399 (δ CH$_3$), 1433 (δ CH$_2$), 1608 (C=C), 2975, 3055, 3080 (all olefinic CH), 3120 (CH of C$_5$H$_5$), other bands 666 to 2925 [6]

k = 15.81, k′ = 0.50 [25]

UV (C$_7$H$_{16}$): 311 (ε = 5400) [114]

96 Fp CD$_2$CH=CH$_2$
    I (5%) [42]

b.p. 40°/10^{-3} Torr [42]
^{1}H NMR (CDCl$_3$): 4.68 (m, =CH$_2$), 5.7 (m, CH) [42]

97 FpCH$_2$CH=CD$_2$
    I (95%) [42]

b.p. 40°/10^{-3} Torr [42]
^{1}H NMR (CDCl$_3$): 2.08 (d, CH$_2$), 5.7 (m, CH) [42]

*98 FpCD$_2$CH=CD$_2$

^{1}H NMR (CDCl$_3$): 4.71 (s, C$_5$H$_5$), 6.07 (br s, CH) [175]
IR (CH$_2$Cl$_2$): 1947, 2000 (both CO) [175]

*99 FpCH$_2$C(CH$_3$)=CH$_2$
    I (88%) [41, 99, 114], (40%) [80], (30%) [42]

amber oil [80], b.p. 45°/5·10^{-4} Torr [42]
^{1}H NMR (CDCl$_3$): 2.14 (s, CH$_3$), 2.50 (br s, –CH$_2$–), 4.66 (s, 1H, =CH$_2$), 5.08 (s, C$_5$H$_5$, 1H, =CH$_2$) [80], 1.79 (s, CH$_3$), 2.06 (s, –CH$_2$–), 4.55 (t, =CH$_2$) [42], 1.78 (s, CH$_3$), 2.13 (s, –CH$_2$–), 4.59 (m, =CH$_2$), 4.61 (s, C$_5$H$_5$) [99]

Table 5 (continued)

No.	compound method of preparation (yield)	properties and remarks
		^{13}C NMR (CS$_2$): 10.3 (d, FeC, J(^{13}C, ^{57}Fe) = 8.82), 24.7 (CH$_3$), 87.4 (C$_5$H$_5$, J(^{13}C, ^{57}Fe) = 2.32), 107.3 (=CH$_2$), 155.2 (>C=), 223.1 (CO, J(^{13}C, ^{57}Fe) = 30.15) [135], see also [186] ^{57}Fe NMR (CS$_2$ at $-30°$, referred to ferrocene): -600.1 [135], see also [186] IR: 1962, 2018 [80], 1950, 1998 [99] in CH$_2$Cl$_2$, 1950, 2010 (all CO) in C$_6$H$_{12}$ [97] UV (C$_7$H$_{16}$): 317 (ε = 6200) [114]
*99a	FpCD$_2$C(CH$_3$)=CH$_2$	obtained as a mixture with No. 99b [99]
*99b	FpCH$_2$C(CH$_3$)=CD$_2$	obtained as a mixture with No. 99a [99]
*100	(E)-FpCH$_2$CH^1=CH^{2}C$_6$H$_5$ I(72%, based on Fp$_2$) [136], (22%) [99], (30%) [42]	yellow solid [136], m.p. 75° [42], amber oil [86] ^{1}H NMR: 2.25 (dd, CH$_2$, J = 7.5, 1), 4.56 (s, C$_5$H$_5$), 6.11 (d, H-2, J = 15), 6.47 (dt, H-1, J = 15, 7.5), 7.15 (m, C$_6$H$_5$) in CS$_2$ [86], see also [99], 2.30 (d, CH$_2$), 6.32 (m, CH=CH, ^{3}J(CH, CH) = 15), 7.29 (m, C$_6$H$_5$) [42], 2.28 (d, CH$_2$, ^{3}J = 7.5), 4.63 (s, C$_5$H$_5$), 6.00 to 6.84 (m, CH=CH), 7.23 (m, C$_6$H$_5$) [136] in CDCl$_3$ IR: 1960, 2010 in CH$_2$Cl$_2$ [86, 99], 1955, 2010 (all CO), 829 (CH of C$_5$H$_5$), 1620 (C=C) in CHCl$_3$ [136]
*101	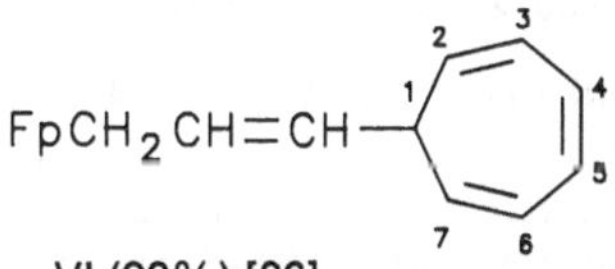 FpCH$_2$CH=CH— VI (98%) [86]	(Z)/(E)-mixture ($\sim$1:2?) [86] amber oil [86] ^{1}H NMR (CS$_2$): 1.92, 2.12 (2 d, (E)- and (Z)-CH$_2$, J = 8), 2.3 (m, C$_7$H$_7$, H-1), 4.53, 4.60 (2 s, (E)- and (Z)-C$_5$H$_5$), 5.1 (m, CH= CH), 5.7 (m, C$_7$H$_7$, H-2,7), 6.0 (m, C$_7$H$_7$, H-3,6), 6.6 (m, C$_7$H$_7$, H-4,5) [86] IR (KBr): 1920, 2000 (both CO) [86]
*102	(Z)-FpCH$_2$CH=CHCl I [136], V (81%) [114]	yellow, very unstable oil [136], red oil [114] ^{1}H NMR (CS$_2$): 2.05 (dd, CH$_2$, J = 9.1), 4.73 (s, C$_5$H$_5$), 5.58 (dt, =CHCl, J = 7.1), 5.90 (dt, CCH=C, J = 7.9) [114] IR: 1968, 2018 in CHCl$_3$ [136], 1967, 2020 in petroleum ether, 1948, 2015 (all CO) in CS$_2$ [114]
103	FpCH$_2$C(CH$_2$Cl)=CH$_2$ I (75%) [141]	prepared from CH$_2$=C(CH$_2$Cl)$_2$ and a deficiency of Na[Fp] [141]

References on pp. 141/7

Table 5 (continued)

No.	compound method of preparation (yield)	properties and remarks
104	(Z)-FpCH$_2$CH=CHBr VI (80%) [86]	orange crystals, m.p. 36 to 37° [86] ^{1}H NMR (CS$_2$): 2.0 (d, CH$_2$, ^{3}J = 9), 4.75 (s, C$_5$H$_5$), 5.7 (d, =CHBr, ^{3}J = 6.5), 6.3 (dt, FeCCH=) [86] IR (CH$_2$Cl$_2$): 1950, 2020 (both CO) [86]
*105	FpCH$_2$C(CH$_3$)=CHBr VI [86]	dark orange oil [86] ^{1}H NMR (CS$_2$): 1.78 (d, CH$_3$, ^{4}J = 1.3), 2.07 (s, CH$_2$), 4.77 (s, C$_5$H$_5$), 5.57 (q, =CH–) [86]
*106	(Z)-FpCH$_2$C^{1}H=C^{2}HOCH$_3$ I [86] VI (97%) [86]	m.p. 40.5 to 41.5° [86], red, m.p. 45 to 47° [188] ^{1}H NMR: 2.08 (d, CH$_2$, J = 9), 3.47 (s, OCH$_3$), 4.6 (m, =CH), 4.68 (s, C$_5$H$_5$), 5.62 (d, =CH, J = 6) in CS$_2$ [86], 2.33 (d, CH$_2$, J = 9), 3.27 (s, OCH$_3$), 4.20 (s, C$_5$H$_5$), 4.75 (td, =C^{1}H–, J = 9, 6), 5.61 (d, =C^{2}H–, J = 6) in C$_6$D$_6$ [188] ^{13}C NMR (C$_6$D$_6$): –4.4 (FeC), 58.9 (OCH$_3$), 85.6 (C$_5$H$_5$), 118.3 (C-1), 141.7 (C-2), 217.6 (CO) [188] IR: 1941, 1996 (both CO), 1660 (C=C) (neat) [86], 1947, 2001 (both CO), 1640 (C=C) in CHCl$_3$ [188] UV (C$_7$H$_{16}$): 337 (ε = 7100) [114]
*107	(E)-FpCH$_2$CH=CHOCH$_3$ I [86]	red oil [86] ^{1}H NMR (CS$_2$): 2.11 (dd, CH$_2$, ^{3}J = 9.0), 3.37 (s, OCH$_3$), 4.59 (s, C$_5$H$_5$), 4.87 (dt, FeCCH=), 6.15 (dt, =CHO, ^{3}J = 12, ^{4}J = 1) [86] IR (CS$_2$): 1950, 2010 (both CO) [86]
*108	FpCH$_2$C(OCH$_3$)=CH$_2$ VI (95%) [175]	amber solid, m.p. 24 to 26° [139, 142], orange oil [175] ^{1}H NMR (CDCl$_3$): 2.00 (s, FeC), 3.52 (s, OCH$_3$), 3.70 (br s, 1H, =CH$_2$), 3.84 (d, 1H, =CH$_2$, J = 2), 4.72 (s, C$_5$H$_5$) [175], almost the same data are given in [139, 142] IR: 1958, 2005 (both CO), 1625, 1656 (both C=C) in CCl$_4$ [139], 1955, 2030 (both CO) in CH$_2$Cl$_2$ [175]
*109	FpCH$_2$C(OC$_2$H$_5$)=CH$_2$	—
110	(Z)-FpCH$_2$CH=CHOC$_6$H$_5$ VI (72%) [114]	red oil [114] ^{1}H NMR (CS$_2$): 2.12 (d, CH$_2$, ^{3}J = 9.1), 4.67 (s, C$_5$H$_5$), 5.03 (dt, 1H, FeCCH=), 6.07 (d, 1H, =CH-O, ^{3}J = 5.5), 6.87 (m, 3H, C$_6$H$_5$), 7.10 (m, 2H, C$_6$H$_5$) [114] IR (Skelly B): 1960, 2015 (both CO), 1597 (C=C) [114]

References on pp. 141/7

Table 5 (continued)

No.	compound method of preparation (yield)	properties and remarks
111	(E)-FpCH$_2$CH=CHSO$_2$CH$_3$ VI (90%) [86]	^{1}H NMR (CDCl$_3$): 2.02 (d, CH$_2$, ^{3}J = 9.5 to 10), 2.87 (s, CH$_3$), 4.80 (s, C$_5$H$_5$), 6.05 (d, =CHS, ^{3}J = 15), 7.25 (dt, FeCCH=) [86] IR (neat): 1960, 2010 (both CO), 1125, 1300 (both SO$_2$) [86]
*112	(E)-FpCH$_2^1$CH^2=CH^{3}Si(CH$_3$)$_3$ I (65%) [103]	wax, purified by chromatography [103] ^{1}H NMR (CDCl$_3$): 0.23 (s, Si(CH$_3$)$_3$), 2.30 (2 d, H-1), 4.80 (s, C$_5$H$_5$), 5.44 (2 t, H-3), 6.40 (2 t, H-2), J(H-1,2) = 8, J(H-1,3) < 0.5, J(H-2,3) = 18 [103] IR: 1961, 2010 (both CO) [103]

No. 113. FpCH$_2$ structure with dioxolane group

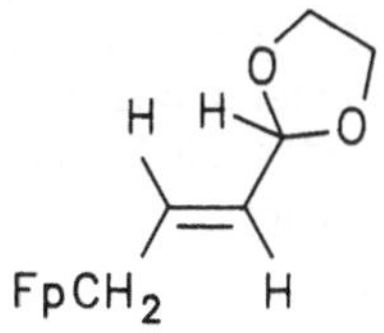

prepared according to [86] (Method VI) as cited in [101]; for the reaction with tropyliumiron tricarbonyl, see on p. 71 [101]

No. 114. CH$_3$, FpCH$_2$ structure with dioxolane group

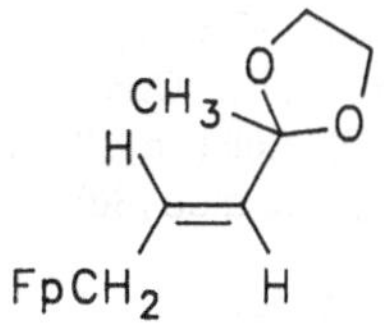

attempted reaction with 2-ethoxycarbonylcyclopent-2-enone is briefly mentioned in [175]

No. *115. n-C$_5$H$_{11}$, FpCH$_2$CH=CH structure with dioxolane group VI (83%) [86]

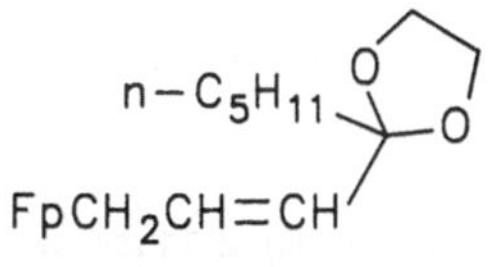

amber oil, (Z)/(E)-mixture (~1:3)
^{1}H NMR (CS$_2$), (E)-compound: 2.06 (d, FeCH$_2$, ^{3}J = 8), 3.74 (s, 4H, OCH$_2$), 4.66 (s, C$_5$H$_5$), 5.12 (d, =CHCO, ^{3}J = 15), 5.95 (dt, FeCCH=) [86]
IR (neat): 1940, 2010 (both CO) [86]

No. *116. C$_6$H$_5$, FpCH$_2$ structure with dioxolane group VI (90%) [86]

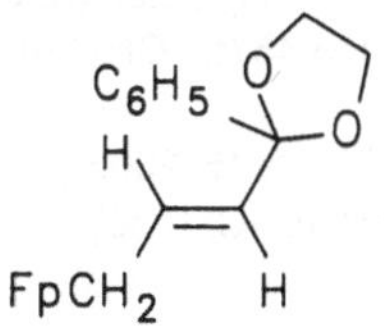

yellow crystals, m.p. 51 to 53° [86]
^{1}H NMR (CS$_2$): 1.95 (d, CH$_2$, ^{3}J = 8), 3.80 (m, 4H, OCH$_2$), 4.53 (s, C$_5$H$_5$), 5.32 (d, =CHCO), 5.8 (dt, FeCCH, ^{3}J = 15), 7.25 (m, C$_6$H$_5$) [86]
IR (KBr): 1940, 2000 (both CO) [86]

*117 FpCH$_2$CH^1=CH^{2}CH$_3$
I (94%) [99],
VI (91%) [86]

deep amber oil [6, 86], (E)/(Z)-mixture
b.p. 45°/5 × 10^{-4} Torr [42]
^{1}H NMR: 1.3 (m, CH$_3$), 1.9 (m, CH$_2$), 4.50 (s, C$_5$H$_5$), 4.56 (s, C$_5$H$_5$), 5.2 (m, H-1) in THF-d^8 [131], 1.53 (dm, CH$_3$), 2.0 (m, CH$_2$), 4.58 (s, C$_5$H$_5$, (E)-isomer), 4.65 (s, C$_5$H$_5$, (Z)-isomer), 4.8 to 5.9 (m, 2H, CH=CH) in

References on pp. 141/7

Table 5 (continued)

No.	compound method of preparation (yield)	properties and remarks
*117 (continued)		CS_2 [86], 1.56 (d, CH_3), 2.12 (d, CH_2), 5.43 (m, CH=CH) in $CDCl_3$ [42], 1.57 (d, CH_3, J(CH_3, H-2) = 5), 2.12 (d, CH_2, J(CH_2, H-1) = 6), 4.56 (s, C_5H_5), 5.43 (complex, CH=CH) as pure liquid [6] IR: 1951, 2009 in C_6H_{12} [97], 1945, 2005 in CH_2Cl_2 [99], 1940, 2000 [86], 1950, 2016 (all CO), 717 (δ =CH– out-of-plane, (Z)-isomer), 826, 1000, 1015, 1363, 1422 (all C_5H_5), 960 (δ =CH– out-of-plane, (E)-isomer), 1377, 1454 (both δ CH_3), 1437 (δ CH_2), 1643 (C=C), 2860, 2925, 2960 (all ν CH of CH_2, CH_3), 3010 (ν olefinic CH), 3115 (ν CH of C_5H_5), other bands at 763, 841, 900, 1056, 1080, 1114, 1160, 1260, 1300, 1401, 1681, 1710 (neat) [6] k = 15.88, k' = 0.53 [25]
*118	(E)-FpCH$_2$CH=CHCH$_3$ I[86, 99]	^{1}H NMR (CS_2): 1.53 ("dm", CH_3, J = 6.0), 2.07 ("dm", CH_2, J = 7.5), 4.57 (s, C_5H_5), 5.06 (dq, CH=, J = 15.0, 6.0), 5.68 ("5 q", CH=, J = 15.0, 7.5, 1.0) [99], see also [86]
119	FpCHDCH=CHCH$_3$ I [42]	b.p. 45°/5 × 10^{-4} Torr, mixture of (Z)/(E)-isomers [42] ^{1}H NMR ($CDCl_3$): 1.56 (d, CH_3), 2.12 (d, CHD), 5.43 (m, CH=CH) [42]
*120	FpCH$_2$CH=C(CH$_3$)$_2$ I (88%) [136], (52%) [99]	amber oil [136] ^{1}H NMR: 1.57 (br s, 6H, CH_3), 2.14 (br, d, CH_2, ^{3}J = 9.0), 4.58 (s, C_5H_5), 5.34 (t, –CH=) in CS_2 [99], 1.63 (s, 6H, CH_3), 2.36 (d, CH_2, ^{3}J = 9), 4.71 (s, C_5H_5), 5.45 (t, –CH=) in $CDCl_3$ [136] IR (neat): 1950, 2005 [99], 1945, 2005 (all CO), 825 (CH of C_5H_5), 1640 (C=C) [136]
121	FpCH$_2$CH=CHCH$_2$C$_6$H$_5$ VI (99%) [120]	(E)/(Z)-mixture [120] ^{1}H NMR (CS_2): 2.07 (d, FeCH$_2$, J = 7.0, (E)-isomer), 2.17 (d, FeCH$_2$, J = 9.0, (Z)-isomer), 3.20 (d, 2H, CH$_2$-aryl, J = 6.0, (E)-isomer), 3.33 (d, 2H, CH$_2$-aryl, J = 8.0, (Z)-isomer), 4.50, 4.55 (both: s, 5H, C_5H_5), 4.9 to 6.1 (m, 4H, (Z)- and (E)-CH=CH) [120] IR (neat): 1955, 2010 (both CO) [120]

Table 5 (continued)

No.	compound method of preparation (yield)	properties and remarks
*122	(E)-FpCH$_2$CH=CHCH$_2$C$_6$H$_5$	oil [120] ^{1}H NMR (CS$_2$): 2.08 (d, FeCH$_2$, J=7.0), 3.18 (d, 2H, CH$_2$-aryl, J=5.5), 4.50 (s, C$_5$H$_5$), 5 to 6 (m, 2H, CH=CH) [120] IR (neat): 1942, 2000 (both CO), other band 960 [120]
*123	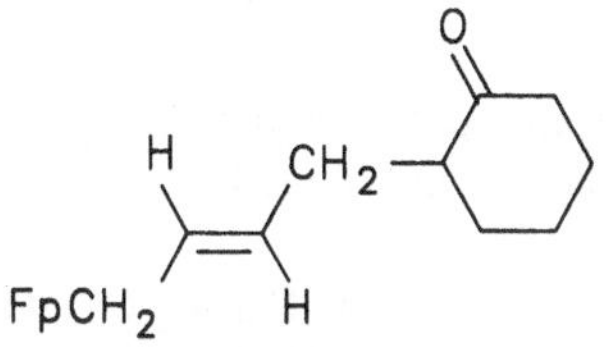	oil [120] ^{1}H NMR (C$_6$D$_6$): 1.1 to 2.4 (m, 13H, >CH-, -CH$_2$-), 4.40 (s, C$_5$H$_5$), 5.50 (dt, -CH=, J= 17, 10), 5.94 (dt, -CH=) [120] IR (neat): 1950, 1995 (both CO) [120]
*124	FpCH$_2$CH=CClCH$_3$ I (30%) [42]	m.p. 23° [42] ^{1}H NMR (CDCl$_3$): 2.02 (s, CH$_3$), 2.10 (d, CH$_2$, partly overlapped by CH$_3$), 5.7 (t, =CH-) [42]
*125	(E)-FpCH$_2$CH=CHCH(OCH$_3$)$_2$ VI (84%) [86]	amber oil [86] ^{1}H NMR (CS$_2$): 2.05 (d, CH$_2$, ^{3}J=8), 3.17 (s, 6H, OCH$_3$), 4.65 (s, C$_5$H$_5$), 4.7 (d, OCHO, ^{3}J=5), 5.15 (dd, =CHCO, J(CH=, CH=)= 15), 6.0 (dt, FeCCH) [86] IR (neat): 1950, 2020 (both CO) [86]
*126	(E)-FpCH$_2$CH^1=CH^{2}CHO	amber oil [86] ^{1}H NMR (CS$_2$): 2.12 (d, CH$_2$, ^{3}J=9.5), 4.74 (s, C$_5$H$_5$), 5.77 (dd, H-2, J(H-1,2)=14.5), 7.04 (dt, H-1), 9.35 (d, CHO, ^{3}J=7.5) [86], 2.15 (d, CH$_2$), 4.75 (s, C$_5$H$_5$), 5.82 (m, H-2, J(H-1,2)=15), 6.94 (m, H-1, J(CH$_2$, H-1)=9), 9.29 (d, CHO, ^{3}J=7.5) [110], see also [96] IR (neat): 1940, 2000 (both CO), 1648 (C=O) [96, 110], 1950, 2010 (both CO), 1592 (C=C), 1653 (C=O) [86]
*127	(E)-FpCH$_2$CH=CHCOC$_6$H$_5$	red oil, crude product [86] ^{1}H NMR (CD$_3$NO$_2$): 2.25 (d, CH$_2$, ^{3}J=9), 4.91 (s, C$_5$H$_5$), 7.27 (d, CH=, J(CH=, CH=)=15), 7.2 to 8 (m, 6H, C$_6$H$_5$, CH=) [86]
*128	(E)-FpCH$_2$CH=CHCOOCH$_3$	amber oil [86] ^{1}H NMR (CS$_2$): 2.0 (d, CH$_2$, ^{3}J=9), 3.55 (s, OCH$_3$), 4.70 (s, C$_5$H$_5$), 5.45 (d, =CH, J(CH=, CH=)=15), 7.16 (dt, =CH) [86] IR (neat): 1950, 2000 (both CO), 1690 (C=O) [86]

                    References on pp. 141/7

Table 5 (continued)

No.	compound method of preparation (yield)	properties and remarks
*129	(E)-$FpCH_2CH=CHCONHCOCCl_3$	orange solid [99] ^1H NMR ($CDCl_3$): 2.13 (d, CH_2, $^3J=10$), 4.79 (s, C_5H_5), 6.58 (d, CH=, $J(CH=, CH=)=15$), 7.7 (m, CH=), 9.03 (br s, NH) [99] IR (KBr): 1960, 2020 (both CO), 1560 (C=C), 1650, 1710 (both C=O) [99]
130	$FpCH_2CH=C(CH_3)SCH_3$	prepared from [Fp=$C(SCH_3)CH_3$]CF_3SO_3 and lithium divinylcuprate [198]
131	$FpCH_2CH=C(CH_3)SC_6H_5$	prepared from [Fp=$C(SC_6H_5)CH_3$]CF_3SO_3 and lithium divinylcuprate [198]
132	$FpCH_2CH=CHC_2H_5$ VI (79%) [120]	(E)/(Z)-mixture (1:1) [120] ^1H NMR (CS_2): 0.93 (t, CH_3, J=7.0), 1.9 (m, =CCH_2C), 2.1 (2d, $FeCH_2$, J=7.0, 9.0), 4.60, 4.65 (2s, C_5H_5), 4.8 to 6.0 (m, CH=CH) [120] IR (neat): 1946, 2008 (both CO) [120]
*133	(E)-$FpCH_2CH=CHC_2H_5$	^1H NMR (CS_2): 0.92 (t, CH_3, J=7), 1.83 (m, =CCH_2C, J=7), 2.06 (d, $FeCH_2$, J=7), 4.56 (s, C_5H_5), 5 to 6 (m, CH=CH) [120] IR (neat): 1950, 2000 (both CO), other band 960 [120]
*134	(E)-$FpCH_2CH=C(C^1H_3)CH_2C^2H_3$	^1H NMR (CS_2): 0.92 (t, C^2H_3, J=7.5), 1.53 (s, C^1H_3), 1.87 (q, =CCH_2C, J=7), 2.12 (d, $FeCH_2$, $^3J=9$), 4.58 (s, C_5H_5), 5.33 (m, -CH=) [120] IR (neat): 1942, 1995 (both CO) [120]
*135	(E)-$FpCH_2CH=CHCOCH_3$	orange crystals, m.p. 34.5 to 36° [86] ^1H NMR (CS_2): 2.0 (s, CH_3), 2.05 (d, CH_2, $^3J=9$), 4.7 (s, C_5H_5), 5.7 (d, O=CCH=, $^3J=15$), 7.0 (dt, FeCCH=) [86] IR (CH_2Cl_2): 1950, 2010 (both CO), 1625 (C=O) [86]
*136	(E)-$FpCH_2CH=CHCOCH(C_6H_5)_2$	orange oil [175] ^1H NMR (CS_2): 1.92 (d, CH_2, $^3J=10$), 4.45 (s, C_5H_5), 5.10 ("s", CH=), 5.80 (d, CH=, $J(CH=, CH=)=15$), 7.15 (s, 11H) [175] IR (neat): 1960, 2010 (both CO), 1650 (C=O) [175]

Table 5 (continued)

No.	compound method of preparation (yield)	properties and remarks
*137	(E)–FpCH$_2$CH=CHCOCH(CH$_3$)$_2$	orange oil [99] ^{1}H NMR (CD$_3$NO$_2$): 0.99 (d, 6H, CH$_3$, J = 7), 2.05 (d, CH$_2$, ^{3}J = 9), 2.69 (sept, >CHC=O), 4.69 (s, C$_5$H$_5$), 5.82 (d, =CHC=O, ^{3}J = 15.0), 7.03 (dt, =CHCFe) [99] IR (neat): 1950, 2010 (both CO), 1570 (C=C) 1640 (C=O) [99]
*138	(E)–FpCH$_2$CH=CHCOCH(CH$_3$)C$_6$H$_5$	orange oil [175] ^{1}H NMR (CS$_2$): 1.30 (d, CH$_3$, J = 7), 1.97 (d, CH$_2$, ^{3}J = 10), 3.75 (q, O=CCH<), 4.42 (s, C$_5$H$_5$), 4.60 (s, CH=), 5.63 (d, CH=, J(CH=, CH=) = 15), 7.17 (s, C$_6$H$_5$) [175] IR (CH$_2$Cl$_2$): 1960, 2010 (both CO), 1640 (C=O) [175]
*139	FpCH$_2$C(OCH$_3$)=CHCH(COOCH$_3$)CH(COOCH$_3$)$_2$	^{1}H NMR (CDCl$_3$): 1.80 (d, 1H, CH$_2$), 2.17 (d, 1H, CH$_2$, J = 9,5), 3.43 (s, OCH$_3$), 3,68, 3.70, 3.73 (3s, 9H, COOCH$_3$), 3.74 to 4.10 (m, 3H, =CHCHCH), 4.80 (s, C$_5$H$_5$) [173] IR (CHCl$_3$): 1955, 2000 (both CO), 1625 (C=C), 1735 (C=O) [173] mass spectrum: [M − 2 CO]$^+$ (49), 252 (100), 152 (59), 1122(?) (33), 121 (62), 111 (24), 59 (27), no parent peak [173]
*140	FpCH$_2$C(OCH$_3$)=CHC(COOCH$_3$)$_2$CH(COOCH$_3$)$_2$	^{1}H NMR (CDCl$_3$): 1.99 (s, CH$_2$), 3.42 (s, OCH$_3$), 3.77 (s, 12H, COOCH$_3$), ∼3.77 (s, =CH–), 4.45 (s, CH(CO)$_2$), 4.83 (s, C$_5$H$_5$) [173] IR (CHCl$_3$): 1955, 2000 (both CO), 1740 (C=O) [173]
*141	FpCH$_2$C(OCH$_3$)=CHC(COOC$_2$H$_5$)$_2$CH(COOC$_2$H$_5$)$_2$	—
142	(E)–FpCH$_2$CH=CHCOCH(C$_2$H$_5$)C$_6$H$_5$	prepared like No. 136 (yield 10%) [175] orange oil [175] ^{1}H NMR (CS$_2$): 0.5 to 1.3 (m, C$_2$H$_5$), 1.90 (d, FeCH$_2$, ^{3}J = 9), 3.50 (t, O=CCH<, ^{3}J = 7), 4.40 (s, 6H, C$_5$H$_5$, CH=), 5.62 (d, CH=, J(CH=, CH=) = 15), 7.17 (s, C$_6$H$_5$) [175] IR (neat): 1960, 2010 (both CO), 1640 (C=O) [175]

Table 5 (continued)

No.	compound method of preparation (yield)	properties and remarks
*143	(Z)-FpCH$_2$C(OCH$_3$)=CHCOCH(CN)C$_4$H$_9$-t	crude product is an (E)/(Z)-mixture [173] yellow crystals, m.p. 104 to 105° [173] ^{1}H NMR (CDCl$_3$): 1.16 (s, C$_4$H$_9$-t), 2.55 (s, CH$_2$), 3.18 (s, O=CCH<), 3.70 (s, OCH$_3$), 4.90 (s, C$_5$H$_5$), 5.46 (s, O-C=CH) [173] IR (CHCl$_3$): 1966, 2008 (both CO), 1646 (C=O), 2230 (CN), other band 1520 [173] mass spectrum: [M−2 CO]$^+$ (21), 186 (58), 139 (36), 121 (54), 99 (100), 59 (43), 57 (46), 56 (46), no parent peak [173]
*144	FpCH$_2$CH=CHCH$_2$CH=CH$_2$	unstable amber oil [155] ^{1}H NMR (CDCl$_3$): 2.10 (d, FeCH$_2$, J=8), 2.68 (t, CCH$_2$C, J=6), 4.70 (s, C$_5$H$_5$), 4.8 to 6 (several m) [155]
*145	FpCH$_2$CH=CHC(COOCH$_3$)=CHCOOCH$_3$	amber oil [172] ^{1}H NMR (CDCl$_3$): 2.14 (br d, FeCH$_2$, ^{3}J~5), 3.74 (s, 6H, COOCH$_3$), 4.55 to 5.15 (CH=CH), 4.95 (s, C$_5$H$_5$), 6.87 (br s, =CHCO) [172] IR (CCl$_4$): 1944, 2008 (both CO) [172]
*146	(Z,Z)-FpCH$_2$C(OCH$_3$)=CH^{1}C(COOCH$_3$)=CH^{2}COOCH$_3$	amber oil [173] ^{1}H NMR (CDCl$_3$): 2.69 (s, FeCH$_2$), 3.62 (s, 6H, COOCH$_3$, OCH$_3$), 3.70 (s, 3H, COOCH$_3$), 4.89 (s, C$_5$H$_5$), 5.48 (d, H-1, J=2.5), 6.29 (d, H-2) [173] IR (CH$_2$Cl$_2$): 1960, 2005 (both CO), 1625 (C=C), 1718 (C=O) [173] mass spectrum: [M−2 CO]$^+$ (15), 177 (18), 149 (20), 122 (20), 121 (100), 96 (19), 95 (23), 56 (32), no parent peak [173]

structures of the type FeC$_n$C=C (n = 1 to 5)

*147	FpCH$_2$CH$_2$CH=CH$_2$ I (80%) [148], (77%) [154], (60%) [45], (75%) [47]	amber oil [45, 148] ^{1}H NMR: 1.45 (m, FeCH$_2$), 2.15 (m, CCH$_2$C=), 4.70 (s, C$_5$H$_5$), 4.4 to 5.2 (m, 7H, C$_5$H$_5$, =CH$_2$), 5.48 to 6.2 (m, =CH-) [154], 1.50 (t, FeCH$_2$), 2.18 (q, CCH$_2$C=), 4.75 (s, C$_5$H$_5$), 4.9 (m, =CH$_2$), 5.8 (m, =CH-) [47, 148] in CDCl$_3$, 1.47 (t, FeCH$_2$, ^{3}J=8.2), 2.23 (q, (CCH$_2$C=) J=7.0), 4.52 (s, C$_5$H$_5$), 5.02 (m, =CH$_2$), 5.93 (m, =CH-) in CCl$_4$ [123] or C$_6$H$_6$ [137], 1.4 (m, FeCH$_2$), 2.1 (m, CCH$_2$C=), 4.6

Table 5 (continued)

No.	compound method of preparation (yield)	properties and remarks
		(s, C_5H_5), 4.9 (m, =CH_2), 5.9 (m, =CH–) in C_6D_6 [45] IR: 1960, 2008 in petroleum ether [154], 1940, 2000 (all CO), 1633 (C=C) neat or mull [45] UV: 350 ($\varepsilon = 765$) [47] mass spectrum: [M]$^+$ (6) [154]
*148	$FpCH_2CH_2C(CH_3)=CH_2$	—
*149	$FpCH_2C(CH_3)_2CH=CH_2$ I (10%) [45]	amber oil [45] ^{1}H NMR (C_6D_6): 1.4 (s, 6H, CH_3), 1.9 (s, $FeCH_2$), 4.4 (s, C_5H_5), 5.1 (complex, =CH_2), 6.1 (complex, =CH–) [45] IR (neat or mull): 1944, 1998 (both CO), 1640 (C=C) [45]
150	$FpCH_2CH(C_6H_5)CH=CH_2$	presumably present in small amounts in the reaction mixture of [Fp(CH_2=CHCHCH$_2$)]$^+$ and C_6H_5MgCl [120]
*151	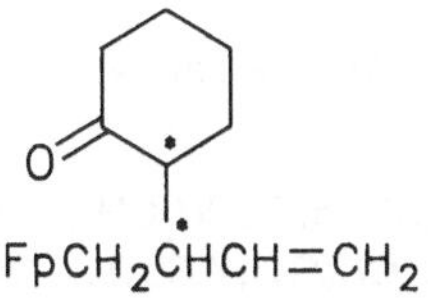 $FpCH_2CHCH=CH_2$	oil [120] ^{1}H NMR: 1.1 to 2.4 (m, 12H, CH, CH_2), 4.38 (s, C_5H_5), 4.7 to 4.8 (m, 3H, CH=CH_2) in C_6D_6(?) [120], 4.76, 4.78 (both C_5H_5) in CS_2 [124] IR (KBr): 1946, 2004 (both CO) [120]
*152	$FpCH^1H^2CH^3(CH=CH_2)CH^4(COOCH_3)_2$	m.p. 40 to 42° [120] ^{1}H NMR (CS_2): 0.88 (t, H–1, J(H–1,2) = J(H–1,3) = 10), 1.7 (dd, H–2, J(H–1,2) = 10, J(H–2,3) = 2.5), 2.5 (m, H–3), 3.24 (d, H–4, J(H–3,4) = 9), 3.54, 3.62 (both: s, OCH_3), 4.70 (s, C_5H_5), 4.7 to 5.5 (m, 3H, CH=CH_2) [120] IR (neat): 1940, 2005 (both CO) [120]
*153	(E)-$FpCH_2CH_2C(COOC_2H_5)=CHC_6H_5$	m.p. 67 to 69° [120] ^{1}H NMR ($CDCl_3$): 1.33 (t, CH_3, J = 7), 1.50 (m, $FeCH_2$), 2.63 (m, CH_2C=), 4.25 (q, OCH_2), 4.75 (s, C_5H_5), 7.39 (br s, C_6H_5), 7.41 (s, =CH–) [120] IR (CH_2Cl_2): 1946, 2004 (both CO), 1635 (C=C), 1701 (C=O) [120]
154	(E)-$FpCH_2CH_2CH=CHCH_3$ I (75%) [47, 148]	^{1}H NMR ($CDCl_3$): 1.40 (m, $FeCH_2$), 1.65 (d, CH_3), 2.10 (m, CH_2C=), 4.75 (s, C_5H_5), 5.5 (m, CH=CH, ^{3}J = 17) [47, 148] UV: 350 ($\varepsilon = 765$) [47] IR (C_5H_{12}): ~1960, ~2000 (both CO) [148]

References on pp. 141/7

Table 5 (continued)

No.	compound method of preparation (yield)	properties and remarks
*155	Fp(CH₂)₃CH=CH₂ I (75%) [47, 148] V (69%) [155]	yellow oil [155], amber oil [148] ^{1}H NMR: 1.47 (m, CH₂CH₂C=), 2.02 (m, FeCH₂), 4.65 (s, C₅H₅), 4.87 (m, =CH₂), 5.70 (m, –CH=) in CS₂ [120, 155], 1.55 (m, FeCH₂CH₂), 2.12 (m, CH₂C=), 4.75 (s, C₅H₅), 4.80 (m, 1H, =CH₂), 5.05 (m, 1H, =CH₂), 5.7 (m, –CH=) in CDCl₃ [47, 148] IR (neat): 1946, 2004 (both CO) [120, 155], 1639 (C=C) [155]
156	FpCH(CH₃)CH₂CH₂CH=CH₂ V (38%) [155]	1:1 mixture with No. 157, spectra of the mix- ture: ^{1}H NMR (CS₂): 0.75 to 2.67 (m, 8H, CH, CH₂, CH₃), 1.30 ("d", FeCH, ^{3}J = 7), 4.63, 4.67 (2s, C₅H₅), 4.92 (m, =CH₂), 5.58 (m, –CH=) [155] IR (neat): 1942, 2000 (both CO), 1639 (C=C) [155]
157	FpCH₂CH(CH₃)CH₂CH=CH₂	1:1 mixture with No. 156, see above
158	Fp(CH₂)₃C(CH₃)=CH₂ I (92%) [148]	amber oil [148] ^{1}H NMR (CDCl₃): 1.50 (m, FeCH₂CH₂), 1.69 (m, CH₃), 2.00 (m, CH₂C=), 4.68 (s, C₅H₅), 4.68 (m, =CH₂) [148] IR (C₅H₁₂): ∼1960, ∼2000 (both CO) [148]
159	FpCH(C₆H₅)CH₂CH₂CH=CH₂ V (75%) [155]	1:3 mixture with No. 160, spectra of the mix- ture: ^{1}H NMR (CS₂): 1.6 to 2.7 (m, 5H, –CH<, CH₂), 4.40, 4.50 (2s, C₅H₅), 4.83 (m, =CH₂), 5.60 (m, –CH=), 7.03, 7.13 (2s, C₆H₅) [155] IR (neat): 1949, 2000 (both CO), 1639 (C=C) [155]
160	FpCH₂CH(C₆H₅)CH₂CH=CH₂	mixture with No. 159, see above
161	Fp(CH₂)₃CH=CHCH₃ I (70%) [148]	amber oil [148] ^{1}H NMR (CDCl₃): 1.42 (m, CH₃), 1.62 (m, FeCH₂CH₂), 2.00 (m, CH₂C=), 4.68 (s, C₅H₅), 5.40 (m, CH=CH) [148] IR (C₅H₁₂): ∼1960, ∼2000 (both CO) [148]
162	Fp(CH₂)₄CH=CH₂ I (91%) [148]	amber oil [148] ^{1}H NMR (CDCl₃): 1.45 (m, FeCH₂CH₂CH₂), 2.05 (m, CH₂C=), 4.67 (s, C₅H₅), 4.75 to 5.05 (m, =CH₂), 5.60 (m, –CH=) [148] IR (C₅H₁₂): ∼1960, ∼2000 (both CO) [148]

References on pp. 141/7

Table 5 (continued)

No.	compound method of preparation (yield)	properties and remarks
*163	FpCH(CHO)CH$_2$CH$_2$CH=CH$_2$ V (73%) [155]	yellow oil [155] ^{1}H NMR (CD$_3$NO$_2$): 1.17 to 2.78 (m, 5H, CH, CH$_2$), 4.90 (s, m, 7H, C$_5$H$_5$, =CH$_2$), 5.80 (m, –CH=), 9.27 (d, CHO, J=2.5) [155] IR (CS$_2$): 1970, 2008 (both CO), 1650 (C=C) [155]
164	FpCH(COOCH$_3$)CH(CH$_3$)CH$_2$CH=CH$_2$ V [155]	–
165	FpCH(COOCH$_2$C^{1}H$_3$)CH(C^{2}H$_3$)- CH$_2$CH=CH$_2$ V (40%) [155]	^{1}H NMR (CS$_2$): 0.95 (d, C^{2}H$_3$, J=6), 1.2 (t, C^{1}H$_3$, J=7), 1.3 to 2.3 (m, 4H, –CH<, –CH$_2$–), 3.94 (q, OCH$_2$, J=7), 4.72 (s, m, 7H, C$_5$H$_5$, =CH$_2$), 5.33 (m, –CH=) [155] IR (neat): 1972, 2020 (both CO), 1684, 1942 [155]

structures of the type FeC$_n$C=C=C (n = 0 or 2)

No.	compound	properties and remarks
*166	FpCH=C=CH$_2$ II (30%) [36], (52%) [69], VI (70%) [134]	b.p. 60°/5 × 10^{-3} Torr [36] ^{1}H NMR (CS$_2$): 3.94 (d, =CH$_2$, J=6.5), 4.8 (s, C$_5$H$_5$), 4.98 (t, –CH=), [69], 3.95 (d, =CH$_2$), 4.89 (t, –CH=, J=6.5) [36], see also [34] IR (neat): 1960, 2020 [36], 1970, 2020 (all CO), 1905 (C=C=C) [69]
167	FpCH^1=C=CH^{2}CH$_3$ II (30%) [36]	b.p. 80°/5 × 10^{-3} Torr [36] ^{1}H NMR (CS$_2$): 1.47 (q, CH$_3$, ^{5}J(H-1, CH$_3$)=3.2, ^{3}J (H-2, CH$_3$)=6.8), 4.40 (oct, H-2), 4.98 (oct, H-1, ^{4}J(H-1, 2)=5.8) [36] IR (neat): 1955, 2010 (both CO) [36]
*168	FpC(CF$_3$)=C=CF$_2$ II (13%) [23]	yellow–brown crystals, m.p. 39.5 to 40° (from pentane) [23, 24] ^{19}F NMR: 60.3 (t, CF$_3$, J(CF$_3$, CF$_2$)=4.0), 110.8 (q, CF$_2$) [23] IR (C$_6$H$_{12}$): 2003, 2049 (both CO), 1992 (ν_{as} C=C=C), other bands 650, 698, 762, 838, 849, 872, 1001, 1017, 1123, 1158, 1202, 1215, 1231, 1360, 1394, 1407 [23]
169	FpC(CH$_2$OH)=C=CH$_2$ II (66%) [164]	bright yellow needles, m.p. 75 to 76° [164] ^{1}H NMR (CDCl$_3$): 2.35 (t, OH, J=5.5, signal disappeared on addition of D$_2$O), 3.92 (dt, CH$_2$O, J=4.0, 5.5, signal collapsed to a triplet with J=4.0 on addition of D$_2$O), 4.28 (t, =CH$_2$, J=4.0), 4.88 (s, C$_5$H$_5$) [164], 2.15 (OH), 3.90 (CH$_2$O), 4.30 (=CH$_2$), 4.85 (C$_5$H$_5$) [204] IR (CHCl$_3$): 1975, 2027 (both CO) [164]

Table 5 (continued)

No.	compound method of preparation (yield)	properties and remarks
170	$FpC(=C=CH_2)CH_2OCH_3$ II (77%) [118]	amber oil [118] ^1H NMR (CS_2): 3.2 (s, OCH_3), 3.74 (t, CH_2, $J=2.5$), 3.95 (t, CH_2, $J=2.5$), 4.79 (s, C_5H_5) [118] IR (neat): 1960, 2020 (both CO) [118]
*171	$FpC[CH_2N(C_2H_5)_2]=C=C(C_6H_5)_2$	^1H NMR ($CDCl_3$): 1 and 2.5 (C_2H_5), 3.2 ($FeCCH_2$), 4.70 (C_5H_5), 7.1 to 7.4 (C_6H_5) [153] IR ($CHCl_3$): 1980, 2010 (both CO) [153]
172	$FpCH_2CH_2CH=C=CH_2$ I (75%) [44, 148], see also [62]	amber oil, air-sensitive [148] ^1H NMR: 1.45 (m, $FeCH_2$), 2.05 (m, $CH_2C=$), 4.66 (s, C_5H_5), 4.66 (m, $=CH_2$), 5.05 (m, $=CH-$) [148] IR: $\sim$1940, $\sim$2000 in C_5H_{12} [148], 1950, 2005 (all CO) [44]
*173	$FpCH_2CH_2C(CH_3)=C=CH_2$ I (60%) [44, 148], see also [62]	amber oil, air-sensitive [148] IR: 1950, 2005 (both CO) [44]
*174	$FpC[CH_2N(C_2H_5)_2]=C=C(CH_3)C_6H_5$	^1H NMR ($CDCl_3$): 1 and 2.6 (C_2H_5), 2.0 ($=CCH_3$), 3.2 (CH_2N), 4.70 (C_5H_5), 7.1 to 7.4 (C_6H_5) [153] IR ($CHCl_3$): 1980, 2010 (both CO) [153]
175	$FpCH_2CH_2CH=C=CHCH_3$ I (85%) [44, 148], see also [62]	amber oil, air-sensitive [148] ^1H NMR: 1.55 (m, $FeCH_2$, CH_3), 2.07 (m, $CH_2C=$), 4.66 (s, C_5H_5), 5 (m, 2H, $HC=C=CH$) [148] IR: $\sim$1940, $\sim$2000 in C_5H_{12} [148], 1950, 2005 (all CO) [44]

supplement

*176	$FpCH=C=O$	very unstable [144] ^1H NMR ($C_6D_5CD_3$, 10% C_6D_6): -0.16 (CH), 4.06 (C_5H_5) [144]
*177	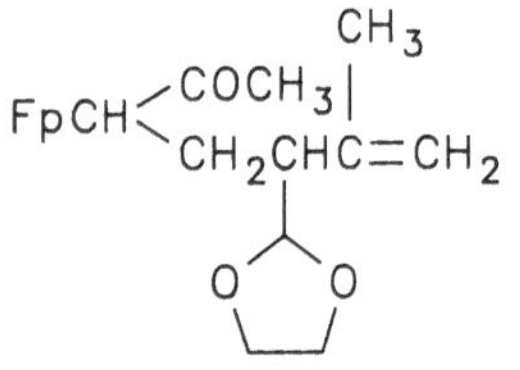 VI [192]	—
178	$FpCH(SC_6H_5)(CH_2)_4CH=CH_2$ I [199]	—

Table 5 (continued)

No.	compound method of preparation (yield)	properties and remarks

*179   FpC[=C(CN)$_2$]C(C$_6$H$_5$)=C(CN)$_2$ · 0.125 CH$_2$Cl$_2$

yellow crystals (from CH$_2$Cl$_2$/C$_5$H$_{12}$), m.p.
>202° (dec.) [210]
^{1}H NMR: 4.95 (s, C$_5$H$_5$), 5.30 (s, 0.25 H,
CH$_2$Cl$_2$), 7.60 (m, 5 H, C$_6$H$_5$) in CDCl$_3$ [210],
5.13 (s, C$_5$H$_5$) [211]
^{13}C NMR (CD$_2$Cl$_2$/CH$_2$Cl$_2$): 86.9 (s, C$_5$H$_5$),
110.5, 112.3, 115.6 (all: s, CN), 130.7 (m,
C$_6$H$_5$), 209.0, 211.0 (both: s, CO) [210]
IR: 2005, 2050 (both CO), 2226 (CN) in CH$_2$Cl$_2$,
662, 695, 741, 772, 826, 865, 1494, 1540
(C=C) in Nujol [210], 2005, 2050 (both CO)
in Nujol(?) [211]
mass spectrum: [M+H]$^+$ (45), [M]$^+$ (38),
[M+H−CO]$^+$ (9), 363 (9), [M−2CO]$^+$
(100), [M−2CO−CN]$^+$ (48),
[M+H−C$_2$(CN)$_4$]$^+$ (90),
[M+H−2CO-C$_6$H$_5$]$^+$ (48), 268 (26),
[Fe(C$_2$C$_6$H$_5$)C$_2$(CN)$_4$)]$^+$ (24) [210], [M]$^+$,
[M−CO]$^+$, [M−2CO]$^+$,
[M−2CO−C$_5$H$_5$]$^+$ (?),
[Fe(C$_2$C$_6$H$_5$)C$_2$(CN)$_4$]$^+$ [211]

*180   (Z,Z)-Fp(CH=CH)$_2$Br

red-brown needles (from pentane), m.p. 75
to 78° [213]
^{1}H NMR (C$_6$D$_6$): 4.0 (s, C$_5$H$_5$), 5.9 to 7.8 (m,
4 H, C$_4$H$_4$Br) [213]
IR (KBr?): 1970, 2000 (both CO) [213]

*Further information:

C$_5$H$_5$Fe(CO)$_2$CH=CH$_2$ (Table 5, No. 1). Formation of the compound by treatment of FpCl
with a THF solution of CH$_2$=CHMgCl (20% yield) [9], treatment of FpI with CH$_2$=CHMgX
(X = halogen) or LiCH=CH$_2$ (25 to 40% yield) [174], see also [198], or from [Fp(CH$_2$=CH$_2$)]$^+$
and CH$_2$=CHMgCl (25% yield) [120] is briefly mentioned. The compound forms in a 2%
yield by direct irradiation at room temperature for 18 h of a mixture obtained from Na[Fp]
and CH$_2$=CHCOCl in THF at −78 °C (via FpCOCH=CH$_2$). Workup by chromatography on
Al$_2$O$_3$ and subsequent sublimation always afford the complex contaminated with ferrocene
as indicated by ^{1}H NMR [8]. According to [174], where no details of reaction conditions
are given, this procedure gives overall yields of 60 to 90%.

In FpCH=CH$_2$ the stereoisomeric methylene protons may be unambiguously assigned
from their vicinal coupling constants. The proton H-3, cis to the Fp group, and therefore
closer to the C$_5$H$_5$-Fe axis in all conformations of the Fp-ligand bond, is found to be more
highly shielded than the trans proton H-2 [99]. A comparison of the ^{1}H NMR spectrum with
that of CH$_2$=CH$_2$ shows that the α-proton H-1 is more influenced by the iron atom (strong

shift to lower fields) than the β-protons H-2 and H-3 [19]. A comparison of the ^{1}H NMR spectra of FpCH=CH$_2$ and FpCOCH=CH$_2$ indicates that a vinyl group bonded directly to a transition metal may be distinguished from those bonded to a carbon atom with NMR [8].

FpCH=CH$_2$ reacts with in situ generated Fp$^+$ to give the bimetallic complex [Fp$_2$CHCH$_2$]$^+$. No more details are given [168].

(E)-C$_5$H$_5$Fe(CO)$_2$CH=CHC$_6$H$_5$ (Table 5, No. 2) has been obtained by UV irradiation of (E)-FpCOCH=CHC$_6$H$_5$ in THF for 9 h [48, 49], see also [154], or in acetone (reaction monitored by IR) [159] and workup by chromatography on Al$_2$O$_3$ with CH$_2$Cl$_2$/hexane (1:4) [49] or toluene [159] as eluent. Further purification by recrystallization from CH$_2$Cl$_2$/hexane [159] or by TLC on Al$_2$O$_3$ [48, 49] gives the product in a 34 or 15% yield, respectively. Decarbonylation of (E)-FpCOCH=CHC$_6$H$_5$ also takes place upon treatment with a stoichiometric amount of Rh(P(C$_6$H$_5$)$_3$)$_3$Cl in THF overnight. Similar workup as described above (however with toluene/CH$_2$Cl$_2$ (5:1) as eluent) affords a 14% yield of (E)-FpCH=CHC$_6$H$_5$ [159]. A yield of 32% results from the reaction of (E)-FpCOCH=CHC$_6$H$_5$ with (CH$_3$)$_3$NO (1.1 equivalents) [154].

The coupling constant J(H, H) = 16 Hz of the olefinic protons is the only evidence for the (E)-configuration [49]. The assignments of the olefinic protons are contradictory in [48, 49] and [154].

The compound is unstable and decomposes at room temperature to give products which include the decomposition products of ferrocene [49].

C$_5$H$_5$Fe(CO)$_2$C(C$_6$H$_5$)=CH$_2$ (Table 5, No. 3) has been prepared by the reaction of FpI with CH$_2$=C(C$_6$H$_5$)MgX (X = halogen) or LiC(C$_6$H$_5$)=CH$_2$ (25 to 40% yields) or by acylation of Na[Fp] with CH$_2$=C(C$_6$H$_5$)COCl followed by photochemical decarbonylation (90% yield) [174]. The resulting product is dissolved in CH$_2$Cl$_2$ and purified by flash chromatography with pentane/CH$_2$Cl$_2$ (4:1) as eluent. A detailed description of the chromatographic technique, which has been developed especially for air-sensitive compounds, is given [191].

C$_5$H$_5$Fe(CO)$_2$C(C$_6$H$_5$)=CHC$_6$H$_5$ (Table 5, No. 4) is formed in low yield along with the main product Fp$_2$ by stirring a mixture of [Fp(C$_6$H$_5$C≡CC$_6$H$_5$)]BF$_4$ and 1.25 equivalents of NaBH$_3$CN in THF at 0 °C for 1 h. After solvent removal at 0 °C the residue is extracted with benzene, and the filtered and concentrated extracts are chromatographed on Al$_2$O$_3$. Elution of the first band with benzene and subsequent evaporation of the solvent afford the compound as an oil [149].

C$_5$H$_5$Fe(CO)$_2$CH=CHC$_6$F$_4$H-4 (Table 5, No. 5) has been prepared by treatment of Na[Fp] with C$_6$F$_5$C≡CH in THF at room temperature for 20 h. After filtration and solvent removal under reduced pressure the residue is dissolved in CH$_2$Cl$_2$ and chromatographed on Florisil. Elution with light petroleum (b.p. 40 to 60 °C)/benzene (6:1) gives a fraction which, after evaporation and sublimation at 100 °C/10^{-2} Torr, affords the product in a 37% yield. The mechanism of formation is discussed [17].

According to [20] the ^{19}F NMR spectrum consists of two resonances (both at δ = 138.4 ppm) which analyze as an AA′XX′ system.

The mass spectrum shows the molecular ion at m/e = 352 and peaks due to loss of two CO groups at m/e = 324 and 296 [17].

C$_5$H$_5$Fe(CO)$_2$CF=CF$_2$ (Table 5, No. 6) is isolated in a 5% yield from the reaction mixture containing Na[Fp] and CF$_2$=CFCl (Method I) by solvent removal at 0 °C/30 Torr and crystallization of the residue from light petroleum (b.p. 40 to 60 °C) at −78 °C [13], see also [12]. A yield of 20% is obtained when Na[Fp] is allowed to react with CF$_2$=CF$_2$ instead of CF$_2$=CFCl in THF at −196 °C followed by warming the mixture to −78 °C and after 4 h, allowing to reach room temperature overnight. Subsequent workup by chromatography on Florisil using benzene/light petroleum, b.p. 60 to 80 °C (3:1), as eluent gives the product [10]. The compound forms in a 4% yield by treatment of Fp_2Hg with $CF_2=CF_2$ at 130 °C/15 atm for 4 h followed by chromatography and repeated low temperature crystallization from pentane [11].

According to [11] the ^{19}F NMR spectrum shows resonances at $\delta = 84.1$ ppm (1F, quartet, peak separation 115 Hz) and in the region 130 to 140 ppm (2F, 8 bands) and may be assigned to an ABX system.

In [10] it is reported that the compound appears to be indefinitely stable on exposure to air, but it decomposes in solution. It is not decomposed by water and sublimes very easily. However, according to [11] the complex is unstable in air and to light and distils in vacuum at 60 °C with decomposition.

Below 1900 cm^{-1} the IR spectrum exhibits four main bands in regions attributable to the C=C stretch, the $=CF_2$ asymmetric stretch, the $=CF$ stretch, and the $=CF_2$ symmetric stretch vibrations [10].

$C_5H_5Fe(CO)_2C(OCH_3)=CH_2$ (Table 5, No. 7) has been prepared in 87% yield by addition of $KOC(CH_3)_3$ in THF to a solution of $[Fp=C(OCH_3)CH_3]BF_4$ (1.05:1 mole ratio) in CH_2Cl_2 at −78 °C, evaporation of the solvent under vacuum below 0 °C, extraction of the residue with hexane, filtration through Celite and solvent removal. The compound is obtained in 55% yield as a mixture with $FpC(OCH_3)_2CH_3$ (39%) and $FpCOCH_3$ (6%) by treatment of $[Fp=C(OCH_3)CH_3]BF_4$ in CH_2Cl_2 with a suspension of $NaOCH_3$ in THF at −78 °C. Extraction of the reaction residue with hexane gives an oil containing the aforementioned substances as shown by ^1H NMR spectroscopy. Attempted separation of $FpC(OCH_3)_2CH_3$, either by vacuum sublimation (60 °C/10^{-5} Torr) or by chromatography on Al_2O_3, led to decomposition [179]. A 1:1 mixture of $FpC(OCH_3)=CH_2$ and $FpC(CH_3)_2OCH_3$ results from the reaction of $[Fp=C(OCH_3)CH_3]BF_4$ with $LiCH_3$ in CH_2Cl_2 at −78 °C. The compound also forms in small, variable amounts along with $FpC(CH_3)_2OCH_3$ (45 to 50%) on treatment of $[Fp=C(OCH_3)CH_3]BF_4$ with $LiCu(CH_3)_2$ in $CH_2Cl_2/O(C_2H_5)_2$ at −78 °C [176, 201].

$C_5H_5Fe(CO)_2C(OLi)=CH_2$ (Table 5, No. 8) is formed by addition of $FpCOCH_3$ in THF to $LiN(Si(CH_3)_3)_2$ in the same solvent at −63 °C [197]. Its formation from $FpCOCH_3$ and $LiN(C_3H_7-i)_2$ in THF is briefly mentioned in [183].

The enolate is reasonably stable and persists in solution up to ~ -50 °C. It may be precipitated from THF by addition of ether, hexane, or other solvents of low polarity [197].

The compound reacts with electrophilic reagents such as XR^1 ($R^1 = CH_3$, i-C_3H_7, n-C_4H_9, $CH_2CH=CH_2$, $CH_2C(CH_3)=CH_2$, $CH_2C_6H_5$, $CH_2C_6H_4OCH_3$-4; X = Br, I, OSO_2CF_3), with ethyloxirane in the presence of $BF_3 \cdot O(C_2H_5)_2$, or with propanal in THF at −78 to −20 °C to give the products $FpCOCH_2R^1$, $FpCOCH_2CH_2CH(OH)C_2H_5$, and $FpCOCH_2CH(OH)C_2H_5$, respectively (cf. Preparation Method V in Section 1.5.2.3.16.1.7 on p. 8). The use of hexamethylphosphoric triamide or tetramethylethylenediamine as a cosolvent leads to faster rates of reaction but to no improvements in the yields of the products. When less reactive alkylating agents (e.g. primary iodides and tosylates) are employed, decomposition of the enolate becomes competitive with the substitution reactions. Obtained as the final products in these cases are the compounds resulting from alkylation at the iron nucleus [197].

$(E)-C_5H_5Fe(CO)_2CH=CH(CH_2)_3CH=CH_2$ (Table 5, No. 9) has been prepared from 1-iodo-hepta-1,6-diene with 2 mole LiC_4H_9-t followed by 1 mole FpI [174], see also [199].

For the reaction with HBF_4, see "General Remarks" (cyclopropanation reactions) on p. 66.

References on pp. 141/7

C₅H₅Fe(CO)₂CH=C(CN)₂ $C_5H_5Fe(CO)_2CH=C(CN)_2$ (Table **5**, No. **10**) has been purified after chromatography on Al_2O_3 (Method I) by recrystallization from mixtures of CH_2Cl_2 and hexane [67].

Studies of the ^{13}C NMR data of $FpCH=C(CN)_2$ and related compounds revealed that introduction of a transition metal group into $(NC)_2C=CHCl$ causes a large downfield shift (50 to 60 ppm) of the resonance of the olefinic carbon atom to which the metal is bonded. A correlation between the ^{13}C NMR chemical shifts and the $\nu(CN)$ frequencies in cyanoolefins and their transition metal derivatives is pointed out [76]. The ^1H NMR and IR data are discussed in connection with the bonding character of the title compound in [67].

The compound is relatively air–stable. It can be sublimed under vacuum (0.1 Torr) at 100 °C with some decomposition. This instability upon such vacuum sublimation prevents the measurement of a satisfactory mass spectrum. Instead trivial decomposition products such as $Fe(C_5H_5)_2$ are obtained [67].

(E)-C₅H₅Fe(CO)₂CH=CHCH₃ $(E)-C_5H_5Fe(CO)_2CH=CHCH_3$ (Table **5**, No. **11**). As described for $(E)-FpCH=CHC_6H_5$ (see No. 2), the compound has been obtained by photochemical and chemical (with $Rh(P(C_6H_5)_3)_3Cl$) decarbonylation of $(E)-FpCOCH=CHCH_3$. However, the product is further purified by vacuum sublimation onto a dry ice/acetone cooled probe. The yields of the resulting slightly impure oil are 57 and 15%, respectively [159].

C₅H₅Fe(CO)₂C(CH₃)=CH₂ $C_5H_5Fe(CO)_2C(CH_3)=CH_2$ (Table **5**, No. **12**) has been prepared in two steps by the reaction of Na[Fp] with $CH_2=C(CH_3)COCl$ (1:1 mole ratio) in THF at 0 °C followed by photochemical decarbonylation of $FpCOC(CH_3)=CH_2$ formed in hexane/toluene for 2 h, 69% yield [176], see also [174]. After solvent removal in vacuum the crude product is purified by distillation at 39 to 42 °C/10^{-3} Torr, 67% yield [201]. The compound has also been obtained by heating $FpC(CH_3)_2OCH_3$ in benzene at 60 to 65 °C for 8 h, evaporation of the solvent under vacuum and distillation of the oily residue at 50 to 55 °C/10^{-3} Torr, 80% yield [176, 201]. Another method, consisting of the reaction of FpI with $CH_2=C(CH_3)MgX$ (X = halogen) or $LiC(CH_3)=CH_2$ (25 to 40% yield), is mentioned in [174], see also [198].

In contrast to $FpC(C_6H_5)=CH_2$ (see No. 3), which requires a special procedure for purification due to its air–sensitivity, $FpC(CH_3)=CH_2$ is sufficiently stable to permit routine operations in the air [174].

C₅H₅Fe(CO)₂C(CH₃)=CHC₆H₅ $C_5H_5Fe(CO)_2C(CH_3)=CHC_6H_5$ (Table **5**, No. **13**). The oil remaining after concentration of the pentane eluate (Method VII) solidifies within 1 h. The ^1H NMR spectrum of the solid indicated a ~4:1 mixture of geometric isomers [91], see also [57]. Several more chromatographic treatments on Al_2O_3, discarding the very front of the eluted yellow band, and two crystallizations from pentane, discarding the more soluble isomer, eventually gave yellow–orange needles of the title compound as a ~20:1 mixture of isomers [91].

(Z)-C₅H₅Fe(CO)₂CH=CHCF₃ $(Z)-C_5H_5Fe(CO)_2CH=CHCF_3$ (Table **5**, No. **14**) has been prepared by treatment of Na[Fp] with $CF_3C\equiv CH$ in THF at −78 °C for 5 h and for a further 5 h at room temperature. After filtration and solvent removal under reduced pressure, the residue is dissolved in CH_2Cl_2 and chromatographed on Florisil with light petroleum (b.p. 40 to 60 °C)/benzene as eluent. The resultant yellow–orange oil is purified by distillation at 40 °C/10^{-3} Torr, 92% yield. The mechanism of formation is discussed [17]. The compound forms in a 5% yield upon irradiation of $FpSn(CH_3)_3$ in the presence of an excess of $CF_3C\equiv CH$ in hexane at 25 °C for 30 h in a sealed tube and subsequent workup after removal of the reaction volatiles by chromatography on SiO_2 [40], see also [31]. For the formation as mixture with $FpC(CF_3)=CH_2$, see No. 15.

The compound is formulated as the (Z)–isomer on the basis of the coupling constants $J(H,H)$ and $J(CF_3,H)$. The spectral data are compared with those of $Re(CO)_5CH=CHCF_3$ [17].

The compound decomposes at room temperature even in vacuum [17]. The mass spectrum shows the molecular ion $[M]^+$, the fragments $[M-CO]^+$ and $[M-2CO]^+$ [17] and ions resulting from elimination of FeF_2 or C_5H_5FeF from the latter fragment. The last step indicates that fluorine transfer to the iron atom may occur during the fragmentation [60]. Also observed are peaks at $m/e = 149$, 121, and 95, characteristic of the Fp group [17].

$C_5H_5Fe(CO)_2C(CF_3)=CH_2$ (Table **5**, No. **15**) is obtained as a mixture with 30% (Z)-FpCH= $CHCF_3$ by the reaction of FpH, prepared from Fp_2, with $CF_3C\equiv CH$ in benzene at room temperature for 4 d in a sealed Pyrex bulb. After removal of gaseous material the product is chromatographed on Florisil with benzene/light petroleum, b.p. 60 to 80 °C (1:1), as eluent. Further purification is performed by repeated distillation at 50 °C/10^{-3} Torr into a receiver cooled to -78 °C giving the above mentioned mixture as a yellow liquid in a 10% yield [26, 27].

The high resolution mass spectrum showing the molecular ion at $m/e = 272$ has been found to be very similar to that reported [17] for $FpCH=CHCF_3$ [26].

(E)-$C_5H_5Fe(CO)_2CF=CFCF_3$ (Table **5**, No. **16**). After chromatography (Method I) the compound is purified by repeated sublimation at 40 °C/0.1 Torr onto a cooled (0 °C) probe [2]. Although Method I has been carried out with $CF_2=CFCF_2Cl$, the compound contains only a perfluoroprop-1-enyl and not a perfluoroallyl group [2, 3]. A higher yield is obtained (56%) when $CF_2=CFCF_3$ is used in place of $CF_2=CFCF_2Cl$ (4.6%) [10].

The assignment of the ^{19}F NMR resonances is made by comparison with spectra of fluorocarbons and is supported by the spin coupling constants, which also suggest the (E)-configuration of the compound [4].

Below 1900 cm^{-1} the IR spectrum shows the same feature as $FpCF=CF_2$ (see No. 6) [10].

The compound is readily soluble in organic solvents [2]. It appears to be indefinitely stable on exposure to air. However, it decomposes in solution, but is not decomposed by water. The compound sublimes very easily [10].

$C_5H_5Fe(CO)_2CF=CClCF_3$ (Table **5**, No. **17**) is predominantly obtained as the (Z)-isomer from the reaction of Na[Fp] with $CF_3CCl=CFCl$ (Method I). The compound has been characterized solely by its spectral properties. The tentative identification of a small percentage (7%) of (E)-isomer rests on the observation of a weak doublet in the ^{19}F NMR spectrum in the region expected for a CF_3-C=C group, and with a splitting consistent with a trans-$CF_3C=CF$ arrangement. The quartet, which should be observed for the CF group, could not be found, perhaps because of signal-to-noise ratio problems. Method I also gives only the rearranged complex and no allyl compound, indicating that $FpCF=CClCF_3$ is thermodynamically the more stable product [30].

$C_5H_5Fe(CO)_2C(CH_2I)=CHC_6H_5$ (Table **5**, No. **18**). Concentration of the CH_2Cl_2 eluate (Method VII) gives a brown tar, characterized as the title compound by IR spectroscopy. The unstable material decomposes spontaneously at room temperature to FpI (proved by mass spectrometry at 60 °C) which is also the main product from the reaction when using Method VII [91].

$C_5H_5Fe(CO)_2C(CH_2OH)=CH_2$ (Table **5**, No. **19**) has also been obtained by addition of an ether solution of $FpC(CH_2OCOCH_3)=CH_2$ to a suspension of $LiAlH_4$ in the same solvent at 0 °C. After 30 min the reaction is quenched successively with H_2O, 15% NaOH, and H_2O. The mixture is filtered and the ether solution is separated and dried. After solvent removal the residue is taken up in petroleum ether and chromatographed on Al_2O_3 with ether/petroleum ether (3:2 to 4:1) to give 35% of the product. This was further purified by crystallization from ether/petroleum ether [134].

Treatment of the compound with LiC_4H_9-t at $-20\,°C$, followed by methylation of the anion formed with $[O(CH_3)_3]BF_4$ gives a $3:1$ mixture of $FpC(CH_2OCH_3)=CH_2$ and XC [122].

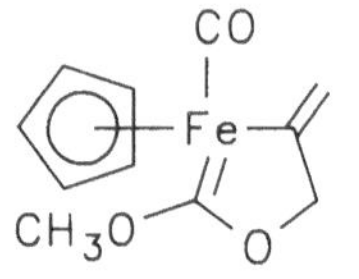

XC

$C_5H_5Fe(CO)_2C(CH_2OH)=CHC_6H_5$ (Table **5**, No. **20**). The oil remaining after concentration of the CH_2Cl_2 eluate (Method VII) is crystallized from CH_2Cl_2 by slow addition of pentane and concentration of the mixture [91]. The compound has also been obtained by hydration of $FpCH_2C\equiv CC_6H_5$ in pentane on Al_2O_3. Elution with CH_2Cl_2 and concentration of the eluate gives an oil which converts into a solid by washing with pentane, 15% yield [91].

$C_5H_5Fe(CO)_2C(CH_2OCOCH_3)=CH_2$ (Table **5**, No. **25**) is reduced to $FpC(CH_2OH)=CH_2$ upon treatment with $LiAlH_4$ in ether at $0\,°C$ for 30 min [134].

(Z)$-C_5H_5Fe(CO)_2C(CH_2OCOCH_3)=CHC_6H_5$ (Table **5**, No. **26**) has been prepared by stirring a solution of $FpCH_2C\equiv CC_6H_5$ in glacial acetic acid for 70 min, dilution with pentane, and neutralization with aqueous $NaHCO_3$. The pentane layer is dried ($MgSO_4$), filtered, and chromatographed on Al_2O_3 with pentane as eluent. Concentration of the eluate gives an oil (25% yield) contaminated with a material (as indicated by ^{1}H NMR spectroscopy) which could not be removed by further chromatography. The compound probably has the (Z)-configuration. This is expected from the formation reaction, and spectroscopic data are not contradictory [91].

$C_5H_5Fe(CO)_2C(=CHC_6H_5)CH_2N(C_2H_5)_2$ (Table **5**, No. **27**). Attempts to crystallize the oil prepared according to Method VII have been unsuccessful [91].

$C_5H_5Fe(CO)_2C(N(CH_3)_2)=C(CH_3)_2$ (Table **5**, No. **29**). The reaction of Na[Fp] with $(CH_3)_2C=CClN(CH_3)_2$ (Method I) gives an air-sensitive, brown liquid which was shown by its ^{1}H NMR spectrum to be a mixture. Attempts to separate this mixture by chromatography failed, owing to decomposition on the column. Vacuum distillation (45 to $55\,°C/0.5$ Torr for 2 d) removed most of the impurities. However, repeated distillation (43 to $47\,°C/0.5$ Torr) introduced a second component, suggested (^{1}H NMR) to be XCI. Integration of the C_5H_5 resonances indicated the presence of 70% No. 29 and 30% XCI [90], see also [89].

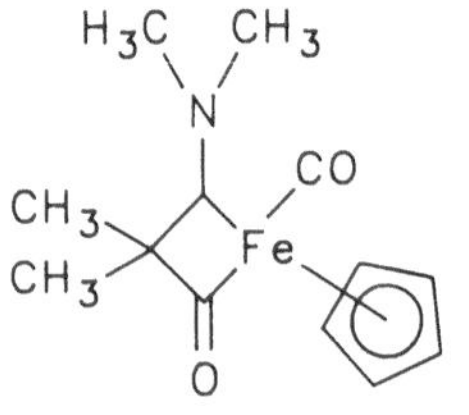

XCI

(E)$-C_5H_5Fe(CO)_2CH=CHCOC_6H_5$ (Table **5**, No. **31**) is purified after chromatography (Method I) by recrystallization from petroleum ether (b.p. 40 to $60\,°C$) at $-70\,°C$ [16].

References on pp. 141/7

Studies of the reaction with cyclopentadiene have shown that even heating an equimolar mixture of the reactants at 80 to 90 °C in benzene for several hours does not lead to the formation of an adduct. It is assumed that the electron-donating properties of the Fp group prevents the occurrence of the expected Diels–Alder reaction. In addition, the steric hindrance of the Fp moiety may also deactivate the olefinic bond of the dienophile [29].

(E)-$C_5H_5Fe(CO)_2CH=CHCOC_6H_4CH_3$-4 (Table 5, No. **32**) is purified like No. 31.

Addition of $[O(C_2H_5)_3]BF_4$ to the compound in CH_2Cl_2 leads to color change of the mixture from yellow to dark red. Further treatment with $C_6H_5NH_2$ followed by addition of $N(C_2H_5)_3$ or ethanolic NaOH affords the anil (E)-FpCH=CHC(=NC_6H_5)C_6H_4CH_3-4 (No. 34) along with Fp_2. It is assumed that the cationic complex $[Fp=CHCH=C(OC_2H_5)C_6H_4CH_3-4]^+$ is first formed (although it could not be isolated), because the reaction with aniline does not take place without prior ethylation of FpCH=CHCOC_6H_4CH_3-4 [140].

(E)-$C_5H_5Fe(CO)_2CH=CHCOC_6H_4Br$-4 (Table 5, No. **33**). The ^{1}H NMR spectrum shows the protons of the phenyl ring as a multiplet, characteristic of a system of the AA'BB' type [61].

(Z)-$C_5H_5Fe(CO)_2C(OCOCH_3)=CHCH_3$ (Table 5, No. **39**) is formed by treatment of FpC≡CCH_3 with anhydrous acetic acid in refluxing cyclohexane [28].

The compound is rapidly hydrolized to the acyl complex $FpCOC_2H_5$ by ethanolic KOH [28].

(E)- and **(Z)-$C_5H_5Fe(CO)_2CF=C(CF_3)COOC_2H_5$** (Table 5, Nos. **42** and **43**). The reaction mixture resulting from Na[Fp] and $CF_2=C(CF_3)COOC_2H_5$ (Method I) is worked up by evaporation of the solvent, extraction of the residue with hexane and thin layer chromatography of the concentrated extracts on Al_2O_3 with benzene/petroleum ether (2:1) as eluent. Two products, which are the (Z)- and (E)-isomers, are isolated in equal amounts [127].

The spin-spin coupling constant $(J(CF_3, F) = 21.1$ Hz) indicates that the compound is the (E)-isomer [127], and an X-ray analysis confirms this structure [146].

The (E)-compound crystallizes in the monoclinic system, space group $P2_1/a - C_{2h}^5$ (No. 14) with a = 14.629(2), b = 12.018(1), c = 8.2800(7) Å, and β = 97.23(1)°; Z = 4, d_c = 1.67 and d_m = 1.65 g · cm^{-3}. The Fe atom has a "piano-stool" coordination. The Fe–C(vinyl) σ-bond is shortened in comparison with the sum of single-bond covalent radii. This shortening is due to the participation of a vicinal F atom in hyperconjugation promoting an increase in the order of the Fe–C bond as result of the donor reaction d_π–σ*, with which there is a transfer of electrons from unbonded d orbitals of the metal atom to σ*-antibonding orbitals of the C–F bond. The participation of the vinyl F atom in the hyperconjugation leads, in addition, to a certain increase in the =C–F bond length to 1.40(2) Å, in comparison with the remaining three bonds (mean value 1.32(2) Å). The structure is shown in **Fig. 2**, p. 120 [146].

(E)- and **(Z)-$C_5H_5Fe(CO)_2CH=CHCONH_2$, (E)-$C_5H_5Fe(CO)_2CH=CHCON(CH_2)_5$** (Table 5, Nos. **44** to **46**). Treatment of these compounds with $[O(C_2H_5)_3]BF_4$ (1:1 mole ratio) at room temperature for 30 min leads to ethylation giving the imidates $[FpCH=CHC(OC_2H_5)NR_2^1]BF_4$ ($R_2^1 = H_2$ or $(CH_2)_5$), see Section 1.5.2.3.19 [166].

(Z)-$C_5H_5Fe(CO)_2C(CF_3)=CHSCH_3$ (Table 5, No. **47**) has been obtained in 60% yield by UV irradiation of compound XCII in THF for 300 h and workup by chromatography on Florisil with CH_2Cl_2/hexane as eluent [145].

The coupling constant $^4J(F, H) = 2.1$ Hz for CF_3 and H is consistent with a (Z)-configuration [145].

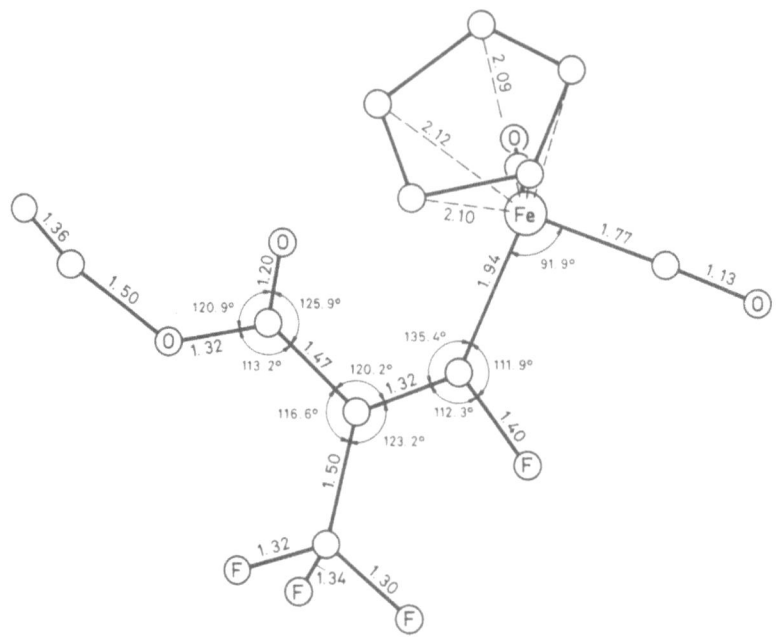

Fig. 2. Molecular structure of (E)−C₅H₅Fe(CO)₂CF=C(CF₃)COOC₂H₅ (No. 42) with selected
bond lengths (in Å) and angles.

The compound is stable in the air in the solid state [145]. Heating in the presence
of an excess CF₃C≡CH affords XCIII and photolysis of No. 47 alone or in the presence
of an excess CF₃C≡CH gives XCIV [145].

$$\text{XCII} \qquad \text{XCIII} \qquad \text{XCIV}$$

C₅H₅Fe(CO)₂C(CH₃)=CHCH₃ (Table **5**, No. **50**) is formed in low yield from [Fp(CH₃C≡
CCH₃)]BF₄ and NaBH₃CN by the same procedure as already described for FpC(C₆H₅)=CHC₆H₅
[149], see No. 4.

C₅H₅Fe(CO)₂C(C₂H₅)=CH₂ (Table **5**, No. **51**) has been obtained in 65% yield from [Fp(CH₂=
C=CH₂)]X and LiCu(CH₃)₂ under conditions as described for (E)−FpCH₂CH=CHC₂H₅ [120],
see No. 133.

(Z)−C₅H₅Fe(CO)₂C(CF₃)=CHCH₃ and **(Z)−C₅H₅Fe(CO)₂C(CF₃)=CHCH₂C₆H₅** (Table **5**, Nos. **52**
and **53**) are formed along with pyran-derived complexes (figured in Table 5, see under

Nos. 52 and 53) and XCV on UV irradiation of FpR ($R = R^1 = CH_3$, $CH_2C_6H_5$) in the presence of $CF_3C\equiv CH$ in hexane for 14 to 15 d (cf. Method IVa). After solvent removal in vacuum the residues are chromatographed on Al_2O_3. Elution of the yellow bands with hexane affords chromatographically inseparable mixtures of No. 52 with the methyl-substituted pyran-complex (ratio 1:7.5), or of No. 53 with the benzyl-substituted pyran-complex (ratio 1:8). Repeated attempts to separate these materials (e.g. crystallization from hexane at −78 °C) have been unsuccessful [158].

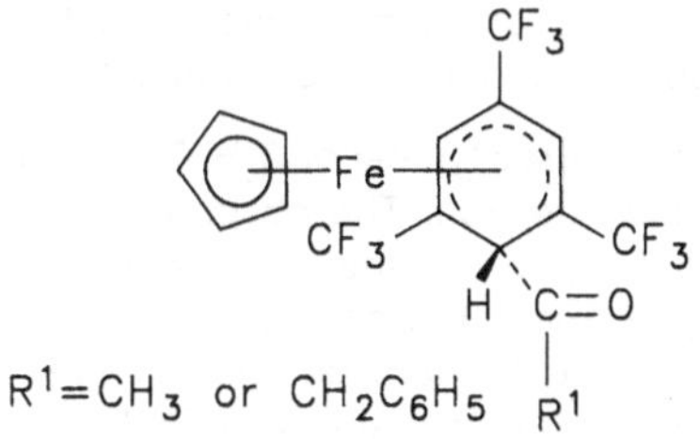

$R^1 = CH_3$ or $CH_2C_6H_5$

XCV

The spectroscopic data for the mixtures given in Table 5 have been analyzed in terms of the respective structures. On the basis of the coupling constants (NMR spectra) a (Z)-structure is proposed for Nos. 52 and 53 [158].

(E)-$C_5H_5Fe(CO)_2C(CF_3)$=$CHCF_3$ (Table 5, No. **54**). The reactions of $FpCH_2CH=CHCH_3$ or $FpCH_2CH=C(CH_3)_2$ with $CF_3C\equiv CCF_3$ in $CHCl=CCl_2$ at 60 to 65 °C for 12 or 1.5 h, respectively, afford a mixture of products including No. 54, which could be separated by chromatography on Al_2O_3 [129].

The assignment of an (E)-orientation is based on a comparison of the values $J(H, CF_3)$ and $J(CF_3, CF_3)$ for No. 54 and related compounds [129].

The compound rapidly sublimes at 45 °C/0.1 Torr [129].

(Z)-$C_5H_5Fe(CO)_2C(CF_3)$=$CHCF_3$ (Table 5, No. **55**) forms along with the main product XCVI and some $FpC(CF_3)=C(CF_3)Sn(CH_3)_3$ on irradiation of $FpSn(CH_3)_3$ in the presence of an excess of $CF_3C\equiv CCF_3$ in hexane at 76 °C for 40 h in a sealed tube and workup of the mixture by removal of the reaction volatiles, extraction of the residue with hexane and chromatography of the extracts on SiO_2 with a hexane/benzene mixture as eluent [40].

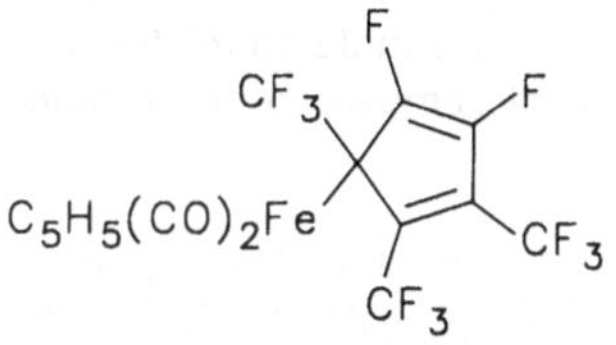

XCVI

The ^1H NMR spectrum shows a singlet (C_5H_5) and a quartet of quartets with $J_1 = 8.5$ and $J_2 = 1.5$ Hz. The ^{19}F NMR spectrum exhibits a pair of quartets and another overlapping pair of quartets. This is consistent with a (Z)-configuration with geminal CF_3-H coupling of 8.5 Hz, cis CF_3-H coupling of 1.5 Hz and trans CF_3-CF_3 coupling of 2 Hz. For the (E)-structure the trans CF_3-H coupling should be close to zero. Resonance data are not given [40].

References on pp. 141/7

(E)-C₅H₅Fe(CO)₂C(CF₃)=CFCF₃ (Table 5, No. **56**) has been obtained in 71% yield by the reaction of FpI with CF₃CF=C(CF₃)Ag (1:1.25 mole ratio) in CH₂Cl₂ for 5 h, filtration of the precipitated AgI, solvent removal from the filtrate at 25 °C/25 Torr, and isolation of the product from the residue by sublimation at 40 °C/0.05 Torr. A yield of only 11% results from the analogous reaction in THF (1 h) [59].

The CF₃-CF₃ coupling [⁵J(F,F)] has not been observed. It is ∼1.7 Hz for related compounds, whereas compounds with a cis-configuration of the CF₃ groups exhibit a ⁵J(F,F) of 12 to 15 Hz. Thus, the coupling constants indicate that the two CF₃ groups are in a trans position about the carbon–carbon double bond [59].

The mass spectrum shows features typical of C₅H₅Fe(CO)₂ derivatives, particularly those with fluorocarbon groups bonded to iron. The observed metastable ions indicate that the molecular ion can fragment to [C₅H₄Fe]⁺ by the successive loss of its two CO groups followed by elimination of a neutral C₄F₆ fragment (presumably CF₃C≡CCF₃) to give [C₅H₅FeF]⁺ which can then undergo loss of a neutral HF molecule producing [C₅H₄Fe]⁺. All fragments, their relative intensities and the metastable ions are given [59].

(Z)-C₅H₅Fe(CO)₂C(CF₃)=C(CF₃)SCF₃ and **(Z)-C₅H₅Fe(CO)₂C(CF₃)=C(CF₃)SC₆F₅** (Table 5, Nos. **57** and **58**). The coupling constants ⁵J(F,F) for the two CF₃ groups, 15.8 Hz for No. 57 and 16.0 Hz for No. 58, are comparable with values for other CF₃C=CCF₃ derivatives with cis-structure for both CF₃ groups rather than for a trans-structure (2 Hz) [87].

The ν(CO) regions of the IR spectra of Nos. 57 and 58 are more complicated than would be expected for simple dicarbonyl species and in particular the spectrum of (Z)-FpC(CF₃)= C(CF₃)SCF₃ is strongly temperature dependent. The spectra are consistent with the presence of rotational isomers due to preferred conformations of the C(CF₃)=C(CF₃)SR¹ (R¹ = CF₃ or C₆F₅) ligands [87].

C₅H₅Fe(CO)₂C(CF₃)=C(CF₃)Sn(CH₃)₃ (Table 5, No. **61**). The mass spectrum shows the molecular ion and fragments resulting from successive loss of the CH₃ groups [15].

(E)-C₅H₅Fe(CO)₂CH=CHC(CH₃)₂OH (Table 5, No. **62**) has been prepared by stirring (E)-FpCH=CHCOCH₃ and LiCH₃ (1:1.15 mole ratio) in ether at 0 °C for 1 h, then quenching the mixture with H₂O, evaporation of the ether layer, and extraction of the oily residue with toluene. Chromatography of the concentrated extracts on Al₂O₃ with hexane and ether as eluents gives the product in 62% yield [201], see also [193].

(Z)-C₅H₅Fe(CO)₂C(CH₂OH)=CHCH₃ (Table 5, No. **63**) forms by stirring a CH₂Cl₂ solution of FpCH₂C≡CCH₃ and CF₃C≡CCF₃ (1:1.2 mole ratio) for 5 h under a dry ice condenser and workup of the mixture after solvent removal by chromatography on Al₂O₃ (6% H₂O) with CH₂Cl₂ as eluent [129].

C₅H₅Fe(CO)₂C(=CH₂)CH(OH)CH₃ (Table 5, No. **64**) is obtained as 1:2 mixture with FpCH(CH₃)COCH₃ on treatment of (E)-[Fp(CH₂=C=CHCH₃)]BF₄ with D₂O in acetone and subsequent chromatography on Al₂O₃ [134].

(Z)-C₅H₅Fe(CO)₂C(CH₂OCH₃)=CHCH₃ (Table 5, No. **65**). The addition of methoxide in methanol to the allene salts [Fp(CH₂=C=CHR¹)]BF₄ (cf. Method VII) gives principally (when R¹ = C₆H₅), or exclusively (R¹ = CH₃, (Z)-compound) the ketones FpCH₂COCH₂R¹ [57, 91]. However, when R¹ = CH₃ and the reagents are carefully dried, (Z)-FpC(CH₂OCH₃)=CHCH₃ may be obtained in a 92% yield. When the reaction is carried out in 10% aqueous methanol, FpCH₂COC₂H₅ becomes the major product. The chemical shift of the vinylic proton is in accord with the (Z)-configuration [104].

References on pp. 141/7

A mixture of (Z)- and (E)-FpC(CH$_2$OCH$_3$)=CHCH$_3$ and FpCH$_2$COC$_2$H$_5$ (ratio 1:1:4, identified by ^{1}H NMR) is obtained when an equilibrium mixture of XCVII and XCVIII (X=BF$_4$) is dissolved in 10% aqueous methanol at 0 °C followed by addition of Na$_2$CO$_3$. The resulting solution is stirred at 0 °C/1 h, then poured into water and extracted with ether [104].

Treatment of (Z)-FpC(CH$_2$OCH$_3$)=CHCH$_3$ with HBF$_4$ · O(C$_2$H$_5$)$_2$ in anhydrous ether at 0 °C regenerates XCVII (X=BF$_4$) [104].

XCVII   XCVIII

C$_5$H$_5$Fe(CO)$_2$C(=CHCH$_3$)CH$_2$N(C$_2$H$_5$)$_2$ (Table **5**, No. **67**). The oil obtained on concentration of the pentane eluate (Method VII) could not be induced to crystallize [91].

(E)-C$_5$H$_5$Fe(CO)$_2$CH=CHCOCH$_3$ (Table **5**, No. **68**) is purified after chromatography (Method I) by recrystallization from petroleum ether (b.p. 40 to 60 °C) at −70 °C [16].

The compound reacts with 2,4-(O$_2$N)$_2$C$_6$H$_3$NHNH$_2$ in alcohol at room temperature (3.5 h) to give the hydrazone FpCH=CHC(CH$_3$)=NNHC$_6$H$_3$(NO$_2$)$_2$-2,4 [39]. Boiling the compound with the more nucleophilic 4-O$_2$NC$_6$H$_4$NHNH$_2$ in glacial acetic acid and leaving the mixture at room temperature for 1 d afford 1-(4-nitrophenyl)-3-methylpyrazole. The presumably first-formed 4-nitrophenylhydrazone could not be isolated [39].

Treatment of No. 68 with C$_6$H$_5$SH in boiling benzene for 9 h or with C$_6$H$_5$SNa, formed in situ from C$_6$H$_5$SH and NaOCH$_3$ in CH$_3$OH, in boiling CH$_3$OH for 8 h causes formation of Fp$_2$ and CH$_3$COCH=CHSC$_6$H$_5$, whereas (C$_5$H$_5$Fe(CO)SC$_6$H$_5$)$_2$ and traces of CH$_3$COCH=CHSC$_6$H$_5$, but no Fp$_2$, result from the reaction with C$_6$H$_5$SSC$_6$H$_5$ in boiling benzene for 42 h. The latter reaction is assumed to proceed by a radical mechanism [39].

Addition of [O(C$_2$H$_5$)$_3$]BF$_4$ to No. 68 in CH$_2$Cl$_2$ leads to color change from yellow to dark red. However, it has not been possible to isolate the ethylation product [Fp=CHCH= C(OC$_2$H$_5$)CH$_3$]$^+$ due to its facile conversion into the starting material [140], cf. similar behavior of (E)-FpCH=CHCOC$_6$H$_4$CH$_3$-4 (No. 32).

Addition of LiCH$_3$ to (E)-FpCH=CHCOCH$_3$ in ether gives, after aqueous workup, (E)-FpCH= CHC(CH$_3$)$_2$OH [193, 201], see No. 62.

C$_5$H$_5$Fe(CO)$_2$C(=CH$_2$)COCH$_3$ (Table **5**, No. **69**) has been obtained in 96% yield from the reaction of [Fp(CH$_2$=C=C(OC$_6$H$_5$)CH$_2$Fp)]PF$_6$ with NaOCH$_3$ in CH$_3$OH, filtration through Celite after 1 min, evaporation of the solvent, extraction of the residue with CH$_2$Cl$_2$, filtration through Al$_2$O$_3$ and solvent removal [133]. The compound forms in 21% yield by addition of saturated aqueous NaOH to a suspension of [Fp$_2$(CH$_2$=C=C=CH$_2$)][PF$_6$]$_2$ in CH$_2$Cl$_2$ and stirring the mixture for 14 h. After drying with anhydrous MgSO$_4$ the filtered and concentrated solution is chromatographed on Al$_2$O$_3$ with CH$_2$Cl$_2$/petroleum ether (1:1) as eluent. The corresponding reaction of [Fp$_2$(CH$_2$=C=C=CH$_2$)][PF$_6$]$_2$ with saturated aqueous NaHCO$_3$ for 3 d gives FpC(=CH$_2$)COCH$_3$ (15% yield) and Fp(=CH$_2$)COCH$_2$Fp (30%). Treatment of the latter compound in CH$_2$Cl$_2$ with saturated aqueous NaOH converts it into FpC(=CH$_2$)COCH$_3$ (1:1 mixture with the starting material after 25 h as indicated by ^{1}H NMR) [133]. The compound also forms by treatment of [Fp$_2$(CH$_2$=C=C=CH$_2$)]$^{2+}$ with excess Cl$^-$ followed by chromatography of the reaction mixture on Al$_2$O$_3$ or by the reaction of FpC(=CH$_2$)CBr=CH$_2$ with H$_2$O [98].

References on pp. 141/7

C₅H₅Fe(CO)₂CH=CHC(CH₃)=NNHC₆H₃(NO₂)₂-2,4 $\mathbf{C_5H_5Fe(CO)_2CH=CHC(CH_3)=NNHC_6H_3(NO_2)_2}$-2,4 (Table 5, No. **70**) precipitates immediately after addition of (E)-FpCH=CHCOCH₃ in alcohol to 2,4-(O₂N)₂C₆H₃NHNH₂. After 3.5 h it is filtered off, washed with water until neutral, then with methanol and petroleum ether, and dried in a stream of nitrogen, 82% yield [39].

The hydrazone is stable in air in the solid state [39].

$\mathbf{C_5H_5Fe(CO)_2C(COOCH_3)=CHCOOCH_3}$ (Table 5, No. **71**) has been obtained from FpH and CH₃OCOC≡CCOOCH₃ in benzene within 30 min. Possible reaction mechanisms are discussed [196].

$\mathbf{C_5H_5Fe(CO)_2C(C_2H_5)=CHC_2H_5}$ (Table 5, No. **72**) is formed in low yield from [Fp(C₂H₅C≡CC₂H₅)]BF₄ and NaBH₃CN by the same procedure as already described for FpC(C₆H₅)=CHC₆H₅ [149], see No. 4.

(E)-$\mathbf{C_5H_5Fe(CO)_2C(COOCH_3)=CHCH_2C(CH_3)_2OH}$ (Table 5, No. **73**) is formed in small amounts along with the lactones IC and C by trans-addition of isobutene to [Fp(CH≡CCOOCH₃)]⁺ generated by the reaction of [Fp(GH₂=C(CH₃)₂)]BF₄ with CH≡CCOOCH₃ in refluxing CH₂Cl₂. No further details are given [150].

IC   C

(Z)-$\mathbf{C_5H_5Fe(CO)_2C(=CHCH_3)CH_2CH(COOCH_3)_2}$ (Table 5, No. **74**) has been prepared by addition of LiCH(COOCH₃)₂, generated by treatment of CH₂(COOCH₃)₂ with Li[N(Si(CH₃)₃)₂] in THF at −78 °C, to a suspension of XCVII (see p. 123) in THF at −78 °C. Warming to room temperature over a period of 2 to 3 h, removal of the solvent under reduced pressure, and chromatography of the residue on Al₂O₃ with ether or CH₂Cl₂ as eluents give the product in 88% yield. The assigned structure of No. 74 follows from trans–addition to the coordinated double bond.

(E)-$\mathbf{C_5H_5Fe(CO)_2C(=CHCH_3)CH_2CH(COOCH_3)_2}$ (Table 5, No. **75**) is obtained along with No. 74, when a mixture of XCVII and XCVIII (see p. 123), formed by thermal equilibration of XCVII, is treated with LiCH(COOCH₃)₂ [120].

(E)-$\mathbf{C_5H_5Fe(CO)_2CH=CHC(CH_3)=CH_2}$ (Table 5, No. **77**) has also been obtained in 27% yield by photochemical decarbonylation of (E)-FpCOCH=CHC(CH₃)=CH₂ in ether/benzene at −5 °C [190], see also [198].

$\mathbf{C_5H_5Fe(CO)_2C(=CH_2)CH=CH_2}$ (Table 5, No. **78**) has been prepared by addition of [Fp(CH₂=CHC(Fp)=CH₂)]PF₆, formed by protonation of FpCH₂C≡CCH₂Fp with HPF₆ in ether, to a mixture of Fp₂ in THF and sodium amalgam. After 0.5 h the solvent is evaporated, the residue dissolved in petroleum ether and chromatographed on Al₂O₃, eluting with benzene/petroleum ether (1:1), 69% yield [102]. The compound is formed in 9% yield along with the main product FpCH₂C≡CCH₂Fp and small amounts of FpCH₂C≡CCH₃ by the reaction of Na[Fp] with XCH₂C≡CCH₂X (X=Cl, OSO₂C₆H₅) in THF at −78 °C. After warming to 24 °C workup of the reaction mixture is carried out by addition of petroleum ether (b.p. 30 to 60 °C), filtration through Celite, evaporation of the solvent and extraction of the residue with petroleum ether. The last two steps are repeated, followed by solvent removal and short–path distillation of the residue at 70 °C/10^{-3} Torr. Separation from FpCH$_2$C≡CCH$_3$, with which the title compound is occasionally contaminated, is possible by chromatography on Al$_2$O$_3$. In a similar experiment BrCH$_2$C≡CCH$_2$Br also afforded a 9% yield of FpC(=CH$_2$)CH=CH$_2$ but no detectable amounts of FpCH$_2$C≡CCH$_2$Fp were formed. The yield was increased to 30% when a large excess of t-butanol was added to the reaction mixture prior to addition of C$_6$H$_5$SO$_3$CH$_2$C≡CCH$_2$OSO$_2$C$_6$H$_5$. In a similar experiment, but with BrCH$_2$C≡CCH$_2$Br, a 1:1 mixture of FpC(=CH$_2$)CH=CH$_2$ and FpCH$_2$C≡CCH$_3$ in a combined yield of 31% is obtained while the analogous reaction with ClCH$_2$C≡CCH$_2$Cl gives only 10% of No. 78. In the presence of methanol, ethanol, or isopropanol C$_5$H$_5$Fe(CO)(η^3-(E)-[CH$_2$CHC(CH$_3$)COOR^1]) (R^1=CH$_3$, C$_2$H$_5$, i-C$_3$H$_7$) was isolated in addition to FpC(=CH$_2$)CH=CH$_2$ and FpCH$_2$C≡CCH$_3$. Mechanisms for all these reactions are discussed. A mixture of No. 78 and FpCH$_2$C≡CCH$_3$ (identified by ^{1}H NMR) is formed in a 13% total yield by addition of butatriene to a THF solution of FpH [102].

The H-1 multiplet resonance, which is obscured by the chemical shift of the C$_5$H$_5$ protons (δ=4.76 ppm) when the ^{1}H NMR spectrum is measured in CS$_2$, is observable when C$_6$D$_6$ is used as solvent. In this solvent the C$_5$H$_5$ resonance is shifted to δ=4.20 ppm whereas the butadienyl resonances are shifted downfield [102].

The compound is soluble in petroleum ether. It can be short-path distilled (10^{-3} Torr) without extensive decomposition [102].

FpC(=CH$_2$)CH=CH$_2$ is converted to a mixture of C$_5$H$_5$Fe(CO)(η^3-(E)-[CH$_2$CHC(CH$_3$)COOCH$_3$]) and Fp$_2$ by treatment with Na[Fp] or NaOCH$_3$ in methanol for 10 h [102].

FpC(=CH$_2$)CH=CH$_2$ forms Diels–Alder adducts with a variety of symmetrical electrophilic dienophiles, e.g. (NC)$_2$C=C(CN)$_2$ or CF$_3$C≡CCF$_3$ (see "General Remarks" on p. 79). However, it failed to react with dimethyl maleate and asymmetric electrophilic reagents such as ketene, ethoxymethylenemalonitrile(?), or cyclopent-2-enone at temperatures below 110 °C [200].

FpC(=CH$_2$)CH=CH$_2$ reacts with CH$_3$OOCC≡CCOOCH$_3$ in refluxing CH$_2$ClCH$_2$Cl for 1.5 h to give the cycloadduct CI. The rate constants for this reaction have been determined as a function of the polarity of the solvent. However, at 70 °C only a small variation in the second-order rate constant on going from nonpolar cyclohexane (k=1.10 × 10^3) to moderately polar CH$_3$NO$_2$ (k=1.4 × 10^3) or CH$_3$CN (k=1.7 × 10^3 M^{-1}·s^{-1}) could be observed (monitored by ^{1}H NMR spectroscopy). Thus, a nonpolar Diels–Alder mechanism is supposed. Kinetic experiments also show that FpC(=CH$_2$)CH=CH$_2$ is at least 100 times more reactive toward CH$_3$OOCC≡CCOOCH$_3$ than the electron-rich 2-trimethylsiloxybuta-1,3-diene [200], see also [141].

$$C_5H_5(CO)_2Fe \quad \text{—} \quad \overset{COOCH_3}{\underset{COOCH_3}{}}$$

CI

CII

CIII

The reaction of FpC(=CH$_2$)CH=CH$_2$ with nitrosobenzene at 24 °C for 20 h gives a mixture of the isomers CII and CIII, the relative yields of which are essentially invariant in solvents of different polarity: cyclo-C$_6$H$_{12}$ (ratio 3:5), CH$_3$CN (3:5), or CH$_2$Cl$_2$ (2:3). Thus, it is assumed that each of the isomers is formed via the same reaction pathway, most likely the Diels–Alder reaction [200], see also [141].

(E)-C$_5$H$_5$Fe(CO)$_2$CF=CFCF=CF$_2$ (Table 5, No. **79**). The ether–soluble material of the reaction residue (Method I) is worked up by chromatography on Florisil with light petroleum (b.p. 40 to 60 °C) as eluent [22].

The resonances observed in the ^{19}F NMR spectrum are multiplets of equal intensity, a pattern which is expected for five spin–coupled nonequivalent F atoms. The stereochemical distinction is made on the basis of the difference in the coupling constants J(F–1,2). The value of 41 Hz is in the range for a cis–position of F–1 and F–2 [22].

(E,E)-C$_5$H$_5$Fe(CO)$_2$CH=CHCH=CHCl (Table 5, No. **80**) has been obtained in 95% yield by refluxing complex CIV in toluene for 1 h, cooling to 24 °C and evaporation of the solvent (thermal isomerization) [110], see also [96]. For the hydrolysis of CIV, see No. 126 [110].

CIV

C$_5$H$_5$Fe(CO)$_2$C(=CH$_2$)CBr=CH$_2$ (Table 5, No. **81**). Small amounts of the compound have been obtained from [Fp$_2$(CH$_2$=C=C=CH$_2$)]$^{2+}$ and [N(C$_2$H$_5$)$_3$CH$_2$C$_6$H$_5$]Br or by bromination of FpCH$_2$C≡CCH$_2$Fp [98].

The compound is thermally unstable. It was difficult to obtain in pure form. Its identity was established by spectroscopic examination of a sample that was about 90% pure [98].

The Br ligand is labile as evidenced by rapid formation of AgBr (along with [Fp(CH$_2$=C=C=CH$_2$)]$^+$ in low yield) on treatment of the compound with AgPF$_6$ in CH$_2$Cl$_2$. The reaction with H$_2$O gives FpC(=CH$_2$)COCH$_3$, see Section 1.5.2.3.16.1.7 [98].

(E,E)-C$_5$H$_5$Fe(CO)$_2$CH=CHCH=CHOCH$_3$ (Table 5, No. **82**) is formed in 16% yield by the reaction of the binuclear cyclobutadiene complex [Fp(μ-C$_4$H$_4$)Fp][PF$_6$]$_2$ with a mixture of methanol and NaHCO$_3$ for 15 h. After solvent removal the residue is taken up in benzene and filtered through Celite. The oil remaining after evaporation of benzene is chromatographed twice on Al$_2$O$_3$ with petroleum ether as eluent [111], see also [82, 116].

The (E,E)-stereochemistry of the compound is based on comparison of the coupling constants of the ^{1}H NMR spectrum with the values for other 1,4-disubstituted 1,3-butadienes. The NMR shift reagent tris(6,6,7,7,8,8,8-heptafluoro-2,2-dimethyl-3,5-octanedionato)europium was used to assign the protons of the ^{1}H NMR spectrum [111].

(E,E)-C$_5$H$_5$Fe(CO)$_2$CH=CHCH=CHCH$_3$ (Table 5, No. **85**) has been obtained in 38% yield by photochemical decarbonylation of (E,E)-FpCOCH=CHCH=CHCH$_3$ in ether/benzene at −5 °C [190], see also [198].

C$_5$H$_5$Fe(CO)$_2$C(=C(CF$_3$)$_2$)C(CF$_3$)=CF$_2$ (Table 5, No. **87**) is obtained in 22% yield when the reaction of Na[Fp] with (CF$_3$)$_2$C=C=C(CF$_3$)$_2$ (Method I) is carried out at −70 °C/1.5 h and further at room temperature for 1 h and at 50 to 60 °C/3 h [105, 106], see also [68, 117], whereas only 7% of the product is formed at 0 °C/3 h and further at room temperature for 2 h [94], see also [117]. In the first case the oil remaining after chromatography on Al₂O₃ is purified by crystallization from pentane at −70 °C [105, 106]. In the second case the oil is worked up by thin-layer chromatography on SiO₂ with petroleum ether/benzene (10:1) to separate the title compound from complex CV also formed followed by recrystallization of the product from petroleum ether [94]. The formation mechanism is discussed [105, 108].

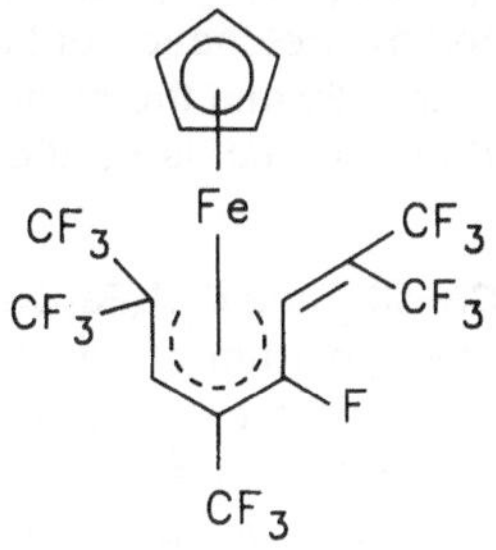

CV

Assignment of the ^{19}F NMR signals given in Table 5 has been done using the double resonance method [106] and spin decoupling experiments [105]. The ^{19}F NMR spectrum (relative to 100% CF₃COOH) shows a quadruplet at −22.7 and three multiplets at −20.9, 3.1, and 11.0 ppm [68]. The structure of No. 87 is also based on the elemental analysis, the IR and NMR data [68, 105].

The solid compound is stable and readily soluble in common organic solvents. Solutions do not decompose in air over a period of several hours [105, 106].

The mass spectrum shows the characteristic feature of carbonyl complexes containing a fluorinated ligand. Remarkable is that under the influence of electron bombardment (30 eV) the heaviest ion has m/e = 442 which corresponds to the removal of one CO group [106].

Treatment with Br₂, HCl, or HgCl₂ at room temperature in inert solvents does not lead to cleavage of the Fe–R bond. When refluxed with Br₂ in CCl₄, 50% of the starting material is recovered after 6 h [105, 106].

Sodium amalgam in THF transforms the compound at 50 to 60 °C to an unidentified fluorinated substance which shows no CO bands in its IR spectrum [105, 106].

UV irradiation of No. 87 in the presence of buta-1,3-diene or cyclohexa-1,3-diene in benzene or pentane, respectively, for 7 h leads to elimination of the CO groups and formation of the complexes CVII (R¹ = H, R² = CH=CH₂; R¹/R² = (CH₂)₂CH=CH), in which the diene reactants function as monodentate ligands. As indicated by IR measurements, the reactions

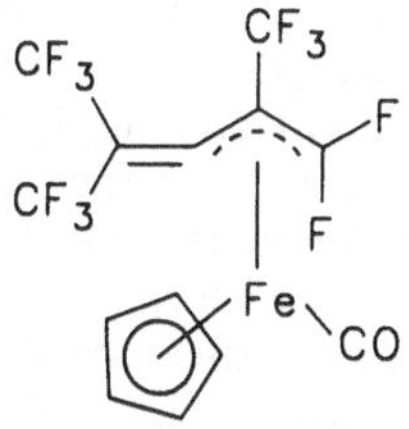

CVI

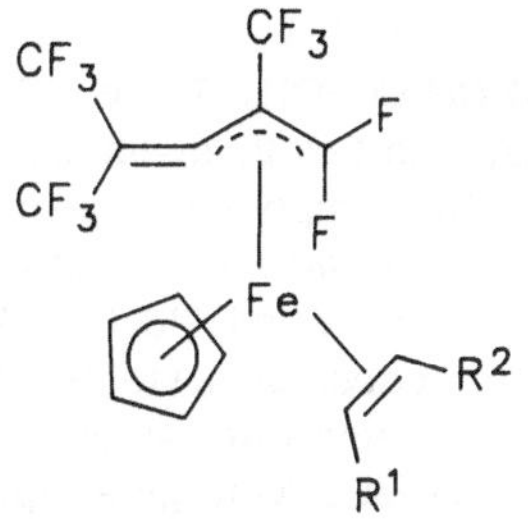

CVII

References on pp. 141/7

proceed via the η^3-allylidene complex CVI in which the remaining CO group is slowly replaced by the respective ligand. From the corresponding reaction with cyclopentadiene, however, no stable compound could be isolated [125].

The spectral data and chemical behavior are compared with the properties of $Re(CO)_5$-$C(=C(CF_3)_2)C(CF_3)=CF_2$ [105, 106].

(Z)-$C_5H_5Fe(CO)_2C(=CHC\equiv CCH_3)CH_2N(C_2H_5)_2$ (Table 5, No. **88**) has been prepared from $[Fp(CH_2=C=CHC\equiv CCH_3)]BF_4$, formed by protonation of $FpCH_2C\equiv CC\equiv CCH_3$ with HBF_4 in ether at $-40\,°C$, and cooled $NH(C_2H_5)_2$. After 30 min the mixture is warmed to room temperature, excess $NH(C_2H_5)_2$ evaporated and the product is purified by chromatography on Al_2O_3 with ether as eluent, 56% yield [195].

$C_5H_5Fe(CO)_2C(CN)=C(CN)CH_2CH=CH_2$ and **$C_5H_5Fe(CO)_2C(CN)=C(CN)CH(CH_3)CH=CH_2$** (Table 5, Nos. **89** and **90**). The ^1H NMR spectrum of compound No. 90 indicates that the allyl fragment underwent 1,3 rearrangement in the course of the insertion (Method IVc). The multiplicity of the methine proton ($\delta = 3.85$ ppm) derives from an overlapping doublet of quartets with both coupling constants about equal, $^3J(H,H) \sim {}^3J(H,CH_3) \sim 6.5$ Hz. This assignment has been substantiated by proton decoupling experiments [129].

The values of $\nu(CO)$ for the compounds are appreciably higher (30 to 49 cm^{-1}) than those for the parent η^1-allyl complexes, suggesting insertion of an electron-withdrawing group into the Fe-C σ-bond [129].

The compounds are rather unstable solids that are sparingly soluble in saturated hydrocarbons but increasingly soluble in benzene, CH_2Cl_2, and acetone [129].

$C_5H_5Fe(CO)_2C(COOCH_3)=C(COOCH_3)CH_2CH=CH_2$ (Table 5, No. **91**) is formed in 9% yield along with $FpCH_2CH=CHC(COOCH_3)=CHCOOCH_3$ (3%) and 1,2-$(CH_3COO)_2$-4-Fp-cyclopentene (42%) by treatment of $FpCH_2CH=CH_2$ with $CH_3OOCC\equiv CCOOCH_3$ (1:2 mole ratio) in dimethylformamide at room temperature for 90 h. The residue obtained after solvent removal is worked up by chromatography on Al_2O_3, eluting with mixtures of ether/light petroleum (b.p. 40 to 60 °C) [172], see also [143].

$C_5H_5Fe(CO)_2CH_2CH=C_6H_8O$ (Table 5, No. **93**). Alkylation with $[O(CH_3)_3]BF_4$ gives CVIII [161].

CVIII

$C_5H_5Fe(CO)_2CH_2CH=CH_2$ (Table 5, No. **95**) prepared according to Method I has been further purified by short-path distillation at 50 °C/10^{-2} Torr [131] or at 45 °C/10^{-3} Torr [6], see also [78]. The compound is also obtained from Na[Fp] and $HC_6F_4OCH_2CH=CH_2$-4 in THF and subsequent workup by chromatography of a CH_2Cl_2 extract of the evaporated reaction mixture on Florisil [18, 32]. A combination of Methods I and VI is described in [172]. The crude product (96% yield) resulting from the reaction of Na[Fp] with $CH_2=CHCH_2Cl$ (THF, 0 °C to room temperature), is dissolved in ether and treated with 40% aqueous HBF_4 in acetic anhydride at 0 °C to give $[Fp(CH_2=CHCH_3)]BF_4$, from which $FpCH_2CH=CH_2$ is then regenerated by addition of $N(C_2H_5)_3$ in CH_2Cl_2 (0 °C to room temperature). After solvent removal the residue is extracted with pentane and the extracts are filtered and concentrated, affording the allyl complex in quantitative yield [172].

The He(I) photoelectron spectrum is represented graphically in [156]. The measured vertical ionization potentials (IP) and empirical assignments are given below:

IP (in eV)	assignment (No. of IP's)
7.97	Fe 3d (3)
8.54	Fe–C σ (1)
9.78	π C_3H_5 (1)
10.24	Fe–π C_5H_5 (3)

A comparison with the spectrum of $FpCH_3$ shows that the lowest-lying ion states resulting from ionization of molecular orbitals with large Fe 3d character move to lower energy in $FpCH_2CH=CH_2$. Extended-Hückel MO calculations revealed that the orbitals in $FpCH_2CH=CH_2$, which are primarily olefinic in character, have substantial amounts of metal d character. A good correlation between the cyclic voltammetric oxidation potentials ($E_{1/2} = 0.90$ eV for $FpCH_2CH=CH_2$, measured in THF/0.05 M $[N(C_4H_9-n)_4]BF_4$ vs. SCE) and the energies of the lowest ion states (from UV photoelectron spectra) of $FpCH_3$, $FpCH_2CH=CH_2$, and FpC_5H_5 has been noted [156].

The orbital interaction diagram for Fp^+ and $[CH_2=CHCH_2]^-$ is analyzed by Extended-Hückel MO calculations [151].

The compound is soluble in the common organic solvents [6]. From titration in ethanol/water (6:1) with HPF_6 an apparent pK_a of 5.0 has been obtained [157]. For polarographic data, see "General Remarks" on p. 83.

$FpCH_2CH=CH_2$ decomposes readily in air [6].

Addition of Br_2 in CH_2Cl_2 to $FpCH_2CH=CH_2$ in the same solvent at -78 °C leads to immediate precipitation of $[Fp(CH_2=CHCH_2Br)]Br$ [86]. The reaction with N-bromopyridinium bromide followed by deprotonation with $N(C_2H_5)_3$ giving exclusively $FpCH_2CH=CHBr$ (cf. Method VI) is mentioned in [77].

Alkylation of $FpCH_2CH=CH_2$ to give the salts $[Fp(CH_2=CHCH_2R^1)]X$ (CIX), where $X=BF_4$ or $SbCl_6$, is effected in CH_2Cl_2 by treatment with either $[O(CH_3)_3]BF_4$ ($R^1=CH_3$) for 90 min, [cyclo C_7H_7]BF_4 ($R^1=$ tropyl) for 15 min, or [cyclo C_3Cl_3]$SbCl_6$ ($R^1=$ trichlorocyclopropenyl; fast reaction) [86]. Acylation with formation of complexes CIX where $R^1=COCH_3$ or COC_6H_5 has been accomplished by employing acyl cations generated from CH_3COCl or C_6H_5COCl and $AgSbF_6$ in CH_3NO_2 at -30 °C. CX is obtained from the reaction with 2-phenyl-1,3-dioxolen-2-ium hexafluorophosphate in CH_2Cl_2 at -30 to 0 °C for 30 min. 2-n-Amyl-1,3-dioxolen-2-ium fluoborate gives CXI after 20 min at -20 °C. No. 95 gives CIX ($R^1=$ $HC(OCH_3)_2$) with $[HC(OCH_3)_2]PF_6$ in CH_2Cl_2 at -78 to $+20$ °C (30 min) [86].

$$\left[C_5H_5(CO)_2Fe \!-\!\! \overset{\|}{\diagdown}_{R^1} \right]^+ X^- \qquad \left[C_5H_5(CO)_2Fe \!-\!\! \overset{\|}{\diagdown}\underset{O}{\overset{R^2\ O}{\diagup}} \right]^+ X^-$$

CIX

CX: $R^2=C_6H_5$; $X=[PF_6]$

CXI: $R^2= n-C_5H_{11}$; $X=[BF_4]$

References on pp. 141/7

Carboxylation takes place when $FpCH_2CH=CH_2$ is treated with $[C(OCH_3)_3]BF_4$ in CH_2Cl_2 at 0 °C for 30 min. The intermediately formed cation $[Fp(CH_2=CHCH_2C(OCH_3)_3)]^+$ is hydrolytically very sensitive and can be directly converted to the ester $[Fp(CH_2=CHCH_2COOCH_3)]BF_4$ [86].

$FpCH_2CH=CH_2$ reacts with α,α'-dibromo ketones $BrR^1R^2CCOCR^3R^4Br$, and $Fe_2(CO)_9$ (formation of oxyallyl-iron cations) in refluxing CH_2Cl_2 for 12 h to give the cyclohexanones CXII. The cyclization proceeds smoothly with tetrasubstituted ($R^1=R^2=R^3=R^4=CH_3$), trisubstituted ($R^1=R^2=R^4=CH_3$, $R^3=H$) and α,α'-disubstituted ketones ($R^1=R^3=CH_3$, C_2H_5, C_3H_7-i, C_6H_5, $R^2=R^4=H$), whereas 1,1,3,3-tetrabromoacetone, monosubstituted ketones ($R^1=R^2=R^4=H$, $R^3=C_2H_5$) and 2,7-dibromocycloheptanone failed to produce any cyclic products [203]. The reaction of No. 95 with the ketenes $R^5R^6C=C=O$ ($R^5=CH_3$, $R^6=C_6H_5$; $R^5=C_2H_5$, $R^6=C_6H_5$; $R^5=R^6=C_6H_5$) in CH_2Cl_2 at room temperature overnight affords the condensation products (E)-$FpCH_2CH=CHCOCHR^5R^6$ (derived by proton transfer within a dipolar intermediate) in low yields [175].

For the reaction with the ketene immonium salt CXVII (p. 134), see No. 137 on p. 134 [99].

The lower reactivity of $FpCH_2CH=CH_2$ in cycloaddition reactions giving CXIII with electrophiles $R^1CH=C(CN)_2$, where $R^1=C_6H_5$ or C_2H_5O, compared with the reactivities of $C_5H_5Fe(CO)(^2D)CH_2CH=CH_2$ is pointed out in [157]; ^2D is, for instance, the bicyclic phosphite $P(OCH_2)_3CCH_3$.

$C_5H_5Fe(CO)_2CD_2CH=CD_2$ (Table 5, No. **98**) has been prepared according to Method VI from $[Fp(CD_2=CHCD_3)]BF_4$ and $N(C_2H_5)_3$ in CH_2Cl_2 at 0 °C (0.5 h) to 25 °C (0.5 h). The reaction residue is taken up in ether/petroleum ether and filtered through a Celite/Al_2O_3 mixture. Removal of the solvent leaves the product in 83% yield [175].

$C_5H_5Fe(CO)_2CH_2C(CH_3)=CH_2$ (Table 5, No. **99**) is further purified after chromatography (Method I) by distillation under vacuum [42]. Because of the decomposition of the compound under the chromatographic conditions, either distillation [131] or short-path high-vacuum distillation [78] were suggested. A variation of Method I is described where $[N(C_4H_9-n)_4][Fp]$, prepared from $FpSi(CH_3)_3$ and $[N(C_4H_9-n)_4]F$ in THF at −77 K, is used in place of Na[Fp]. The yield of 63% has been determined by ^1H NMR spectroscopy [184].

The compound reacts with Br_2 in CH_2Cl_2 at −78 °C (15 min) followed by addition of $HBF_4 \cdot O(C_2H_5)_2$ to give $[Fp(CH_2=C(CH_3)CH_2Br)]^+$ [86], the starting material for the preparation of $FpCH_2C(CH_3)=CHBr$ (see Method VI). The reaction with the 1,3-dioxolylium cation in CH_2Cl_2 at −20 °C yielding CXIV is briefly mentioned in [192].

References on pp. 141/7

$C_5H_5Fe(CO)_2CD_2C(CH_3)=CH_2$ and $C_5H_5Fe(CO)_2CH_2C(CH_3)=CD_2$ (Table 5, Nos. **99a** and **99b**). A mixture of these compounds has been obtained by deprotonation of the labelled complex $[Fp(CD_2=C(CH_3)_2)]^+$ with N-methylmorpholine (cf. Method VI) at room temperature or at 0 °C followed by washing with water, extraction into Skelly-B, and vacuum distillation of the residue. A rapid equilibration is observed in $FpCD_2C(CH_3)=CH_2$ compared to a similar, but slower isomerization of $FpCH_2CH=CD_2$ (No. 97, see "General Remarks" on p. 63) [99].

(E)-$C_5H_5Fe(CO)_2CH_2CH=CHC_6H_5$ (Table 5, No. **100**). Different yields are obtained by Method I depending on the allyl reactant used. Thus, while $C_6H_5CH=CHCH_2Cl$ gives a 72% yield, $C_6H_5CH=CHCH_2OSO_2C_6H_5$ affords only 22% of No. 100 [136]. After the product has been further purified by chromatography on Al_2O_3 and distillation under vacuum, only 10% is obtained from $C_6H_5CH=CHCH_2Cl$ and 30% from $CH_2=CHCHClC_6H_5$ [42]. The compound is formed in 89% yield (determined by ^{1}H NMR spectroscopy) when $[N(C_4H_9-n)_4][Fp]$, prepared from $FpSi(CH_3)_3$ and $[N(C_4H_9-n)_4]F$ in THF at -77 K, is used in place of Na[Fp] for Method I [184].

$C_5H_5Fe(CO)_2CH_2CH=CHC_7H_7$ (Table 5, No. **101**). The predominant component of the mixture obtained by Method VI is presumed to be the (E)-isomer [86].

(Z)-$C_5H_5Fe(CO)_2CH_2CH=CHCl$ (Table 5, No. **102**). It is not clear which isomer has been obtained (Method I) in [136]. According to these authors extensive decomposition prevented a meaningful ^{1}H NMR spectrum from being recorded [136], but see [114].

$C_5H_5Fe(CO)_2CH_2C(CH_3)=CHBr$ (Table 5, No. **105**). The ^{1}H NMR spectrum shows that it is a single stereoisomer of as yet unknown stereochemistry [86].

$C_5H_5Fe(CO)_2CH_2CH=CHOCH_3$ (Table 5, Nos. **106** and **107**). Although starting with (E)-$CH_3OCH=CHCH_2Cl$ (Method I), a 2:1 mixture of (Z)- and (E)-$FpCH_2CH=CHOCH_3$ is obtained in 50% yield after chromatography on Al_2O_3 in petroleum ether. Separation of the isomers is achieved on Al_2O_3 using ether/petroleum ether (1:19) to elute the (Z)-isomer (No. 106), and ether/petroleum ether (1:9) to elute the (E)-isomer (No. 107) [86], see also [114]. The (Z)-compound has been prepared in 36% yield by addition of $CH_2=CHMgBr$ in THF to a suspension of $[Fp=CHOCH_3]PF_6$ in the same solvent at -23 °C. After warming to room temperature and solvent removal, the residue is extracted with hexane and further purified by thin layer chromatography on SiO_2 using hexane/ether (3:1) and recrystallization from pentane. (Z)-$FpCH_2CH=CHOCH_3$ is also obtained by treatment of $[Fp=CHOCH_3]PF_6$ with LiCH=CHI$_2$ in ether at -70 °C and workup as aforementioned, 28% yield [188]. A preparation for No. 106 involving treatment of Na[Fp] with methoxymethyloxirane followed by reaction with 48% HBF_4 and treatment of the resulting $[Fp(CH_2=CHCH_2OCH_3)]^+$ with $N(C_2H_5)_3$ is reported in [114].

No. 106 proved to be resistent to mild acid hydrolysis and was recovered in almost quantitative yield after treatment with 3% aqueous HCl at reflux after 30 min [86].

For the reaction of No. 106 with $CH_3OOCC≡CCOOCH_3$, see "General Remarks" on p. 78.

Attempted isomerization of No. 107 by treatment with $[N(C_2H_5)_4]BF_4$ in CH_2Cl_2 at 0 °C for 15 min to give No. 106 failed [86].

$C_5H_5Fe(CO)_2CH_2C(OCH_3)=CH_2$ (Table 5, No. **108**) has been obtained in 40% yield along with $FpCH_2C(OCH_3)_2CH_3$ (~1:2 mixture) by the reaction of Na[Fp] with $BrCH_2C(OCH_3)_2CH_3$ in THF at 0 °C/1 h and at room temperature/3 h. It is separated from the mixture by treatment with a 40% aqueous solution of HBF_4 in acetic anhydride at 0 °C followed by treatment of the salt $[Fp(CH_2=C(OCH_3)CH_3)]BF_4$, formed in 75% yield, with 1,4-diazabicyclo[2.2.2]oc-

References on pp. 141/7

tane in THF at 0 °C/30 min and at room temperature/1 h. The latter step proceeds quantitatively [139], see also [142].

The compound No. 108 is extremely air- and moisture-sensitive, being readily hydrolyzed to the ketone $FpCH_2COCH_3$ [139, 142]. Normal chromatographic purification procedures cannot be employed because such treatment leads to exclusive formation of $FpCH_2COCH_3$ [139].

For the reaction with $CH_3OOCC\equiv CCOOCH_3$ to give (Z,Z)-$FpCH_2C(OCH_3)=CHC(COOCH_3)=$ $CHCOOCH_3$ see No. 146, and "General Remarks" on p. 78. No products were isolated from the reaction of $FpCH_2C(OCH_3)=CH_2$ in dimethylformamide with methyl acrylate, acrylonitrile, (E)-NCCH=CHCN, methyl cinnamate, $O_2NCH=CHCOOC_2H_5$, $C_6H_5CH=CHNO_2$, maleic anhydride, p-benzoquinone, $CH\equiv CCOOCH_3$, or $(CH_3)_3SiC\equiv CSi(CH_3)_3$, even by using excess reactant and long reaction times [173].

$C_5H_5Fe(CO)_2CH_2C(OC_2H_5)=CH_2$ (Table 5, No. **109**) forms, along with compound CXV, by treatment of $[Fp(CH_2=C(OC_2H_5)CH_3)]^+$ with lithium 5-methylcyclohex-1-enolate in THF at -78 °C for 3.5 h [170].

CXV

(E)-$C_5H_5Fe(CO)_2CH_2CH=CHSi(CH_3)_3$ (Table 5, No. **112**). The (E)-configuration of the starting material $ClCH_2CH=CHSi(CH_3)_3$ (Method I) is retained in the product as indicated by the CH=CH coupling constant of 18 Hz [103].

$C_5H_5Fe(CO)_2CH_2CH=CH-C_3H_4O_2(C_5H_{11}-n)$ (Table 5, No. **115**) gives a compound with a strong IR absorption at 1715 cm^{-1} (unconjugated demetallated ketone) on treatment with 5% HCl at room temperature for 24 h [86].

(E)-$C_5H_5Fe(CO)_2CH_2CH=CH-C_3H_4O_2(C_6H_5)$ (Table 5, No. **116**). The oil obtained by Method VI crystallizes from hexane [86].

The compound resists hydrolysis in a 10% aqueous dioxane solution of HCl at room temperature [86].

$C_5H_5Fe(CO)_2CH_2CH=CHCH_3$ (Table 5, No. **117**) is obtained in 94% yield as 1:2 mixture of (Z)- and (E)-isomers when $CH_2=CHCHClCH_3$ is used as starting material for Method I [99]. However, when the same reaction is carried out on a commercial mixture of crotyl chloride containing ~20% of $CH_2=CHCHClCH_3$ and a 1:1 ratio of (Z)- and (E)-$CH_3CH=$ $CHCH_2Cl$, the isomers are formed in the ratio 3:4 in 84% yield [99], see also [42]. Chromatography of either of the above mixtures on Al_2O_3 employing petroleum ether as eluent leads to a partial separation of isomers, the forerun consisting of the (E)-isomer in 90% purity [172], see also [88]. Similarly, the (E)-isomer may be largely separated from the (Z)-isomer (3:2 mixture, obtained by Method VI) by chromatography on Al_2O_3 and elution with pentane [86]. The deprotonation of (Z)-$[Fp(CH_3CH=CHCH_3)]^+$ with $N(C_2H_5)_3$ (cf. Method VI) in CH_2Cl_2 at room temperature (30 min) and workup of the mixture by extraction of the reaction residue with ether and solvent removal from the filtered extracts give a caramel-colored oil identified by ^{1}H NMR spectroscopy as a 1:1 mixture of (Z)- and (E)-isomers, 97% yield. The reaction mechanism is discussed [99]. FpCH$_2$CH=CHCH$_3$ has also been obtained as an isomeric mixture by treatment of a solution of FpCl in THF with an excess of NaBH$_4$ (formation of FpH) for 20 min, during which a stream of butadiene gas is passed through the solution. After addition of light petroleum (b.p. 30 to 40 °C) and water, the petroleum layer which separates is worked up by chromatography and distillation. This mixture of isomers decomposes at ~60 °C [6].

(E)-C$_5$H$_5$Fe(CO)$_2$CH$_2$CH=CHCH$_3$ (Table 5, No. **118**) has been prepared according to Method I from Na[Fp] and (E)-CH$_3$CH=CHCH$_2$X (X = OSO$_2$C$_6$H$_5$ [86], or OSO$_2$C$_6$H$_4$CH$_3$-4 [99]).

C$_5$H$_5$Fe(CO)$_2$CH$_2$CH=C(CH$_3$)$_2$ (Table 5, No. **120**). The use of (CH$_3$)$_2$C=CHCH$_2$Br in place of the respective chloride (Method I) may be the reason for the lower yield (52%) obtained by [99].

Due to its instability, the compound could not be purified by short-path high vacuum distillation at 10^{-6} Torr [78].

Treatment with (C$_6$H$_5$)$_2$C=C=O in CH$_2$Cl$_2$ at room temperature for 21 h leads to recovery of the starting material and not the expected [3+2]cycloaddition product [126].

(E)-C$_5$H$_5$Fe(CO)$_2$CH$_2$CH=CHCH$_2$C$_6$H$_5$ (Table 5, No. **122**) has been obtained in 55% yield from [Fp(CH$_2$=CHCH=CH$_2$)]$^+$ and C$_6$H$_5$MgCl. A small amount of FpCH$_2$CH(C$_6$H$_5$)CH=CH$_2$ may also be present in the reaction mixture [120].

(E)-C$_5$H$_5$Fe(CO)$_2$CH$_2$CH=CHCH$_2$-(C$_6$H$_9$O) (Table 5, No. **123**) has been prepared by the reaction of [Fp(CH$_2$=CHCH=CH$_2$)]$^+$ with 1-pyrrolidinocyclohexene (1:1 mole ratio) in CH$_3$CN at 0 °C for 30 min, followed by hydrolysis of the intermediately formed iminium salt by brief heating with aqueous NaOH on a steam bath. Extraction into ether and chromatography on Al$_2$O$_3$ gives the product in 26% yield [120].

On the basis of ^{1}H NMR and IR data, the (E)-structure is assigned to the compound. Its exclusive formation suggests that the complex is formed directly from the butadiene complex by conjugate addition of the enamine [120].

C$_5$H$_5$Fe(CO)$_2$CH$_2$CH=C(Cl)CH$_3$ (Table 5, No. **124**) is purified after chromatography (Method I) by distillation under vacuum [42].

(E)-C$_5$H$_5$Fe(CO)$_2$CH$_2$CH=CHCH(OCH$_3$)$_2$ (Table 5, No. **125**) is rapidly and quantitatively hydrolyzed in THF by 1% HCl to give (E)-FpCH$_2$CH=CHCHO (No. 126) [86].

(E)-C$_5$H$_5$Fe(CO)$_2$CH$_2$CH-CHCHO (Table 5, No. **126**) has been obtained in 95% yield by hydrolysis of (E)-FpCH$_2$CH=CHCH(OCH$_3$)$_2$ in THF with 1% HCl, extraction with CH$_2$Cl$_2$, solvent removal from the extracts, and chromatography of the residue on Al$_2$O$_3$ with ether as eluent [86]. The compound also forms by hydrolysis of a benzene solution of CIV (see p. 126) on Al$_2$O$_3$. Elution with benzene, then with 5% methanol in benzene followed by evaporation of the solvent gives No. 126 in 75% yield [110], see also [96, 116]. A possible mechanism for this formation is discussed [110].

(E)-C$_5$H$_5$Fe(CO)$_2$CH$_2$CH=CHCOC$_6$H$_5$ (Table 5, No. **127**). As described for (E)-FpCH$_2$CH=CHCOCH$_3$ (No. 135) the compound has been prepared by treatment of FpCH$_2$CH=CH$_2$ with C$_6$H$_5$CO$^+$ and subsequent deprotonation of the cation formed, [Fp(CH$_2$=CHCH$_2$COC$_6$H$_5$)]$^+$, with N(C$_2$H$_5$)$_3$. The (E)-isomer was the exclusive product. For reaction conditions, see No. 135 [86].

(E)-C$_5$H$_5$Fe(CO)$_2$CH$_2$CH=CHCOOCH$_3$ (Table 5, No. **128**) has been prepared by the reaction of FpCH$_2$CH=CH$_2$ with [C(OCH$_3$)$_3$]BF$_4$ in CH$_2$Cl$_2$ at 0 °C for 30 min, addition of ether and treatment of the precipitated [Fp(CH$_2$=CHCH$_2$COOCH$_3$)]BF$_4$ with N(C$_2$H$_5$)$_3$ in CH$_2$Cl$_2$ for 15 min (cf. Method VI). Solvent removal, extraction of the residue with ether, chromatography of the concentrated extracts on Al_2O_3 and elution with ether/hexane (1:1) afford the product in 54% yield [86].

(E)-$C_5H_5Fe(CO)_2CH_2CH=CHCONHCOCCl_3$ (Table 5, No. **129**) has been prepared by treatment of $FpCH_2CH=CH_2$ with CCl_3CONCO (1:1 mole ratio) in CH_2Cl_2 at room temperature for 1 h and crystallization of the residue from CH_2Cl_2/hexane after solvent removal, 40% yield [99].

(E)-$C_5H_5Fe(CO)_2CH_2CH=CHC_2H_5$ (Table 5, No. **133**) has been prepared by addition of Li-Cu(CH_3)_2 in ether to $[Fp(CH_2=CHCH=CH_2)]^+$ in THF at $-50\,°C$. After 1 h the solvent is removed, the residue taken up in petroleum ether and chromatographed on Al_2O_3, 62% yield [120].

The almost exclusive (E)-stereochemistry (less than 2% of the (Z)-isomer is present) is evidenced by IR data and confirmed by ^1H NMR decoupling experiments [120].

(E)-$C_5H_5Fe(CO)_2CH_2CH=C(CH_3)C_2H_5$ (Table 5, No. **134**) has been obtained in 72% yield from $[Fp(CH_2=CHC(CH_3)=CH_2)]^+$ and $LiCu(CH_3)_2$ under conditions described for (E)-$FpCH_2CH=CHC_2H_5$, but at $-78\,°C$ [120], see No. 133.

The structure is deduced from the ^1H NMR spectrum and from the stereochemistry of the adduct resulting from the reaction with $(NC)_2C=C(CN)_2$ (see "General Remarks" on p. 79) [120].

(E)-$C_5H_5Fe(CO)_2CH_2CH=CHCOCH_3$ (Table 5, No. **135**) has been prepared by acylation of $FpCH_2CH=CH_2$ at $-30\,°C$, employing CH_3CO^+ generated from CH_3COCl and $AgSbF_6$ in CH_3NO_2, followed by in situ deprotonation of the cation formed, $[Fp(CH_2=CHCH_2COCH_3)]^+$, by treatment with $N(C_2H_5)_3$ in CH_3NO_2 at $0\,°C$ for 15 min (cf. Method VI). Solvent removal, extraction of the residue with ether, and chromatography on Al_2O_3 with ether as eluent give an oil which is crystallized from CS_2 at $-45\,°C$. The product from this reaction was exclusively the (E)-isomer [86].

(E)-$C_5H_5Fe(CO)_2CH_2CH=CHCOCH(C_6H_5)_2$ and **(E)-$C_5H_5Fe(CO)_2CH_2CH=CHCOCH(CH_3)C_6H_5$** (Table 5, Nos. **136** and **138**) have been prepared by stirring a mixture of $FpCH_2CH=CH_2$ and $(C_6H_5)_2C=C=O$, or $C_6H_5C(CH_3)=C=O$, respectively (1:1.2 mole ratio) in CH_2Cl_2 at room temperature overnight. After solvent removal the residue is taken up in ether and chromatographed on Al_2O_3 with ether/petroleum ether (3:7) as eluent. The yields are 30 and 11%, respectively [175].

(E)-$C_5H_5Fe(CO)_2CH_2CH=CHCOCH(CH_3)_2$ (Table 5, No. **137**) has been obtained in 19% yield by treatment of the ammonium salt CXVI with 1% aqueous oxalic acid in acetone for 12 h, extraction with ether and chromatography on Al_2O_3. CXVI is formed by addition of $FpCH_2CH=CH_2$ to the keteneimmonium salt CXVII in CH_2Cl_2 at $-60\,°C$. The filtrate of the reaction mixture is concentrated and finally triturated with ether at room temperature [99].

CXVI   CXVII

C$_5$H$_5$Fe(CO)$_2$CH$_2$C(OCH$_3$)=CHCH(COOCH$_3$)CH(COOCH$_3$)$_2$ (Table 5, No. **139**) has been obtained in 28% yield by the reaction of FpCH$_2$C(OCH$_3$)=CH$_2$ with (CH$_3$OOC)$_2$C=CHCOOCH$_3$ (1:2 mole ratio) in CH$_2$Cl$_2$ at room temperature for 20 h, solvent removal and chromatography on Al$_2$O$_3$ with ether/light petroleum (2:3) as eluent [173], see also [143]. For solvent dependence and formation of other products, see the corresponding reaction in "General Remarks" on p. 80.

Hydrolysis of the compound to give FpCH$_2$COCH$_2$CH(COOCH$_3$)CH(COOCH$_3$)$_2$ takes place during purification by chromatography or by treatment of the crude product with 4-CH$_3$C$_6$H$_4$SO$_3$H in aqueous THF at room temperature for 6 h [143, 173].

C$_5$H$_5$Fe(CO)$_2$CH$_2$C(OCH$_3$)=CHC(COOCH$_3$)$_2$CH(COOCH$_3$)$_2$ (Table 5, No. **140**) is obtained in 13% yield by the reaction of FpCH$_2$C(OCH$_3$)=CH$_2$ with (CH$_3$OOC)$_2$C=C(COOCH$_3$)$_2$ in DMF for 70 h and subsequent chromatography on Al$_2$O$_3$ [173].

The compound is partly hydrolyzed to FpCH$_2$COCH$_2$C(COOCH$_3$)$_2$CH(COOCH$_3$)$_2$ during purification by chromatography. Also, treatment of the crude product with 4-CH$_3$C$_6$H$_4$SO$_3$H in aqueous THF for 6 h causes hydrolysis [173].

C$_5$H$_5$Fe(CO)$_2$CH$_2$C(OCH$_3$)=CHC(COOC$_2$H$_5$)$_2$CH(COOC$_2$H$_5$)$_2$ (Table 5, No. **141**) is formed in the reaction of FpCH$_2$C(OCH$_3$)=CH$_2$ with (C$_2$H$_5$OOC)$_2$C=C(COOC$_2$H$_5$)$_2$ in dimethylformamide during 48 h. However, purification by chromatography on Al$_2$O$_3$ leads to hydrolysis. Thus, elution with ether/light petroleum (3:1) gives a mixture of FpCH$_2$COCH$_2$C(COOC$_2$H$_5$)$_2$CH-(COOC$_2$H$_5$)$_2$ and FpCH$_2$COCH$_3$ [173].

(Z)-C$_5$H$_5$Fe(CO)$_2$CH$_2$C(OCH$_3$)=CHCOCH(CN)(C$_4$H$_9$-t) (Table 5, No. **143**) has been prepared by addition of t-C$_4$H$_9$C(CN)=C=O in benzene to FpCH$_2$C(OCH$_3$)=CH$_2$ in the same solvent at room temperature, solvent removal after 2 h and chromatography of the residue on Al$_2$O$_3$ with ether/light petroleum (1:3) as eluent, 63% yield. Because of the signals observed in the ^{1}H NMR spectrum of the crude product, the existence of two isomers is assumed. However, chromatography causes isomerization, resulting in formation of the (Z)-isomer, which is thought to be the more stable due to the possible dipolar interactions between methoxy- and ketone groups [173].

Treatment with [NH$_4$]$_2$[Ce(NO$_3$)$_6$] in methanol at room temperature leads to replacement of the Fp moiety by a proton to give (Z)-CH$_3$OC(CH$_3$)=CHCOCH(CN)C$_4$H$_9$-t [173].

C$_5$H$_5$Fe(CO)$_2$CH$_2$CH=CHCH$_2$CH=CH$_2$ (Table 5, No. **144**) is formed in small amounts along with ferrocene and FpC$_5$H$_5$ by treatment of [Fp$_2$(CH$_2$=CHCH$_2$CH$_2$CH=CH$_2$)][BF$_4$]$_2$ with N(C$_4$H$_9$-n)$_3$ in CH$_3$NO$_2$ and after solvent removal with NaI in acetone. It is isolated from the other products by chromatography on Al$_2$O$_3$ [155].

C$_5$H$_5$Fe(CO)$_2$CH$_2$CH=CHC(COOCH$_3$)=CHCOOCH$_3$ (Table 5, No. **145**). For its formation see further information on FpC(COOCH$_3$)=C(COOCH$_3$)CH$_2$CH=CH$_2$ (No. 91) [172], also [142, 143].

(Z,Z)-C$_5$H$_5$Fe(CO)$_2$CH$_2$C(OCH$_3$)=CHC(COOCH$_3$)=CHCOOCH$_3$ (Table 5, No. **146**) has been prepared by addition of CH$_3$OOCC≡CCOOCH$_3$ to FpCH$_2$C(OCH$_3$)=CH$_2$ (2:1 mole ratio) in dimethylformamide at room temperature, solvent removal after 3 h, and chromatography of the residue on Al$_2$O$_3$ with ether/light petroleum (2:3) as eluent, 63% yield [173], see also [142, 143].

The compound does not hydrolyze during chromatography as is the case with related compounds, presumably because of extended conjugation of the enol ether [173].

C$_5$H$_5$Fe(CO)$_2$CH$_2$CH$_2$CH=CH$_2$ (Table 5, No. **147**) is formed together with FpCH$_2$C$_3$H$_5$-cyclo (3:7 mixture) by the reaction of Na[Fp] with cyclo-C$_3$H$_5$CH$_2$I in THF at 0 °C. An ESR spectrum

References on pp. 141/7

of the mixture shows the presence of the radical $CH_2=CHCH_2\dot{C}H_2$, revealing that this reaction proceeds through free radicals as intermediates. By contrast, the analogous reaction of Na[Fp] with cyclo-$C_3H_5CH_2Br$ yields almost exclusively $FpCH_2C_3H_5$-cyclo [123, 137].

The electron-impact mass spectrum (70 eV) of No. 147 (=FpR) shows the following ions (relative intensity): $[M]^+$ (8), $[M-CO]^+$ (7), $[M-2CO]^+$ (100), $[M-2CO-H_2]^+$ (22), $[M-2CO-2H_2]^+$ (25), $[M-R]^+$ (18), $[M-R-CO]^+$ (28), $[M-R-2CO]^+$ (93), $[M-C_4H_6]^+$ (2), $[M-C_4H_6-CO]^+$ (8), $[M-C_4H_6-2CO]^+$ (63), $[Fe]^+$ (76), $[M-RH-CO]^+$ (75), $[Fe(C_5H_5)(C_4H_7)]^+$ (100). Remarkable is the base peak $[C_5H_5FeC_4H_7]^+$, which may be the methallyl complex $[\eta^5\text{-}C_5H_5Fe\text{-}\eta^3\text{-}C_3H_4CH_3]^+$ [162].

Treatment of the compound with a suspension of $[C(C_6H_5)_3]BF_4$ in anhydrous ether for 12 h gives a yellow-orange precipitate of $[Fp(CH_2=CHCH=CH_2)]BF_4$, the IR spectrum of which indicates that iron is coordinated to only one double bond of the butadiene [45].

The compound reacts with a CH_2Cl_2 solution of $FpFBF_3$, prepared from FpI and $AgBF_4$, at 22 °C to give the dinuclear salt $[Fp(CH_2=CHCH_2CH_2Fp)]BF_4$ [202].

On cooling, the compound behaves similarly to $FpCH_2C(CH_3)_2CH=CH_2$, affording a small amount of CXVIII [45], see No. 149.

CXVIII: $R^1=H$

CXIX: $R^1=CH_3$

$C_5H_5Fe(CO)_2CH_2CH_2C(CH_3)=CH_2$ (Table 5, No. **148**) reacts regiospecifically with free radicals ($Cl_3C^\cdot$, $Br_3C^\cdot$, $4\text{-}CH_3C_6H_4SO_2^\cdot$) and with electrophiles like CF_3COOH at the δ-carbon of the but-3-enyl ligand, causing displacement of the metal and formation of cyclopropylcarbinyl compounds. Thus, $FpCH_2CH_2C(CH_3)=CH_2$ reacts with CF_3COOH in $CDCl_3$ at ambient temperature within a few seconds to give the cyclopropane CXX ($R^1=H$) and $[Fp][CF_3COO]$. The same type of products CXX ($R^1=CCl_3$ or CBr_3) and FpX (X=Cl or Br, respectively) result from the reaction with CCl_3SO_2Cl and CBr_4 in CH_2Cl_2, which proceeds more slowly. In the corresponding reaction with $4\text{-}CH_3C_6H_4SO_2I$ only CXX ($R^1=4\text{-}CH_3C_6H_4SO_2$) could be isolated. A radical chain mechanism is proposed [160].

CXX

$C_5H_5Fe(CO)_2CH_2C(CH_3)_2CH=CH_2$ (Table 5, No. **149**) prepared by Method I is purified by chromatography on Al_2O_3 and then by distillation in vacuum [45].

It has been observed that the compound becomes partly crystalline after 30 days at -6 °C. Chromatography of an ether solution of the crystals on Al_2O_3 lead to separation of Fp_2 and an orange crystalline compound, for which the IR data suggest the acyl derivative CXIX [45].

The compound decomposes slowly and continuously and is readily oxidized. The mass spectrum shows the molecular ion and fragments resulting from successive loss of CO. A peak at m/e=83 is assigned to the ion $[CH_2C(CH_3)_2CH=CH_2]^+$ [45].

$C_5H_5Fe(CO)_2CH_2CH(C_6H_9O)CH=CH_2$ (Table 5, No. 151) has been prepared from $[Fp(CH_2=CHCH=CH_2)]^+$ and 1-pyrrolidinocyclohexene as described for (E)-$FpCH_2CH=CHCH_2-C_6H_9O$ (No. 123), but in a somewhat modified way. Thus, the compound has been isolated in 56% yield by low-temperature hydrolysis of the reaction mixture and subsequent rapid chromatography on Al_2O_3 [120].

The structure of No. 151 is proposed on the basis of ^{1}H NMR and IR data [120]. The ^{1}H NMR spectrum in CS_2 shows two C_5H_5 proton resonances, indicative of the presence of the two anticipated diastereomers in approximately equal proportion [124]. This is also mentioned in [120] although only one signal from the C_5H_5 protons is observed.

The compound converts to CXXI on brief warming in CH_3CN at 60 to 80 °C. The reaction proceeds with high diastereospecificity. Thus, although two additional centers of chirality are created in this conversion, only three of a possible eight diastereomeric pairs are formed, deduced from the ^{13}C NMR spectrum. Steric factors controlling this reaction are discussed [120, 124].

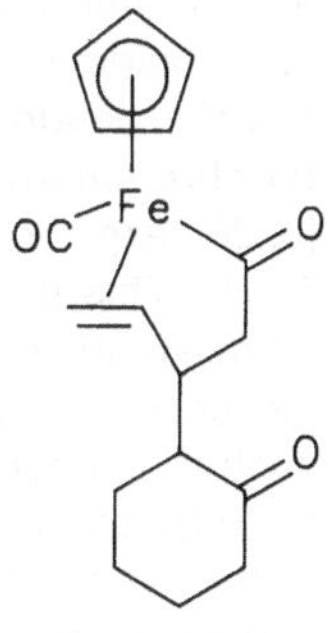

CXXI

$C_5H_5Fe(CO)_2CH_2CH(CH=CH_2)CH(COOCH_3)_2$ (Table 5, No. 152). Under the reaction conditions described for (Z)-$FpC(=CHCH_3)CH_2CH(COOCH_3)_2$ (No. 74) the compound has been obtained in 86% yield by addition of $LiCH(COOCH_3)_2$ to $[Fp(CH_2=CHCH=CH_2)]^+$. The structure is derived from the ^{1}H NMR data [120].

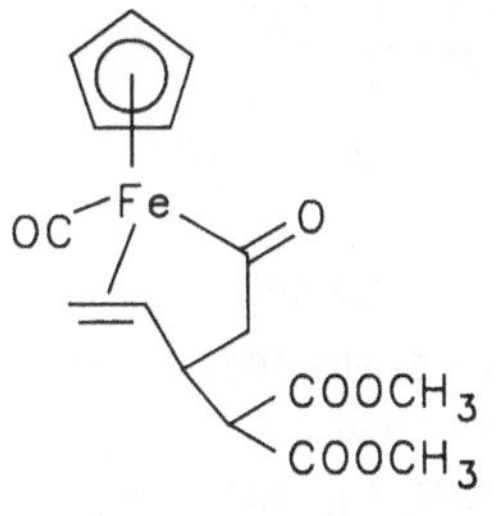

CXXII

Prolonged heating in THF led to recovery of the compound in high yield. Only a small amount of CXXII resulting from ligand transfer and olefin capture by the coordinatively unsaturated intermediate could be isolated [120].

References on pp. 141/7

(E)-C$_5$H$_5$Fe(CO)$_2$CH$_2$CH$_2$C(COOC$_2$H$_5$)=CHC$_6$H$_5$ (Table **5**, No. **153**) has been obtained by re-fluxing the ylid FpCH$_2$CH$_2$C(COOC$_2$H$_5$)=P(C$_6$H$_5$)$_3$ (see "Organoiron Compounds" B12, 1984, p. 282) and C$_6$H$_5$CHO (1:10 mole ratio) in benzene for 5.5 h (Wittig reaction), evaporation of the solvent from the cooled mixture, extraction of the residue with ether and chromatography on Al$_2$O$_3$ eluting with ether, 46% yield [120]. The (E)–structure is tentatively assigned to the product [120].

C$_5$H$_5$Fe(CO)$_2$(CH$_2$)$_3$CH=CH$_2$ (Table **5**, No. **155**) is also obtained by exposure of [Fp(CH$_2$=CH(CH$_2$)$_3$Fp)]$^+$ to an acetone solution of NaI at room temperature for 15 min, 69% yield [71]. The compound is formed in 14% yield along with Fp$_2$ by addition of CH$_2$=CHCH$_2$MgCl to [Fp(CH$_2$=CH$_2$)]$^+$ [120]. The compound is mentioned in connection with an investigation of transition metal complexes possessing alkenyl ligands in [33].

C$_5$H$_5$Fe(CO)$_2$CH(CHO)CH$_2$CH$_2$CH=CH$_2$ (Table **5**, No. **163**). Treatment of the compound with [O(CH$_3$)$_3$]BF$_4$ yields the vinyl ether complex [Fp(CH$_3$OCH=CHCH$_2$CH$_2$CH=CH$_2$)]$^+$ as a (Z)/(E)-mixture (5:1) [155].

C$_5$H$_5$Fe(CO)$_2$CH=C=CH$_2$ (Table **5**, No. **166**). The reaction product of Na[Fp] and HC≡CCH$_2$Br was first claimed to be FpCH$_2$C≡CH (see Section 1.5.2.3.16.1.12, Table 6, No. 7) [5], then FpC≡CCH$_3$ (see Section 1.5.2.3.16.1.12, Table 6, No. 4) [28], and later on FpCH=C=CH$_2$ (No. 166) [34, 36]. The FpCH$_2$C≡CH structure is not supported by the ^{1}H NMR spectrum. In [5] a singlet with 3H is explained by the coincidence of the CH$_2$ and ≡CH protons at $\delta =$ 1.8 ppm. At any rate, the compound primarily obtained is rearranged to give FpC≡CCH$_3$ by "acid" [34], "anhydrous acids" or by chromatography on silica gel or "acid washed Al$_2$O$_3$" [28]. The products studied in [5, 28] are isolated by chromatography (twice on "Al$_2$O$_3$") and should therefore be FpC≡CCH$_3$. This is supported by the respective ^{1}H NMR spectra on comparison with a sample of FpC≡CCH$_3$ obtained [212] from CH$_3$C≡CMgBr and FpCl. An exact structural decision for the unrearranged product being FpCH=C=CH$_2$ has not been given. The suggestion of this structure is based on the position of CH ($\delta =$ 4.89 ppm [34, 36]) and CH$_2$ ($\delta =$ 3.95 [36], 3.97 ppm [34]), and on the coupling constant J = 6.5 Hz in the ^{1}H NMR spectrum [34, 36]. This is in good agreement with the values of the product prepared by Method II [69]. The IR spectrum is not helpful in determining the possible structures, because the allenic band is in the range of the terminal CO bands [34].

Protonation of No. 166 with HPF$_6$·O(C$_2$H$_5$)$_2$ at -20 °C affords [Fp(HC≡CCH$_3$)]PF$_6$, which rapidly reacts with water to give a 2:1 mixture of FpCH$_2$COCH$_3$ and FpCOC$_2$H$_5$ [69]. The reaction with N–carbomethoxysulfonylamine yields the oxathiazepine derivative CXXIII [69].

CXXIII

C$_5$H$_5$Fe(CO)$_2$C(CF$_3$)=C=CF$_2$ (Table **5**, No. **168**) is reasonably stable in air, but solutions in organic solvents slowly decompose [23].

C$_5$H$_5$Fe(CO)$_2$C(CH$_2$N(C$_2$H$_5$)$_2$)=C=C(C$_6$H$_5$)$_2$ and **C$_5$H$_5$Fe(CO)$_2$C(CH$_2$N(C$_2$H$_5$)$_2$)=C=C(CH$_3$)C$_6$H$_5$** (Table **5**, Nos. **171** and **174**) have been prepared at -80 °C by addition of NH(C$_2$H$_5$)$_2$ to [Fp(CH$_2$=C=C=C(C$_6$H$_5$)R^1)]BF$_4$, obtained from FpCH$_2$C≡CCR^1(OH)C$_6$H$_5$ (R^1 = CH$_3$ or C$_6$H$_5$) and HBF$_4$ in ether at -50 °C [153].

C$_5$H$_5$Fe(CO)$_2$CH$_2$CH$_2$C(CH$_3$)=C=CH$_2$ (Table **5**, No. **173**) could not be purified by chromatography on Al$_2$O$_3$ due to decomposition on the column. It is isolated from the reaction mixture

(Method I) by filtration through Celite, evaporation of the solvent, extraction of the residue with pentane and solvent removal [148].

$C_5H_5Fe(CO)_2CH=C=O$ (Table 5, No. 176) is formed by thermolysis (30 °C) of CXXIV in $C_6D_5CD_3$ or CD_3COCD_3 as a rather unstable compound, which can be trapped by methanolysis to give $FpCH_2COOCH_3$. It has been identified by its ^1H NMR spectrum. However, a complete characterization of this ketenyl complex was not possible due to its instability [144].

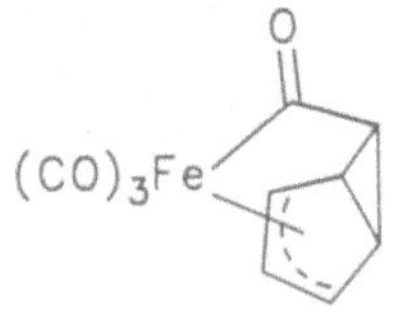

CXXIV

$C_5H_5Fe(CO)_2CH_2C(CH_3)=CH-C_3H_5O_2$ and $C_5H_5Fe(CO)_2CH(COCH_3)CH_2CH(C_3H_5O_2)C(CH_3)=CH_2$ (Table 5, Nos. 92 and 177). Both compounds are produced as follows [192]:

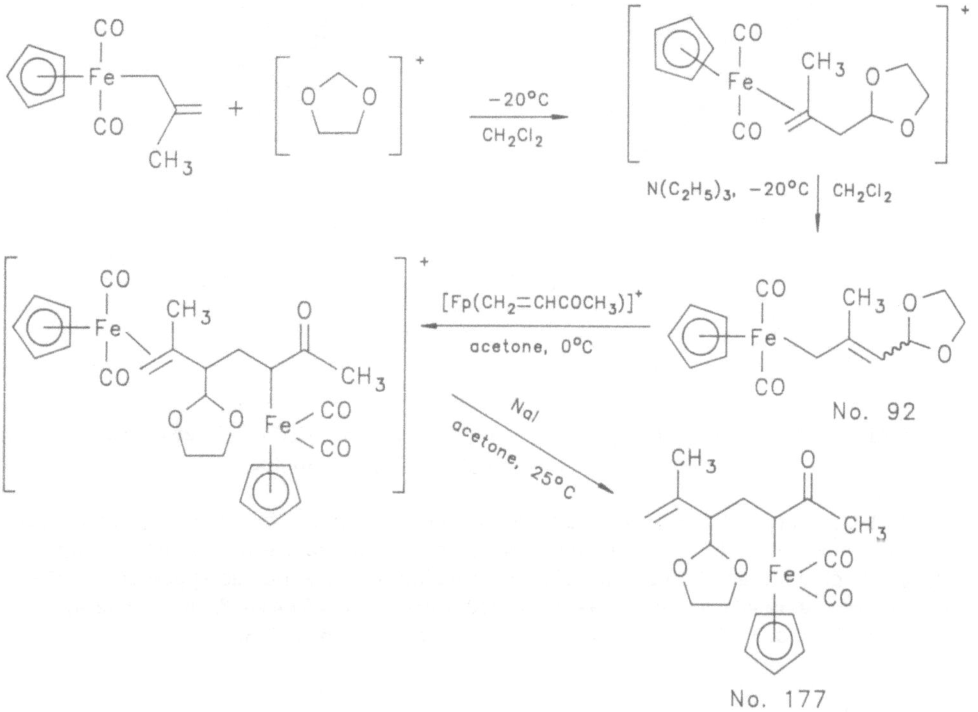

No. 92

No. 177

The reaction of No. 177 with CH_3Li at -78 °C followed by treatment with HCl gives $(CH_3)_2C=CHCH_2CH(C(CH_3)=CH_2)-C_3H_5O_2$ [192].

$C_5H_5Fe(CO)_2C[=C(CN)_2]C(C_6H_5)=C(CN)_2 \cdot 0.125\ CH_2Cl_2$ (Table 5, No. 179) is prepared from $FpC\equiv CC_6H_5$ and $(NC)_2C=C(CN)_2$ (1:2 mole ratio) in CH_2Cl_2 at room temperature. After 90 min the filtrate is diluted with C_2H_5OH and concentrated. On cooling to -50 °C the compound crystallizes, yield 77% [210]. A similar reaction is said to give a cyclobutenyl derivative, yield 82% (see Section 1.5.2.3.16.3, Table 8, No. 21) [211]. This compound is believed to be No. 179 because the cyclobutenyl compound could be obtained using nonpolar solvents instead of CH_2Cl_2 and has different structural and spectroscopic parameters [210].

References on pp. 141/7

The compound crystallizes in the triclinic space group P$\bar{1}$ – C$_i^1$ (No. 2) with a = 9.378(2), b = 13.874(3), c = 7.935(4) Å, α = 92.92(3)°, β = 101.57(2)°, γ = 108.78(1)°; Z = 2, d$_c$ = 1.420 g/cm³. The structure is shown in **Fig. 3** [210].

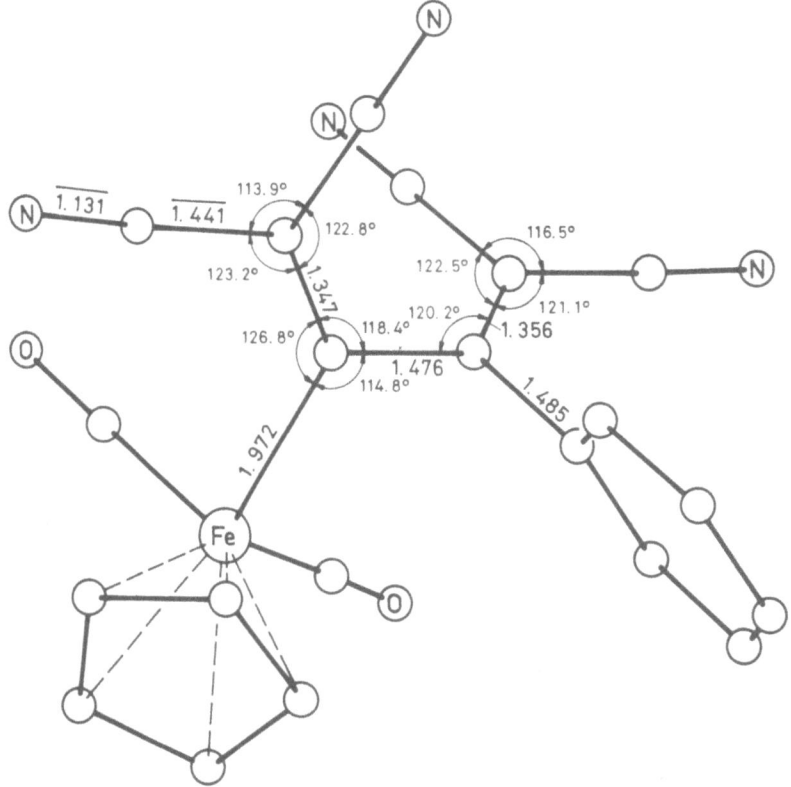

Fig. 3. Molecular structure of C$_5$H$_5$Fe(CO)$_2$C[=C(CN)$_2$]C(C$_6$H$_5$)=C(CN)$_2$ · 0.125 CH$_2$Cl$_2$ (No. 179) with selected bond lengths (in Å) and angles.

(Z,Z)–C$_5$H$_5$Fe(CO)$_2$(CH=CH)$_2$Br (Table 5, No. **180**) is prepared by stirring FpI and Br(CH= CH)$_2$Li in ether at −78 °C for 2 h, warming to room temperature and chromatography on Al$_2$O$_3$ with pentane as eluent, yield 23%. It crystallizes in the triclinic space group P1 – C$_1^1$ (No. 1) with a = 7.402(3), b = 7.499(2), c = 10.574(7) Å, α = 91.41(4)°, β = 103.42(4)°, γ = 99.97(2)°; Z = 2, d$_c$ = 1.83 g/cm³. The structure is shown in **Fig. 4** [213].

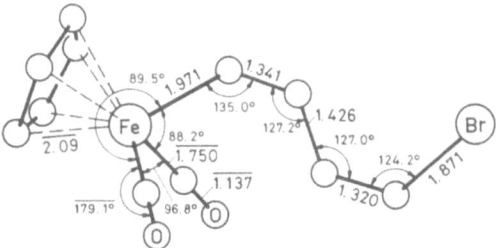

Fig. 4. Molecular structure of (Z,Z)–C$_5$H$_5$Fe(CO)$_2$(CH=CH)$_2$Br (No. 180) with selected bond lengths (in Å) and angles.

References:

[1] Green, M.L.H.; Nagy, P. (Proc. Chem. Soc. **1961** 378).
[2] King, R.B.; Stafford, S.L.; Treichel, P.M.; Stone, F.G.A. (J. Am. Chem. Soc. **83** [1961] 3604/8).
[3] King, R.B.; Treichel, P.M.; Stone, F.G.A. (Proc. Chem. Soc. **1961** 69/70).
[4] Pitcher, E.; Stone, F.G.A. (Spectrochim. Acta **17** [1961] 1244/8).
[5] Ariyaratne, J.K.P.; Green, M.L.H. (J. Organometal. Chem. **1** [1963/64] 90/3).
[6] Green, M.L.H.; Nagy, P.L.I. (J. Chem. Soc. **1963** 189/97).
[7] Ariyaratne, J.K.P.; Green, M.L.H.; Nagy, P.L.I. (Proc. Chem. Soc. **1963** 107).
[8] King, R.B.; Bisnette, M.B. (J. Organometal. Chem. **2** [1964] 15/37).
[9] Green, M.L.H.; Ishaq, M.; Mole, T. (Z. Naturforsch. **20b** [1965] 598).
[10] Jolly, P.W.; Bruce, M.I.; Stone, F.G.A. (J. Chem. Soc. **1965** 5830/7).

[11] Green, M.L.H.; Stear, A.N. (Z. Naturforsch. **20b** [1965] 812).
[12] Jolly, P.W.; Stone, F.G.A. (Chem. Commun. **1965** 85/6).
[13] Bruce, M.I.; Jolly, P.W.; Stone, F.G.A. (J. Chem. Soc. A **1966** 1602/6).
[14] Green, M.; Mayne, N.; Stone, F.G.A. (Chem. Commun. **1966** 755/6).
[15] Bichler, R.E.J.; Booth, M.R.; Clark, H.C. (Inorg. Nucl. Chem. Letters **3** [1967] 71/4).
[16] Nesmeyanov, A.N.; Rybin, L.V.; Rybinskaya, M.I.; Ustynyuk, Yu.A. (Zh. Obshch. Khim. **37** [1967] 1587/91; J. Gen. Chem. [USSR] **37** [1967] 1505/8).
[17] Bruce, M.I.; Harbourne, D.A.; Waugh, F.; Stone, F.G.A. (J. Chem. Soc. A **1968** 895/8).
[18] Bruce, M.I.; Davies, C.H. (Inorg. Nucl. Chem. Letters **4** [1968] 675/7).
[19] Bayer, E.; Breitmeyer, E.; Schurig, V. (Chem. Ber. **101** [1968] 1594/1600).
[20] Bruce, M.I. (J. Chem. Soc. A **1968** 1459/64).

[21] Andrianov, V.G.; Struchkov, Yu.T. (Chem. Commun. **1968** 1590/2).
[22] Green, M.; Mayne, N.; Stone, F.G.A. (J. Chem. Soc. A **1968** 902/5).
[23] Goodfellow, R.J.; Green, M.; Mayne, N.; Rest, A.J.; Stone, F.G.A. (J. Chem. Soc. A **1968** 177/80).
[24] Green, M.; Mayne, N.; Stone, F.G.A. (unpublished results from Bruce, M.I.; Stone, F.G.A., Preparat. Inorg. React. **4** [1968] 177/235, 191).
[25] King, R.B. (Inorg. Chim. Acta **2** [1968] 454/8).
[26] Harbourne, D.A.; Stone, F.G.A. (J. Chem. Soc. A **1968** 1765/71).
[27] Harbourne, D.A.; Stone, F.G.A. (unpublished results from Bruce, M.I.; Stone, F.G.A., Preparat. Inorg. React. **4** [1968] 177/235, 191).
[28] Jolly, P.W.; Pettit, R. (J. Organometal. Chem. **12** [1968] 491/5).
[29] Nesmeyanov, A.N.; Rybin, L.V.; Rybinskaya, M.I.; Kaganovich, V.S.; Ustynyuk, Yu.A.; Leshcheva, I.F. (Zh. Obshch. Khim. **38** [1968] 1471/6; J. Gen. Chem. [USSR] **38** [1968] 1424/7).
[30] Goldwhite, H.; Rowsell, D.G.; Valdez, C. (J. Organometal. Chem. **12** [1968] 133/41).

[31] Bichler, R.E.J.; Clark, H.C. (4th Intern. Conf. Organometal. Chem., Bristol 1969, Abstr. O 4).
[32] Bruce, M.I.; Davies, C.H. (J. Chem. Soc. A **1969** 1077/81).
[33] Fahey, D.R. (Diss. Univ. North Dakota 1969; Diss. Abstr. Intern. B **30** [1969] 1589).
[34] Johnson, M.D.; Mayle, C. (J. Chem. Soc. D **1969** 192).
[35] Nesmeyanov, A.N.; Rybinskaya, M.I.; Rybin, L.V.; Kaganovich, V.S.; Ustynyuk, Yu.A.; Leshcheva, I.F. (Izv. Akad. Nauk SSSR Ser. Khim. **1969** 1100/3; Bull. Acad. Sci. USSR Div. Chem. Sci. **1969** 1004/6).
[36] Roustan, J.L.; Cadiot, P. (Compt. Rend. C **268** [1969] 734/6).
[37] Rybinskaya, M.I.; Rybin, L.V.; Nesmeyanov, A.N. (4th Intern. Conf. Organometal. Chem., Bristol 1969, Abstr. K 5).

[38] Struchkov, Yu.T.; Andrianov, V.G.; Bir'Yukov, B.P.; Gusev, A.I.; Semion, V.A. (4th Intern. Conf. Organometal. Chem., Bristol 1969, Abstr. A 15).

[39] Rybin, V.L.; Kaganovich, V.S.; Rybinskaya, M.I. (Izv. Akad. Nauk SSSR Ser. Khim. **1969** 360/2; Bull. Acad. Sci. USSR Div. Chem. Sci. **1969** 311/3).

[40] Bichler, R.E.J.; Booth, M.R.; Clark, H.C. (J. Organometal. Chem. **24** [1970] 145/58).

[41] Giering, W.P.; Rosenblum, M. (J. Organometal. Chem. **25** [1970] C71/C73).

[42] Merour, J.Y.; Cadiot, P. (Compt. Rend. C **271** [1970] 83/5).

[43] Merour, J.Y. (Compt. Rend. C **271** [1970] 1397/9).

[44] Benaim, J.; Merour, J.Y.; Roustan, J.L. (Tetrahedron Letters **1971** 983/6).

[45] Green, M.L.H.; Smith, M.J. (J. Chem. Soc. A **1971** 3220/3).

[46] Giering, W.P.; Rosenblum, M. (J. Am. Chem. Soc. **93** [1971] 5299/301).

[47] Merour, J.Y.; Charrier, C.; Roustan, J.L.; Benaim, J. (Compt. Rend. C **273** [1971] 285/8).

[48] Nesmeyanov, A.N.; Rybinskaya, M.I.; Rybin, L.V.; Kaganovich, V.S.; Petrovskii, P.V. (J. Organometal. Chem. **31** [1971] 257/67).

[49] Nesmeyanov, A.N.; Rybin, L.V.; Rybinskaya, M.I.; Kaganovich, V.S.; Petrovskii, P.V. (Izv. Akad. Nauk SSSR Ser. Khim. **1971** 2733/8; Bull. Acad. Sci. USSR Div. Chem. Sci. **1971** 2592/6).

[50] Ross, D.A. (Diss. Ohio State Univ., Columbus 1970; Diss. Abstr. Intern. B **31** [1971] 3905/6).

[51] Jacobson, S.E.; Wojcicki, A. (J. Am. Chem. Soc. **93** [1971] 2535/7).

[52] Thomasson, J.E.; Robinson, P.W.; Ross, D.A.; Wojcicki, A. (Inorg. Chem. **10** [1971] 2130/7).

[53] Su, S.R.; Wojcicki, A. (J. Organometal. Chem. **31** [1971] C34/C36).

[54] Su, S.R. (Diss. Ohio State Univ., Columbus 1971; Diss. Abstr. Intern. B **32** [1972] 6283).

[55] Cutler, A.; Fish, R.W.; Giering, W.P.; Rosenblum, M. (J. Am. Chem. Soc. **94** [1972] 4354/5).

[56] Ciappenelli, D.J.; Cotton, F.A.; Kruczynski, L. (J. Organometal. Chem. **42** [1972] 159/62).

[57] Lichtenberg, D.W.; Wojcicki, A. (J. Am. Chem. Soc. **94** [1972] 8271/3).

[58] King, R.B.; Saran, M.S. (J. Am. Chem. Soc. **94** [1972] 1784/5).

[59] King, R.B.; Zipperer, W.C. (Inorg. Chem. **11** [1972] 2119/25).

[60] Müller, J. (Angew. Chem. **84** [1972] 725/37; Angew. Chem. Intern. Ed. Engl. **11** [1972] 653/65).

[61] Nesmeyanov, A.N.; Rybinskaya, M.I.; Petrovskaya, E.A.; Popova, T.V. (Izv. Akad. Nauk SSSR Ser. Khim. **1972** 1646/7; Bull. Acad. Sci. USSR Div. Chem. Sci. **1972** 1588/9).

[62] Roustan, J.L.; Benaim, J.; Charrier, C.; Merour, J.Y. (Tetrahedron Letters **1972** 1953/6).

[63] Yamamoto, Y.; Wojcicki, A. (Inorg. Nucl. Chem. Letters **8** [1972] 833/6).

[64] Yamamoto, Y.; Wojcicki, A. (J. Chem. Soc. Chem. Commun. **1972** 1088/9).

[65] Giering, W.P.; Raghu, S.; Rosenblum, M.; Cutler, A.; Ehntholt, D.; Fish, R.W. (J. Am. Chem. Soc. **94** [1972] 8251/3).

[66] Churchill, M.R.; Ni Chang, S.W.Y. (J. Am. Chem. Soc. **95** [1973] 5931/8).

[67] King, R.B.; Saran, M.S. (J. Am. Chem. Soc. **95** [1973] 1811/7).

[68] Nesmeyanov, A.N.; Kolobova, N.E.; Znobina, G.K.; Anisimov, K.N.; Zlotina, I.B.; Bargamova, M.D. (Izv. Akad. Nauk SSSR Ser. Khim. **1973** 2168; Bull. Acad. Sci. USSR Div. Chem. Sci. **1973** 2127).

[69] Raghu, S.; Rosenblum, M. (J. Am. Chem. Soc. **95** [1973] 3060/2).

[70] Nesmeyanov, A.N.; Rybinskaya, M.I.; Kaganovich, V.S.; Popova, T.V.; Petrovskaya, E.A. (Izv. Akad. Nauk SSSR Ser. Khim. **1973** 2087/9; Bull. Acad. Sci. USSR Div. Chem. Sci. **1973** 2031/3).

[71] Rosan, A.; Rosenblum, M.; Tancrede, J. (J. Am. Chem. Soc. **95** [1973] 3062/4).

[72] Yamamoto, Y.; Wojcicki, A. (Inorg. Chem. **12** [1973] 1779/88).

[73] Cohen, L.; Giering, W.P.; Kenedy, D.; Magatti, C.V.; Sanders, A. (J. Organometal. Chem. **65** [1974] C57/C60).

[74] Chen, L.S.; Su, S.R.; Wojcicki, A. (J. Am. Chem. Soc. **96** [1974] 5655/6).

[75] King, R.B. (Ann. N.Y. Acad. Sci. **239** [1974] 171/9).

[76] Gansow, O.A.; Burke, A.R.; King, R.B.; Saran, M.S. (Inorg. Nucl. Chem. Letters **10** [1974] 291/5).

[77] Nicholas, K. (unpublished results from Rosenblum, M., Accounts Chem. Res. **7** [1974] 122/8).

[78] Magatti, C.V.; Giering, W.P. (J. Organometal. Chem. **73** [1974] 85/92).

[79] King, R.B.; Hodges, K.C. (J. Am. Chem. Soc. **96** [1974] 1263/4).

[80] Su, S.R.; Wojcicki, A. (Inorg. Chim. Acta **8** [1974] 55/60).

[81] Tancrede, J.M.P. (Diss. Brandeis Univ. 1973; Diss. Abstr. Intern. B **34** [1974] 3177).

[82] Sanders, A.; Giering, W.P. (J. Am. Chem. Soc. **96** [1974] 5247/8).

[83] Raghu, S. (unpublished results from Rosenblum, M., Accounts Chem. Res. **7** [1974] 122/8).

[84] Faller, J.W.; Johnson, B.V.; Dryja, T.P. (J. Organometal. Chem. **65** [1974] 395/400).

[85] King, R.B. (Organotransition Met. Chem. Proc. 1st Japan. Am. Semin., Honolulu 1974 [1975], pp. 37/47).

[86] Cutler, A.; Ehntholt, D.; Lennon, P.; Nicholas, K.; Marten, D.F.; Madhavarao, M.; Raghu, S.; Rosan, A.; Rosenblum, M. (J. Am. Chem. Soc. **97** [1975] 3149/57).

[87] Davidson, J.L.; Sharp, D.W.A. (J. Chem. Soc. Dalton Trans. **1975** 2283/7).

[88] Cutler, A.R. (Diss. Brandeis Univ. 1975; Diss. Abstr. Intern. B **36** [1975] 233).

[89] Hodges, K.C. (Diss. Univ. Georgia 1974; Diss. Abstr. Intern. B **35** [1975] 4822).

[90] King, R.B.; Hodges, K.C. (J. Am. Chem. Soc. **97** [1975] 2702/12).

[91] Lichtenberg, D.W.; Wojcicki, A. (J. Organometal. Chem. **94** [1975] 311/26).

[92] Lichtenberg, D.W.; Wojcicki, A. (Inorg. Chem. **14** [1975] 1295/301).

[93] Nesmeyanov, A.N.; Kolobova, N.E.; Zlotina, I.B.; Solodova, M.Ya.; Anisimov, K.N. (Tezisy Dokl. 12th Vses. Chugaevskoe Soveshch. Khim. Kompleksn. Soedin. Novosibirsk, USSR, 1975, Vol. 3, pp. 474/5; C.A. **86** [1977] No. 5580).

[94] Nesmeyanov, A.N.; Kolobova, N.E.; Zlotina, I.B.; Nekrasov, Yu.S.; Sizoi, V.F.; Anisimov, K.N. (Dokl. Akad. Nauk SSSR **224** [1975] 604/6; Dokl. Chem. Proc. Acad. Sci. USSR **220/225** [1975] 568/9).

[95] Wojcicki, A.; Dizikes, L.J. (7th Intern. Conf. Organometal. Chem., Venice, Italy 1975, Abstr. S1C).

[96] Sanders, A.; Giering, W.P. (J. Am. Chem. Soc. **97** [1975] 919/21).

[97] Faller, J.W.; Johnson, B.V. (J. Organometal. Chem. **88** [1975] 101/13).

[98] Bauch, T.E.; Konowitz, H.; Giering, W.P. (J. Organometal. Chem. **114** [1976] C15/C18).

[99] Cutler, A.; Ehntholt, D.; Giering, W.P.; Lennon, P.; Raghu, S.; Rosan, A.; Rosenblum, M.; Tancrede, J.; Wells, D. (J. Am. Chem. Soc. **98** [1976] 3495/507).

[100] Dizikes, L.J.; Wojcicki, A. (Inorg. Chim. Acta **20** [1976] L29/L30).

[101] Genco, N.; Marten, D.; Raghu, S.; Rosenblum, M. (J. Am. Chem. Soc. **98** [1976] 848/9).

[102] Bauch, T.E.; Giering, W.P. (J. Organometal. Chem. **114** [1976] 165/74).

[103] Pannell, K.H.; Lappert, M.F.; Stanley, K. (J. Organometal. Chem. **112** [1976] 37/48).

[104] Lennon, P.; Madhavarao, M.; Rosan, A.; Rosenblum, M. (J. Organometal. Chem. **108** [1976] 93/109).

[105] Nesmeyanov, A.N.; Kolobova, N.E.; Zlotina, I.B.; Lokshin, B.V.; Leshcheva, I.F.; Znobina, G.K.; Anisimov, K.N. (J. Organometal. Chem. **110** [1976] 399/44).

[106] Nesmeyanov, A.N.; Kolobova, N.E.; Zlotina, I.B.; Lokshin, B.V.; Leshcheva, I.F.; Znobina, G.K.; Anisimov, K.N. (Izv. Akad. Nauk SSSR Ser. Khim. **1976** 1124/9; Bull. Acad. Sci. USSR Div. Chem. Sci. **1976** 1093/7).

[107] Nesmeyanov, A.N.; Rybinskaya, M.I.; Rybin, L.V.; Petrovskaya, E.A.; Svoren, V.A. (Izv. Akad. Nauk SSSR Ser. Khim. **1976** 1592/6; Bull. Acad. Sci. USSR Div. Chem. Sci. **1976** 1511/4).

[108] Nesmeyanov, A.N.; Aleksandrov, G.G.; Bokii, N.G.; Zlotina, I.B.; Struchkov, Yu.T.; Kolobova, N.E. (J. Organometal. Chem. **111** [1976] C9/C12).

[109] Sanders, A.; Bauch, T.; Magatti, C.V.; Lorenc, C.; Giering, W.P. (J. Organometal. Chem. **107** [1976] 359/75).

[110] Sanders, A.; Giering, W.P. (J. Organometal. Chem. **104** [1976] 67/78).

[111] Sanders, A.; Giering, W.P. (J. Organometal. Chem. **104** [1976] 49/65).

[112] Williams, J.P. (Diss. Ohio State Univ., Columbus 1975; Diss. Abstr. Intern. B **36** [1976] 3950).

[113] Davidson, J.L.; Green, M.; Stone, F.G.A.; Welch, A.J. (J. Chem. Soc. Dalton Trans. **1976** 2044/53).

[114] Fish, R.W.; Giering, W.P.; Marten, D.; Rosenblum, M. (J. Organometal. Chem. **105** [1976] 101/18).

[115] Stone, F.G.A. (AD-AO-32905 [1976] 1/29; C.A. **87** [1977] No. 23353).

[116] Sanders, A. (Diss. Boston. Univ. Grad. School 1975; Diss. Abstr. Intern. B **36** [1976] 2814/5).

[117] Aleksandrov, G.G.; Andrianov, V.G.; Bokii, N.G.; Struchkov, Yu.T. (3rd Intern. Semin. Cryst. Chem. Coord. Organometal. Compounds Proc. Trzebieszowice, Pol., 1977, pp. 130/3; C.A. **93** [1980] No. 95373).

[118] Foxman, B.; Marten, D.; Rosan, A.; Raghu, S.; Rosenblum, M. (J. Am. Chem. Soc. **99** [1977] 2160/5).

[119] Bauch, T.E.; Konowitz, H.; Kerber, R.C.; Giering, W.P. (J. Organometal. Chem. **131** [1977] C7/C10).

[120] Lennon, P.; Rosan, A.M.; Rosenblum, M. (J. Am. Chem. Soc. **99** [1977] 8426/39).

[121] Dizikes, L.J.; Wojcicki, A. (J. Organometal. Chem. **137** [1977] 79/90).

[122] Klemarczyk, P.; Price, T.; Priester, W.; Rosenblum, M. (J. Organometal. Chem. **139** [1977] C25/C28).

[123] Krusic, P.J.; Fagan, P.J.; San Filippo, J. (J. Am. Chem. Soc. **99** [1977] 250/2).

[124] Lennon, P.; Priester, W.; Rosan, A.; Madhavarao, M.; Rosenblum, M. (J. Organometal. Chem. **139** [1977] C29/C33).

[125] Kolobova, N.E.; Zlotina, I.B.; Ivanova, I.I. (Izv. Akad. Nauk SSSR Ser. Khim. **1977** 2130/1; Bull. Acad. Sci. USSR Div. Chem. Sci. **1977** 1973/4).

[126] Chen, L.S.; Lichtenberg, D.W.; Robinson, P.W.; Yamamoto, Y.; Wojcicki, A. (Inorg. Chim. Acta **25** [1977] 165/72).

[127] Nesmeyanov, A.N.; Zlotina, I.B.; Khomutov, M.A.; Leshcheva, I.F.; Kolobova, N.E.; Anisimov, K.N. (Izv. Akad. Nauk SSSR Ser. Khim. **1977** 705/6; Bull. Acad. Sci. USSR Div. Chem. Sci. **1977** 642/3).

[128] Nesmeyanov, A.N.; Kolobova, N.E.; Zlotina, I.B.; Ivanova, L.V.; Anisimov, K.N. (Izv. Akad. Nauk SSSR Ser. Khim. **1977** 707/8; Bull. Acad. Sci. USSR Div. Chem. Sci. **1977** 644/6).

[129] Williams, J.P.; Wojcicki, A. (Inorg. Chem. **16** [1977] 3116/24).

[130] Waterman, P.S.; Giering, W.P. (J. Organometal. Chem. **155** [1978] C47/C50).

[131] Cotton, J.D. (J. Organometal. Chem. **159** [1978] 465/76).

[132] Chen, L.S.; Su, S.R.; Wojcicki, A. (Inorg. Chim. Acta **27** [1978] 79/89).

[133] Bauch, T.E.; Giering, W.P. (J. Organometal. Chem. **144** [1978] 335/49).

[134] Klemarczyk, P.; Rosenblum, M. (J. Org. Chem. **43** [1978] 3488/93).

[135] Koridze, A.A.; Astakhova, N.M.; Petrovskii, P.V.; Lutsenko, A.I. (Dokl. Akad. Nauk SSSR **242** [1978] 117/20; Dokl. Chem. Proc. Acad. Sci. USSR **238/243** [1978] 416/8).

[136] Downs, R.L.; Wojcicki, A. (Inorg. Chim. Acta **27** [1978] 91/103).

[137] Filippo, J.S.; Silbermann, J.; Fagan, P.J. (J. Am. Chem. Soc. **100** [1978] 4834/42).

[138] Bell, P.B. (Diss. Ohio State Univ., Columbus 1978; Diss. Abstr. Intern. B **39** [1979] 4878).

[139] Abram, T.S.; Baker, R. (Syn. React. Inorg. Metal-Org. Chem. **9** [1979] 471/7).

[140] Nesmeyanov, A.N.; Petrovskaya, S.A.; Rybin, L.V.; Rybinskaya, M.I. (Izv. Akad. Nauk SSSR Ser. Khim. **1979** 2045/9; Bull. Acad. Sci. USSR Div. Chem. Sci. **1979** 1887/90).

[141] Waterman, P.S. (Diss. Boston Univ. Grad. School 1979; Diss. Abstr. Intern. B **40** [1979] 2200).

[142] Abram, T.S.; Baker, R. (J. Chem. Soc. Chem. Commun. **1979** 267/8).

[143] Abram, T.S.; Baker, R.; Exon, C.M. (Tetrahedron Letters **1979** 4103/6).

[144] Aumann, R.; Wörmann, H. (Chem. Ber. **112** [1979] 1233/51).

[145] Petillon, F.Y.; Le Floch-Perennou, F.; Guerchais, J.E.; Sharp, D.W.A. (J. Organometal. Chem. **173** [1979] 89/106).

[146] Andrianov, V.G.; Struchkov, Yu.T.; Zlotina, I.B.; Khomutov, M.A. (Koord. Khim. **5** [1979] 1872/6; Soviet J. Coord. Chem. **5** [1979] 1462/5).

[147] Fabian, B.D.; Labinger, J.A. (J. Am. Chem. Soc. **101** [1979] 2239/40).

[148] Roustan, J.L.; Merour, J.Y.; Charrier, C.; Benaim, J.; Cadiot, P. (J. Organometal. Chem. **168** [1979] 61/86).

[149] Reger, D.L.; Coleman, C.J.; McElligott, P.J. (J. Organometal. Chem. **171** [1979] 73/84).

[150] Samuels, S.B.; Berryhill, S.R.; Rosenblum, M. (J. Organometal. Chem. **166** [1979] C9/C12).

[151] Schilling, B.E.R.; Hoffmann, R.; Faller, J.W. (J. Am. Chem. Soc. **101** [1979] 592/8).

[152] Fadel, S.; Weidenhammer, K.; Ziegler, M.L. (Z. Anorg. Allgem. Chem. **453** [1979] 98/106).

[153] Benaim, J.; Giulieri, F. (J. Organometal. Chem. **202** [1980] C9/C14).

[154] Laycock, D.E.; Hartgerink, J.; Baird, M.C. (J. Org. Chem. **45** [1980] 291/9).

[155] Lennon, P.J.; Rosan, A.; Rosenblum, M.; Tancrede, J.; Waterman, P. (J. Am. Chem. Soc. **102** [1980] 7033/8).

[156] Fabian, B.D.; Fehlner, T.P.; Hwang, L.-S.J.; Labinger, J.A. (J. Organometal. Chem. **191** [1980] 409/13).

[157] Rosenblum, M.; Waterman, P.S. (J. Organometal. Chem. **187** [1980] 267/75).

[158] Bottrill, M.; Green, M.; O'Brien, E.; Smart, L.S.; Woodward, P. (J. Chem. Soc. Dalton Trans. **1980** 292/8).

[159] Quinn, S.; Shaver, A. (Inorg. Chim. Acta **38** [1980] 243/5).

[160] Bury, A.; Johnson, M.D.; Stewart, M.J. (J. Chem. Soc. Chem. Commun. **1980** 622/3).

[161] Chang, T.C.T.; Rosenblum, M.; Samuels, S.B. (J. Am. Chem. Soc. **102** [1980] 5930/1).

[162] Stone, J.A.; Laycock, D.E.; Lin, M.; Baird, M.C. (J. Chem. Soc. Dalton Trans. **1980** 2488/92).

[163] Bell, P.B.; Wojcicki, A. (Inorg. Chem. **20** [1981] 1585/92).

[164] Marten, D.F. (J. Org. Chem. **46** [1981] 5422/5).

[165] Lee, M.T.; Giering, W.P.; Magnuson, R.H. (Abstr. Papers 182nd Natl. Meeting Am. Chem. Soc., New York 1981, INOR 111).

[166] Rybin, L.V.; Petrovskaya, E.A.; Batsanov, A.S.; Struchkov.; Yu.T.; Rybinskaya, M.I. (J. Organometal. Chem. **212** [1981] 95/105).

[167] Bodnar, T.; Cutler, A.R. (J. Organometal. Chem. **213** [1981] C31/C36).

[168] Cutler, A.R.; Bodnar, T. (Abstr. Papers 182nd Natl. Meeting Am. Chem. Soc., New York 1981, INOR 110).

[169] Chang, T.C.T.; Foxman, B.M.; Rosenblum, M.; Stockman, C. (J. Am. Chem. Soc. **103** [1981] 7361/2).

[170] Chang, T.C.T.; Rosenblum, M. (J. Org. Chem. **46** [1981] 4103/5).

[171] Priester, W.; Rosenblum, M.; Samuels, S.B. (Syn. React. Inorg. Metal-Org. Chem. **11** [1981] 525/37).

[172] Abram, T.S.; Baker, R.; Exon, C.M.; Rao, V.B. (J. Chem. Soc. Perkin Trans. I **1982** 285/94).

[173] Abram, T.S.; Baker, R.; Exon, C.M.; Rao, V.B.; Turner, R.W. (J. Chem. Soc. Perkin Trans. I **1982** 301/6).

[174] Kremer, K.A.M.; Kuo, G.-H.; O'Connor, E.J.; Helquist, P.; Kerber, R.C. (J. Am. Chem. Soc. **104** [1982] 6119/21).

[175] Bucheister, A.; Klemarczyk, P.; Rosenblum, M. (Organometallics **1** [1982] 1679/84).

[176] Casey, C.P.; Miles, W.H.; Tukada, H.; O'Connor, J.M. (J. Am. Chem. Soc. **104** [1982] 3761/2).

[177] Casey, C.P.; Miles, W.H.; Tukada, H.; O'Connor, J.M. (Abstr. Papers 184th Natl. Meeting Am. Chem. Soc., Kansas City 1982, ORGN 198).

[178] Baker, R.; Exon, C.M.; Rao, V.B.; Turner, R.W. (J. Chem. Soc. Perkin Trans. I **1982** 295/300).

[179] Casey, C.P.; Tukada, H.; Miles, W.H. (Organometallics **1** [1982] 1083/4).

[180] Leung, T.W.; Christoph, G.G.; Wojcicki, A. (Inorg. Chim. Acta **76** [1983] L281/L282).

[181] Lee, M.T. (Diss. Boston Univ. Grad. School 1983; Diss. Abstr. Intern. B **44** [1983] 1458).

[182] Lennon, P.; Rosenblum, M. (J. Am. Chem. Soc. **105** [1983] 1233/41).

[183] Liebeskind, L.S.; Welker, M.E. (Organometallics **2** [1983] 195/7).

[184] Marten, D.F.; Wilburn, S.M. (J. Organometal. Chem. **251** [1983] 71/8).

[185] Zulu, S.J. (Diss. Boston Univ. Grad. School 1983; Diss. Abstr. Intern. B **44** [1983] 493).

[186] Koridze, A.A.; Astakhova, N.M.; Petrovskii, P.V. (J. Organometal. Chem. **254** [1983] 345/60).

[187] Baker, R.; Rao, V.B.; Erdik, E. (J. Organometal. Chem. **243** [1983] 451/60).

[188] Casey, C.P.; Miles, W.H. (J. Organometal. Chem. **254** [1983] 333/7).

[189] Fabian, B.D.; Labinger, J.A. (Organometallics **2** [1983] 659/64).

[190] Kuo, G.-H.; Helquist, P.; Kerber, R.C. (Organometallics **3** [1984] 806/8).

[191] Kremer, K.A.M.; Helquist, P. (Organometallics **3** [1984] 1743/5).

[192] Rosenblum, M.; Bucheister, A.; Chang, T.C.T.; Cohen, M.; Marsi, M.; Samuels, S.B.; Scheck, D.; Sofen, N.; Watkins, J.C. (Pure Appl. Chem. **56** [1984] 129/36).

[193] Casey, C.P.; Miles, W.H. (Organometallics **3** [1984] 808/9).

[194] Baker, R.; Keen, R.B.; Morris, M.D.; Turner, R.W. (J. Chem. Soc. Chem. Commun. **1984** 987/8).

[195] Giulieri, F.; Benaim, J. (J. Organometal. Chem. **276** [1984] 367/76).

[196] Ferguson, S.B.; Sanderson, L.J.; Shackleton, T.A.; Baird, M.C. (Inorg. Chim. Acta **83** [1984] L45/L47).

[197] Brinkman, K.; Helquist, P. (Tetrahedron Letters **26** [1985] 2845/8).

[198] Kuo, G.-H. (Diss. New York State Univ. 1985; Diss. Abstr. Intern. B **46** [1985] 1925).

[199] Iyer, R.S.; Kuo, G.-H.; Helquist, P. (J. Org. Chem. **50** [1985] 5898/900).

[200] Waterman, P.S.; Belmonte, J.E.; Bauch, T.E.; Belmonte, P.A.; Giering, W.P. (J. Organometal. Chem. **294** [1985] 235/50).

[201] Casey, C.P.; Miles, W.H.; Tukada, H. (J. Am. Chem. Soc. **107** [1985] 2924/31).

[202] Bodnar, T.W.; Cutler, A.R. (Organometallics **4** [1985] 1558/65).

[203] Hegedus, L.S.; Holden, M.S. (J. Org. Chem. **50** [1985] 3920/3).

[204] Benaim, J.; L'Honore, A. (J. Organometal. Chem. **202** [1980] C53/C57).

[205] Denisovich, L.I.; Gubin, S.P.; Chapovskii, Yu.A. (Izv. Akad. Nauk SSSR Ser. Khim. **1967** 2378/84; Bull. Acad. Sci. USSR Div. Chem. Sci. **1967** 2271/5).

[206] Denisovich, L.I.; Gubin, S.P. (J. Organometal. Chem. **57** [1973] 109/19).

[207] Jacobson, S.E.; Wojcicki, A. (J. Am. Chem. Soc. **95** [1973] 6962/70).

[208] Williams, J.P.; Wojcicki, A. (Inorg. Chem. **16** [1977] 2506/12).

[209] Watkins, J.C.; Rosenblum, M. (Tetrahedron Letters **25** [1984] 2097/100).

[210] Bruce, M.I.; Duffy, D.N.; Liddell, M.J.; Snow, M.R.; Tiekink, E.R.T. (J. Organometal. Chem. **335** [1987] 365/78).

[211] Davison, A.; Solar, J.P. (J. Organometal. Chem. **166** [1979] C13/C17).

[212] Abu Salah, O.M.; Bruce, M.I. (J. Chem. Soc. Dalton Trans. **1974** 2302/4).

[213] Ferede, R.; Noble, M.; Cordes, A.W.; Allison, N.T.; Lay, J., Jr. (J. Organometal. Chem. **339** [1988] 1/6).

1.5.2.3.16.1.12 $C_5H_5Fe(CO)_2R$ Compounds with R = Alkynyl

The compounds listed in Table 6 have been prepared by the following methods.

Method I: Addition of Na[Fp] in THF to an excess of $BrCH_2C\equiv CR^1$ in the same solvent at 0 °C/2 h ($R^1 = CH_3$ or C_6H_5) gives $FpCH_2C\equiv CR^1$ [12], see also [13, 22]. (With $BrCH_2C\equiv CH$, the reaction mixture is cooled in a dry ice/acetone bath; for the structure of the product with $R^1 = H$, see Section 1.5.2.3.16.1.11, Table 5, No. 166.) After stirring at room temperature for 1 h, solvent and excess $BrCH_2C\equiv CH$ are removed under vacuum and the residue is extracted with ether/light petroleum, b.p. 30 to 40 °C (1:1). The concentrated extract is chromatographed on Al_2O_3 with ether/light petroleum (1:1) as eluent. Repeated chromatography of the concentrated eluate and recrystallization from light petroleum of the residue remaining after solvent removal gives the product with $R^1 = H$ [2]. No. 8 is purified by chromatography on Al_2O_3 with C_5H_{12} as eluent and No. 10 by sublimation at 30 °C/5×10^{-3} Torr [12]. Workup of the products resulting from Na[Fp] and $BrCH_2CH_2C\equiv CR^2$ ($R^2 = H$, CH_3, C_6H_5) in THF at room temperature after 2 h is performed by filtration through Celite, evaporation of the solvent and chromatography on Al_2O_3 with pentane as eluent [60]. The preparation of compounds 10 and 14 from Na[Fp] and $ClCH_2C\equiv CCH_3$ or $ClCH_2C\equiv CC\equiv CCH_3$ is briefly mentioned in [15] and [71], respectively. Also, the reaction of Na[Fp] with $ClCH_2C\equiv CC(OH)R^3C_6H_5$ ($R^3 = CH_3$ or C_6H_5) in THF to give Nos. 11 to 13 is reported without further details [62], see also [61].

Method II: Treatment of FpCl with THF solutions of $R^1C\equiv CMgBr$ at -78 °C ($R^1 = CH_3$) or -60 °C ($R^1 = n\text{-}C_4H_9$, C_6H_5) and, after warming to room temperature, hydrolysis with 5% aqueous NH_4Cl [32] or ~ 2% aqueous HCl [8], respectively. Workup is carried out by chromatography on Al_2O_3 (No. 6) [8] and presumably on Florisil (No. 4) [32]. After hydrolysis $FpC\equiv CC_6H_5$ separates as an oil and slowly solidifies. It is filtered, dissolved in ether and precipitated by addition of light petroleum. For further purification it can be recrystallized from ethanol and sublimed [8].

Method III: Addition of FpCl in ether (No. 2) or THF (No. 3) to $LiC\equiv CR^1$ ($R^1 = C_6H_5$ or C_6F_5) in hexane/ether or THF, respectively, at -78 °C. After stirring at room temperature for 15 h, the solvent is removed under reduced pressure and the residue

References on pp. 160/2

is chromatographed on Florisil with light petroleum/benzene as eluent. The product is purified by sublimation twice at 120 to 140 °C/10^{-2} Torr (No. 3) or by sublimation at 130 to 140 °C/0.1 Torr, fractional crystallization from ether at -10 °C followed by sublimation at 110 °C/0.1 Torr (No. 2) [5].

Similarly $FpC\equiv CCF_3$ has been obtained from FpI and $LiC\equiv CCF_3$ in THF at -78 °C/ 1 h and at room temperature/4 h. After solvent removal the residue is worked up by sublimation at 100 °C/10^{-3} Torr, recrystallization of the sublimate from light petroleum/benzene and resublimation [5].

Compounds No. 2 [5, 8], 3, 5 [5], and 6 [8] are stable in air. Nos. 2 and 6 decompose only slowly in $CHCl_3$ [8]. Compound No. 4 is air sensitive [9]. Compound No. 7 decomposes slowly in air in the solid state [2, 9]. Solutions are stable when kept under N_2 but are readily oxidized on exposure to air [2]. For the cyclic voltammograms of $FpCH_2C\equiv CC_6H_5$ and $FpCH_2C\equiv CCH_3$, see [67].

Treatment of $FpC\equiv CCH_3$ (but see No. 4 for the structure) with anhydrous HCl in light petroleum for 1 min results in the immediate formation of a deep yellow precipitate, assumed to be $[Fp(CH_2=C=CH_2)]Cl$ [2], see also [3, 4], which converts into deep red crystalline FpCl on further treatment with HCl for 5 min. Similarly, shaking a solution of $FpC\equiv CCH_3$ in light petroleum with $HSbCl_6$ in concentrated HCl affords $[Fp(CH_2=C=CH_2)]SbCl_6$ [2]. Attempts to protonate the compound with anhydrous $HClO_4$ in benzene or HBF_4 in ether led either to decomposition or the formation of intractable oils, respectively [2]. $FpC\equiv CCH_3$ is decomposed by "anhydrous HBF_4 and $HSbCl_6$", giving salts of the cation $[FpCO]^+$ [9].

Whereas decomposition of $FpC\equiv CC_6H_5$ to FpCl occurs on treatment with concentrated aqueous HCl or HCl in organic solvents [8], an acid catalyzed addition of H_2O to the acetylide to give $FpCOCH_2C_6H_5$ takes place in the reaction with 5% aqueous HCl in THF during 4 h [32] or with $HBF_4 \cdot O(CH_3)_2$ in anhydrous methanol [49]. Similarly $FpC\equiv CCH_3$ reacts with 10% HCl in ethanol to yield $FpCOC_2H_5$ [9]. It is assumed that these reactions proceed via the vinylidenecarbene cation $[FpC=CHR^1]^+$, which can be trapped with nucleophiles [9, 49, 50]. Evidence for the formation of such an intermediate is seen in the reaction of $FpC\equiv CCH_3$ with anhydrous acetic acid in refluxing cyclohexane, which leads to the vinylacetate complex $FpC(OCOCH_3)=CHCH_3$ [9]. Also, the formation of the phosphonium salt $[FpC(P(C_6H_5)_3)=CHC_6H_5]X$ ($X = ClO_4$ or BF_4) from $FpC\equiv CC_6H_5$ and $HClO_4$ or HBF_4 in acetic anhydride in the presence of $P(C_6H_5)_3$ at -30 °C to room temperature supports this assumption [58]. In the absence of an appropriate nucleophile, however, the cation reacts with a second molecule of $FpC\equiv CC_6H_5$ to give the cation I. This is the case, when $FpC\equiv CC_6H_5$ is treated with $HBF_4 \cdot O(CH_3)_2$ in CH_2Cl_2 at -78 °C [49], or with $HClO_4$ or HBF_4 in acetic anhydride at -70 °C to room temperature [58]. In contrast, the reaction of $FpC\equiv CC_6H_5$ with equimolar amounts of $HClO_4$ or HBF_4 in benzene, acetone, or methanol at room temperature results in formation of $[FpCO]X$, the yields increase from 9% in benzene to 79% in methanol [58].

$$\left[\begin{array}{c} C_5H_5(CO)_2Fe \quad\quad C_6H_5 \\ \diagdown \quad\quad\quad\quad | \\ \rule{0pt}{0pt}\quad\quad\quad\quad\quad H \\ C_6H_5 \quad\quad\quad Fe(CO)_2C_5H_5 \end{array} \right]^+$$

I

Action of ethanolic HCl on $FpCH_2C\equiv CH$ affords a mixture of $FpCH_2COCH_3$ and some $FpCOC_2H_5$. However, treatment of $FpCH_2C\equiv CH$ with anhydrous acids (e.g. acetic acid in CH_3NO_2 at room temperature) or more efficiently by chromatography on SiO_2, acid washed Al_2O_3, or Florisil causes isomerization to $FpC\equiv CCH_3$. In both cases $[Fp(CH_2=C=CH_2)]^+$ is assumed as the intermediate [9]. Isolation of such intermediates, i.e., $[Fp(CH_2=C=CHR^1)]^+$ ($R^1 = CH_3$, C_6H_5), was possible in the cases where $FpCH_2C\equiv CR^1$ was allowed to react with aqueous 48% HBF_4 in acetic anhydride at 0 °C [25, 38], see also [20, 34], or in ether at -30 °C [44], or with anhydrous $HClO_4$ in benzene [18, 60], or presumably with $HPF_6 \cdot O(C_2H_5)_2$ at -20 °C [30]. The protonation reactions yield a single stereoisomer, to which the syn-structure II is assigned. The high stereospecificity is interpreted in terms of protonation at $\equiv CR^1$, a concerted process with trans–periplanar participation of the organometallic group [30, 44].

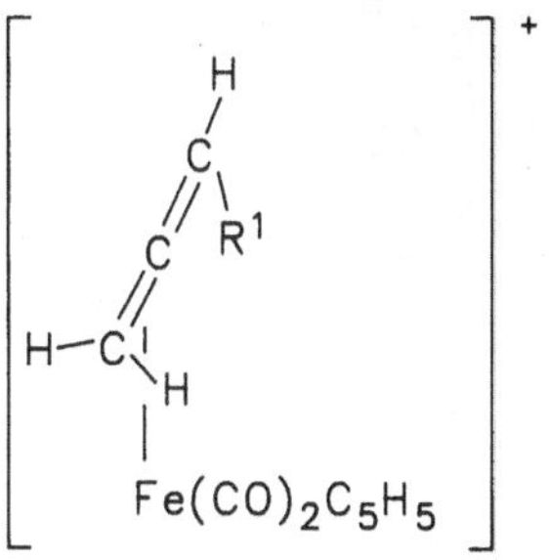

II

$[Fp(CH_2=C=CHC\equiv CCH_3)]BF_4$ could be isolated from the reaction of $FpCH_2(C\equiv C)_2CH_3$ with $HBF_4 \cdot O(C_2H_5)_2$ in ether at -40 °C to room temperature [71]. Whereas treatment of $FpCH_2C\equiv CCR^1(OH)C_6H_5$ ($R^1 = CH_3$ or C_6H_5) with anhydrous HBF_4 in ether at -50 °C affords $[Fp(CH_2=C=C=CR^1C_6H_5)]BF_4$ [61], $FpCH_2C\equiv CC(CH_3)_2OH$ is reported to react with aqueous HBF_4 in THF to give the cationic dihydrofuran complex III [62].

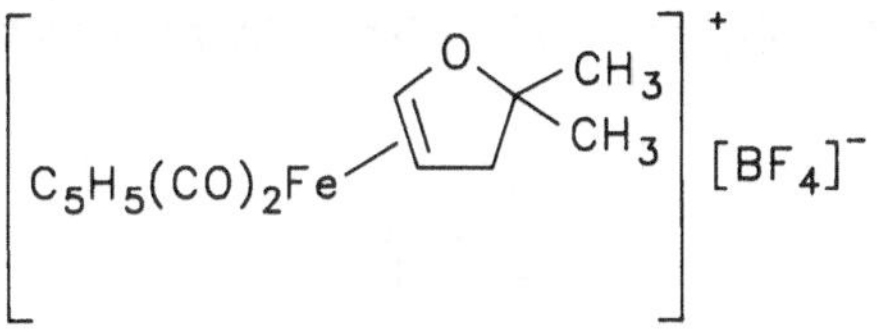

III

The reaction of $FpC\equiv CR^1$ with CuX (1:1 mole ratio) in refluxing acetone for 100 min ($R^1 = CH_3$, $X = Cl$), 5.5 h ($R^1 = C_6H_5$, $X = Cl$), or 20 h ($R^1 = C_6H_5$, $X = Br$) gives the yellow dimeric adduct $[FpC\equiv CR^1CuX]_2$ (IV), in which the iron acetylide group is π–bonded to copper [32].

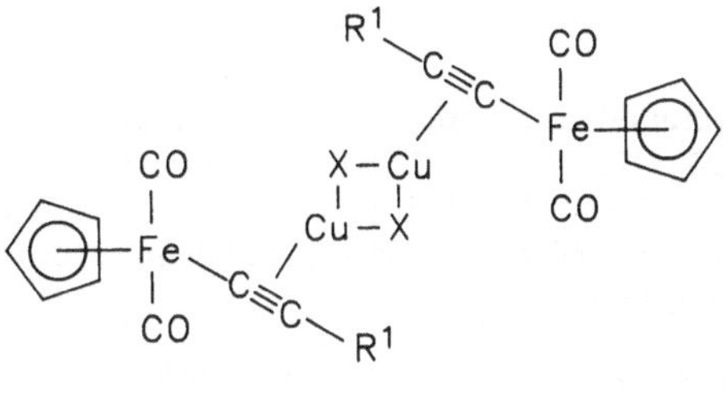

IV

References on pp. 160/2

Addition of a solution of $Co_2(CO)_8$ in benzene to $FpC\equiv CC_6H_5$ (1:1 mole ratio) in the same solvent at room temperature over a period of 90 min and stirring for further 30 min leads to formation of black crystals of Vb; the structure is from an X-ray analysis [72]. In contrast, in an earlier report, structure VI has been given as a plausible one, whereas structure Va has been assigned to the product resulting from the analogous reaction of $FpC\equiv CCH_3$ [26].

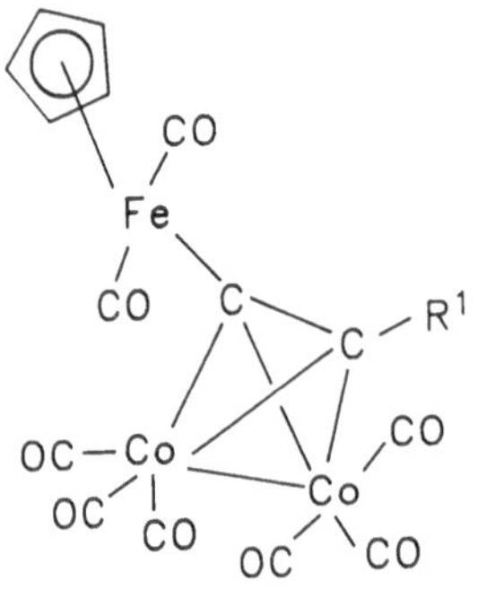

V

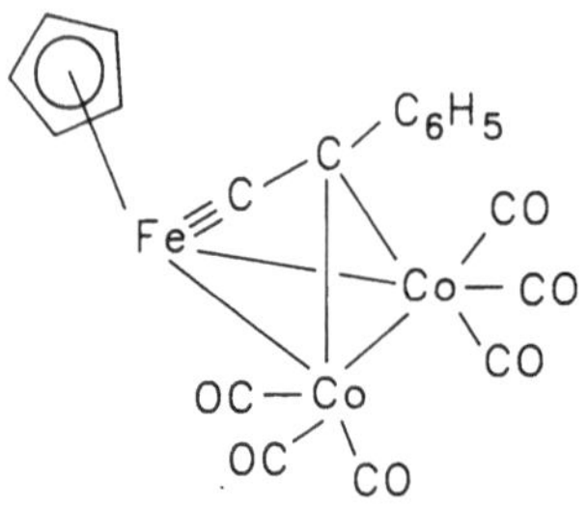

a: $R^1 = CH_3$
b: $R^1 = C_6H_5$

VI

The reaction of $FpCH_2C\equiv CCH_3$ or $FpCH_2C\equiv CSi(CH_3)_3$ in CH_2Cl_2 with the trimethylsilanyl-oxytropylium irontricarbonyl or di-n-butylboranyloxytropylium irontricarbonyl (prepared from tropeneiron tricarbonyl and $(CH_3)_3SiOSO_2CF_3$ or $(n-C_4H_9)_2BOSO_2CF_3$ in CH_2Cl_2) at $-78\,°C$ for 2 h followed by warming to room temperature, refluxing for 2 h and quenching the reaction with absolute ethanol and solid K_2CO_3 at room temperature for 12 h affords the ketohydroazulene complex VII where $R^1 = CH_3$ or H, respectively ($Si(CH_3)_3$ group is lost during reaction). Workup of the reaction mixtures from $FpCH_2C\equiv CCH_3$ by chromatography on SiO_2 leads to VII ($R^1 = CH_3$) and $FpC_6H_3(CH_3)_2$-2,5 presumably due to the presence of a trace of acid [70].

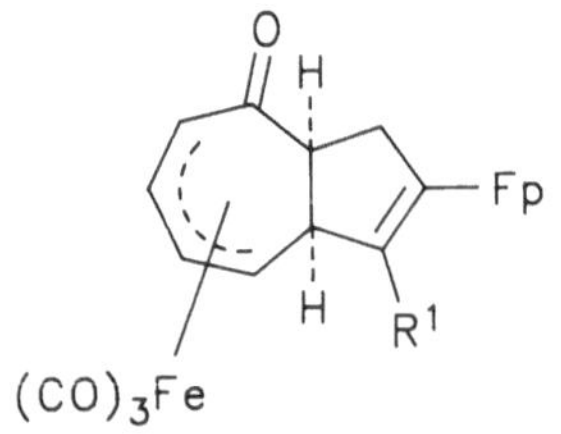

VII

For the following cycloaddition reactions of the alkynyl compounds with the electrophilic reagents SO_2 [22], SO_3 [20, 27, 29], C_6H_5NSO [45], R^1SO_2NSO, $S[NSO_2CH_3]_2$ [74], 4-$CH_3C_6H_4SO_2NCO$ [54, 63], $ClSO_2NCO$ [31], $(C_6H_5)_2C=C=O$, $t-C_4H_9C(CN)=C=O$ [45], CF_3COCF_3 [59], and $(NC)_2C=C(CN)_2$ [35, 59] a two-step mechanism is proposed, which proceeds via a dipolar metal-η^2-allene intermediate.

The reactions of $FpCH_2C\equiv CR^1$ ($R^1 = CH_3$, C_6H_5) with liquid SO_2 under reflux lead, after some hours, to insertion of SO_2 and formation of the allenic sulfinates $FpOSOCR^1=C=CH_2$

[13]. However, later studies show that $FpCH_2C{\equiv}CR^1$ reacts with SO_2 in pentane or CH_2Cl_2 at 25 °C/5 to 30 min to give a yellow precipitate of the sultine complex VIII (n=1). The same product is also obtained when liquid SO_2 is condensed onto $FpCH_2C{\equiv}CCH_3$ at −75 °C followed by warming to −10 °C after 30 min [22], see also [15, 19, 21, 51], whereas under similar conditions $FpCH_2CH_2C{\equiv}CCH_3$ inserts SO_2 yielding $FpSO_2CH_2CH_2C{\equiv}CCH_3$ [22]. Holding $FpC{\equiv}CCH_3$ in liquid SO_2 at reflux for 12 to 48 h gives no detectable amount of an SO_2-containing product. The reaction with SO_2 also failed when it was carried out in a pressure bottle at 25 to 30 °C for 24 to 48 h [22], see also [21]. The lack of reactivity is ascribed to the relatively strong Fe−C bond in this compound resulting from possible π-bonding between the metal and the unsaturated σ-bonded hydrocarbon ligand [22].

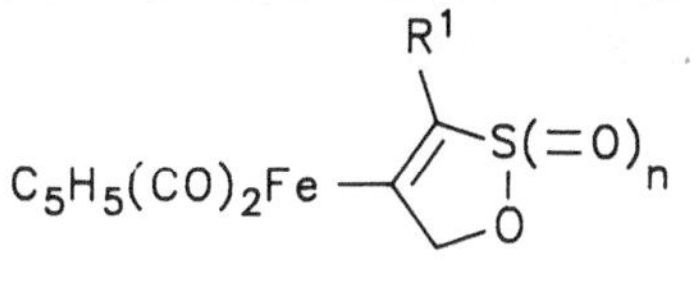

VIII

The reaction of $FpCH_2C{\equiv}CR^1$ ($R^1=CH_3$, C_6H_5) in $Cl_2C{=}CCl_2$ with SO_3 dissolved in the same solvent at room temperature for 0.5 h affords the sultone complex VIII, n=2 [29]. The same product is also obtained when the reaction is carried out with $SO_3 \cdot$ dioxane either in CH_2Cl_2 at 0 °C/10 min [20, 29] or at −70 °C [27], or in $ClCH_2CH_2Cl$ at −30 °C during less than 2 h [27].

Keeping a solution of $FpCH_2C{\equiv}CCH_3$ in neat C_6H_5NSO at room temperature for 48 h leads to formation of the 1:1 adduct IX, $R^1=CH_3$, which separates as a brown precipitate. In contrast, no precipitation was observed after $FpCH_2C{\equiv}CC_6H_5$ and C_6H_5NSO had been stored for 90 h. However, treatment of the reaction mixture with benzene and pouring it into pentane affords also a brown precipitate of IX, $R^1=C_6H_5$ [17, 45].

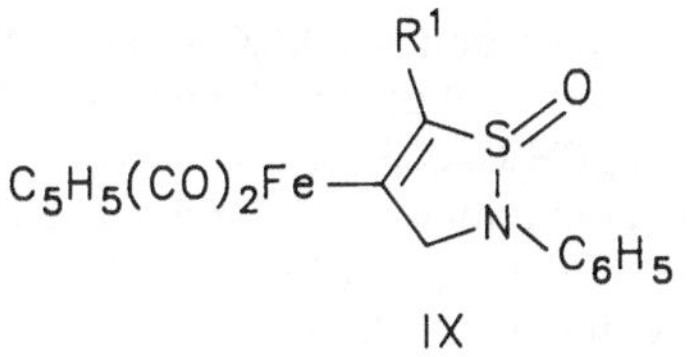

IX

Treatment of $FpCH_2C{\equiv}CR^1$ ($R^1=CH_3$, C_6H_5) with R^2SO_2NSO ($R^2=CH_3$, C_6H_5) (1:1 mole ratio) in CH_2Cl_2 at room temperature for 30 min causes formation of the [3+2]cycloadducts X as yellow solids [74]. Under similar conditions $FpCH_2C{\equiv}CC_6H_5$ reacts with 4-$CH_3C_6H_4SO_2NSO$ to give a yellow glass, to which, due to its spectroscopic properties, structure X ($R^2=4$-$CH_3C_6H_4$) has also been assigned [45]. The analogous reaction of $FpCH_2C{\equiv}CC_6H_5$ with $S(NSO_2CH_3)_2$ yields the cycloadduct XI [66, 74].

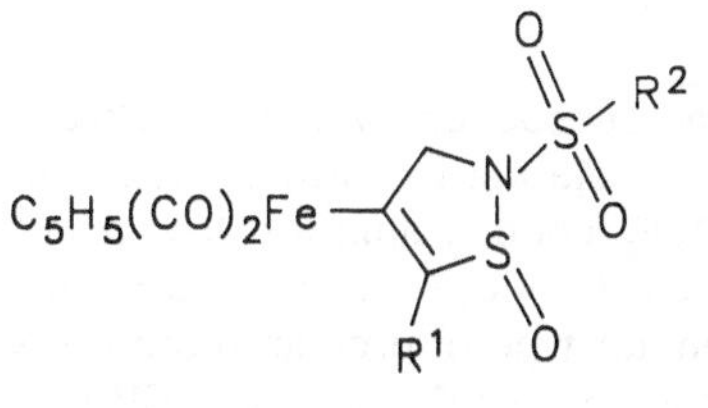

X

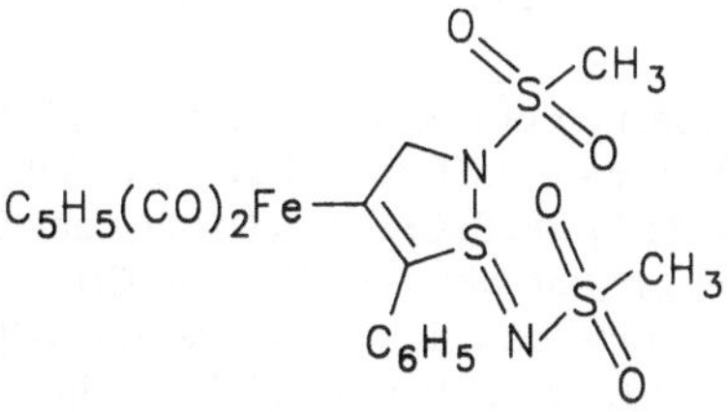

XI

References on pp. 160/2

Kinetic studies on the reactions of $FpCH_2C\equiv CR^1$ ($R^1 = CH_3$, C_6H_5) with $4-CH_3C_6H_4SO_2NCO$ in CH_2Cl_2 at 25 °C to produce the [3+2]cycloadducts XII [1, 28, 54], revealed that the cycloaddition is first order in each reactant. The second-order rate constants 6.3×10^{-2} $M^{-1} \cdot s^{-1}$ for $FpCH_2C\equiv CCH_3$ and 2.1×10^{-3} $M^{-1} \cdot s^{-1}$ for $FpCH_2C\equiv CC_6H_5$ have been obtained. The lower reactivity of $FpCH_2C\equiv CC_6H_5$ is explained by steric and electronic properties of the substituent C_6H_5. An approximately twofold increase in k_2 was observed for the reaction of $FpCH_2C\equiv CC_6H_5$ with $4-CH_3C_6H_4SO_2NCO$ on going from CH_2Cl_2 to the more polar solvent CH_3CN. Although this small solvent effect indicates a relatively nonpolar transition state, which would be consistent with a concerted process, the authors do not exclude the aforementioned two-step dipolar pathway as a possible reaction mechanism [63]. Comparable rates of cycloaddition of $4-CH_3C_6H_4NCO$ have been found for the structurally related complexes $FpCH_2C\equiv CR^1$ and $FpCH_2CH=CHR^1$ [54, 63].

XII

XIII

Addition of $ClSO_2NCO$ in benzene to $FpCH_2C\equiv CR^1$ (3:2 mole ratio) in the same solvent ($R^1 = C_6H_5$) or in CH_2Cl_2/benzene (2:1) at 10 °C/30 min ($R^1 = CH_3$) or at 25 °C ($R^1 = C_6H_5$) results in formation of XIII. Under comparable experimental conditions the less electrophilic C_6H_5NCO does not react with $FpCH_2C\equiv CCH_3$. Only the unreacted compound could be isolated after chromatography of the reaction residue on Al_2O_3 [31]. The cycloaddition reaction of $FpCH_2C\equiv CC_6H_5$ with CCl_3CONCO occurs at conveniently measurable rates, but the product decomposes gradually in solution [63].

Keeping $FpC\equiv CR^1$ ($R^1 = CH_3$, C_6H_5) and $(C_6H_5)_2C=C=O$ (1:1 mole ratio) in benzene at room temperature for 1 d affords the cyclobutenonyl complex XIV [64]. Addition of $(C_6H_5)_2C=C=O$ in benzene to a solution of $FpCH_2C\equiv CR^2$ ($R^2 = CH_3$, C_6H_5) (1.4:1 mole ratio) in CH_2Cl_2 at −40 °C followed by stirring at room temperature for 5 h gives the cycloadduct XV. Similarly, the reaction of $FpCH_2C\equiv CR^3$ with $t-C_4H_9C(CN)=C=O$ (1:1.7 mole ratio) in benzene at room temperature for 3 h yields compound XVI [45].

XIV

XV

XVI

The reaction of $FpC\equiv CC_6H_5$ with CF_3COCF_3 and subsequent workup by chromatography on Al_2O_3 gives a mixture of XVII and XVIII, which could not be further separated [59]. The compounds $FpCH_2C\equiv CR^1$ react with neat CF_3COCF_3 under reflux for 45 min ($R^1 = CH_3$) or 9 h ($R^1 = C_6H_5$) or with CF_3COCF_3 in CH_2Cl_2 for 0.5 h ($R^1 = C_6H_5$) under a dry ice condenser to yield the cycloaddition products XIX [39]. A similar type of product (Formula XX) results from the reaction of $FpCH_2C\equiv CC_6H_5$ with an excess of $CCl_2FCOCClF_2$ in CH_2Cl_2 at 25 °C/2 to 3 h [63].

References on pp. 160/2

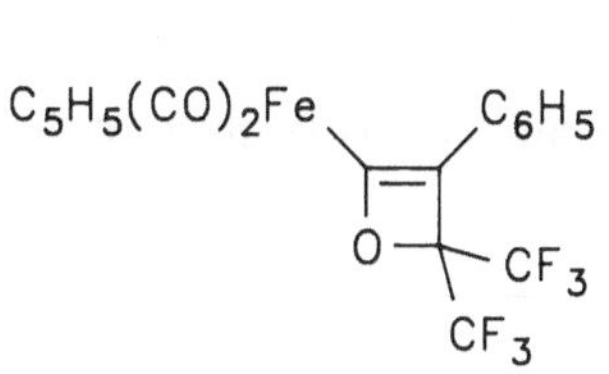

XVII

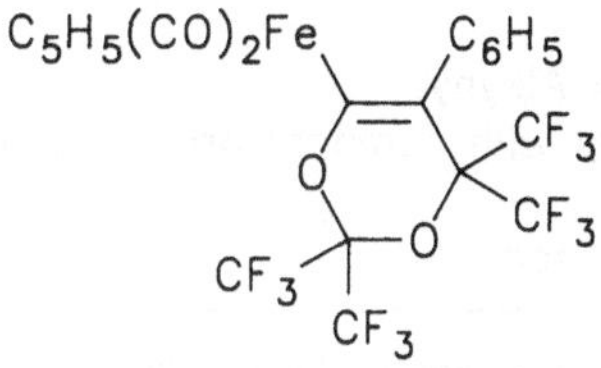

XXVIII

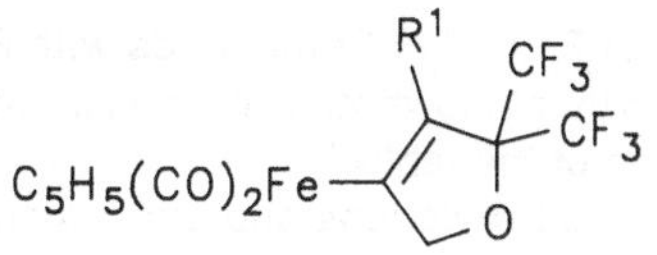

XIX

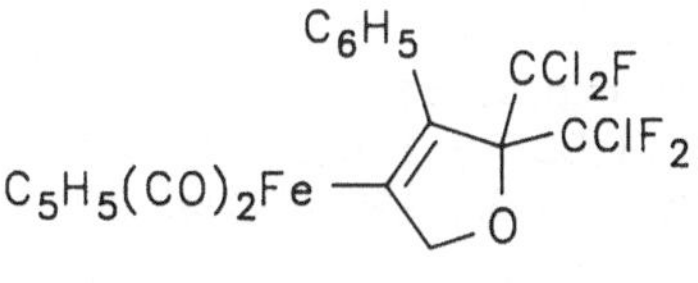

XX

FpC≡CC$_6$H$_5$ reacts rapidly with (NC)$_2$C=C(CN)$_2$ in CH$_2$Cl$_2$ to give the cyclobutenyl complex XXI. The analogous reaction in ether affords a different product. Thus, a green solution, assumed to be due to a charge-transfer complex, is formed on addition of ether to a 1:1 mixture of the solids. The color fades and, after a few minutes, the yellow solid XXII precipitates and converts into XXI over a period of ~1 h as indicated by ^{1}H NMR spectroscopy. This is also observed during attempted recrystallization or heating of XXII in the solid state. It is suggested that XXII is formed by electrophilic attack on the alkyne [59].

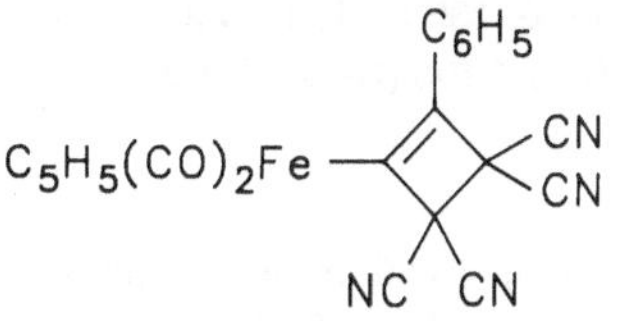

XXI

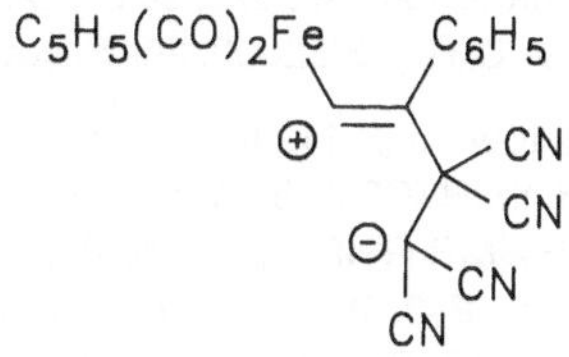

XXII

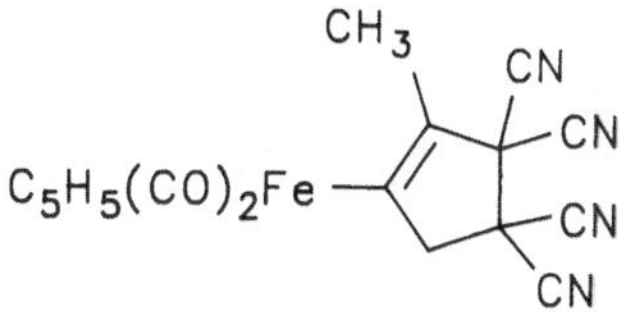

XXIII

FpCH$_2$C≡CCH$_3$ and (NC)$_2$C=C(CN)$_2$ (1:1 mole ratio) react in less than 60 s in CH$_3$CN, in about 60 s in THF or benzene, and in approximately 30 min in pentane at 25 °C to yield XXIII. This rate dependence of the reactions on the polarity of the solvent supports the proposal of a mechanism involving the intermediacy of a dipolar metal-allene complex [23, 24, 35]. Consistent with such a mechanism is also the observed lack of reactivity of (NC)$_2$C=C(CN)$_2$ toward FpCH$_2$CH$_2$C≡CCH$_3$ (3:1 mole ratio) in CH$_2$Cl$_2$ or THF at 25 °C. After 3 d only unreacted material (50%) and some insoluble decomposition material could be found [35].

In the presence of catalytic amounts of AlBr$_3$, FpC≡CCH$_3$ reacts with cyclohex-2-enone in CH$_2$Cl$_2$ at −78 to 0 °C/2 h to give the [2+2]cycloadduct XXIV, whereas under similar conditions FpCH$_2$C≡CCH$_3$ affords the [3+2]cycloadduct XXV, to which, on the basis of spectroscopic data, a cis-hydrindenone structure is assigned [65].

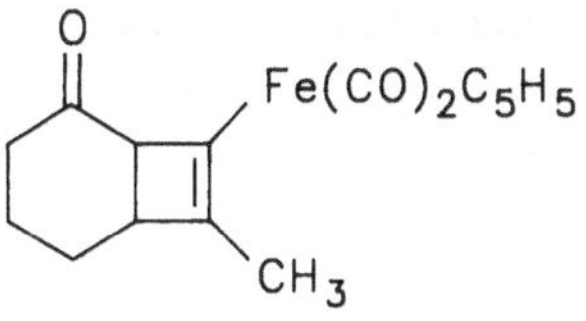

XXIV

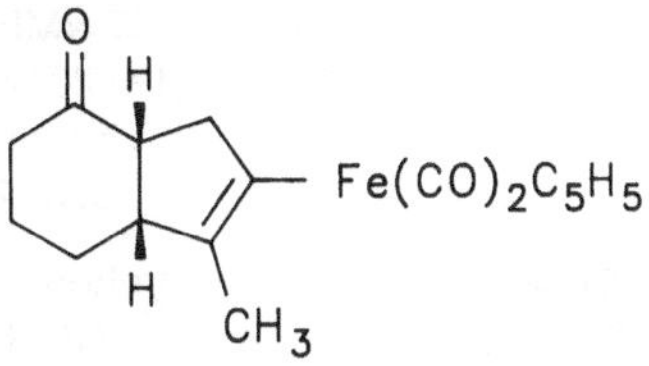

XXV

References on pp. 160/2

Table 6

$C_5H_5Fe(CO)_2R$ Compounds with R = Alkynyl.

Further information on compounds with numbers preceded by an asterisk is given at the end of the table.

For abbreviations and dimensions, see p. X.

No.	compound method of preparation (yield)	properties and remarks
1	FpC≡CH II [16]	yellow, m.p. 55 to 60° [16] ^{1}H NMR $(CDCl_3)$: 1.30 (s, CH), 4.92 (s, C_5H_5) [16] IR (Nujol): 1965, 1998, 2050 (all CO), 3290 (≡CH) [16]

*2	FpC≡CC$_6$H$_5$
	II (60%) [8]
	III (3.5%) [5]
		yellow, m.p. 120 to 122° (dec.) [5, 8], 114 to 117° (dec.) [69]
		^{1}H NMR: 4.85 (s, C_5H_5), 7.00 (m, C_6H_5) in CS_2 [5], 5.05
		   (s, C_5H_5), 7.3 (m, C_6H_5) in $CDCl_3$ [8]
		^{13}C NMR (CH_2Cl_2): 85.6 (C_5H_5, $J(^{13}C, ^{57}Fe) = 2.3$), 89.8
		   (d, FeC, $J(^{13}C, ^{57}Fe) = 19.5$), 116.3 (d, ≡C-aryl,
		   $J(^{13}C, ^{57}Fe) = 3.0$), 125.3, 128.1, 131.2 (all C_6H_5), 213.0
		   (CO, $J(^{13}C, ^{57}Fe) = 27.8$) [68]
		^{57}Fe NMR $(CH_2Cl_2, 30°$, referred to ferrocene): −515.7 [68]
		IR: 2007, 2047 (both CO), 2121 (C≡C), 838 (C_5H_5), other
		   bands at 746, 970, 981, 1027, 1130, 1368, 1420, 1490, 1509
		   in C_6H_{12} [5], 2000, 2045 (both CO), 2105 (C≡C) [8, 69]

*3	FpC≡CC$_6$F$_5$
	III (18%) [5]
		yellow, m.p. 112 to 113° [5]
		^{1}H NMR (CS_2): 5.05 (s, C_5H_5) [5]
		^{19}F NMR (THF, 35°): 140.8 (m, F-2,6), 162.0 (m, F-4), 165.6
		   (m, F-3,5), $\vert J(F-2,4) \vert = 4.4$, $\vert J(F-3,4) \vert = 20.45$ [5, 6]
		IR: 1997, 2041 (both CO), 2113 (C≡C) in C_6H_{12}, 836 (C_5H_5),
		   1488 and 1595 (both C_6F_5), other bands 693, 1422, 2850,
		   2925, 3080 in CCl_4 [5]

*4	FpC≡CCH$_3$
	I (15%) [2]
	II (45%) [32]
		yellow crystals, m.p. 95 to 96° [9, 32]
		^{1}H NMR (CS_2): 1.79 (s, CH_3), 4.93 (s, C_5H_5) [9], 1.8 (s),
		   4.90 (s) [2]
		IR (C_6H_{12} and CS_2): 1999, 2048 (both CO), 2140 (C≡C), other
		   bands at 831, 1001, 1015, 1370, 2845, 2900 [9]
		k = 16.64, k′ = 0.39 [10]

*5	FpC^1≡C^{2}CF$_3$
	III (34%) [5]
		yellow, m.p. 126 to 129° [5], 128° [56]
		^{1}H NMR: 5.08 (s, C_5H_5) in $CDCl_3$ [5], 4.98 (s, C_5H_5) [56]
		^{13}C NMR $(CDCl_3)$: 85.5 (C_5H_5), 101.1 (C-2, $^{2}J(F,C) = 47.8$),
		   104.3 (C-1, $^{3}J(F,C) = 7.8$), 111.5 (q, CF_3, $^{1}J(F,C) = 254.0$),
		   210.9 (CO) [56]
		^{19}F NMR: 46.1 (s) in $CDCl_3$ [5], 45.8 (s) [56]
		IR: 2012, 2956 (both CO), 825 (C_5H_5), 2147 (C≡C), other
		   bands at 758, 1113, 1247 in C_6H_{12} [5], 2000, 2050 (both
		   CO), 2130 (C≡C) in Nujol [56]

6	FpC≡CC$_4$H$_9$-n II [8]	yellow liquid [8] ^{1}H NMR $(CDCl_3)$: 0.7 to 2.4 (3 m, n-C_4H_9), 5.0 (s, C_5H_5) [8] IR: 1930, 2045 (both CO), 2125 (C≡C) [8]

References on pp. 160/2

Table 6 (continued)

No.	compound method of prepara- tion (yield)	properties and remarks
*7	$FpCH_2C{\equiv}CH$ I (34%) [9]	amber oil [9] ^1H NMR (CS_2): 3.97 (d, CH_2, J = 6.5), 4.76 (s, C_5H_5), 4.89 (q, CH) [9] IR $(C_6H_{12}$ and $CS_2)$: 1980, 2032 (both CO), 3315 (≡CH), other bands at 626, 804, 831, 1002, 1015, 1050, 1140, 1422, 2988, 3045 [9]
*8	$FpCH_2C{\equiv}CC_6H_5$ I (30%) [12]	orange crystals [38], m.p. 79° [12] ^1H NMR $(CDCl_3)$: 1.85 (s, CH_2), ~7 (m, C_6H_5), 4.7 to 4.8 (s, C_5H_5) [12] IR: 1960, 2016 in $CHCl_3$ [13], 1960, 2020 (all CO), ~2200 (C≡C) in $CDCl_3$ [12], 1958, 2006 (both CO) in THF [46]
9	$FpCH_2C{\equiv}CSi(CH_3)_3$	for the reaction with [di–n–butylboranyloxytropylium iron- tricarbonyl]$^+$ see p. 150
*10	$FpCH_2C{\equiv}CCH_3$ I (50%) [15, 22], (30%) [12]	yellow crystals, m.p. 46 to 48° [15, 22], 50° [12] ^1H NMR: 1.58 (quint, CH_2, J~3), 1.68 (sext, CH_3), 4.7 to 4.8 (s, C_5H_5) in CS_2 [12], 1.6 to 2.0 (m, CH_2, CH_3), 4.76 (s, C_5H_5) in $CDCl_3$ [15, 22] IR: 1975, 2025 in C_6H_{14} [22], 1950, 2000 (all CO), 2200 (C≡C) in KBr [12] R: 2210 (C≡C) [12]
11	$FpCH_2C{\equiv}CC(OH)(C_6H_5)_2$ I [61]	—
12	$FpCH_2C{\equiv}CC(OH)(CH_3)_2$ I [62]	—
13	$FpCH_2C{\equiv}CC(OH)(C_6H_5)CH_3$ I [61]	—
14	$FpCH_2C{\equiv}CC{\equiv}CCH_3$ I [71]	^1H NMR $(CDCl_3)$: 1.5 (CH_2), 1.8 (CH_3), 4.6 (C_5H_5) [71] IR $(CHCl_3)$: 1945, 2000 (both CO) [71]
15	$(Z)-FpC(=CHC{\equiv}CCH_3)CH_2N(C_2H_5)_2$	see Section 1.5.2.3.16.1.11, Table 5, No. 88 [71]
16	$FpCH_2CH_2C{\equiv}CH$ I (65%) [60]	m.p. 35° [60] ^1H NMR $(CDCl_3)$: 1.55 (m, $FeCH_2$), 2.00 (t, ≡CH), 2.27 (m, $CH_2C{\equiv}$, J = 2), 4.65 (s, C_5H_5) [60] IR (C_5H_{12}): ~1960, ~2000 (both CO) [60]
*17	$FpCH_2CH_2C{\equiv}CC_6H_5$ I (65%) [60]	m.p. 55° [60] ^1H NMR $(CDCl_3)$: 1.70 (m, $FeCH_2$), 2.50 (m, $CH_2C{\equiv}$), 4.85 (s, C_5H_5), 7.30 (m, C_6H_5) [60] IR (C_5H_{12}): ~1960, ~2000 (both CO) [60]

References on pp. 160/2

Table 6 (continued)

No.	compound method of prepara- tion (yield)	properties and remarks
18	$FpCH_2CH_2C{\equiv}CCH_3$ I (60%) [22], (43%) [60]	red [22], m.p. 47° [60] ^{1}H NMR (CDCl$_3$): 1.58 (m, FeCH$_2$), 1.78 (t, CH$_3$), 2.27 (m, CH$_2$C≡, J = 2), 4.77 (s, C$_5$H$_5$) [60] IR: 1955, 2005 in Nujol [22], ~1960, ~2000 (all CO) in C$_5$H$_{12}$ [60]

* Further information:

$C_5H_5Fe(CO)_2C{\equiv}CC_6H_5$ (Table 6, No. 2) has also been prepared by addition of $C_6H_5C{\equiv}CH$ to a suspension of FpCl (6:5 mole ratio) and CuI as catalyst in deoxygenated $N(C_2H_5)_3$. After stirring for 2 h the solvent is removed under vacuum and the residue is extracted with ether. Addition of hexane to the extract precipitates the product in a 74% yield [69].

The compound crystallizes in the monoclinic space group $P2_1/n$ ($P2_1/c$) $-C_{2h}^5$ (No. 14) with a = 9.473(2), b = 9.796(2), c = 13.887(3) Å, β = 109.70(2)°; Z = 4, d_c = 1.53, and d_m = 1.51 g/cm^3. The structure is shown in **Fig. 5**. The compound consists of discrete monomeric molecules in which the phenylethynyl group is linearly σ-bonded to the Fp group [33].

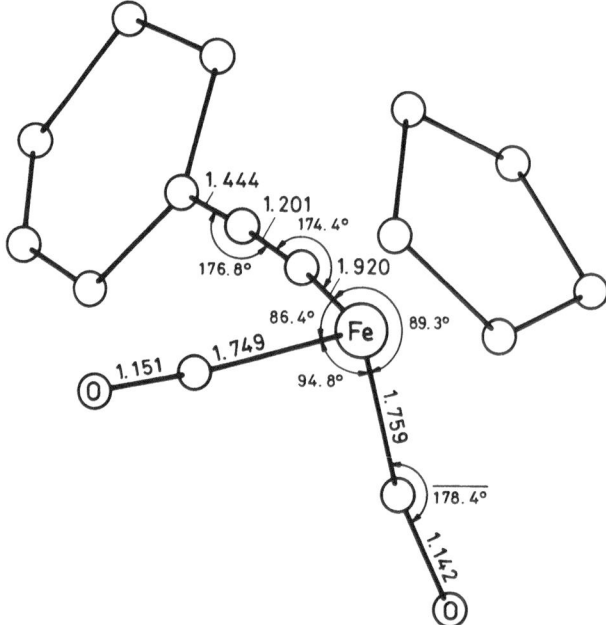

Fig. 5. Molecular structure of $FpC{\equiv}CC_6H_5$ (No. 2) with selected bond lengths (in Å) and angles.

The mass spectrum exhibits the molecular ion and fragments resulting from successive loss of the two CO groups [5]. Also mentioned is the ion $[FeC{\equiv}CC_6H_5]^+$. The most prominent metastable peak (m/e 124) corresponds to the reaction $[C_5H_5FeC{\equiv}CC_6H_5]^+ \rightarrow [C_5H_5C{\equiv}CC_6H_5]^+ + Fe$ [8].

Attempts to hydrogenate the substance failed, even when Adam's catalyst was used [8].

Treatment of $FpC\equiv CC_6H_5$ with $P(C_6H_5)_3$ at 160 °C proceeds with gas evolution to give $C_5H_5Fe(CO)(P(C_6H_5)_3)C\equiv CC_6H_5$ [8].

The compound reacts with $Fe_2(CO)_9$ (1:1 mole ratio) in benzene at room temperature for 12 h [32] or in refluxing benzene [26] to yield the trinuclear complex $C_5H_5Fe_3(C_2C_6H_5)(CO)_7$ (see "Organoiron Compounds" C7, 1986, p. 146).

Addition of a solution of $FpC\equiv CC_6H_5$ in benzene to $RuCo_2(CO)_{11}$ (1.05:1 mole ratio) in hexane at room temperature over a period of 1 h causes formation of the tetranuclear cluster $RuCo_2Fe(C_2C_6H_5)C_5H_5(CO)_{10}$ (XXVI) [73].

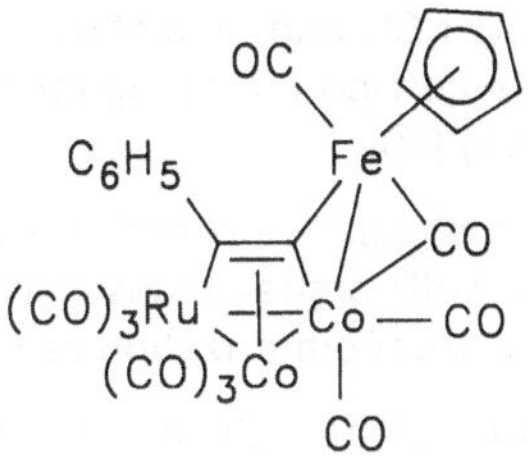

XXVI

The reaction of $FpC\equiv CC_6H_5$ with CH_3OSO_2F in CH_2Cl_2 gives, after metathesis with NH_4PF_6, compound XXVII. This alkylation follows the same pathway as the protonation, see p. 148 [49].

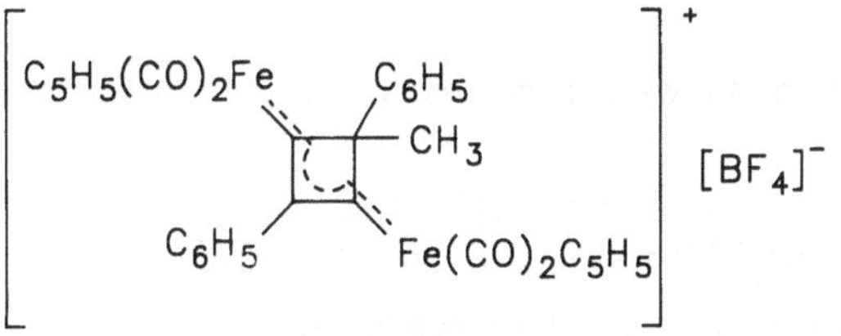

XXVII

Addition of $C_6H_5CH_2COCl$ in benzene to a solution of $FpC\equiv CC_6H_5$ (10:1 mole ratio) and $N(C_2H_5)_3$ in the same solvent at 0 °C and stirring the mixture at room temperature for 5 h leads to formation of 3-Fp substituted 2,4-diphenylcyclobut-2-enone [64].

$C_5H_5Fe(CO)_2C\equiv CC_6F_5$ (Table **6**, No. 3). Compared with $C_6F_5C\equiv CH$, the ring fluorines of $FpC\equiv CC_6F_5$ resonate at higher field. This is explained by the interaction of the filled metal d-orbitals with the acetylenic orbitals, with the result that the latter cannot accept as much π-electron density from the ring π-orbitals as in the parent fluorocarbon. The resulting increase in electron density in the ring shields the fluorines and causes the observed higher resonances [6].

The mass spectrum shows the molecular ion and fragments formed by the successive loss of the two CO groups [5].

$C_5H_5Fe(CO)_2C\equiv CCH_3$ (Table **6**, No. 4). For its formation from Na[Fp] and $HC\equiv CCH_2Br$, see Section 1.5.2.3.16.1.11, Table 5, No. 166.

References on pp. 160/2

The compound has also been obtained by addition of acetic acid to $FpCH_2C \equiv CH$ in CH_3NO_2. After 1 h at room temperature the solvent is removed under high vacuum and the product is sublimed and recrystallized from petroleum ether. Isomerization of $FpCH_2C \equiv CH$ during chromatography on SiO_2, eluting with ether/petroleum ether (4:1) (60% yield) or less efficient, on acid washed Al_2O_3 or Florisil, also leads to formation of $FpC \equiv CCH_3$. From the reaction of FpBr with $NaC \equiv CCH_3$ in liquid ammonia, the compound has been isolated in only 1% yield [9]. As shown by IR spectroscopy a mixture of $FpC \equiv CCH_3$ and $FpCH_2C \equiv CH$ (major product) is formed by treatment of Na[Fp] with $4\text{-}HC_6F_4OCH_2C \equiv CH$ in THF [11], see also [7].

$C_5H_5Fe(CO)_2C \equiv CCF_3$ (Table 6, No. 5) has been obtained in a 16% yield by the reaction of $FpSCH_3$ with $HC \equiv CCF_3$ (1:1.1 mole ratio) in THF at 293 K for 1 d followed by chromatography on Florisil, elution with hexane/CH_2Cl_2 and crystallization from the same solvents. The yield decreases when the reaction is carried out at higher temperature (6% yield at 333 K) or under UV irradiation (1% after 48 h) [56].

The mass spectrum shows the molecular ion and the fragments $[M-CO]^+$ and $[M-2CO]^+$ [5]. According to [56] peaks with higher masses than the molecular ion are also observed. They are believed to result from dimerization or intermolecular interaction [56].

Treatment of the compound with HCl has given a product which showed an IR band at ~ 1600 cm^{-1}. It is suggested that an addition of HCl across the triple bond has occurred. The reaction with I_2 in benzene afforded only a 20% yield of FpI and a small amount of benzene-insoluble material. The majority of the compound was recovered unchanged [5].

$C_5H_5Fe(CO)_2CH_2C \equiv CH$ (Table 6, No. 7) has been obtained as major product along with $FpC \equiv CCH_3$ by the reaction of Na[Fp] with $4\text{-}HC_6F_4OCH_2C \equiv CH$ in THF as shown by IR spectroscopy [11], see also [7].

The mass spectrum shows the molecular ion and the fragments $[M-CO]^+$ and $[M-2CO]^+$ [7, 11].

For the assumed formation from Na[Fp] and $HC \equiv CCH_2Br$, and the isomerization to give $FpC \equiv CCH_3$, see Section 1.5.2.3.16.1.11, Table 5, No. 166.

When $FpCH_2C \equiv CH$ is refluxed with a suspension of $Fe_2(CO)_9$ (1:1.35 mole ratio) in benzene for 1 h, a mixture of products is formed, from which $Fe_3(CO)_{12}$, Fp_2, $C_5H_5Fe_2(CO)_6C_3H_3$, and $C_5H_5Fe_3(CO)_8C_3H_3$ could be isolated by thin layer chromatography on SiO_2 [55].

$C_5H_5Fe(CO)_2CH_2C \equiv CC_6H_5$ (Table 6, No. 8) has been obtained in a 30% yield by addition of $N(C_2H_5)_3$ to a suspension of $[Fp(CH_2=C=CHC_6H_5)]BF_4$ in CH_2Cl_2 at 25 °C. Solvent evaporation affords a red-orange gum. This is extracted with pentane, and the extract is filtered through Al_2O_3 and concentrated [38].

The reaction of $FpCH_2C \equiv CC_6H_5$ with $HgCl_2$ in THF at 25 °C for 55 h in a sealed flask proceeds by cleavage of the Fe–C bond and formation of FpCl and $C_6H_5C \equiv CCH_2HgCl$. Another possible mechanism consists of the electrophilic attack of $HgCl_2$ on the triple bond, leading to an η^2-alleneiron zwitterion in equilibrium with the corresponding ionic species, $[Fp(CH_2=C=C(C_6H_5)HgCl)]^+Cl^-$, as intermediates. The rate of the reaction has been monitored by IR spectroscopy in the $\nu(CO)$ region at 25 °C under pseudo-first-order conditions. However, the reaction was too slow to use a wide enough range of concentrations of $HgCl_2$ for accurate measurement of the rate dependence on $[HgCl_2]$. Based on comparable reactions, the cleavage is assumed to be third-order overall, obeying the rate expression $-d[FpCH_2C \equiv CC_6H_5]/dt = k_3[FpCH_2C \equiv CC_6H_5][HgCl_2]^2$, where $k_3 = 1.8 \times 10^{-5}$ M$^{-2} \cdot$ s^{-1} [46], see also [40].

$FpCH_2C\equiv CC_6H_5$ is regiospecifically coupled through the γ-carbon of the propargyl ligand to afford the dinuclear complex XXVIII when oxidized by $AgPF_6$ in CH_2Cl_2 or CH_2ClCH_2Cl at -78 to $+24\,°C$. A mechanism involving cation radicals is discussed [52], see also [57].

Hydration to $FpC(CH_2OH)=CHC_6H_5$ is effected by eluting a solution of the compound in pentane over a column of Al_2O_3 during 1 h [38].

The compound reacts with $(E)-CF_3C(CN)=C(CN)CF_3$ in benzene at room temperature during 5 min to yield a diastereomeric mixture of the [3+2]cycloaddition products XXIX (9%) and XXX (91%). The amount of XXIX increases with longer reaction times. The corresponding reaction with $(Z)-CF_3C(CN)=C(CN)CF_3$ is complicated by isomerization to $(E)-CF_3C(CN)=C(CN)CF_3$ under the reaction conditions. The results are interpreted as being consistent with a two-step mechanism involving a zwitterionic intermediate [37, 43, 47].

XXVIII

XXIX   XXX

The reaction of the compound with excess $CF_3C\equiv CCF_3$ in CH_2ClCH_2Cl at $60\,°C$ for 24 h in a bomb or for 10 h under a dry ice condenser gives, among several low-yield products, a 2:1 adduct of the metal complex and the alkyne. The spectroscopic data of this orange solid indicate that both molecules of $CF_3C\equiv CCF_3$ are bonded to the propargyl fragment, however the structure of the newly formed ligand has not been elucidated [48].

The olefins $Cl_2C=CCl_2$, $(C_6H_5)_2C=C(CN)_2$, and $2-ClC_6H_4CH=C(CN)_2$ show no apparent reaction with $FpCH_2C\equiv CC_6H_5$ at $25\,°C$ [63].

2,3-Dichloro-5,6-dicyano-1,4-benzoquinone reacts rapidly with $FpCH_2C\equiv CC_6H_5$ in $CHCl_3$ at room temperature and even at $-78\,°C$ to give a cycloadduct (structure not given). The less electrophilic tetrachloro-1,4-benzoquinone undergoes very slow reaction with $FpCH_2C\equiv CC_6H_5$ at $25\,°C$, whereas 1,4-benzoquinone does not react under these conditions [63].

$C_5H_5Fe(CO)_2CH_2C\equiv CCH_3$ (Table 6, No. 10) has also been obtained by treatment of [Fp-{(Z)-CH_2=C=CHCH_3}]BF_4 with $N(C_6H_{11}-cyclo)_2C_2H_5$ in CH_2Cl_2 at $0\,°C$ for 30 min [53]. Occasionally, $FpCH_2C\equiv CCH_3$ forms in small amounts, in addition to $FpC(=CH_2)CH=CH_2$ and the main product $FpCH_2C\equiv CCH_2Fp$, when Na[Fp] is allowed to react with $XCH_2C\equiv CCH_2X$ (X = Cl, $OSO_2C_6H_5$). Small amounts of $FpCH_2C\equiv CCH_3$ and $FpC(=CH_2)CH=CH_2$ are also produced by the reaction of Na[Fp] with $BrCH_2C\equiv CCH_2Br$ or by bubbling butatriene through a THF solution of FpH [41].

References on pp. 160/2

The compound reacts with $SnCl_2$ (1:2.2 mole ratio) in THF during 24 h to give $FpSnCl_3$ [36].

Addition of $FpCH_2C\equiv CCH_3$ to a solution of Na[Fp] in THF and CH_3OH results in formation of Fp_2 and $(E)-C_5H_5Fe(CO)[\eta^3-CH_2=CHC(CH_3)COOCH_3]$ as indicated by ^1H NMR spectroscopy after 18 h [41].

Treatment of the compound with tropyliumiron tricarbonyl in CH_3NO_2 at room temperature for 15 min affords XXXI [42]. For the formation of the corresponding ketohydroazulene complex see p. 150 [70].

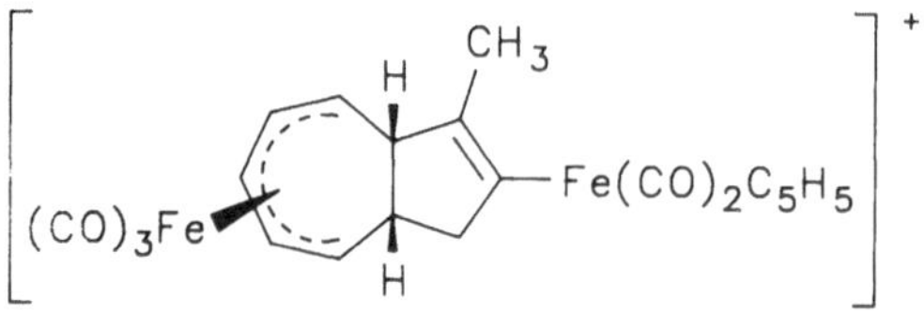

XXXI

The reaction with a slight excess of $CF_3CC\equiv CCF_3$ in CH_2Cl_2 for 5 h under a dry ice condenser, followed by chromatography on Al_2O_3 (6% H_2O), gives Fp_2, a mixture of inseparable iron carbonyls and $FpC(CH_2OH)=CHCH_3$ [48].

No reaction has been observed when $C_2H_5OOCN=NCOOC_2H_5$ in benzene was added to $FpCH_2C\equiv CCH_3$ in CH_2Cl_2 at room temperature and stirred for 28 h. The lack of reactivity is ascribed to the relatively poor electrophilic properties of the azodicarboxylate toward the coordinated alkynyl fragment [45].

The reaction with $CH_3O_2CNSO_2$ to give the isothiazoline derivative XXXII is briefly mentioned in [30].

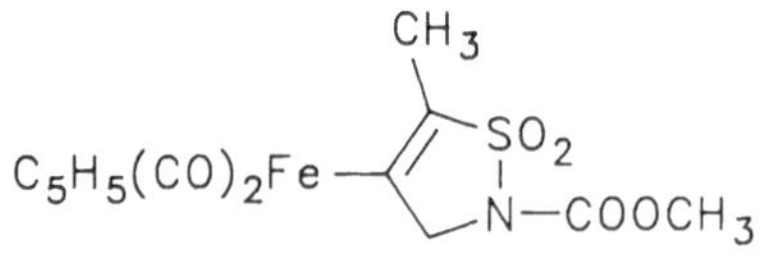

XXXII

$C_5H_5Fe(CO)_2CH_2CH_2C\equiv CC_6H_5$ (Table 6, No. 17) reacts with $P(C_6H_5)_3$ in CH_3CN at 60 °C during 45 h to give the acyl complex $C_5H_5Fe(CO)(P(C_6H_5)_3)COCH_2CH_2C\equiv CC_6H_5$ [60].

References:

[1] Cotton, F.A.; Liehr, A.D.; Wilkinson, G. (J. Inorg. Nucl. Chem. **1** [1955] 175/86).
[2] Ariyaratne, J.K.P.; Green, M.L.H. (J. Organometal. Chem. **1** [1963/64] 90/3).
[3] Ariyaratne, J.K.P.; Green, M.L.H. (Chem. Eng. News **41** [1963] 46).
[4] Ariyaratne, J.K.P.; Green, M.L.H.; Nagy, P.L.I. (Proc. Chem. Soc. **1963** 107).
[5] Bruce, M.I.; Harbourne, D.A.; Waugh, F.; Stone, F.G.A. (J. Chem. Soc. A **1968** 356/9).
[6] Bruce, M.I. (J. Chem. Soc. A **1968** 1459/64).
[7] Bruce, M.I.; Davies, C.H. (Inorg. Nucl. Chem. Letters **4** [1968] 675/7).
[8] Green, M.L.H.; Mole, T. (J. Organometal. Chem. **12** [1968] 404/6).
[9] Jolly, P.W.; Pettit, R. (J. Organometal. Chem. **12** [1968] 491/5).
[10] King, R.B. (Inorg. Chim. Acta **2** [1968] 454/8).

[11] Bruce, M.I.; Davies, C.H. (J. Chem. Soc. A **1969** 1077/81).

[12] Roustan, J.L.; Cadiot, P. (Compt. Rend. C **268** [1969] 734/6).

[13] Roustan, J.L.; Charrier, C. (Compt. Rend. C **268** [1969] 2113/6).

[14] Johnson, M.D.; Mayle, C. (J. Chem. Soc. D **1969** 192).

[15] Churchill, M.R.; Wormald, J.; Ross, D.A.; Thomasson, J.E.; Wojcicki, A. (J. Am. Chem. Soc. **92** [1970] 1795/6).

[16] Kim, P.J.; Masai, H.; Sonogashira, K.; Hagihara, N. (Inorg. Nucl. Chem. Letters **6** [1970] 181/5).

[17] Robinson, P.W.; Wojcicki, A. (J. Chem. Soc. D **1970** 951/2).

[18] Benaim, J.; Mérour, J.Y.; Roustan, J.L. (Compt. Rend. C **272** [1971] 789/91).

[19] Churchill, M.R.; Wormald, J. (J. Am. Chem. Soc. **93** [1971] 354/9).

[20] Lichtenberg, D.W.; Wojcicki, A. (J. Organometal. Chem. **33** [1971] C77/C79).

[21] Ross, D.A. (Diss. Ohio State Univ. 1970; Diss. Abstr. Intern. B **31** [1971] 3905/6).

[22] Thomasson, J.E.; Robinson, P.W.; Ross, D.A.; Wojcicki, A. (Inorg. Chem. **10** [1971] 2130/7).

[23] Su, S.R.; Wojcicki, A. (J. Organometal. Chem. **31** [1971] C34/C36).

[24] Su, S.R. (Diss. Ohio State Univ. 1971; Diss. Abstr. Intern. B **32** [1972] 6283).

[25] Lichtenberg, D.W.; Wojcicki, A. (J. Am. Chem. Soc. **94** [1972] 8271/3).

[26] Yasufuku, K.; Yamazaki, H. (Bull. Chem. Soc. Japan **45** [1972] 2664).

[27] Roustan, J.L.; Mérour, J.Y.; Benaim, J.; Charrier, C. (Compt. Rend. C **274** [1972] 537/40).

[28] Giering, W.P.; Raghu, S.; Rosenblum, M.; Cutler, A.; Ehntholt, D.; Fish, R.W. (J. Am. Chem. Soc. **94** [1972] 8251/3).

[29] Lichtenberg, D.W.; Wojcicki, A. (Inorg. Chim. Acta **7** [1973] 311/4).

[30] Raghu, S.; Rosenblum, M. (J. Am. Chem. Soc. **95** [1973] 3060/2).

[31] Yamamoto, Y.; Wojcicki, A. (Inorg. Chem. **12** [1973] 1779/88).

[32] Abu Salah, O.M.; Bruce, M.I. (J. Chem. Soc. Dalton Trans. **1974** 2302/4).

[33] Goddard, R.; Howard, J.; Woodward, P. (J. Chem. Soc. Dalton Trans. **1974** 2025/7).

[34] Lichtenberg, D.W. (Diss. Ohio State Univ. 1973; Diss. Abstr. Intern. B **34** [1974] 5372).

[35] Su, S.R.; Wojcicki, A. (Inorg. Chim. Acta **8** [1974] 55/60).

[36] Magatti, C.V.; Giering, W.P. (J. Organometal. Chem. **73** [1974] 85/92).

[37] Williams, J.P.; Wojcicki, A. (Inorg. Chim. Acta **15** [1975] L21/L22).

[38] Lichtenberg, D.W.; Wojcicki, A. (J. Organometal. Chem. **94** [1975] 311/26).

[39] Lichtenberg, D.W.; Wojcicki, A. (Inorg. Chem. **14** [1975] 1295/301).

[40] Wojcicki, A.; Dizikes, L.J. (7th Intern. Conf. Organometal. Chem., Venice, Italy, 1975, Abstr. S1C).

[41] Bauch, T.E.; Giering, W.P. (J. Organometal. Chem. **114** [1976] 165/74).

[42] Genco, N.; Marten, D.; Raghu, S.; Rosenblum, M. (J. Am. Chem. Soc. **98** [1976] 848/9).

[43] Williams, J.P. (Diss. Ohio State Univ. 1975; Diss. Abstr. Intern. B **36** [1976] 3950).

[44] Foxman, B.; Marten, D.; Rosan, A.; Raghu, S.; Rosenblum, M. (J. Am. Chem. Soc. **99** [1977] 2160/5).

[45] Chen, L.S.; Lichtenberg, D.W.; Robinson, P.W.; Yamamoto, Y.; Wojcicki, A. (Inorg. Chim. Acta **25** [1977] 165/72).

[46] Dizikes, L.J.; Wojcicki, A. (J. Organometal. Chem. **137** [1977] 79/90).

[47] Williams, J.P.; Wojcicki, A. (Inorg. Chem. **16** [1977] 2506/12).

[48] Williams, J.P.; Wojcicki, A. (Inorg. Chem. **16** [1977] 3116/24).

[49] Davison, A.; Solar, J.P. (J. Organometal. Chem. **155** [1978] C8/C12).

[50] Davison, A.; Selegue, J.P. (J. Am. Chem. Soc. **100** [1978] 7763/5).

[51] Chen, L.S.; Su, S.R.; Wojcicki, A. (Inorg. Chim. Acta **27** [1978] 79/89).

[52] Waterman, P.S.; Giering, W.P. (J. Organometal. Chem. **155** [1978] C47/C50).

[53] Klemarczyk, P.; Rosenblum, M. (J. Org. Chem. **43** [1978] 3488/93).

[54] Bell, P.B. (Diss. Ohio State Univ. 1978; Diss. Abstr. Intern. B **39** [1979] 4878).

[55] Aleksandrov, G.G.; Skripkin, V.V.; Kolobova, N.E.; Struchkov, Yu.T. (Koord. Khim. **5** [1979] 1479/83; Soviet J. Coord. Chem. **5** [1979] 1148/52).

[56] Petillon, F.Y.; Le Floch-Perennou, F.; Guerchais, J.E.; Sharp, D.W.A. (J. Organometal. Chem. **173** [1979] 89/106).

[57] Waterman, P.S. (Diss. Boston Univ. Grad. School 1979; Diss. Abstr. Intern. B **40** [1979] 2200).

[58] Kolobova, N.E.; Skripkin, V.V.; Aleksandrov, G.G.; Struchkov, Yu.T. (J. Organometal. Chem. **169** [1979] 293/300).

[59] Davison, A.; Solar, J.P. (J. Organometal. Chem. **166** [1979] C13/C17).

[60] Roustan, J.L.; Mérour, J.Y.; Charrier, C.; Benaim, J.; Cadiot, P. (J. Organometal. Chem. **168** [1979] 61/86).

[61] Benaim, J.; Giulieri, F. (J. Organometal. Chem. **202** [1980] C9/C14).

[62] Benaim, J.; L'Honore, A. (J. Organometal. Chem. **202** [1980] C53/C57).

[63] Bell, P.B.; Wojcicki, A. (Inorg. Chem. **20** [1981] 1585/92).

[64] Hong, P.; Sonogashira, K.; Hagihara, N. (J. Organometal. Chem. **219** [1981] 363/9).

[65] Bucheister, A.; Klemarczyk, P.; Rosenblum, M. (Organometallics **1** [1982] 1679/84).

[66] Leung, T.W.; Christoph, G.G.; Wojcicki, A. (Inorg. Chim. Acta **76** [1983] L281/L282).

[67] Zulu, S.J. (Diss. Boston Univ. Grad. School 1983; Diss. Abstr. Intern. B **44** [1983] 493).

[68] Koridze, A.A.; Astakhova, N.M.; Petrovskii, P.V. (J. Organometal. Chem. **254** [1983] 345/60).

[69] Bruce, M.I.; Humphrey, M.G.; Matisons, J.G.; Roy, S.K.; Swincer, A.G. (Australian J. Chem. **37** [1984] 1955/61).

[70] Watkins, J.C.; Rosenblum, M. (Tetrahedron Letters **25** [1984] 2097/100).

[71] Giulieri, F.; Benaim, J. (J. Organometal. Chem. **276** [1984] 367/76).

[72] Bruce, M.I.; Duffy, D.N.; Humphrey, M.G. (Australian J. Chem. **39** [1986] 159/63).

[73] Roland, E.; Bernhardt, W.; Vahrenkamp, H. (Chem. Ber. **119** [1986] 2566/81).

[74] Leung, T.W.; Christoph, G.G.; Galucci, J.; Wojcicki, A. (Organometallics **5** [1986] 846/53).

1.5.2.3.16.2 $C_5H_5Fe(CO)_2R$ Compounds with R = Aryl

The following methods have been used to prepare the compounds listed in Table 7, pp. 168/79.

Method I: Reaction of Na[Fp] with $C_6F_5R^1$ (R^1 = CH=CH$_2$ [36], see also [23], Si(CH$_3$)$_3$, Si(C$_6$H$_5$)$_2$C$_6$F$_5$, Si(CH$_3$)$_2$(CH$_2$)$_3$Si(CH$_3$)$_3$ [64], or $C_6F_4R_2^2$ (R^2 = H, CN [9]) in THF at room temperature, or with $C_6F_5R^1$ (R^1 = H, F, CF$_3$ [5], CH$_3$ [36]), or 1,2,3,4-$C_6H_2F_4$ [5] under reflux for several hours. In some cases the reactants are mixed at 0 °C (reaction with $C_6F_4(COOC_2H_5)_2$-1,2 [9]), or at −70 to −78 °C ($C_6F_5SR^3$ [24], 1,3,5-$C_3F_3Cl_3$ [65]) before allowing the mixtures to warm to room temperature. Workup is carried out by chromatography on Florisil, eluting with CHCl$_3$ (Nos. 55 and 56 [9]), hexane (Nos. 45 and 47 [64]), acetone (No. 46) [64], light petroleum, b.p. 40 to 60 °C (No. 25) [36], or mixtures of benzene/light petroleum (Nos. 27 [36], 39 [24], 48, and 49 [9]); or on Al$_2$O$_3$, eluting with benzene (Nos. 24, 26, 50 [5], and 54 [65]). The compounds are purified by recrystallization from light petroleum (No. 25 [36]; followed by sublimation, No. 27 [36]), from benzene (No. 55 [9]), THF (No. 54 [65]), or by extraction of the eluate residue with pentane, cooling to −78 °C, and sublimation of the precipitate at 80 °C/0.1 Torr (Nos. 24, 26, and 50 [5]). Significant additional amounts of

FpC_6F_4H-4 (No. 24) and $FpC_6F_4CF_3-4$ (No. 26) are obtained by sublimation of the insoluble material remaining after pentane extraction [5]. In some cases the compounds are first sublimed at 80 °C/0.2 Torr (Nos. 48 and 49) or at 140 to 160 °C/0.01 Torr (No. 56) before they are recrystallized from light petroleum (Nos. 49 and 56) or light petroleum/benzene (No. 48) [9]. Special workup methods for FpC_6F_5 (No. 20) and $FpC_6F_4SCH_3-4$ (No. 38) are given under "Further information".

For the preparation of FpR ($R = C_6F_4Br-2$, C_6F_4Br-3, C_6H_4Br-4, C_6F_4H-2, C_6F_4Cl-3), Li[Fp] in hexane/ether is allowed to react with the respective $C_6F_4Br_2$ species, $1-HC_6F_4Br-2$, or $C_6F_4Cl_2-1,3$ at -78 °C, then at room temperature for 12 h followed by chromatography on Al_2O_3 with benzene as eluent and recrystallization from pentane, hexane, or ethanol [59].

Method II: Addition of FpX ($X = Cl$ or I) to RLi (1:1 mole ratio), prepared from $n-C_4H_9Li$ and the corresponding halobenzene in ether at -78 °C [59] or -78 to -110 °C [85], followed by stirring the mixture for several hours either at the low temperatures [85] or at room temperature [59]. After filtration and evaporation of the solvent, the residue is worked up at room temperature by chromatography on Al_2O_3 with benzene [59] or ether/hexane [85] as eluent. For the preparation of $FpC_6H_4CH_3-2$ or $FpC_6H_4CH_2C_6H_5-2$ the entire procedure is carried out at room temperature [85].

Method III: UV irradiation of $FpCOC_6H_5$ in benzene for 14 h followed by solvent removal from the filtered reaction mixture, dissolution of the residue in pentane, chromatography on Al_2O_3 with pentane as eluent and recrystallization from pentane at -78 °C [6]. The same procedure has been used to prepare compounds No. 3, 5 [63], 11 (elution with benzene) [55], and No. 15 [63, 76]. A higher yield of FpC_6H_5 (75% instead of 17% [6]) is obtained when the crude product resulting from Na[Fp] and C_6H_5COCl is directly photolyzed without isolation [87]. $FpC_6H_4CH_3-4$ (No. 6) has also been prepared in this way [58]. Photochemical decarbonylation of $FpCOC_6H_4F-3$ and $FpCOC_6H_4F-4$ has been carried out in THF during 10 h. Workup of the reaction residue is performed as described above, however with benzene as solvent. The products are recrystallized from hexane at -78 °C [45].

Method IVa: Reaction of Na[Fp] or K[Fp] with onium salts. Treatment of Na[Fp] with $[IR_2']I$ in THF [12, 18, 56] or tetrahydrofurfural [33] at -60 to -50 °C for 1 h and at room temperature for 0.5 to 3 h, after solvent removal, chromatography on Al_2O_3 with benzene/petroleum ether (1:9) as eluent. A lower yield of FpC_6H_5 (18% in contrast to 34% [12]) has been obtained when $[I(C_6H_5)_2]BF_4$ in THF is used in place of $[I(C_6H_5)_2]I$ (elution with benzene/hexane, 1:9) [19], see also [8].

Method IVb: Reaction of Na[Fp] or K[Fp] with onium salts under similar conditions as above. Na[Fp] also reacts with $[S(C_6H_5)_3]BF_4$ in THF to give FpC_6H_5 [8, 12, 19]. However, only traces of FpC_6H_5 (the main product being Fp_2) result from the corresponding reaction with the diazonium salt $[C_6H_5N_2]BF_4$ [19]. An extremely rapid reaction occurs upon addition of K[Fp] in THF to $[P(C_6H_5)_4]Br$ in the same solvent at -70 °C and warming to room temperature. FpC_6H_5 obtained after chromatography on Al_2O_3 with benzene/petroleum ether (1:9) as eluent is contaminated with $P(C_6H_5)_3$. Only repeated chromatography of a hexane solution of this fraction (elution with hexane) gives the pure product. The analogous reaction of

References on pp. 189/91

K[Fp] with $[As(C_6H_5)_4]Cl$ gives similar yields as obtained with $[P(C_6H_5)_4]Br$ [73]. The mechanisms of the above reactions are discussed [19, 73].

Method V: Reaction of FpBr with CuR, formed by addition of CuCl to RMgBr in ether/THF, at 0 °C for 1 h and at 20 °C for 4 h. After solvent removal, chromatography on Al_2O_3 with 1:1 benzene/petroleum ether as eluent [79].

The effects of R and X ligands upon the chemical shifts of the C_5H_5 protons of FpR including FpC_6H_5 and FpX (X = halogen) compounds are predominantly inductive [57]. A good linear correlation between the Hammett σ-constants and the chemical shift of the C_5H_5 protons as well as the symmetric and antisymmetric $\nu(CO)$ frequencies (for FpC_6H_5 and FpC_6F_5, see [22]) has been found for a series of meta- and para-substituted derivatives of FpC_6H_4X (X = CH_3, F, Cl, CN, OCH_3, $COOCH_3$). The small differences in values ($\Delta\delta$ = 0.07 ppm, $\Delta\nu_s(CO)$ = 12 cm^{-1}, $\Delta\nu_{as}(CO)$ = 15 cm^{-1}) for groups ranging from CN (σ_R = +0.63) to OCH_3 (σ_R = −0.27) suggest very little resonance interaction between these para-substituents and the other ligands, CO and C_5H_5, bonded to the metal [37].

The carbonyl chemical shift parameters of FpX compounds including FpC_6H_5, FpC_6H_4Cl-4, and FpC_6F_5 are observed to be linearly dependent on the Taft σ_I-values of the X substituent, the carbonyl stretching frequencies, and the reductive half-wave potentials. These relationships are explained by considering variation in $\delta(CO)$ to be principally determined by changes in the paramagnetic screening term [55].

The inductive (σ_I) and resonance (σ_R) constants of the Fp group have been determined on the basis of the ^{19}F chemical shifts of FpC_6H_4F-3 and FpC_6H_4F-4 by the Taft equation: σ_I = −0.25, σ_R = −0.29. The negative values indicate that the organic moiety is a strong σ- and π-donor. Studies on related 3- and 4-fluorophenyl transition metal complexes revealed that only the inductive parameters varied greatly, whereas the resonance parameters were almost invariant. It is concluded that the inductive interaction, i.e., σ-bonding, predominates in the iron-aryl bond and that little π-interaction is involved [45, 51, 56, 57], although another interpretation supports significant π-acceptance by the aryl ligand [37].

The ^{57}Fe Mössbauer spectra have been studied for a series of compounds, all showing a well-resolved quadrupole doublet spectrum. The observed asymmetry in the areas of the two quadrupole lines is attributed to preferential crystal orientation. A systematic increase in the isomer shift has been found for the 2-halophenyl derivatives FpR in the order R = C_6H_5 < C_6H_4I-2 < C_6H_4Br-2 < C_6H_4Cl-2 < C_6H_4F-2 < C_6F_5, indicating a decrease in s-electron density on iron. The observed trend is explained by increased σ-donation by the metal as the electronegativity of the halogen increases. The alkyl-substituted aryl groups, $C_6H_4CH_3$-2, $C_6H_4CH_2C_6H_5$-2, should be better σ-donors than the halogenated compounds and hence should have smaller isomer shifts due to a larger electron density at the iron. However, the values found are either larger (R = $C_6H_4CH_3$-2) or equal (R = $C_6H_4CH_2C_6H_5$-2) to the isomer shifts of the monohalogenated aryl complexes. Thus, a direct interaction between the alkyl groups and the iron is suggested as a mechanism for the anomalously low s-electron density on the metal. A support of this theory is seen in the good correlation between the isomer shifts and the carbonyl stretching force constants for all compounds studied except for $FpC_6H_4CH_3$-2 and $FpC_6H_4CH_2C_6H_5$-2. No obvious correlation seems to exist between the observed quadrupole splitting and the aryl group [91].

A systematic increase in the stretching force constants have been found for FpR in the order R = $C_6H_4CH_2C_6H_5$-2 ~ $C_6H_4CH_3$-2 < C_6H_5 < C_6H_4I-2 < C_6H_4Br-2 ~ C_6H_4Cl-2 < C_6H_4F-2 < C_6F_5. This order follows the increasing electron withdrawing properties of the ortho-phenyl substituent, as given by the Hammett-Taft inductive constants, σ_I, but is the inverse of the corresponding resonance constants, σ_R [91].

The fluoroaryl compounds No. 24, 25, 27, 28, 29, 33, 36, and 37 are yellow or yellow-orange crystalline solids, which can be sublimed in vacuum, and which are air-stable in the solid state, although solutions decompose more or less rapidly in the presence of oxygen [36]. In general, perfluorophenyl derivatives show remarkable thermal and oxidative stability compared to their hydrocarbon analogs [7]. The mass spectra of the aforementioned fluoroaryl compounds, of No. 53 [36] and of Nos. 20, 24, 48, 49, and 50 [25], are characterized by the presence of ions which correspond to $[M]^+$, $[M-CO]^+$, and $[M-2CO]^+$. The CO-free ions of the latter series of fluoroaryl compounds $C_5H_5Fe(CO)_2R$ are reported to subsequently eliminate either the metal as FeF or FeF_2 with concomitant ligand transfer to give fragments such as $[C_5H_5R-F]^+$, or they first lose HF forming $[C_5H_4FeR-F]^+$ and then FeF, FeF_2, or a second molecule of HF. In each case, the ion $[C_5H_5R-F]^+$ fragments further with loss of C_2H_2, presumably from C_5H_5, a process which is confirmed by the appropriate meta-stable peaks. Other characteristic ions are $[C_3H_3]^+$, $[C_5H_5]^+$, $[C_5H_6]^{+\cdot}$, $[C_3H_3Fe]^+$, $[C_5H_5Fe]^+$, $[C_5H_5FeF]^+$, $[FeF]^+$, and $[(C_5H_5)_2Fe]^+$. Complete lists of the fragments and their relative intensities, metastable peaks, and fragmentation schemes are given. The data reveal that differing substitution positions have no apparent effect on the breakdown routes. Comparison with the parent fluoroaromatic compounds shows that a completely different fragmentation pattern results when a transition metal is σ-bonded to the aromatic ring [25]. The mass spectra of compounds No. 20, 24, 26, and 50 are also described in a similar way in [38], see also [16]. In addition to the general features, the mass spectrum of $FpC_6F_4CF_3$-4 (No. 26) exhibits the ions $[C_6F_nC_5H_5]^+$ (n=3, 4, or 5) and $[C_6F_4H]^+$ which may arise by a CF_2 elimination from $[C_7F_nC_5H_5]^+$ (n=5, 6, or 7) and $[C_7F_6H]^+$, respectively. The absence of the fragments $[C_6F_nC_5H_5]^+$ (n=3, 4, or 5) in the mass spectra of the other FpR derivatives suggests that the CF_2 group comes from the CF_3 substituent on the aromatic ring [38].

Similar fragmentation patterns as those described above for some FpR compounds [25] have been observed for compounds No. 59 to 63 and 64 to 70 [40].

The polarographic reduction of FpR in CH_3CN or $HCON(CH_3)_2/0.1$ N $[N(C_2H_5)_4]ClO_4$ takes place in an irreversible two-electron step, giving the anions $[Fp]^-$ and R^- [15, 22, 62]. The observation of two waves in the polarogram of FpC_6F_5 in CH_3CN is thought to result from a one-electron step, which leads initially to the possible products $[Fp]^-$ and the radical $R^\cdot$. A point of inflection that appears on the wave of FpC_6H_5 and FpC_6H_4Cl-4 in $HCON(CH_3)_2$ is attributed to the instability of the compounds in this solvent [15]. The following half-wave potentials have been measured:

compound	$E_{1/2}$ (in V vs. saturated calomel electrode)	
	in CH_3CN	in $HCON(CH_3)_2$
FpC_6H_5	−1.96 [15, 22, 62]	−1.86, −2.09 [15]
FpC_6H_4Cl-4	−1.92 [15, 22]	−1.82, −2.07 [15]
FpC_6F_5	−1.68, −2.30 [15], see also [22]	

The ease of reduction depends upon the stability of the carbanion formed and varies parallel with its basicity. There is a good linear correlation between the half-wave potentials of FpC_6H_5, FpC_6H_4Cl-4, and related FpR (R=alkyl) complexes, and the pK_a values of the corresponding RH [15], see also [46, 62]. A linear relationship between $E_{1/2}$ and ν(CO) of FpR with R=C_6H_5, C_6H_4Cl-4, and C_6F_5 is pointed out in [22].

The polarographic reduction of FpC_6F_5 in $CH_3OCH_2CH_2OCH_3/0.1$ M $[N(C_4H_9-n)_4]ClO_4$ gives two waves at $E_{1/2}$ = −2.3 and −3.0 V (vs. 0.001 M $AgClO_4$/Ag). The number of the electrons involved in the first step is n=1.5. The reduction product has not been identified, but is not $[Fp]^-$ [11].

References on pp. 189/91

Fe–C bond cleavage takes place upon treatment of FpR with $CHCl_2COOH$ to give FpO-$COCHCl_2$ and RH. Kinetic studies carried out under pseudo-first-order conditions (tenfold excess of acid) in CH_2Cl_2 (R = C_6H_5, $C_6H_4CH_3$-4, C_6H_4F-4, C_6H_4Cl-4), benzene, and CH_2ClCH_2Cl (R = C_6H_5) at 25 °C show that the reaction is first-order in FpR. However, it was not possible to determine the concentration of the $CHCl_2COOH$ monomer and the order of the reaction in acid, because the "dimerization constant" for $CHCl_2COOH$ was known only in benzene, and not in the other solvents employed. The rate data, measured in CH_2Cl_2, reveal that the observed rate constants, k, increase as the σ^+-value for the para substituent on the phenyl ring becomes more negative, i.e., C_6H_4Cl-4 < C_6H_5 < C_6H_4F-4 < $C_6H_4CH_3$-4. A comparison of the values of k obtained at similar concentrations of the acid indicates that the reaction is somewhat faster in CH_2Cl_2 and CH_2ClCH_2Cl than in benzene. The corresponding reaction of FpC_6H_5 with CF_3COOH in CH_2Cl_2 was too rapid to be followed by the techniques employed for the above reactions (IR or ^1H NMR spectroscopy). Thus, with 3.2×10^{-2} M acid in a tenfold excess over FpC_6H_5 in CH_2Cl_2 at 25 °C, the cleavage of the Fe–C bond by $CHCl_2COOH$ and CF_3COOH has half-lives of 16 and <2.5 min, respectively [83].

The ready cleavage of $FpC_6H_4R^1$-4 (R^1 = H, CHO, $COCH_3$) by dry HCl yielding FpCl and $C_6H_5R^1$ is briefly mentioned in [37].

The reactions of FpR (R = C_6H_5, C_6H_4Cl-4, or $C_6H_4OCH_3$-4) with excess $HgCl_2$ in THF, acetone, or isopropyl alcohol at 25 °C proceed by three distinct pathways, giving FpCl and RHgCl (reaction 1), FpHgCl and RX (reaction 2), or Hg_2Cl_2 and various substances derived from decomposition of the oxidized FpR (reaction 3). The main products obtained are FpCl (>95%) and RHgCl (>85%) [76], see also [69, 71]. Some FpHgCl (<5%) was detected by IR and ^1H NMR spectroscopy for the reactions of FpC_6H_5 and $FpC_6H_4OCH_3$-4. Traces of Hg_2Cl_2 were observed in the reactions of $FpC_6H_4OCH_3$-4. No evidence of either reaction 2 or reaction 3 was found when R = C_6H_4Cl-4 [76]. Mechanisms leading to these different reaction products were proposed. The rates of the reactions were monitored by IR spectroscopy at 25 °C under pseudo-first-order conditions. Third-order kinetics, first-order in FpR and second-order in $HgCl_2$, were found for reactions 1 and 2. The following rate constants (in $M^{-2} \cdot s^{-1}$) are reported: $k_3 = 4.9 \times 10^{-3}$ (R = $C_6H_4OCH_3$-4) [71, 76], 6.7×10^{-4} (R = C_6H_5), 3.8×10^{-4} (R = C_6H_4Cl-4) [76] in i-C_3H_7OH, and 5.5×10^{-6} (R = $C_6H_4OCH_3$-4) in THF [71, 76]. Support for the electrophilic nature of these cleavage processes is seen in the decrease of the rate constants with increasing Taft polar substituent constants σ^+ of the aryl groups [76].

Kinetic studies on the reactions of FpR with liquid SO_2 to give $FpSO_2R$ have been carried out. The insertion obeys the rate law $-d[FpR]/dt = k[FpR]$ where k contains an indeterminable dependence on the concentration of SO_2. The following rate constants and activation parameters were obtained [63].

compound	$k(s^{-1})$ at −40 °C	ΔH^+ (kcal/mol)	ΔS^+ (cal · mol^{-1} · K^{-1})
$FpC_6H_4OCH_3$-4	1.4×10^{-3}	3.5 ± 0.5	-55 ± 2
$FpC_6H_4CH_3$-2	2.9×10^{-4}	3.4 ± 1.1	-59 ± 2
$FpC_6H_4CH_3$-4	6.6×10^{-5}	7.8 ± 0.4	-43 ± 2
$FpC_6H_4CH_3$-3	1.9×10^{-5} (at −23 °C)	−	−
FpC_6H_5	2.0×10^{-6} (accurate to ±20%)	7.3 ± 0.5	-52 ± 2

With the exception of $k = 3.8 \times 10^{-4}$ s^{-1} for $FpC_6H_4OCH_3$-4 the same data are reported for FpR (R = C_6H_5, $C_6H_4CH_3$-3, $C_6H_4CH_3$-4) in [54]. It is proposed that the reaction proceeds via a transition state I that resembles the classical σ-complex of electrophilic aromatic substitution reactions. This complex may then dissociate into an ion pair, $[Fp]^+[O_2SR]^-$, or form the O-sulfinate FpOS(O)R in a concerted fashion through simultaneous scission of the Fe–aryl bond and formation of the Fe–OS(O) bond, or do both. The O-sulfinate then isomerizes to the S-sulfinate $FpSO_2R$, the final product of the insertion [63]. The existence of an O-bonded sulfinate as intermediate is supported by the relatively facile replacement of RSO_2^- by I^- to give FpI, as is demonstrated with $FpC_6H_4CH_3$-4 and related FpR compounds. Under similar conditions, the S-sulfinate does not react with KI [50, 58]. Mechanisms involving either direct attack of SO_2 on iron accompanied by migration of R onto the SO_2 ligand or initial formation of the acyl complex $C_5H_5Fe(CO)(COR)SO_2$ followed by a rapid transfer of R onto the SO_2 ligand are suggested in [4, 10]. The last two mechanisms would be consistent with the relatively slow insertion reaction of FpC_6H_5 compared to $FpCH_3$ or FpC_2H_5, since in each case, migration of the bulky C_6H_5 group would be involved [4].

$$C_5H_5(CO)_2Fe \quad \overset{\ominus}{O}{-}S \overset{\displaystyle \nearrow}{\underset{\displaystyle O}{}} \quad \underset{\oplus}{\bigcirc}\!\!-\!R^1$$

I

$FpSO_2C_6H_5$ is also formed when SO_2 is bubbled through a solution of FpC_6H_5 in pentane at 27 °C [10]. Only unreacted material could be isolated from the reaction of FpC_6F_5 with liquid SO_2 at −40 °C after 168 h [10].

For the meta- and para-substituted aryl complexes, the observed rate constants decrease with an increasing value of the Hammett σ^+-parameter of the respective substituent [54, 63].

The reactions of FpR (R = $C_6H_4CH_3$-4, C_6H_4F-4, C_6H_4Cl-4) with a slight excess of R^1SO_2N= S=O ($R^1 = CH_3$ or 4-ClC_6H_4) or of $FpC_6H_4CH_3$-4 with CH_3SO_2N=S=NSO_2CH_3 in $CHCl_3$ or CH_2Cl_2 at ambient temperature (generally within 1 h) yield $FpN(SOR)SO_2R^1$ and $CH_3SO_2N(Fp)S(R)$= NSO_2CH_3, respectively. The latter compound (R = $C_6H_4CH_3$-4) was also isolated as by-product from the reaction of $FpC_6H_4CH_3$-4 with concentrated CH_3SO_2NSO, presumably owing to disproportion of CH_3SO_2NSO to $CH_3SO_2NSNSO_2CH_3$ and SO_2 [84].

In contrast to the FpR alkyl compounds (see "Organoiron Compounds" B 12, 1984, p. 204), FpC_6H_5 and FpC_6F_5 do not insert $GeCl_2$ into the Fe–C bond. $FpGeCl_3$ is the product which results from heating FpC_6H_5 with $GeCl_2 \cdot C_4H_8O_2$ ($C_4H_8O_2$ = dioxane) in dioxane at 60 to 70 °C for 3 h (30% yield) or at 100 °C for 2 h (46%). Only traces of $FpGeCl_3$ are obtained from the corresponding reaction of FpC_6F_5 at 100 °C after 3 h [44].

UV irradiation of FpR (R = C_6H_5, $C_6H_4CH_3$-4, C_6H_4Cl-4 [42], C_6F_5 [20], C_6H_4F-3 [56], C_6H_4F-4 [33, 42]) in the presence of various phosphines ^2D such as $P(C_6H_5)_3$, $P(C_6H_4CH_3$-4$)_3$, $P(C_2H_5)_2C_6H_5$, or $P(C_6H_5)_2C_2H_5$ in benzene either at 25 to 30 °C for 5 to 10 h ("150 W lamp") or at 80 °C for 2 to 3 h ("375 W lamp") leads to formation of $C_5H_5Fe(CO)(^2D)R$. When, however, the latter reation is carried out in refluxing benzene with less intense light ("150 W"), two products, $C_5H_5Fe(CO)(^2D)R$ and $C_5H_5Fe(CO)(^2D)COR$, are formed with the former predominating [42], see also [12, 19]. The proportion of these two compounds is reversed when FpC_6H_5 and $P(C_6H_5)_3$ are refluxed for 22 h without irradiation; the aroyl derivative is then the main product [42]. In contrast, heating of FpC_6H_4F-2 in p-xylene at 100 °C for 6 h or of FpC_6H_4Cl-2 in toluene under reflux for 8 h in the presence of $P(C_6H_5)_3$ results in formation of only $C_5H_5Fe(CO)(^2D)R$ [85].

References on pp. 189/91

Whereas photolysis ("375 W lamp") of FpC_6H_4F-4 and $P(C_2H_5)_3$ in benzene at 25 to 30 °C for 3 h affords $C_5H_5Fe(CO)[P(C_2H_5)_3]C_6H_4F$-4, no reaction takes place under similar conditions with the ferrocenyl phosphine $P(C_5H_4FeC_5H_5)_3$. When the mixture is heated at 80 °C for 90 min, $C_5H_5Fe(CO)[P(C_5H_4FeC_5H_5)_3]C_6H_4F$-4 is formed [42]. Also, no reaction occurs on irradiation ("150 W lamp") of FpC_6H_4F-4 with trinaphth-1-ylphosphine in benzene/ tetrahydrofurfural (5:3) at 25 to 30 °C during 13 h. The corresponding reaction with $P(C_6H_5)_2Cl$ in benzene leads to decomposition of FpC_6H_4F-4 [33]. Photolysis of FpC_6H_4CHO-4 in the presence of $P(C_6H_5)_3$ to give $C_5H_5Fe(CO)[P(C_6H_5)_3]C_6H_4CHO$-4 is briefly mentioned in [37].

The photochemical reactions ("150 or 375 W lamp") of FpR (R = C_6H_5, C_6H_4F-4, C_6H_4Cl-4) with $P(OC_6H_5)_3$ (^2D) in benzene at 25 to 30 °C for 2 to 5.5 h afford the exchange products $C_5H_5Fe(CO)[P(OC_6H_5)_3]R$. Refluxing in benzene under more intense irradiation (375 W) for 9 h gives the dimer $[C_5H_5Fe(P(OC_6H_5)_3)_2]_2$, whereas both products are formed under reflux and less intense irradiation ("150 W") during 6.5 to 15 h [18, 42], see also [12]. Both types of products also result from photolysis ("150 W") of FpC_6H_4F-3 and $P(OC_6H_5)_3$ in benzene at room temperature after 15.5 h [56]. $C_5H_5Fe(CO)[P(OC_6H_5)_3]C_6F_5$ was the only product isolated after irradiation ("375 W") of a boiling mixture of FpC_6F_5 and $P(OC_6H_5)_3$ in benzene for 3 h [18]. The product formed in the photochemical reaction ("450 W") of FpC_6H_5 with $P(OC_6H_5)_3$ in benzene at 32 °C for 3.5 h was formulated as the ortho-metalated derivative II. When cyclohexane is used as solvent "at 360 W" and the reaction time is increased by a factor of eight, purportedly the same product together with $[C_5H_5Fe(P(OC_6H_5)_3)_2]_2$ is obtained [75], but according to [70] the latter is in fact III. Another similar type of product, (IV), results from photolysis of FpC_6H_5 and 1-phenoxy-1,2,5-phosphadioxolane in benzene at 28 °C for 4 h [75].

II   III   IV

Table 7
$C_5H_5Fe(CO)_2R$ Compounds with R = Aryl.
Further information on compounds with numbers preceded by an asterisk is given at the end of the table.
For abbreviations and dimensions, see p. X.

No. compound method of preparation (yield)	properties and remarks
*1 FpC_6H_5 III (17%) [6], (19%) [63], (75%) [87] IVa (34%) [12], (18%) [8, 19] IVb (45%) [8, 19, 73] V (55%) [79]	yellow crystals, m.p. 35 to 36° [6, 79], 34° (from hexane at −70°) [8, 12, 19, 66], caramel, m.p. 26 to 30° [1, 2], brown crystals [68] ^1H NMR: 5.30 (C_5H_5), 7.5 (m, C_6H_5) in $CDCl_3$ at 38° [68], 4.81 (C_5H_5) in CH_2Cl_2 [55], 4.73 (s, C_5H_5), 6.78, 7.26 (both complex m, C_6H_5, relative intensities 3:2) in CS_2, 4.75 in $CHCl_3$ [57]

Table 7 (continued)

No. compound method of preparation (yield)	properties and remarks
	$^{13}C\{^1H\}$ NMR: 85.7 (C_5H_5), 122.8 (C-4), 127.5 (C-3,5), 145.0 (C-2,6), 145.5 (C-1), 216.1 (CO) in $CDCl_3$ [91], similar in $CHCl_3$ [75], 86.3 (C_5H_5), 217.1 (CO) in CH_2Cl_2 [55] ^{57}Fe-γ (78 K): $\delta = 0.019$ (Fe), $\Delta = 1.863$, also depicted [91] IR: 1959, 2013 in THF [22], 1965, 2011 in CH_2Cl_2 [55], 1967, 2019 in i-C_3H_7OH [76], 1969, 2024 in CCl_4 [91], 1972, 2023 in C_6H_{12} [54, 63], see also [68], 1943, 1973, 2020 (all CO), 1565 (C_6H_5) in liquid petrolatum, 2020, 1570 (C_6H_5) in C_6H_6 [12], 1930, 1960, 2020 (all CO), 3040, 3105 (both CH), 3900 (CO overtone) in CCl_4, other bands at 675, 695, 728, 760, 828, 840sh, 995, 1010, 1050, 1060 in CS_2, 1360, 1420, 1430, 1470, 1570 in CCl_4 [1], 1969, 2021 (both CO) as halocarbon oil mull, 3010, 3070 (both CH) in KBr, other bands (728 to 1555) in KBr [6] k = 15.96, k' = 0.369 [55], k = 16.07, k' = 0.42 [31], k = 16.10, k' = 0.443 [91] UV (C_6H_{12}): 225 ($\varepsilon = 239\,000$), 354 ($\varepsilon = 1070$) [6]
*2 $C_5H_4DFe(CO)_2C_6H_5$	^1H NMR (CCl_4): 4.6 (s, 4H, C_5H_4D), 6.8 (m, 3H, C_6H_5), 7.4 (m, 2H, C_6H_5) [88] IR (CCl_4): 1969, 2021 (both CO) [88]
3 $FpC_6H_4CH_3$-2 II (20%) [85] III (30%) [63]	yellow crystals, m.p. 68.5° [85], 67 to 69° [63] ^1H NMR: 2.38 (s, CH_3), 4.71 (s, C_5H_5), 6.53 to 6.99 (m, 3H, C_6H_4), 7.29 to 7.57 (m, 1H, C_6H_4) in CCl_4 [63], 2.39 (s, CH_3), 4.76 (s, C_5H_5), 6.6 to 7.0 (m, 3H, C_6H_4), 7.25 to 7.35 (m, 1H, C_6H_4) in $CDCl_3$ [85] $^{13}C\{^1H\}$ NMR ($CDCl_3$): 28.15 (CH_3), 86.1 (C_5H_5), 123.4, 124.6 (C-4,5), 129.3 (C-3), 144.1 (C-1 or C-2), 146.8 (C-6), 148.9 (C-2 or C-1), 215.9 (CO) [91] ^{57}Fe-γ (78 K): $\delta = 0.044$ (Fe), $\Delta = 1.881$ [91] IR: 1968, 2024 in C_6H_{12} [63], 1962, 2017 (all CO) [85, 91], other bands at 832, 1008 [85] in CCl_4 k = 15.99, k' = 0.442 [91] mass spectrum: $[M]^+$, $[M-CO]^+$, $[M-2CO]^+$ [85]
4 $CH_2C_6H_5$ II (24%) [85]	yellow crystals, m.p. 83° [85] ^1H NMR ($CDCl_3$): 4.07 (s, CH_2), 4.59 (s, C_5H_5), 6.47 to 7.6 (m, 9H, aryl) [85] $^{13}C\{^1H\}$ NMR ($CDCl_3$): 46.8 (CH_2), 86.1 (C_5H_5), 123.4, 124.9 (C-4,5), 130.4 (C-3), 144.8 (C-1 or C-2), 147.3 (C-6), 151.4 (C-2 or C-1), 215.9 (CO) [91] ^{57}Fe-γ (78 K): $\delta = 0.028$ (Fe), $\Delta = 1.944$ [91]

References on pp. 189/91

Table 7 (continued)

No.	compound method of preparation (yield)	properties and remarks

4 (continued)

IR (CCl$_4$): 1962, 2016 (both CO) [91], other bands at 834, 1008, 1444, 1450 [85]

k = 15.98, k′ = 0.434 [91]

mass spectrum: [M]$^+$, [M − 2CO]$^+$ [85]

5 FpC$_6$H$_4$CH$_3$–3
III (39%) [63]

m.p. 48 to 49° [63]

^{1}H NMR (CCl$_4$): 2.18 (s, CH$_3$), 4.42 (s, C$_5$H$_5$), 6.47 to 7.37 (m, C$_6$H$_4$) [63]

IR (C$_6$H$_{12}$): 1970, 2027 (both CO) [54, 63]

*6 FpC$_6$H$_4$CH$_3$–4
III [58], 40% [94]
IVa (45%, based on the iodonium salt) [33]

orange, m.p. 78° [58], yellow crystals, m.p. 77 to 78° (from petroleum ether) [94], m.p. 76.5° [33], yellow oil [10]

^{1}H NMR (CS$_2$): 2.13 (s, CH$_3$), 4.64 (s, C$_5$H$_5$), 6.62, 7.10 (both: m, 2H, C$_6$H$_4$) [94]

IR: 1971, 2027 in C$_6$H$_{12}$ [54, 63], 1963, 2112 (all CO), other bands from 558 to 2970 (neat) [10], 1940, 1955 sh, 2000, 2020 (all CO), 781, 827, 835, 848, 999, 1034, 1425, 1470, 3060, 3125 (mulls) [94]

7 FpC$_6$H$_4$F–2
II (80%) [85]

m.p. 61° (from C$_6$H$_{14}$/ether, 19:1) [85]

^{1}H NMR (CDCl$_3$): 4.86 (s, C$_5$H$_5$), 6.6 to 7.1 (m, 3H, C$_6$H$_4$), 7.3 to 7.6 (m, 1H, C$_6$H$_4$) [85]

^{13}C{^{1}H} NMR (CDCl$_3$): 85.4 (C$_5$H$_5$), 113.6 (C–3, J(C,F) = 30.9), 123.7, 125.1 (C–4,5, J(C,F) = 7.8), 127.9 (C–1, J(C,F) = 39.5), 146.5 (C–6, J(C,F) = 13.1), 169.5 (C–2, J(C,F) = 229.6), 215.3 (CO, J(C,F) = 3.9) [91]

^{19}F NMR (CDCl$_3$): 81.2 (m) [85]

^{57}Fe-γ (78 K): δ = 0.039 (Fe), Δ = 1.912 [91]

IR (CCl$_4$): 1974, 2028 (both CO) [85, 91], other bands at 835, 988, 1020, 1430, 1718 [85]

k = 16.17, k′ = 0.436 [91]

mass spectrum: [M]$^+$, [M − CO]$^+$, [M − 2CO]$^+$ [85]

8 FpC$_6$H$_4$F–3
III [45]
IVa (28%) [56]
V (61%) [79]

m.p. 45 to 46° (dec., from benzene/hexane) [79], yellow crystals, m.p. 41° (from heptane at −50°) [56], yellow-gold, m.p. 39 to 41° [45]

^{19}F NMR (referred to C$_6$H$_5$F): +2.37 in CH$_2$Cl$_2$ [45], +2.35 in CHCl$_3$ [56, 57], +1.50 [37]

IR: 1963, 2018 in THF [56], 1975, 2020 (all CO) in CCl$_4$ [45]

9 FpC$_6$H$_4$F–4
III (53%) [45]
IVa (26%) [33]
V (71%) [79]

yellow-gold plates, m.p. 60 to 62° [45], 59 to 60° [79], yellow crystals, m.p. 56° (from hexane at −5°) [33]

^{19}F NMR (referred to C$_6$H$_5$F): +10.95 in CHCl$_3$ [56, 57], +10.92 in CH$_2$Cl$_2$ [45], +10.75 in C$_6$H$_6$ [57], +10.23 [37]

IR: 1961, 2015 in THF [56], 1973, 2018 (all CO) in CCl$_4$ [45]

References on pp. 189/91

Table 7 (continued)

No. compound method of preparation (yield)	properties and remarks
10 FpC$_6$H$_4$Cl-2 II (50%) [85]	m.p. 108° (from C$_6$H$_{14}$/ether, 19:1) [85] ^{1}H NMR (CDCl$_3$): 4.85 (s, C$_5$H$_5$), 6.7 to 7.0 (m, 2H, C$_6$H$_4$), 7.1 to 7.24 (m, 1H, C$_6$H$_4$), 7.48 to 7.64 (m, 1H, C$_6$H$_4$) [85] ^{13}C{^{1}H} NMR (CDCl$_3$): 86.2 (C$_5$H$_5$), 124.8, 125.1 (C-4,5), 128.7 (C-3), 145.9 (C-1), 147.4 (C-6), 148.2 (C-2), 215.1 (CO) [91] ^{57}Fe-γ (78 K): $\delta = 0.033$ (Fe), $\Delta = 2.002$ [91] IR (CCl$_4$): 1973, 2026 (both CO) [91], other bands at 835, 1004, 1194, 1439, 1562 [85] k = 16.15, k' = 0.428 [91] mass spectrum: 290, [M]$^+$, 262, [M−CO]$^+$, 234, [M−2CO]$^+$ [85]
*11 FpC$_6$H$_4$Cl-4 III (26%) [55, 76] IVa (28%) [18]	yellow crystals [18, 55], m.p. 89.5 to 91° (from hexane at 0°) [18, 76], 91 to 92° [52] ^{1}H NMR: 4.80 (s, C$_5$H$_5$), 7.14 (A$_2$B$_2$, C$_6$H$_4$) in CDCl$_3$ [76], 4.84 (C$_5$H$_5$) in CH$_2$Cl$_2$ [55] ^{13}C NMR (CH$_2$Cl$_2$): 86.7 (C$_5$H$_5$), 216.8 (CO) [55] IR: 1963, 2016 in THF [22], 1968, 2014 in CH$_2$Cl$_2$ [55], 1976, 2026 in C$_5$H$_{12}$, 1970, 2022.5 (all CO) in i-C$_3$H$_7$OH [76] k = 16.01, k' = 0.370 [55]
12 FpC$_6$H$_4$Br-2 II (30%) [85] III (40%) [94]	m.p. 110° (from C$_6$H$_{14}$/ether, 19:1) [85], yellow crystals (ether/petroleum ether, 1:1), m.p. 106 to 107° [94] ^{1}H NMR: 4.84 (s, C$_5$H$_5$), 6.4 to 6.9, 7.0 to 7.6 (both: m, 2H, C$_6$H$_4$) in CDCl$_3$ [85], 4.81 (s, C$_5$H$_5$), 6.89 (m, 2H, C$_6$H$_4$), 7.20, 7.40 (both: m, 1H, C$_6$H$_4$) in CS$_2$ [94] ^{13}C{^{1}H} NMR (CDCl$_3$): 86.5 (C$_5$H$_5$), 125.1, 125.4 (C-4, 5), 132.3 (C-3), 141.5 (C-2), 147.7 (C-6), 149.3 (C-1), 215.2 (CO) [91] ^{57}Fe-γ (78 K): $\delta = 0.028$ (Fe), $\Delta = 2.000$ [91] IR: 1973, 2026 (both CO) [85, 91], other bands at 835, 1020, 1030, 1077, 1433 in CCl$_4$ [85], 1950, 1975 sh, 2010 (all CO), 693, 740, 822, 841, 976, 1013, 1065, 1218, 1420, 3055, 3070, 3130 (mulls) [94] k = 16.15, k' = 0.428 [91] mass spectrum: 334, [M]$^+$, 306, [M−CO]$^+$, 278, [M−2CO]$^+$ [85]
13 FpC$_6$H$_4$I-2 II (64%) [85]	m.p. 84.5° (from C$_6$H$_{14}$/ether, 19:1) [85] ^{1}H NMR (CDCl$_3$): 4.84 (s, C$_5$H$_5$), 6.46 (td, 1H, C$_6$H$_4$), 6.81 (td, 1H, C$_6$H$_4$), 7.38 (dd, 1H, C$_6$H$_4$), 7.58 (dd, 1H, C$_6$H$_4$) [85]

References on pp. 189/91

Table 7 (continued)

No.	compound method of preparation (yield)	properties and remarks
13	(continued)	$^{13}C\{^1H\}$ NMR (CDCl$_3$): 87.0 (C$_5$H$_5$), 121.6 (C-2), 125.2, 125.9 (C-4,5), 139.7 (C-3), 147.4 (C-6), 156.3 (C-1), 215.6 (CO) [91] ^{57}Fe-γ (78 K): $\delta = 0.028$ (Fe), $\Delta = 2.013$ [91] IR (CCl$_4$): 1972, 2024 (both CO) [85, 91], other bands at 835, 988, 1020, 1718, 1930 [85] k = 16.12, k′ = 0.420 [91] mass spectrum: [M]$^+$, [M−CO]$^+$, [M−2CO]$^+$ [85]
14	FpC$_6$H$_4$CN-4	for Hammett relations and spectroscopic data, see p. 164 [37]
*15	FpC$_6$H$_4$OCH$_3$-4 III (20%) [63], (40%) [94]	m.p. 63 to 64° [63], yellow crystals, m.p. 60 to 61° [94] ^{1}H NMR: 3.64 (s, CH$_3$), 4.70 (s, C$_5$H$_5$), 6.25 to 6.79, 6.90 to 7.44 (both: m, 2H, C$_6$H$_4$) in CCl$_4$ [63], 3.56 (s, CH$_3$), 4.70 (s, C$_5$H$_5$), 6.47, 7.07 (both: m, 2H, C$_6$H$_4$) in CS$_2$ [94] IR: 1958, 2014 in THF, 1966, 2018 in i-C$_3$H$_7$OH [76], 1971, 2027 in C$_6$H$_{12}$ [54, 63], 1945, 1965 sh, 2000 sh, 2020 (all CO), 702, 809, 826, 990, 1015, 1040, 1090, 1169, 1219, 1252, 1435, 1480, 1580, 3055, 3125 (mulls) [94]
*16	FpC$_6$H$_4$CHO-4	for the reaction with P(C$_6$H$_5$)$_3$, see p. 168 [37]
17	FpC$_6$H$_4$COCH$_3$-4	prepared from FpC$_6$H$_5$ and CH$_3$COCl/AlCl$_3$ at 0° for 1 h [37]
18	FpC$_6$H$_4$COOCH$_3$-4	for Hammett relations and spectroscopic data, see p. 164 [37]
*19	FpC$_6$H$_4$COOC$_2$H$_5$-4	yellow [17]
*20	FpC$_6$F$_5$ I (49%) [5]	golden-yellow crystals, m.p. 144.5 to 145° [7], yellow-orange, m.p. 142 to 143° [5], 139 to 141° [21] ^{1}H NMR: 4.96 (s, C$_5$H$_5$) in CDCl$_3$ at 36° [7], 4.97 (C$_5$H$_5$) in CH$_2$Cl$_2$ [55], 4.98 (C$_5$H$_5$) in CHCl$_3$ [5] $^{13}C\{^1H\}$ NMR: 85.0 (C$_5$H$_5$), 110.3 (m, C-1), 136.5 (m, C-3,5), 137.6 (m, C-4), 150.9 (m, C-2,6), 213.2 (CO) in CDCl$_3$ [91], 86.0 (C$_5$H$_5$), 214.6 (CO) in CH$_2$Cl$_2$ [55] ^{19}F NMR: 106.7 (F-2,6), 161.7 (F-4), 164.3 (F-3,5), J(2,3) = 30.15, J(2,4) ≈ 1.0, J(2,5) = J(3,6) = 8.8, J(2,6), J(3,5) = 3.45, 5.7, J(3,4) = 19.6 in THF at 35° [26], see also [9], 107.3 (d, F-2,6, J = 21), 160.3 (t, F-4, J = 21), 163.6 (t, F-3,5, J = 21) in CHCl$_3$ [5], 28.6 (m), 82.5 (t, J(F,F) = 20.6), 85.4 (m) in C$_6$H$_6$ at 27° referred to CF$_3$COOH [7] ^{57}Fe-γ: $\delta = 0.16$ (Fe/Cr), $\Delta = 1.82$ at 298 K [43], $\delta = 0.083$ (Fe), $\Delta = 1.976$ at 78 K [91]

References on pp. 189/91

Table 7 (continued)

No.	compound method of preparation (yield)	properties and remarks
*20	(continued)	IR: 1997, 2046 in CCl_4 [91], 1987, 2035 in THF [22] or CH_2Cl_2 [55], 1958, 1997, 2045 (all CO) as halocarbon oil mull, 3100 (CH), other bands at 713, 738, 768, 842sh, 854, 955, 962, 1003, 1014, 1019sh, 1045sh, 1050sh, 1055, 1062sh, 1111, 1246, 1262, 1352 sh, 1360, 1435sh, 1450, 1490, 1595, 1620 in KBr [5] $k = 16.42$, $k' = 0.334$ [55], $k = 16.50$, $k' = 0.400$ [91] UV (C_6H_{12}): 356 $(\varepsilon = 942)$ [5]
*21	FpC_6F_4H-2 I (68%) [59]	m.p. 135.5 to 136.5° [59]
22	FpC_6F_4D-2	prepared by stirring of an equimolar mixture of FpC_6F_4Br-2 and n-C_4H_9Li at $-78°$ for 2 h, and subsequent addition of D_2O, yield 100% [61]
*23	FpC_6F_4H-3 II (82%) [59]	m.p. 87.5 to 88° [48, 59]
*24	FpC_6F_4H-4 I (39%) [5]	orange crystals [9], m.p. 128 to 130° [5] ^1H NMR: 4.97 (C_5H_5), 6.66 (tt, C_6F_4H, $J = 7, 10$) in $CHCl_3$, 5.20 (C_5H_5), 6.88 (tt, C_6F_4H, $J = 7, 10$) in CH_3COCH_3 [5] ^{19}F NMR: 109.3 (F-2,6), 142.0 (F-3,5), $J(2,6)$, $J(3,5) = 3.45, 1.2$, $J(2,3) = \pm 29.0$, $J(2,5) = \mp 13.2$, $J(H, F-2) = 7.3$, $J(H, F-3) = 9.35$ in $CDCl_3$ at 35° [9, 26], 108.9 (dd, F-2,6, $J = 7, 15$), 141.5 (dd, F-3,5, $J = 10, 15$) in $CHCl_3$ [5] IR: 1946, 1973, 1986, 2033 (all CO) as halocarbon oil mull, 3075, 3100 (both CH), other bands from 698 to 1626 in KBr [5]
25	$FpC_6F_4CH_3$-4 I (46%) [36]	yellow crystals, m.p. 123 to 124° [36] ^{19}F NMR (THF): 109.3 (F-2,6), 145.9 (F-3,5), $J(2,6)$, $J(3,5) = 0, 4.1$, $J(2,5) = 28.65$, $J(3,6) = 11.9$ [36], see also [26] IR (C_6H_{12}): 1993, 2038 (both CO), other bands from 768 to 1635 [36]
26	$FpC_6F_4CF_3$-4 I (48%) [5], (24%) [27]	yellow-orange, m.p. 119 to 120° [5] ^1H NMR $(CHCl_3)$: 5.01 (C_5H_5) [5] ^{19}F NMR: 104.9 (F-2,6), 144.2 (F-3,5), $J(2,6)$, $J(3,5) = 7.7$, 1.8, $J(2,3) = \pm 29.0$, $J(2,5) = \mp 13.2$, $J(CF_3, F-3) = 21.3$, $J(CF_3, F-2) = 0$ in THF at 35° [26], 56.9 (t, CF_3, $J = 20$), 106.5 (F-2,6), 144.5 (F-3,5) in $CHCl_3$ [5] ^{57}Fe-γ (298 K): $\delta = 0.18$ (Fe/Cr), $\Delta = 1.90$ [43] IR: 1968, 2003, 2054 (all CO) as halocarbon oil mull, 3100 (CH), other bands from 710 to 1630 in KBr [5] UV (C_6H_{12}): 356 $(\varepsilon = 926)$ [5]

References on pp. 189/91

Table 7 (continued)

No.	compound method of preparation (yield)	properties and remarks
27	FpC$_6$F$_4$CH=CH$_2$-4 I (42%) [36]	yellow crystals, m.p. 115 to 117° [23, 36] ^{19}F NMR: 108.6 (F-2,6), 145.6 (F-3,5), J(2,6)=J(3,5)=3.25, J(2,3)=±27.35, J(2,5)=∓12.6 in THF at 35° [26, 36], 110.4 (F-2,6), 148.0 (F-3,5) [28] IR (C$_6$H$_{12}$): 1995, 2040 (both CO), other bands from 675 to 1633 [36]
*28	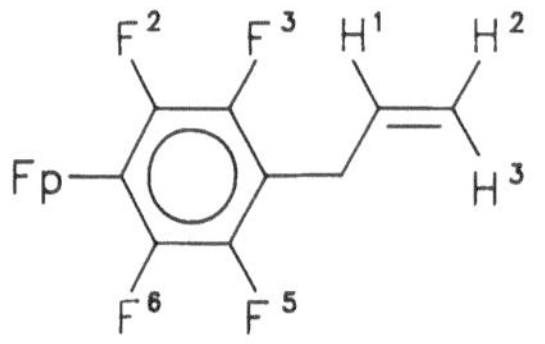	m.p. 108 to 111° [36] ^{1}H NMR: 3.34 (d, CH$_2$), 4.88 (s, 6H, C$_5$H$_5$, H-2), 5.02 (d, H-3), 5.74 (br m, H-1) [36] ^{19}F NMR (THF): 108.8 (F-2,6), 146.9 (F-3,5), J(2,6), J(3,5)=1.6, 4.2, J(2,5)=29.0, J(3,6)=13.6 [36], see also [26] IR (C$_6$H$_{12}$): 1995, 2040 (both CO), other bands from 744 to 1640 [36]
*29	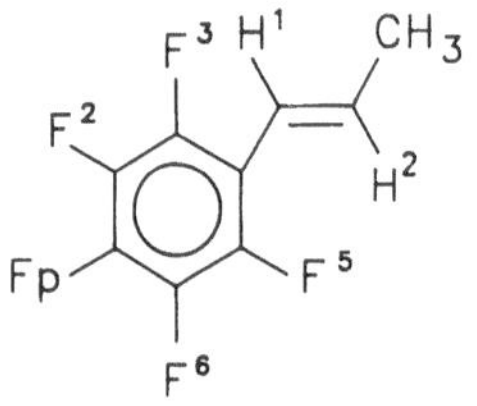	^{1}H NMR: 1.90 (d, CH$_3$), ~4.0 (br, H-2), 6.25 (br s, H-1) [36] ^{19}F NMR (THF): 109.4 (F-2,6), 146.5 (F-3,5), J(2,6), J(3,5)=2.35, 4.3, J(2,5)=27.35, J(3,6)=11.8 [36] IR (C$_6$H$_{12}$): 1976, 2026 (both CO) [36]
*30	FpC$_6$F$_4$Cl-3 I (29%), II (62%) [59]	m.p. 209 to 211° [59]
*31	FpC$_6$F$_4$Br-2 I (61%), II (36%) [59]	m.p. 135.5 to 136.5° [59]
*32	FpC$_6$F$_4$Br-3 I (68%), II (32%) [59], see also [48, 49]	m.p. 209 to 211° (dec.) [59]
*33	FpC$_6$F$_4$Br-4 I (60%), II (69%) [59]	139.5 to 140.5° [59] ^{19}F NMR (THF): 105.2 (F-2,6), 134.0 (F-3,5), J(2,6)=J(3,5)=6.4, J(2,5)=32.2, J(3,6)=14.8 [36] IR (C$_6$H$_{12}$): 1999, 2043 (both CO) [36]
*34	FpC$_6$F$_4$I-2	—
*35	FpC$_6$F$_4$CN-4	orange crystals, m.p. 99 to 101° [9] ^{19}F NMR (THF at 35°): 105.1 (F-2,6), 139.4 (F-3,5), J(2,6), J(3,5)=7.3, 1.55, J(2,3)=±28.5, J(2,5)=∓12.8 [9, 26] IR (C$_6$H$_{12}$): 2000, 2040 (both CO), 2220 (CN), other bands from 723 to 1627 [9]
*36	FpC$_6$F$_4$OCH$_3$-3	^{19}F NMR (THF): 99.7 (F-6), 107.4 (F-2), 156.3 (F-5), 165.5 (F-4) [36] IR (C$_6$H$_{12}$): 1987, 2032 (both CO) [36]

References on pp. 189/91

Table 7 (continued)

No.	compound method of preparation (yield)	properties and remarks
*37	FpC$_6$F$_4$OCH$_3$-4	^{19}F NMR (THF): 108.3 (F-2,6), 158.15 (F-3,5), J(2,6), J(3,5) = 1.0, 5.9, J(2,5) = 29.35, J(3,6) = 9.55 [26, 36] IR (C$_6$H$_{12}$): 1994, 2038 (both CO), other bands from 771 to 1628 [36]
*38	FpC$_6$F$_4$SCH$_3$-4 I (29%) [24]	yellow needles, m.p. 130° (from C$_6$H$_6$/light petroleum), air-stable [24] ^{1}H NMR: 2.79 (CH$_3$), 4.97 (C$_5$H$_5$) [24] ^{19}F NMR (CHCl$_3$): 109 (F-2,6), 137.7 (F-3,5), J(2,3) = J(5,6) = +30.8, J(3,6) = J(2,5) = ∓12.65, J(2,6), J(3,5) = ±3.3, ±4.2 [24] IR (C$_6$H$_{12}$): 1994, 2040 (both CO), other bands from 755 to 1612 [24]
39	FpC$_6$F$_4$SC$_6$H$_5$-4 I (56%) [24]	yellow crystals, m.p. 125° (from C$_6$H$_6$/light petroleum), air-stable [24] ^{1}H NMR: 5.05 (C$_5$H$_5$), 7.03 (C$_6$H$_5$) [24] ^{19}F NMR (CHCl$_3$): 107.3 (F-2,6), 138.4 (F-3,5), J(2,3) = J(5,6) = ±31.2, J(3,6) = J(2,5) = ∓12.5, J(2,6), J(3,5) = ±3.8, ±4.7 [24] IR (C$_6$H$_{12}$): 1999, 2042 (both CO), other bands from 930 to 1200 [24]
*40	FpC$_6$F$_4$COOH-4	yellow, m.p. 182 to 184 (dec.) [41] ^{19}F NMR (CHCl$_3$ or CH$_2$Cl$_2$, referred to CF$_3$COOH): 28.6 (F-3,5), 63.7 (F-2,6) [41] IR (CHCl$_3$): 1995, 2042 (both CO), 1725 (C=O), 2580 to 3350 (OH), other bands from 920 to 1717 [41]
*41	FpC$_6$F$_4$COOC$_2$H$_5$-4	yellow needles, m.p. 127 to 128° [9, 41] ^{19}F NMR (THF at 35°): 108.2 (F-2,6), 142.8 (F-3,5), J(2,6), J(3,5) = 2.95, 5.4, J(2,3) = ±29.7, J(2,5) = ∓13.8 [9, 26] IR (C$_6$H$_{12}$): 2001, 2045 (both CO), other bands from 731 to 1740 [9]
*42	FpC$_6$F$_4$Li-2	not isolated
*43	FpC$_6$F$_4$Li-3	not isolated
*44	FpC$_6$F$_4$Li-4	not isolated
45	FpC$_6$F$_4$Si(CH$_3$)$_3$-4 I (48%) [64]	pale yellow, m.p. 141° [64] ^{19}F NMR (C$_6$H$_6$): 109.3 (d, F-2,6), 130.6 (d, F-3,5), spectrum also displayed [64] IR (C$_6$H$_{14}$): 2001, 2050 (both CO) [64]
46	FpC$_6$F$_4$Si(C$_6$H$_5$)$_2$C$_6$F$_5$-4 I (59%) [64]	yellow, m.p. 205 to 210° (dec.) [64] IR (C$_6$H$_{14}$): 2004, 2050 (both CO) [64]

References on pp. 189/91

Table 7 (continued)

No.	compound method of preparation (yield)	properties and remarks
47	$FpC_6F_4Si(CH_3)_2(CH_2)_3Si(CH_3)_3$-4 I [64]	yellow-orange oil, could not be purified satisfactorily for analysis [64] IR (C_6H_{14}): 2002, 2051 (both CO) [64]

48	$FpC_6F_3H_2$-2,4
	I (52%) [9]
		yellow platelets, m.p. 121 to 122° [9]
		^{19}F NMR (THF at 35°): 112.1 (F-6), 119.5 (F-3), 135.9
		  (F-5), J(F-3,6) = 28.7, J(F-5,6) = 16.1, J(F-3,5) = 1.67,
		  J(H-4, F-6) = 6.15, J(H-4, F-3) = 8.15, J(H-4,
		  F-5) = 10.4, J(H-2, F-6) = 2.4, J(H-2, F-3) = 8.15,
		  J(H-2, F-5) = 1.75 [9, 26]
		IR (C_6H_{12}): 1990, 2032 (both CO), other bands from 753
		  to 1607 [9]

*49	$FpC_6F_3H_2$-2,5
	I (9%) [9]
		orange crystals, m.p. 90 to 91° (from light petroleum)
		  [9]
		^{19}F NMR (THF at 35°): 84.0 (F-6), 143.6 (F-4), 146.4 (F-3),
		  J(F-4,6) = 15.2, J(F-3,6) = 3.33, J(F-3,4) = 20.4, J(H-2,
		  F-6) = 4.35, J(H-2, F-4) = 9.65, J(H-2, F-3) = 11.1,
		  J(H-5, F-6) = 7.2, J(H-5, F-4) = 10.65, J(H-5,
		  F-3) = 6.55 [9, 26]
		IR (C_6H_{12}): 1989, 2032 (both CO), other bands from 650
		  to 1611 [9]

*50	$FpC_6F_3H_2$-3,4
	I (29%) [5]
		yellow-orange, m.p. 73 to 75° [5]
		^1H NMR: 4.93 (C_5H_5), ~6.6 (m) in $CHCl_3$, 5.15 (C_5H_5),
		  ~6.7 (m) in CH_3COCH_3 [5]
		^{19}F NMR: 84.1 (F-2), 101.5 (F-6), 143.8 (F-5) in THF at
		  35° [9, 26], 85.5, 102.6 (both m, F-2,6), 144.4 (fine
		  structure not identified, F-5) in $CHCl_3$ [5]
		^{57}Fe-γ (298 K): δ = 0.15 (Fe/Cr), Δ = 1.94 [43]
		IR: 1933, 1957, 1992, 2041 (all CO) as halocarbon oil
		  mull, 3020 (CH), other bands from 710 to 1548 in KBr
		  [5]

*51	FpC_6F_3(H-3)Cl-5	pale orange crystals, m.p. 151 to 153° [65] ^{19}F NMR (THF, referred to CF_3COOH): −1.1 (F-6), −0.5 (F-2), +41.5 (F-4) [65]

*52	FpC_6F_3(H-2)CN-5	^{19}F NMR (THF at 35°): 76.8 (F-6), 138.4 (F-4), 142.8 (F-3),
		  J(F-4,6) = 14.9, J(F-3,6) = 4.35, J(F-3,4) = 19.7,
		  J(H, F-6) = 6.8, J(H, F-4) = 6.65, J(H, F-3) = 10.8 [26]

*53	FpC_6F_3(H-2)OCH$_3$-5	^{19}F NMR (THF): 79.3 (F-6), 142.7 (F-3), 160.5 (F-4) [36] IR (C_6H_{12}): 1983, 2032 (both CO) [36]
*54	$FpC_6F_3Cl_2$-3,5 I (14%) [65]	pale orange crystals, m.p. >200° (dec.) [65] ^{19}F NMR (THF, referred to CF_3COOH): +2.7 (F-2,6), +43.5 (F-4) [65]

References on pp. 189/91

Table 7 (continued)

No.	compound method of preparation (yield)	properties and remarks
55	FpC$_6$F$_3$(CN)$_2$–3,4 I (17%) [9]	yellow flakes, m.p. 180 to 181° (dec.), sublimes at 160°/0.1 Torr with much decomposition [9] ^{19}F NMR (THF at 35°): 69.8 (F-2), 86.8 (F-6), 136.0 (F-5), J(F-2,6) = 11.5, J(F-2,5) = 7.3, J(F-5,6) = 28.05 [9, 26] IR (CHCl$_3$): 2005, 2048 (both CO), 2235 (CN), other bands from 636 to 1580 [9]
56	FpC$_6$F$_3$(COOC$_2$H$_5$)$_2$–3,4 I (32%) [9]	yellow needles, m.p. 115 to 116° (from light petroleum), stable in air, decomposes partially on sublimation [9] ^{19}F NMR (THF at 35°): 83.3 (F-2), 97.3 (F-6), 144.3 (F-5), J(F-2,6) = 14.5, J(F-2,5) = 3.15, J(F-5,6) = 27.65 [9, 26] IR (CHCl$_3$): 1998, 2045 (both CO), other bands from 843 to 1726 [9]
*57	FpC$_6$F$_3$(Cl-3)COOH-5	yellowish brown crystals, m.p. 181° (dec.) [65] ^{19}F NMR (THF, referred to CF$_3$COOH): −4.2 (F-2), −1.8 (F-6), +40.8 (F-4) [65]
*58	FpC$_6$F$_3$(Cl-3)Li-5	not isolated
*59	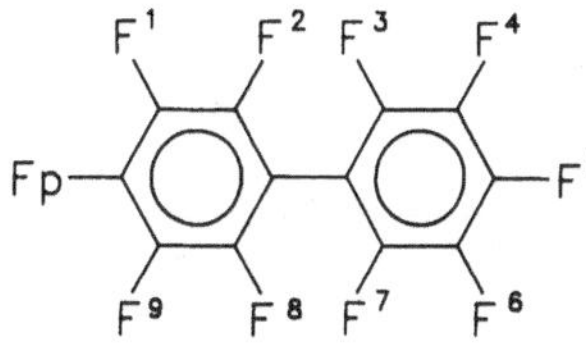	^{19}F NMR (THF): 106.4 (F-1,9), 139.2 (F-2,3,7,8), 153.9 (F-5), 162.9 (F-4,6), relative intensities 1:2:1(?):1 [40]
*60	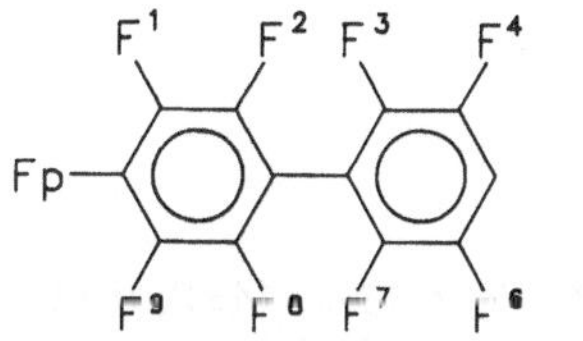	^{19}F NMR (THF): 106.6 (F-1,9), 139.4 (F-2,3,7,8), 142.1 (F-4,6), relative intensities 4:8:4 [40]
*61	FpC$_{12}$H$_2$F$_7$ C$_{12}$H$_{10}$ = biphenyl	—
*62	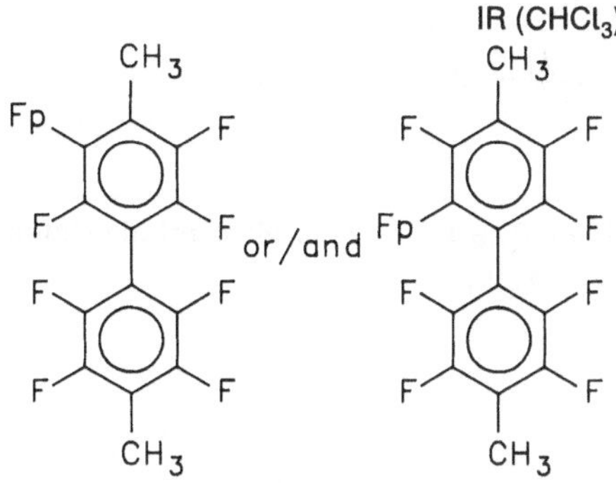	IR (CHCl$_3$): 1979, 2028 (both CO) [40]

References on pp. 189/91

Table 7 (continued)

No.	compound method of preparation (yield)	properties and remarks
*63	$FpC_{12}H(CH_3)_2F_6$ (cf. No. 62, one F substituted by H)	—
*64	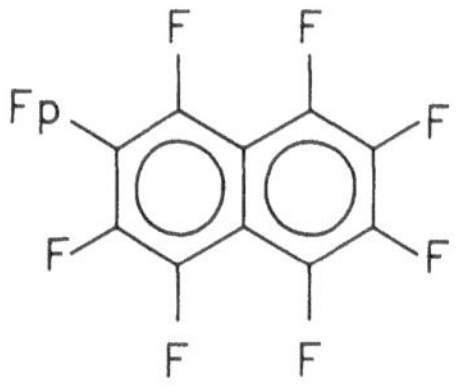	^{19}F NMR (THF): 83.4, 99.4, 146.5, 149.4, 152.8, 160.9, relative intensities 2:2:2:2:2:3.5 [40]
*65	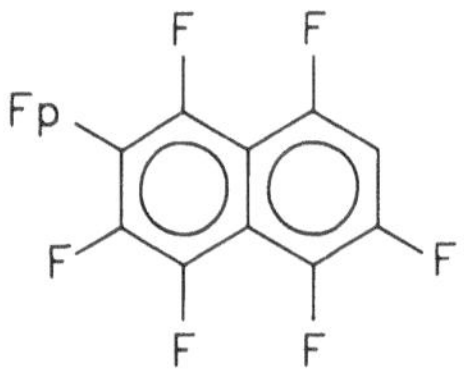 or/and 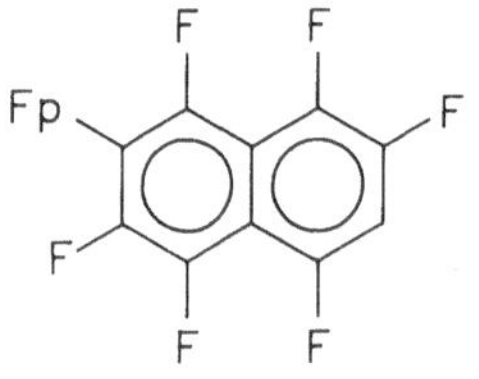	^{19}F NMR (THF): 83.2, 101.6, 119.7, 123.5, 139.8, 153.0, relative intensities 1:1:1:1:1:1 [40]
*66	$FpC_{10}H_2F_5$ ($C_{10}H_7$ = naphth-2-yl)	—
*67	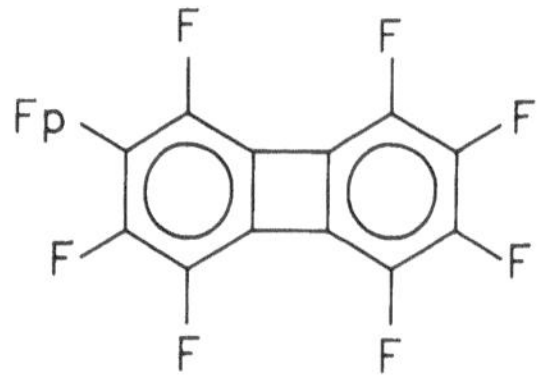	—
*68	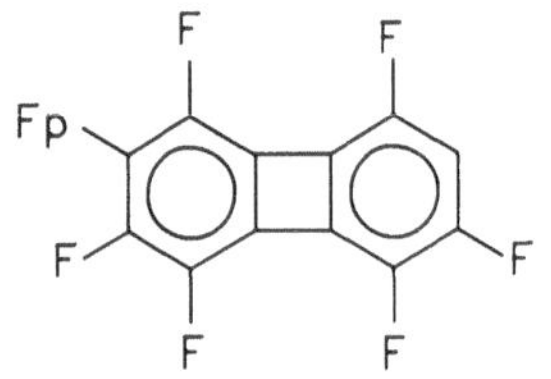 or/and 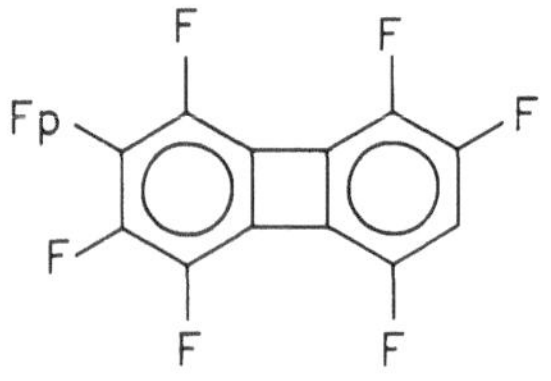	^{19}F NMR (THF): 90.1, 92.9, 123.7, 129.2, 148.2, 154.9, relative intensities 3:3:3:3:4:4 [40]
*69	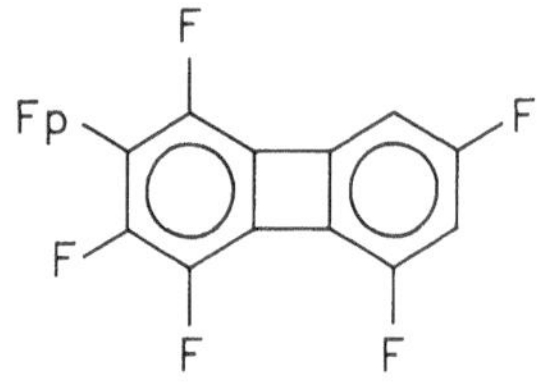	^{19}F NMR (THF): 88.9, 94.3, 124.0, 145.4, relative intensities 3:2.5:3:5 [40]

References on pp. 189/91

Table 7 (continued)

No.	compound method of preparation (yield)	properties and remarks

*70

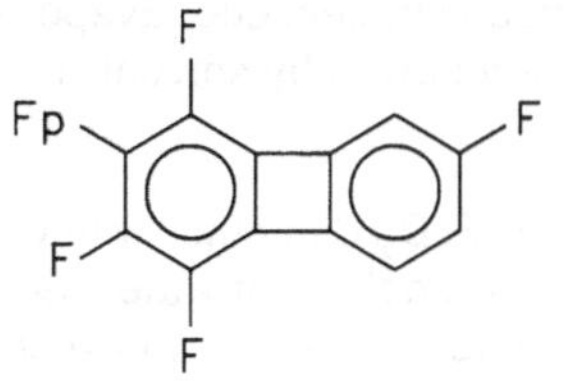

^{19}F NMR (THF): 88.8, 93.1, 95.6, 144.2, relative intensities 1:1:1:1 [40]

supplement

*71 FpC$_6$H$_3$(CH$_3$)$_2$-3,6

dark yellow oil [90]
^{1}H NMR (CDCl$_3$): 2.20 (s, 3H), 2.40 (s, 3H), 4.84 (s, 5H), 6.70 (d, 1H, J=9), 6.95 (d, 1H), 7.45 (s, 1H) [90]
IR (CH$_2$Cl$_2$): 1960, 2010 (both CO), other bands at 1040, 2940, 2970 [90]

*72

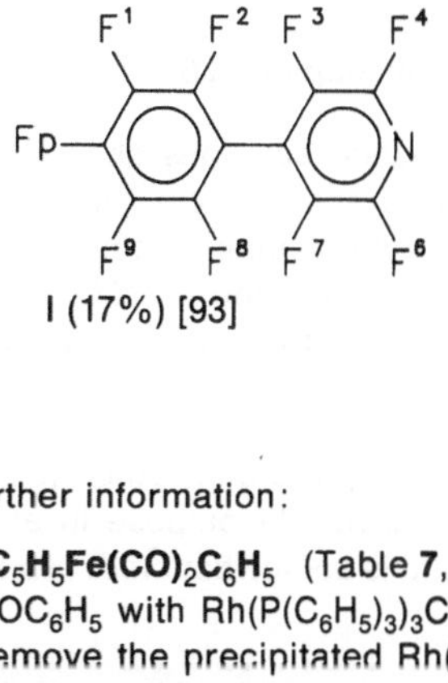

I (17%) [93]

yellow crystals, m.p. 163 to 164°, air stable [93]
^{19}F NMR (CHCl$_3$): 90.6 (F-4,6), 106.3 (F-1,9), 139.6 (F-3,7), 141.0 (F-2,8), J(6,7)=J(3,4)=21.1, J(4,6)= 13.2, J(3,6)=J(4,7)=29.5, J(3,7)=2.2, J(8,9)= J(1,2)=28.8, J(2,8)=2.6, J(1,8)=12.6, J(1,9)=6.4 [93]
IR: 1996, 2046 (both CO) in C$_6$H$_{12}$, other bands at 846, 942, 978, 1019, 1439, 1466, 1645 in CHCl$_3$ [93]

*Further information:

C$_5$H$_5$Fe(CO)$_2$C$_6$H$_5$ (Table 7, No. 1) has also been prepared by decarbonylation of FpCOC$_6$H$_5$ with Rh(P(C$_6$H$_5$)$_3$)$_3$Cl (1:1 mole ratio) in CH$_2$Cl$_2$ during 3 h, followed by filtration to remove the precipitated Rh(P(C$_6$H$_5$)$_3$)$_2$(CO)Cl and chromatography on Al$_2$O$_3$ with hexane as eluent, 40% yield [81], see also [34, 35, 60]. The use of (Rh(P(C$_6$H$_5$)$_3$)$_2$Cl)$_2$ in CH$_3$CN or of Ir(P(C$_6$H$_5$)$_3$)$_3$Cl in benzene gives only 32 and 18% yields, respectively. Only traces result from the corresponding reaction with Ru(P(C$_6$H$_5$)$_3$)$_3$Cl$_2$ in refluxing benzene, while both Pt(P(C$_6$H$_5$)$_3$)$_4$ and Pt(P(C$_6$H$_5$)$_3$)$_3$ in benzene and THF failed to decarbonylate FpCOC$_6$H$_5$ even at reflux [81]. Due to photochemical decarbonylation, FpC$_6$H$_5$ forms in 22% yield, in addition to 2,3-diphenylindenone, when a solution of FpCOC$_6$H$_5$ is irradiated in the presence of C$_6$H$_5$C≡CC$_6$H$_5$ (1:1 mole ratio) in benzene for 4 h. A possible reaction mechanism for the formation of the two products is discussed [82].

FpC$_6$H$_5$ has been obtained in 52% yield by UV irradiation of FpI and Hg(C$_6$H$_5$)$_2$ in benzene for 23 h, solvent removal and chromatography on Al$_2$O$_3$ [17]. The compound forms in 5% yield by treatment of FpI with C$_6$H$_5$MgBr (1:1 mole ratio) in ether at 25 °C for 2 to 3 h, followed by chromatography on Al$_2$O$_3$ (benzene/petroleum ether as eluent) and sublimation at 50 °C/10^{-3} Torr [1], see also [2]. Addition of a 1.5-fold excess of C$_6$H$_5$MgBr in THF to [Fp$_2$I]BF$_4$ in the same solvent affords a 25% yield after stirring at room temperature for 2 to 4 h and workup by chromatography on Al$_2$O$_3$ [67].

References on pp. 189/91

FpC_6H_5 forms along with CO and H_2 on irradiation of Fp_2 and C_6H_5CHO (1:50 mole ratio) in benzene until the red color is converted to a pale yellow–brown (after 4 h). Evaporation of the solvent from the filtered solution and recrystallization of the residue from pentane give the compound in >70% yield. Possible reaction mechanisms are discussed [77].

FpC_6H_5 also forms when $[Fp_2SC_2H_5][B(C_6H_5)_4]$ is either irradiated in THF for 2 h or refluxed in CH_2Cl_2 for 1 h. After solvent removal, the residue is extracted with benzene. Evaporation of the solvent and reextraction of the residue with cyclohexane followed by solvent removal afford the product in 20 and 10% yield, respectively [68].

FpC_6H_5 has also been obtained by treatment of $FpCH_2COC_5H_4FeC_5H_5$ in ether with C_6H_5Li in the same solvent at $-5\,°C$ for 15 min, cooling to $-30\,°C$, and addition of water. Solvent removal from the ether layer and chromatography of the residue on Al_2O_3 with ether as eluent give the compound in an 85% yield [66].

FpC_6H_5 has been isolated in only 2% yield from the reaction of $Na[Fp]$ with C_6H_5I (cf. Method I) at 25 °C after 4 d and workup by chromatography on Al_2O_3 and sublimation [1].

The assignment of the chemical shifts of C-1 and C-4 of the phenyl group is based on the relative peak intensities, the C-1 resonance giving the weakest signal. Distinction between the C-2,6 and C-3,5 resonances has been possible by interpretation of the proton-coupled ^{13}C NMR spectrum. Thus, each member of the C-2,6 doublet shows a triplet fine structure arising from two $^3J(C,H)$ values of about 7 Hz, while each member of the C-3,5 doublet displays a doublet fine structure due to only one $^3J(C,H)$ coupling [75].

The IR spectrum of solid FpC_6H_5 and of concentrated benzene solutions shows three intense bands in the $\nu(CO)$ region. On dilution, the bands at 1943 and 1973 cm^{-1} disappear and the one at 2020 cm^{-1} shows a shoulder. The splitting (no more than 10 cm^{-1}) of the latter band is attributed to symmetrical and asymmetrical vibrations of the two CO groups. The appearance of supplementary bands in the spectrum of the solid is supposed to result from strong intermolecular interactions of the CO ligands [12].

FpC_6H_5 has a camphoraceous odor. It is readily soluble in petroleum ether (b.p. 30 to 60 °C) and in all common organic solvents. The solutions slowly decompose in air but can be preserved in vacuum or in an inert atmosphere. The solid is somewhat more resistant to air than $FpCH_3$, which decomposes within a few hours [1].

The mass spectrum shows the molecular ion, which undergoes stepwise loss of the CO groups giving finally $[C_5H_5FeC_6H_5]^+$. The subsequent dehydrogenation affords $[C_{11}H_8Fe]^+$. However, the fragment $[C_5H_5FeC_6H_5]^+$ may also expel C_2H_2 to form $[C_9H_8Fe]^+$ or the iron atom to give $[C_5H_5C_6H_5]^+$ [132]. The formation of the latter fragment by migration of the C_6H_5 ligand to the C_5H_5 ring is seen as a characteristic rearrangement process observed in the mass spectrum of FpC_6H_5 [92]. Due to pyrolysis, ions of ferrocene, phenylferrocene and diphenylferrocene have also been observed. The fragment $[C_{12}H_{10}]^+$ is assumed to arise from biphenyl, another likely pyrolysis product of FpC_6H_5. The ions $[C_5H_n]^+$ ($n=$ 3, 5, 6) and $[C_3H_3]^+$ are suggested to result from the C_5H_5 ring, and $[C_6H_n]^+$ ($n=5,6$) and $[C_4H_n]^+$ ($n=2,3,4$) from the C_6H_5 ring. All fragments and their relative intensities are given [32].

Photolysis of FpC_6H_5 in pentane or benzene at 15 °C for 8 h yields Fp_2, benzene and diphenyl (ca. 1% in pentane, 26% in benzene). Irradiation in C_6D_6 gives a mixture of $C_{12}D_{10}$, $C_{12}D_5H_5$, and $C_{12}H_{10}$ (4:1:7). Thus, it is concluded that the phenyl group abstracts hydrogen from the solvent to give benzene [80, 86].

References on pp. 189/91

The reduction of FpC_6H_5 with sodium amalgam in dioxane at 70 °C for 3 h followed by addition of glacial CH_3COOH after cooling, affords benzene and Fp_2, which results from the thermal decomposition of FpH at room temperature [15].

Subjection of FpC_6H_5 in ether to 210 atm (3000 lb/in²) of CO pressure at 125 °C for 2 h leads to formation of $FpCOC_6H_5$ [3].

FpC_6H_5 is unaffected by water, but is attacked by acids and bases, yielding the cation $[Fp]^+$ with acids [1].

The compound reacts with I_2 in $CHCl_3$ at 20 °C to give FpI and C_6H_5I [1, 8, 19]. UV irradiation of FpC_6H_5 and $n-C_3F_7I$ (3:10 mole ratio) in benzene at 40 °C for 12 h in a sealed ampule affords FpI. The absence of C_6H_5I as a reaction product is attributed to the supposition that cleavage of the Fe–C σ-bond is accomplished not by the I_2 molecule, but by the radical formed from $n-C_3F_7I$ under irradiation [12].

H/D exchange occurs under basic conditions (C_2H_5OD/C_2H_5ONa in C_6H_6 at 100 °C) on the C_5H_5 ring, rate constant $k = 1.4 \times 10^{-5}$ s^{-1} [53] or 1.05×10^{-5} s^{-1} [89]. Deuteration of the C_5H_5 ligand to give $C_5H_4DFe(CO)_2C_6H_5$ also takes place when FpC_6H_5 in THF is treated with $n-C_4H_9Li$ in hexane at -78 °C for 10 to 15 min, followed by addition of a 10- to 50-fold excess of D_2O [88].

Heating FpC_6H_5 and $HgCl_2$ (1:1 mole ratio) in benzene at 40 to 50 °C for 1.5 h leads to formation of FpCl and C_6H_5HgCl [8, 19], see also p. 166.

Treatment of FpC_6H_5 in THF with $n-C_4H_9Li$ in hexane at -78 °C for 10 to 15 min followed by addition of a 10- to 50-fold excess of $(CH_3)_3SiCl$, $(CH_3)_3SnCl$, CH_3I, or C_2H_5Br leads to substitution on the C_5H_5 ring to give $R^1C_5H_4Fe(CO)_2C_6H_5$ ($R^1 = (CH_3)_3Si$, $(CH_3)_3Sn$, CH_3, C_2H_5) [88]. Similarly $HOCOC_5H_4Fe(CO)_2C_6H_5$ is formed when the reaction mixture resulting from FpC_6H_5 and $n-C_4H_9Li$ is poured onto solid CO_2 [89].

Heating FpC_6H_5 and $Cr(CO)_6$ in diglyme/heptane (3:1) at 138 to 140 °C for 13 h affords $FpC_6H_5Cr(CO)_3$, where the C_6H_5 ligand is σ-bonded to Fe and π-bonded to Cr [74]. Similar products are obtained by the reactions with $L_3M(CO)_3$ (L = CO, C_5H_5N, CH_3CN; M = Cr, Mo, W) as is briefly mentioned in [78].

Quantitative conversion of FpC_6H_5 into $FpC_6H_4COCH_3$-4 is effected during 1 h at 0 °C using the $CH_3COCl/AlCl_3$ complex. Substitution under Vilsmeier conditions ($POCl_3/C_6H_5N(CH_3)CHO$) produces the formyl derivative FpC_6H_4CHO-4, also obtained using $Cl_2CHOC_2H_5$ and $AlCl_3$. In these reactions, no substitution could be detected on the C_5H_5 ring or at the other positions of the phenyl group. A reaction mechanism involving a structure like V ($R^1 = CHO$, CH_3CO) is thought to be consistent with these observations [37].

V

References on pp. 189/91

FpC$_6$H$_5$ reacts with the amphoteric ligands (C$_6$H$_5$)$_2$PN(C$_4$H$_9$-t)AlR$_2^1$ (1:1 mole ratio) in C$_6$D$_6$ during 2 h (R^1=CH$_3$) or 15 min (R^1=C$_2$H$_5$) to give cyclic complexes of the composition C$_5$H$_5$(CO)FeC(C$_6$H$_5$)OAlR$_2^1$N(C$_4$H$_9$-t)P(C$_6$H$_5$)$_2$ (VI). The reactions have been monitored by ^{1}H NMR spectroscopy. VI (R^1=C$_2$H$_5$) showed no tendency to react any further, either in the presence of excess ligand or on standing for 4 d, or heating at 50 °C for several hours. Possible reaction mechanisms are proposed [87].

VI

In contrast to the FpR alkyl compounds (see "Organoiron Compounds" B 12, 1984, p. 204), FpC$_6$H$_5$ does not yield any detectable 1:1 adducts with (NC)$_2$C=C(CN)$_2$ in CH$_2$Cl$_2$ or THF at 25 °C [39, 72].

FpC$_6$H$_5$ can be employed as an additive to improve the burning characteristics of solid propellant compositions [14].

C$_5$H$_4$DFe(CO)$_2$C$_6$H$_5$ (Table **7**, No. **2**) has been prepared by treatment of FpC$_6$H$_5$ in THF with n-C$_4$H$_9$Li in hexane at −78 °C for 10 to 15 min, followed by addition of a 10- to 50-fold excess of D$_2$O. After solvent removal under reduced pressure at room temperature, the residue is chromatographed on Al$_2$O$_3$ with hexane or hexane/benzene (1:1) as eluent, 70% yield [88]. H/D exchange in the C$_5$H$_5$ ligand of FpC$_6$H$_5$ also occurs under basic conditions (C$_2$H$_5$ONa:C$_2$H$_5$OD=1:9 in benzene) at 100 °C in a sealed tube. The reaction is stopped by pouring the mixture into water. For purification, the product is chromatographed on Al$_2$O$_3$, <30% yield [53, 89].

The mass spectrum includes the ions [C$_5$H$_4$DFe]$^+$ and [C$_6$H$_5$]$^+$, suggesting that deuterium is present in the cyclopentadienyl ligand only [88].

C$_5$H$_5$Fe(CO)$_2$C$_6$H$_4$CH$_3$-4 (Table **7**, No. **6**) has been obtained as an oil in <0.5% yield by refluxing Na[Fp] and CH$_3$C$_6$H$_4$I-4 in THF for 12 h and workup by chromatography on Al$_2$O$_3$ with pentane as eluent. The solid, formed by sublimation of the oil at 50 °C/0.1 Torr onto a probe cooled with dry ice and isopropanol, reverts back to an oil on rubbing with a spatula or glass rod. The low yield prevented elemental analysis. The compound has been characterized by conversion to its sulfinate derivative [10].

The pure oil is stable in air for only a few hours. Pentane and benzene dissolve the compound quite readily, however, these solutions also decompose in a few hours when exposed to air [10].

With P(C$_2$H$_5$)$_3$ at 100 °C, the compound yields C$_5$H$_5$Fe(CO)(P(C$_2$H$_5$)$_3$)COC$_6$H$_4$CH$_3$-4 [94].

C$_5$H$_5$Fe(CO)$_2$C$_6$H$_4$Cl-4 (Table **7**, No. **11**) has been obtained in 17% yield by UV irradiation of FpI and Hg(C$_6$H$_4$Cl-4)$_2$ in THF for 40 h. After filtration and solvent removal, the residue is extracted with petroleum ether and the extracts are chromatographed on Al$_2$O$_3$, eluting with petroleum ether and benzene/petroleum ether. The product is recrystallized from hexane [52].

References on pp. 189/91

C₅H₅Fe(CO)₂C₆H₄OCH₃-4 $C_5H_5Fe(CO)_2C_6H_4OCH_3$-4 (Table 7, No. **15**) has also been prepared by decarbonylation of FpCOC₆H₄OCH₃-4 with (Rh(P(C₆H₅)₃)₂Cl)₂ (2:1 mole ratio) in CH₃CN during 20 h. After solvent removal, the residue is chromatographed on Al_2O_3 with hexane/ether as eluent, 49% yield. The use of Rh(P(C₆H₅)₃)₃Cl in CH_2Cl_2 gives only a trace amount of FpC₆H₄OCH₃-4, while Ir(P(C₆H₅)₃)₃Cl or Ir(P(C₆H₅)₃)₂(N₂)Cl in benzene failed to decarbonylate FpCO-C₆H₄OCH₃-4 [81].

C₅H₅Fe(CO)₂C₆H₄CHO-4 (Table 7, No. **16**) has been obtained from FpC₆H₅ under Vilsmeier conditions (POCl₃/C₆H₅N(CH₃)CHO) or by using Cl₂CHOC₂H₅ and AlCl₃ [37].

C₅H₅Fe(CO)₂C₆H₄COOC₂H₅-4 (Table 7, No. **19**) is assumed to form by UV irradiation of FpI and Hg(C₆H₄COOC₂H₅-4)₂ in benzene at room temperature for 44 h. The product, isolated after chromatography on Al_2O_3, was used without further purification for the photolysis reaction with P(C₆H₅)₃ in benzene, which led to the formation of C₅H₅Fe(CO)(P(C₆H₅)₃)C₆H₄-COOC₂H₅-4 after 18 h [17].

C₅H₅Fe(CO)₂C₆F₅ (Table 7, No. **20**) is isolated from the reaction mixture (Method I) by solvent removal, extraction of the residue with CH_2Cl_2, crystallization from pentane and vacuum sublimation [5]. The compound is obtained as a mixture with FpC₆F₄H-4 after the reaction of Na[Fp] with C₆F₅X (X=halogen) in THF (Method I) at room temperature overnight followed by chromatography on Florisil with benzene/light petroleum (b.p. 40 to 60 °C) = 1:9 as eluent and recrystallization from light petroleum. The two products have been identified by their ¹⁹F NMR data [9] and by the mass spectrum of the product from the reaction with C₆F₅Br, which showed the molecular ion and the fragments [M−CO]⁺ and [M−2CO]⁺ of each compound [26]. The total yield ranges from 37% (X=I) to 27% (X=F, Br). The proportion of the two compounds has been determined by ¹⁹F NMR spectroscopy, showing that FpC₆F₅ is the main product in each case (52%, when X=I, 83%, when X=F). In addition to these complexes, the reaction with C₆F₅I gives a third product in a relative yield of ~30%, identified in [27] as Hg(C₆F₅)₂ (resulting from the preparation of Na[Fp] from Fp₂ and sodium amalgam). A possible reaction mechanism is discussed [9].

FpC₆F₅ forms in 4% yield when Na[Fp] is allowed to react with hexafluorobicyclo-[2.2.0]hexa-2,5-diene ("Dewar hexafluorobenzene") in THF at −78 °C. After warming to room temperature it is separated from the main product VII (62%; see Section 1.5.2.3.16.3, Table 8, No. 148) by chromatography on Florisil with light petroleum (b.p. 40 to 60 °C) as eluent [30].

$$C_5H_5(CO)_2Fe$$

VII

FpC₆F₅ is obtained in 3% yield along with Fp₂ and other unidentified products by addition of solid [Fp(C₅H₅N)]PF₆ to a solution of C₆F₅Li in THF at −78 °C and workup at room temperature by chromatography on Al_2O_3 with benzene as eluent. A similar procedure using [FpCH₃CN]PF₆ and C₆F₅Li as reactants gives FpC₆F₅ in 9% yield [21]. The reaction of [FpCO]PF₆ in THF with C₆F₅Li in ether at −78 °C/1 h, then at 0 °C/2 h and subsequent chromatography on Al_2O_3 eluting with 10% benzene in hexane leads to formation of FpC₆F₅ (13% yield) in addition to FpCOC₆F₅ (18%) and traces of a substance assumed to be exo-1-C₆F₅C₅H₅Fe(CO)₃ [13, 21].

FpC₆F₅ is formed in 7% yield by the reaction of FpI with C₆F₅MgBr in ether and workup by chromatography [7].

References on pp. 189/91

The visible spectrum exhibits no maxima; the absorption decreases gradually from 400 to ~580 nm [5].

In an attempt to carbonylate FpC_6F_5 in THF under 1 atm CO at 25 °C for 2.5 h, only the starting material was recovered [21].

$C_5H_5Fe(CO)_2C_6F_4H$-2, $C_5H_5Fe(CO)_2C_6F_4H$-3, and $C_5H_5Fe(CO)_2C_6F_4H$-4 (Table 7, Nos. 21, 23, and 24) have been obtained from the corresponding FpC_6F_4Li derivatives (Nos. 42 to 44) in ether by treatment with water at −78 °C. After warming to room temperature, the organic layer is separated, dried over $MgSO_4$, and filtered. The products remaining after solvent removal are obtained in 93, 87 and 89% yields, respectively [48, 49, 59].

FpC_6F_4H-2 has also been obtained in addition to $FpC_6F_3H_2$-3,4 and FpC_6F_4I-2 from Na[Fp] and $C_6F_4I_2$-1,2 [9, 26], see also No. 34.

The reaction of Li[Fp] with C_6F_5H (Method I) to give FpC_6F_4H-4 in a 63% yield, in contrast to only 39% yield when Na[Fp] is used, is briefly mentioned in [59].

FpC_6F_4H-4 is formed in 20% yield by treatment of Na[Fp] with $C_6F_5NHNH_2$ in THF. After workup by chromatography on Florisil with benzene/light petroleum (b.p. 40 to 60 °C) = 1:4 as eluent, the product is further purified by sublimation at 80 °C/0.1 Torr and crystallization from light petroleum [9].

From the reaction of Na[Fp] with C_6F_5COCl, a mixture of FpC_6F_4H-4 and FpC_6F_5 has been isolated in addition to a third product, FpC_6F_4COFp, (chromatographic separation on Florisil), which, on irradiation in ether/light petroleum, b.p. 40 to 60 °C (1:1) for 17 h also gives small amounts of FpC_6F_4H-4 [47].

A 3:2 mixture of FpC_6F_4H-4 and FpC_6F_4Br-4 (5.5% total yield) has been obtained from the reaction of Na[Fp] with $C_6F_4Br_2$-1,4 in refluxing THF overnight, chromatography on Florisil with benzene/light petroleum as eluent and recrystallization from light petroleum. A comparison of the observed ^{19}F NMR chemical shifts of the mixture with those of the known complex FpC_6F_4H-4 and the values calculated for FpC_6F_4Br-4 (using substituent shielding parameters) suggested the presence of these two compounds [36].

FpC_6F_4H-4 forms in 3% yield from Na[Fp] and C_6F_5CHO [47]. For its formation as mixture with FpC_6F_5 see No. 20 [9].

All three isomers react with n-C_4H_9Li in ether at −78 °C to give the corresponding Li derivatives (Nos. 42 to 44) [48, 59].

$C_5H_5Fe(CO)_2C_6F_4CH_2CH{=}CH_2$-4 and (E)-$C_5H_5Fe(CO)_2C_6F_4CH{=}CHCH_3$-4 (Table 7, Nos. 28 and 29) have been obtained as a 7:1 mixture (24% total yield) from the reaction of Na[Fp] with $C_6F_5CH_2CH{=}CH_2$ in THF and workup by chromatography on Florisil, eluting with 10% benzene/light petroleum (b.p. 40 to 60 °C) and recrystallization from light petroleum. Further crystallization gave pure $FpC_6F_4CH_2CH{=}CH_2$, but a pure sample of (E)-$FpC_6F_4CH{=}CHCH_3$-4 could not be obtained from the mother liquors. A possible mode of formation is discussed [36].

The high-field signal in the ^{1}H NMR spectrum of (E)-$FpC_6F_4CH{=}CHCH_3$-4, assigned to the methyl group is not sharp, indicating an (E)-stereochemistry [36].

$C_5H_5Fe(CO)_2C_6F_4Cl$-3 (Table 7, No. 30). The reaction with n-C_4H_9Li results in incomplete Li/Cl exchange, the yield after hydrolysis of FpC_6F_4H-3 is estimated at 65% from the ^{1}H NMR spectrum of the mixture [59].

References on pp. 189/91

$C_5H_5Fe(CO)_2C_6F_4Br-2$, $C_5H_5Fe(CO)_2C_6F_4Br-3$, and **$C_5H_5Fe(CO)_2C_6F_4Br-4$** (Table 7, Nos. **31**, **32**, and **33**) are obtained by treatment of the respective FpC_6F_4H derivatives (Nos. 21, 23, and 24) with a slight excess of $n-C_4H_9Li$, followed by an excess of Br_2 [59].

$C_5H_5Fe(CO)_2C_6F_4Br-3$ is obtained in only 21% yield (cf. 68% with Method I), when the entire reaction of Li[Fp] with $C_6F_4Br_2$ is carried out at room temperature [59].

For the formation of a mixture of FpC_6F_4Br-4 and FpC_6F_4H-4 from $C_6F_4Br_2-1,4$ and Na[Fp], see under No. 24 [36].

Due to the presence of ^{79}Br and ^{81}Br the mass spectrum of FpC_6F_4Br-4 shows pairs of ions corresponding to $[M]^+$, $[M-CO]^+$, and $[M-2CO]^+$ [36].

All three isomers undergo Li/Br exchange with $n-C_4H_9Li$ to give the corresponding Li derivatives (see Nos. 42 to 44) [48, 49, 59].

$C_5H_5Fe(CO)_2C_6F_4I-2$ (Table 7, No. **34**) has been obtained as a mixture with FpC_6F_4H-2 (2.4:1) and a small amount of $FpC_6F_3H_2-3,4$ by addition of $C_6F_4I_2-1,2$ to Na[Fp] in THF (15 h or 19 h under reflux) at room temperature, chromatography on Florisil with benzene/light petroleum (b.p. 40 to 60 °C) = 1:9, evaporation of the solvent and sublimation at 90 °C/ 0.1 Torr. No separation of the mixture could be achieved by either fractional sublimation, crystallization, or chromatography. A possible reaction mechanism is discussed [9]. $FpC_6F_3H_2-3,4$ could be identified from its known ^{19}F NMR data. A comparison of the other observed chemical shifts of the mixture with the calculated values (using substituent shielding parameters) for FpC_6F_4I-2 and FpC_6F_4H-2 shows good agreement. Further evidence for the presence of all three compounds has been obtained from the mass spectrum, which exhibited the molecular ion and the fragments $[M-CO]^+$ and $[M-2CO]^+$ of each complex [26].

$C_5H_5Fe(CO)_2C_6F_4CN-4$ (Table 7, No. **35**) is formed in 1% yield by refluxing Na[Fp] and C_6F_5CN in THF for 18 h and workup by chromatography on Florisil (elution with ether), sublimation and trituration with ethanol and light petroleum (b.p. 40 to 60 °C) [9]. The ^{19}F NMR spectrum showed the presence of a trace of a second compound, which, by the fine structure of the resonance, is suggested to be $FpC_6F_3(H-2)CN-5$ [9], see also [26].

$C_5H_5Fe(CO)_2C_6F_4OCH_3-3$ and **$C_5H_5Fe(CO)_2C_6F_4OCH_3-4$** (Table 7, Nos. **36** and **37**) have been obtained as a 1:4 mixture (33% yield) by the reaction of Na[Fp] with $C_6F_5OCH_3$ in THF followed by chromatography on Florisil with benzene/light petroleum, b.p. 40 to 60 °C (1:9) as eluent and sublimation at 90 °C/0.1 Torr [29, 36]. A comparison of the chemical shifts observed in the ^{19}F NMR spectrum with the values calculated (using substituent shielding parameters) for the meta- and para-isomers confirmed the assignments [36].

$C_5H_5Fe(CO)_2C_6F_4SCH_3-4$ (Table 7, No. **38**) is isolated from the reaction mixture (Method I) by solvent removal, sublimation of the residue at 70 °C/0.01 Torr onto a probe at −78 °C and recrystallization from benzene/light petroleum (b.p. 40 to 60 °C) [24].

$C_5H_5Fe(CO)_2C_6F_4COOH-4$ (Table 7, No. **40**) has been prepared from Na[Fp] and C_6F_5COOH in $(CH_3OCH_2CH_2)O$. Extraction with $CHCl_3$ leaves a solid, which is dissolved in water and acidified with concentrated H_2SO_4. Extraction of the precipitate with $CHCl_3$, solvent removal, sublimation (to remove C_6F_5COOH) and recrystallization of the residue from ether/light petroleum, b.p. 40 to 60 °C (1:1) give the product in 11% yield. It was not formed when the reaction was carried out in ether or THF, because the precipitation of C_6F_5COONa prevented the acid from reacting with $[Fp]^-$ [41].

An attempt to prepare the anilide, by heating the compound under reflux with $SOCl_2$ for 30 min and treating the resultant mixture with aniline in benzene at 50 °C for 10 min,

References on pp. 189/91

was unsuccessful. Removal of the solvent gave only a tarry residue that contained no metal carbonyl bands in the IR spectrum [41].

$C_5H_5Fe(CO)_2C_6F_4COOC_2H_5-4$ (Table 7, No. **41**) is obtained in 20% yield from Na[Fp] and $C_6F_5COOC_2H_5$ in THF (Method I) at room temperature overnight and subsequent chromatography on Al_2O_3 (elution with ether) [41], whereas only 8% yield results from the corresponding reaction in refluxing THF for 4 h, chromatography on Florisil (elution with ether), sublimation and recrystallization from ethanol/light petroleum (b.p. 40 to 60 °C) [9].

The compound is unstable in solution, soon depositing solid [9]. Various attempts to hydrolyze the ester complex to the corresponding acid under both acidic and basic conditions have been unsuccessful and resulted in decomposition [41].

$C_5H_5Fe(CO)_2C_6F_4Li-2$, $C_5H_5Fe(CO)_2C_6F_4Li-3$, and $C_5H_5Fe(CO)_2C_6F_4Li-4$ (Table 7, Nos. **42**, **43**, and **44**) are formed by addition of n-C_4H_9Li to a solution of the corresponding FpC_6F_4Br (1:1 mole ratio) in ether at −78 °C. The compounds have not been isolated but were allowed to further react with water to give the respective FpC_6F_4H derivatives (see Nos. 21, 23, and 24). The Li intermediates can similarly be obtained by Li/H exchange from FpC_6F_4H. Subsequent treatment with excess Br_2 affords a mixture of the starting material FpC_6F_4H and the corresponding FpC_6F_4Br derivatives [48, 49, 59].

Addition of one equivalent of FpI to the 3- or 4-isomer in ether at −78 °C and stirring the mixture at room temperature for several hours affords 1,3- or 1,4-$Fp_2C_6F_4$, respectively. Attempts to prepare the corresponding 1,2-complex led only to the formation of FpC_6F_4H-2, presumably due to the unfavorable stereochemistry of the bis-iron derivative [48, 49, 59, 61].

In the presence of furan, FpC_6F_4Li-3 and -4 produce the expected Diels–Alder addition products from a "benzyne + furan type" reaction, whereas FpC_6F_4Li-2 gives only FpC_6F_4H-2 [61].

When FpC_6F_4Li-2 in ether is stirred at room temperature for 24 h, again FpC_6F_4H-2 is formed in almost quantitative yield. An analogous Li/H exchange has not been observed for the 3- and 4-derivatives [61].

$C_5H_5Fe(CO)_2C_6F_3H_2-2,5$ (Table 7, No. **49**) is relatively unstable. The crystals become brown in air and solutions decompose within minutes [9].

$C_5H_5Fe(CO)_2C_6F_3H_2-3,4$ (Table 7, No. **50**). The reaction of Na[Fp] with $C_6F_4I_2-1,2$ gives a mixture of compounds (see No. 34), in which $FpC_6F_3H_2-3,4$ could be identified due to its known ^{19}F NMR data [9, 26].

The relatively low symmetry of the substituted benzene ring has prevented interpretation of the fine structures of both the ^1H NMR and ^{19}F NMR resonances [5].

$C_5H_5Fe(CO)_2C_6F_3(H-3)Cl-5$ (Table 7, No. **51**) has been obtained from $FpC_6F_3(Cl-3)Li-5$ (No. 58) in hexane/THF by pouring the mixture onto ice, filtration of the resulting yellow precipitate (92%), and recrystallization from hexane [65].

$C_5H_5Fe(CO)_2C_6F_3(H-2)CN-5$ (Table 7, No. **52**). The reaction of Na[Fp] with C_6F_5CN gives FpC_6F_4CN-4 (No. 35) and a second product [9] of the empirical formula FpC_6F_3HCN. A comparison of the observed ^{19}F NMR chemical shifts with the calculated values (using substituent shielding parameters) for the isomers $FpC_6F_3(H-2)CN-5$ and $FpC_6F_3(H-3)CN-6$ indicates the presence of the former complex. Consistent with this assignment are also the coupling constants, which indicate two meta and one ortho H–F-couplings [26].

$C_5H_5Fe(CO)_2C_6F_3(H-2)OCH_3-5$ (Table 7, No. **53**) is formed in 0.5% yield in addition to $FpCH_3$ by the reaction of Na[Fp] with $4-CH_3OC_6F_4H$ in THF overnight and is isolated by chromatography on Florisil, elution with benzene/light petroleum, b.p. 40 to 60 °C (3:7) and sublimation at 85 °C/0.1 Torr [29, 36]. The structure of the compound has been confirmed by comparison of the observed ^{19}F NMR chemical shifts with the values calculated by means of substituent shielding parameters [36].

$C_5H_5Fe(CO)_2C_6F_3Cl_2$-3,5 (Table 7, No. 54) has been obtained in 25 (X = Br) and 12% (X = I) yields, by Method I, with $3,5-Cl_2C_6F_3X$-1 used in place of $1,3,5-C_3F_3Cl_3$. It may be recrystallized from various organic solvents including $HCON(CH_3)_2$ [65].

The compound is very stable. The crystals gradually darken and decompose above 200 °C without any definite melting or decomposition point [65].

In the mass spectrum $[M]^+$ and $[M+2]^+$ peaks appeared at a 100:65 ratio [65].

No chemical change occurred upon boiling in a methanolic solution either with NaOH or with concentrated HCl "for a long period". Concentrated H_2SO_4, however, hydrolyzed the Fe-C bond even at room temperature giving 1,3-dichloro-2,4,6-trifluorobenzene [65].

One of the Cl atoms can be lithiated with $n-C_4H_9Li$ in hexane/THF at -66 °C to give $FpC_6F_3(Cl-3)Li$-5 (No. 58) [65].

$C_5H_5Fe(CO)_2C_6F_3(Cl-3)COOH$-5 (Table 7, No. 57) has been prepared by bubbling CO_2 for 2 h through a solution of $FpC_6F_3(Cl-3)Li$-5 (see No. 58) in hexane/THF. Hydrolyzation with ice water and acidification with concentrated HCl afford a yellow precipitate, which is filtered off. Dissolution in dilute aqueous NaOH followed by addition of concentrated HCl to the filtered solution gives the product in a 84% yield [65].

$C_5H_5Fe(CO)_2C_6F_3(Cl-3)Li$-5 (Table 7, No. 58) has been obtained by addition of $n-C_4H_9Li$ in hexane to $FpC_6F_3Cl_2$-3,5 (No. 54) in THF at -66 °C. The compound has not been isolated, but was further combined with water or CO_2 to give $FpC_6F_3(H-3)Cl$-5 or $FpC_6F_3(Cl-3)COOH$-5, respectively (see Nos. 51 and 57) [65].

$C_5H_5Fe(CO)_2C_{12}F_9$, $C_5H_5Fe(CO)_2C_{12}HF_8$, and $C_5H_5Fe(CO)_2C_{12}H_2F_7$ (Table 7, Nos. 59, 60, and 61) have been obtained as an inseparable mixture by the reaction of Na[Fp] with decafluorobiphenyl in THF overnight, filtration, evaporation of the solvent, dissolution of the remaining syrup in CH_2Cl_2, and chromatography on Florisil with benzene/light petroleum (b.p. 40 to 60 °C) as eluent. After solvent removal, the residue is purified by sublimation. A similar mixture is formed by carrying out the reaction in refluxing THF. As further product $4,4'-Fp_2C_{12}F_8$ ($C_{12}H_{10}$ = biphenyl) could also be isolated [40].

The ^{19}F NMR spectrum of the mixture was complex, but examination of the spectra of different mixtures enabled the various related resonances to be assigned. The results are consistent with the presence of 4-substituted biphenyl complexes; the number and observed chemical shifts are in fair agreement with those calculated for $FpC_{12}F_9$ and $FpC_{12}HF_8$ [40].

The mass spectrum indicated that the mixture contains $FpC_{12}F_9$ and $FpC_{12}HF_8$ in a ratio of about 1:4. The intensities of the ions associated with $FpC_{12}H_2F_7$ were less than 6% of those of $FpC_{12}HF_8$ ($C_{12}H_{10}$ = biphenyl) [40].

$C_5H_5Fe(CO)_2C_{12}F_7(CH_3)_2$ and $C_5H_5Fe(CO)_2C_{12}HF_6(CH_3)_2$ (Table 7, Nos. 62 and 63). From the reaction of Na[Fp] with 4,4'-dimethyloctafluorobiphenyl in THF only milligram amounts of an orange product could be isolated. The mass spectrum shows ions consistent with the composition $FpC_{12}F_7(CH_3)_2$, as well as a small amount ($\sim$10%) of the hydrogen-substituted complex $FpC_{12}HF_6(CH_3)_2$. There was not sufficient material for the ^{19}F NMR spectrum, thus the position of substitution could not be determined. A further product of the composition $(C_5H_5)_3Fe_3(CO)_5(C_{12}F_6(CH_3)_2)_2$ is obtained as deep purple-black crystals if the reaction is

          References on pp. 189/91

prolonged. After chromatography on Florisil and recrystallization, it was separated from the orange crystals of $FpC_{12}F_7(CH_3)_2$ by hand picking (in all cases $C_{12}H_8(CH_3)_2$ is 4,4'-dimethylbiphenyl) [40].

$C_5H_5Fe(CO)_2C_{10}F_7$, $C_5H_5Fe(CO)_2C_{10}HF_6$, and $C_5H_5Fe(CO)_2C_{10}H_2F_5$ (Table 7, Nos. **64**, **65**, and **66**) have been obtained as a mixture by refluxing Na[Fp] and octafluoronaphthalene in THF overnight followed by chromatography of a CH_2Cl_2 extract on Florisil, elution with benzene/light petroleum, b.p. 40 to 60 °C (3:17), evaporation of the solvent and sublimation at 130 to 140 °C/0.1 Torr. The volatile yellow product was shown by analysis and mass spectrometry to be an approximately 2:1 mixture of $FpC_{10}F_7$ and $FpC_{10}HF_6$. About 2% $FpC_{10}H_2F_5$ was also present [40].

The ^{19}F NMR spectrum showed most of the thirteen resonances expected for a mixture of $FpC_{10}F_7$ and $FpC_{10}HF_6$. The strong deshielding caused by the transition metal group enabled the resonances of the ortho fluorines to be assigned, and the presence of two low-field signals indicated that 2-substitution had occurred; $FpC_{10}F_7$ is assigned to the structure shown in Table 7. Full analysis of the spectrum of $FpC_{10}HF_6$ was not possible, because of its complexity and low intensity. However, the observed chemical shifts compare well with those expected for 2,6- or 2,7-isomers. Because no resonances due to $FpC_{10}H_2F_5$ were observed in the ^{19}F NMR spectrum, nothing else could be concluded concerning this complex [40].

$C_5H_5Fe(CO)_2C_{12}F_7$, $C_5H_5Fe(CO)_2C_{12}HF_6$, $C_5H_5Fe(CO)_2C_{12}H_2F_5$, and $C_5H_5Fe(CO)_2C_{12}H_3F_4$ (Table 7, Nos. **67** to **70**). The reaction of Na[Fp] with octafluorobiphenylene at room temperature for 3 h and subsequent chromatography on Florisil eluting first with light petroleum (b.p. 40 to 60 °C), then with benzene and rechromatography of the combined eluates afford a yellow solid. Sublimation at 110 to 140 °C/0.1 Torr gives a mixture of $FpC_{12}F_7$, $FpC_{12}HF_6$, and $FpC_{12}H_2F_5$. When the reaction mixture was refluxed overnight, no simple substitution product, such as $FpC_{12}F_7$, was present. However, mass spectrometry indicated that another complex, i.e., $FpC_{12}H_3F_4$, had been formed [40].

The assignment of the ^{19}F NMR spectrum was made with a mixture that contained no $FpC_{12}F_7$. The observed resonances indicate the illustrated structures in Table 7 (in all cases $C_{12}H_8$ is biphenylene) [40].

$C_5H_5Fe(CO)_2C_6H_3(CH_3)_2$-**3,6** (Table 7, No. **71**) could be isolated from the reaction of $FpCH_2C \equiv CCH_3$ with the tropylium salt VIII (prepared from tropeneiron tricarbonyl and $(CH_3)_3SiOSO_2CF_3$ or $(n\text{-}C_4H_9)_2BOSO_2CF_3$) in CH_2Cl_2 at -78 °C for 2 h followed by warming to room temperature, refluxing for 2 h, and quenching the reaction with absolute ethanol and solid K_2CO_3 at room temperature for 12 h. After filtration through Celite and solvent removal, the residue is chromatographed on SiO_2 ($CH_3COOC_2H_5$/Skelly = 1:1). The formation of the p-xylene complex is thought to result from the presence of a trace of acid, which reacts with $FpCH_2C \equiv CCH_3$. The main product of this reaction is the ketohydroazulene complex IX [90].

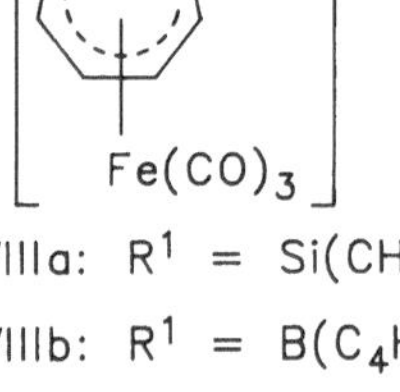

VIIIa: $R^1 = Si(CH_3)_3$

VIIIb: $R^1 = B(C_4H_9\text{-}n)_2$ IX

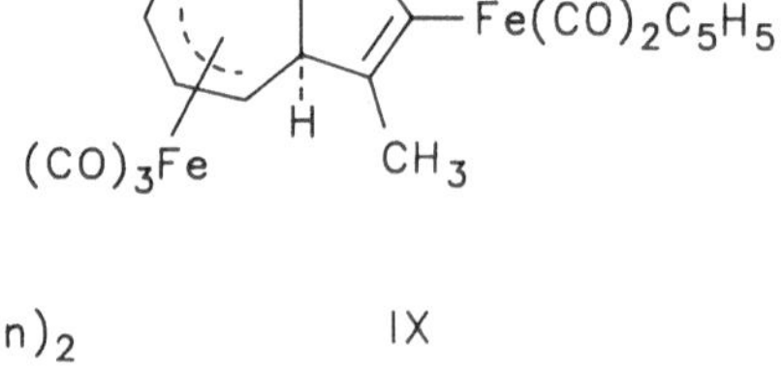

$C_5H_5Fe(CO)_2C_{11}F_8N$ (Table 7, No. 72) prepared from Na[Fp] and 4-pentafluorophenylte-trafluoropyridine (Method I) is further purified after chromatography on Florisil with light petroleum/$CHCl_3$ (3:2) by recrystallization from pentane/$CHCl_3$ [93].

References:

[1] Piper, T.S.; Wilkinson, G. (J. Inorg. Nucl. Chem. **3** [1956/57] 104/24).

[2] Piper, T.S.; Wilkinson, G. (Naturwissenschaften **43** [1956] 15/6).

[3] Closson, R.D.; Coffield, T.H.; Ethyl Corp. (U.S. 3159660 [1962/64]).

[4] Bibler, J.P.; Wojcicki, A. (J. Am. Chem. Soc. **86** [1964] 5051/3).

[5] King, R.B.; Bisnette, M.B. (J. Organometal. Chem. **2** [1964] 38/43).

[6] King, R.B.; Bisnette, M.B. (J. Organometal. Chem. **2** [1964] 15/37).

[7] Rausch, M.D. (Inorg. Chem. **3** [1964] 300/1).

[8] Nesmeyanov, A.N.; Chapovskii, Yu.A.; Makarova, L.G. (Izv. Akad. Nauk SSSR Ser. Khim. **1965** 1310/11; Bull. Acad. Sci. USSR Div. Chem. Sci. **1965** 1286).

[9] Bruce, M.I.; Stone, F.G.A. (J. Chem. Soc. A **1966** 1837/42).

[10] Bibler, J.P.; Wojcicki, A. (J. Am. Chem. Soc. **88** [1966] 4862/70).

[11] Dessy, R.E.; Stary, F.E.; King, R.B.; Waldrop, M. (J. Am. Chem. Soc. **88** [1966] 471/6).

[12] Nesmeyanov, A.N.; Chapovskii, Yu.A.; Lokshin, B.V.; Polovyanyuk, I.V.; Makarova, L.G. (Dokl. Akad. Nauk SSSR **166** [1966] 1125/8; Dokl. Chem. Proc. Acad. Sci. USSR **166/171** [1966] 213/6).

[13] Treichel, P.M.; Shubkin, R.L. (J. Organometal. Chem. **5** [1966] 488/90).

[14] Rifkin, E.B.; Closson, R.D.; Ethyl Corp. (U.S. 3336751 [1964/67]).

[15] Denisovich, L.I.; Gubin, S.P.; Chapovskii, Yu.A. (Izv. Akad. Nauk SSSR Ser. Khim. **1967** 2378/84; Bull. Acad. Sci. USSR Div. Chem. Sci. **1967** 2271/5).

[16] King, R.B. (J. Am. Chem. Soc. **89** [1967] 6368/9).

[17] Nesmeyanov, A.N.; Polovyanyuk, I.V.; Lokshin, B.V.; Chapovskii, Yu.A.; Makarova, L.G. (Zh. Obshch. Khim. **37** [1967] 2015; J. Gen. Chem. [USSR] **37** [1967] 1911/3).

[18] Nesmeyanov, A.N.; Chapovskii, Yu.A.; Ustynyuk, Yu.A. (J. Organometal. Chem. **9** [1967] 345/53).

[19] Nesmeyanov, A.N.; Chapovskii, Yu.A.; Polovyanyuk, I.V.; Makarova, L.G. (J. Organometal. Chem. **7** [1967] 329/37).

[20] Nesmeyanov, A.N.; Chapovskii, Yu.A. (Izv. Akad. Nauk SSSR Ser. Khim. **1967** 2075/7; Bull. Acad. Sci. USSR Div. Chem. Sci. **1967** 1988/90).

[21] Treichel, P.M.; Shubkin, R.L. (Inorg. Chem. **6** [1967] 1328/34).

[22] Nesmeyanov, A.N.; Chapovskii, Yu.A.; Denisovich, L.I.; Lokshin, B.V.; Polovyanyuk, I.V. (Dokl. Akad. Nauk SSSR **174** [1967] 1342/4; Dokl. Chem. Proc. Acad. Sci. USSR **172/177** [1967] 576/8).

[23] Bruce, M.I. (unpublished results from Bruce, M.I.; Stone, F.G.A.; Preparat. Inorg. React. **4** [1968] 177/235, 192).

[24] Cooke, J.; Green, M.; Stone, F.G.A. (J. Chem. Soc. A **1968** 170/3).

[25] Bruce, M.I.; Thomas, M.A. (Org. Mass Spectrom. **1** [1968] 835/46).

[26] Bruce, M.I. (J. Chem. Soc. A **1968** 1459/64).

[27] Bruce, M.I. (J. Organometal. Chem. **14** [1968] 461/4).

[28] Bruce, M.I.; Harbourne, D.A.; Waugh, F.; Stone, F.G.A. (J. Chem. Soc. A **1968** 895/8).

[29] Bruce, M.I.; Davies, C.H. (Inorg. Nucl. Chem. Letters **4** [1968] 675/7).

[30] Cook, D.J.; Green, M.; Mayne, N.; Stone, F.G.A. (J. Chem. Soc. A **1968** 1771/5).

[31] King, R.B. (Inorg. Chim. Acta **2** [1968] 454/8).

[32] King, R.B. (J. Am. Chem. Soc. **90** [1968] 1417/29).

[33] Nesmeyanov, A.N.; Chapovskii, Yu.A.; Polovyanyuk, I.V.; Makarova, L.G. (Izv. Akad. Nauk SSSR Ser. Khim. **1968** 1628/9; Bull. Acad. Sci. USSR Div. Chem. Sci. **1968** 1536/8).

[34] Alexander, J.J.; Wojcicki, A. (J. Organometal. Chem. **15** [1968] P23/P24).

[35] Wojcicki, A.; Alexander, J.J.; Graziani, M.; Thomasson, J.E.; Hartmann, F.A. (New Aspects Chem. Metal Carbonyls Deriv. 1st Intern. Symp. Proc., Venice 1968, Abstr. C6, pp. 1/6).

[36] Bruce, M.I.; Davies, C.H. (J. Chem. Soc. A **1969** 1077/81).

[37] Bolton, E.S.; Knox, G.R.; Robertson, C.G. (J. Chem. Soc. D **1969** 664).

[38] King, R.B. (Appl. Spectrosc. **23** [1969] 137/47).

[39] Wojcicki, A.; Su, S.R.; Hanna, J.A. (4th Intern. Conf. Organometal. Chem., Bristol 1969, Abstr. O5).

[40] Bruce, M.I. (J. Organometal. Chem. **21** [1970] 415/25).

[41] Booth, B.L.; Haszeldine, R.N.; Taylor, M.B. (J. Chem. Soc. A **1970** 1974/8).

[42] Nesmeyanov, A.N.; Makarova, L.G.; Polovyanyuk, I.V. (J. Organometal. Chem. **22** [1970] 707/12).

[43] King, R.B.; Epstein, L.M.; Gowling, E.W. (J. Inorg. Nucl. Chem. **32** [1970] 441/5).

[44] Nesmeyanov, A.N.; Kolobova, N.E.; Anisimov, K.N.; Denisov, F.S. (Dokl. Akad. Nauk SSSR **192** [1970] 813/6; Dokl. Chem. Proc. Acad. Sci. USSR **190/195** [1970] 395/7).

[45] Stewart, R.P.; Treichel, P.M. (J. Am. Chem. Soc. **92** [1970] 2710/8).

[46] Gubin, S.P. (Pure Appl. Chem. **23** [1970] 463/87, 481/4).

[47] Blackmore, T.; Bruce, M.I.; Davidson, P.J.; Iqbal, M.Z.; Stone, F.G.A. (J. Chem. Soc. A **1970** 3153/8).

[48] Cohen, S.C. (J. Organometal. Chem. **30** [1971] C15/C16).

[49] Cohen, S.C. (5th Intern. Conf. Organometal. Chem., Moscow 1971, Vol. 2, Abstr. 397).

[50] Jacobson, S.E.; Reich-Rohrwig, P.; Wojcicki, A. (J. Chem. Soc. D **1971** 1526/7).

[51] Makarova, L.G.; Ustunyuk, N.A.; Polovyanyuk, I.V. (5th Intern. Conf. Organometal. Chem., Moscow 1971, Vol. 1, Abstr. 229).

[52] Nesmeyanov, A.N.; Makarova, L.G.; Vinogradova, V.N. (Izv. Akad. Nauk SSSR Ser. Khim. **1971** 1984/7; Bull. Acad. Sci. USSR Div. Chem. Sci. **1971** 1869/72).

[53] Orlova, T.Yu.; Setkina, V.N.; Makarova, L.G.; Polovyanyuk, I.V.; Kursanov, D.N. (Dokl. Akad. Nauk SSSR **201** [1971] 622/3; Dokl. Chem. Proc. Acad. Sci. USSR **196/201** [1971] 966/7).

[54] Jacobson, S.E.; Wojcicki, A. (J. Am. Chem. Soc. **93** [1971] 2535/7).

[55] Gansow, O.A.; Schexnayder, D.A.; Kimura, B.Y. (J. Am. Chem. Soc. **94** [1972] 3406/8).

[56] Nesmeyanov, A.N.; Makarova, L.G.; Polovyanyuk, I.V. (Izv. Akad. Nauk SSSR Ser. Khim. **1972** 607/9; Bull. Acad. Sci. USSR Div. Chem. Sci. **1972** 567/9).

[57] Nesmeyanov, A.N.; Leshcheva, I.F.; Polovyanyuk, I.V.; Ustynyuk, Yu.A.; Makarova, L.G. (J. Organometal. Chem. **37** [1972] 159/65).

[58] Jacobson, S.E.; Reich-Rohrwig, P.; Wojcicki, A. (Inorg. Chem. **12** [1973] 717/23).

[59] Cohen, S.C. (J. Chem. Soc. Dalton Trans. **1973** 553/5).

[60] Alexander, J.J.; Wojcicki, A. (Inorg. Chem. **12** [1973] 74/6).

[61] Cohen, S.C.; Iorns, T.V.; Mosher, R.S. (J. Fluorine Chem. **3** [1973/74] 233/4).

[62] Denisovich, L.I.; Gubin, S.P. (J. Organometal. Chem. **57** [1973] 109/19).

[63] Jacobson, S.E.; Wojcicki, A. (J. Am. Chem. Soc. **95** [1973] 6962/70).

[64] Haiduc, I.; King, R.B.; Gilman, H. (Rev. Roumaine Chim. **19** [1974] 1709/15).

[65] Ishikawa, N.; Ikubi, S. (Bull. Chem. Soc. Japan **47** [1974] 2621/2).

[66] Nesmeyanov, A.N.; Perevalova, E.G.; Leont'eva, L.I.; Eremin, S.A.; Grigor'eva, O.V. (Izv. Akad. Nauk SSSR Ser. Khim. **1974** 2645/7; Bull. Acad. Sci. USSR Div. Chem. Sci. **1974** 2558/60).

[67] Nesmeyanov, A.N.; Makarova, L.G.; Polovyanyuk, I.V. (Dokl. Akad. Nauk SSSR **217** [1974] 360/1; Dokl. Chem. Proc. Acad. Sci. USSR **214/219** [1974] 509/10).

[68] English, R.B.; Haines, R.J.; Nolte, C.R. (J. Chem. Soc. Dalton Trans. **1975** 1030/3).

[69] Dizikes, L.J.; Wojcicki, A. (J. Am. Chem. Soc. **97** [1975] 2540/2).

[70] Stewart, R.P.; Benedict, J.J.; Isbrandt, L.; Ampulski, R.S. (Inorg. Chem. **14** [1975] 2933/6).

[71] Wojcicki, A.; Dizikes, L.J. (7th Intern. Conf. Organometal. Chem., Venice, 1975, Abstr. S1C).

[72] Su, S.R.; Wojcicki, A. (Inorg. Chem. **14** [1975] 89/98).

[73] Ellis, J.E. (J. Organometal. Chem. **111** [1976] 331/7).

[74] Nesmeyanov, A.N.; Polovyanyuk, I.V.; Makarova, L.G. (Dokl. Akad. Nauk SSSR **230** [1976] 1351/2; Dokl. Chem. Proc. Acad. Sci. USSR **226/231** [1976] 650/1).

[75] Stewart, R.P.; Isbrandt, L.R.; Benedict, J.J.; Palmer, J.G. (J. Am. Chem. Soc. **98** [1976] 3215/9).

[76] Dizikes, L.J.; Wojcicki, A. (J. Am. Chem. Soc. **99** [1977] 5295/303).

[77] Labinger, J.A.; Madhavan, S. (J. Organometal. Chem. **134** [1977] 381/9).

[78] Nesmeyanov, A.N.; Polovyanyuk, I.V.; Makarova, L.G. (8th Intern. Conf. Organometal. Chem., Kyoto, Japan, 1977, Abstr. 1A36).

[79] Zakharkin, L.I.; Kovredov, A.I.; Meiramov, M.G.; Kazantsev, A.V. (Izv. Akad. Nauk SSSR Ser. Khim. **1977** 1673/5; Bull. Acad. Sci. USSR Div. Chem. Sci. **1977** 1544/5).

[80] Alt, H.G.; Herberhold, M.; Rausch, M.D.; Edwards, B.H. (Z. Naturforsch. **34b** [1979] 1070/7).

[81] Kuhlmann, E.J.; Alexander, J.J. (J. Organometal. Chem. **174** [1979] 81/7).

[82] Kolobova, N.E.; Goncharenko, L.V. (Khim. Geterosikl. Soedin. **1979** No. 11, pp. 1461/5; Chem. Heterocycl. Compounds [USSR] **1979** 1173/6).

[83] de Luca, N.; Wojcicki, A. (J. Organometal. Chem. **193** [1980] 359/78).

[84] Severson, R.G.; Leung, T.W.; Wojcicki, A. (Inorg. Chem. **19** [1980] 915/23).

[85] Butler, I.R.; Lindsell, W.E.; Preston, P.N. (J. Chem. Res. S **1981** 185; J. Chem. Res. M **1981** 2573/84).

[86] Bruce, E. (Diss. Univ. Massachusetts 1982; Diss. Abstr. Intern. B **43** [1983] 2554).

[87] Labinger, J.A.; Bonfiglio, J.N.; Grimmett, D.L.; Masuo, S.T.; Shearin, E.; Miller, J.S. (Organometallics **2** [1983] 733/40).

[88] Orlova, T.Yu.; Setkina, V.N.; Sizoi, V.F.; Kursanov, D.N. (J. Organometal. Chem. **252** [1983] 201/4).

[89] Orlova, T.Yu.; Setkina, V.N.; Kursanov, D.N. (J. Organometal. Chem. **267** [1984] 309/12).

[90] Watkins, J.C.; Rosenblum, M. (Tetrahedron Letters **25** [1984] 2097/100).

[91] Butler, L.R.; Lindsell, W.E.; Thomas, M.J.K. (J. Organometal. Chem. **262** [1984] 59/68).

[92] Sizoi, V.; Semenov, V.; Vasyukova, N.; Sukharev, Yu.; Nekrasov, Yu. (12th Intern. Conf. Organometal. Chem., Wien 1985, Abstr. 449).

[93] Green, M.; Taunton-Rigby, A.; Stone, F.G.A. (J. Chem. Soc. A **1968** 2762/5).

[94] Duncan, J.D. (Diss. Balliol College, Oxford 1969, pp. 1/157).

1.5.2.3.16.3 $C_5H_5Fe(CO)_2R$ Compounds with R = Other Carbocycle

Table 8 summarizes compounds where iron is directly bonded to an carbocyclic ligand. The ligands themselves are arranged by increasing ring size and number of double bonds within the ring.

The table does not include the cyclohexanone complexes $C_5H_5Fe(CO)_2C_6H_5R^1R^2R^3R^4O$ (see Formula I), formed by heating $FpCH_2CH=CH_2$, $BrCR^1R^2C(=O)CR^3R^4Br$, and $Fe_2(CO)_9$ in CH_2Cl_2 at reflux for 12 h, because they proved impossible to purify and difficult to handle.

However, the crude products have been used to prepare the corresponding 4-methoxycar-bonylcyclohexanones by oxidation for a period of 4 h with excess $[NH_4]_2[Ce(NO_3)_6]$ in CH_3OH saturated with CO [140]. For reactions of other compounds with $[NH_4]_2[Ce(NO_3)_6]$ see p. 195.

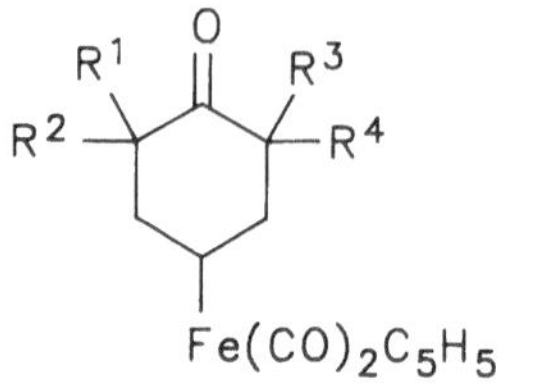

$R^1 = R^3 = CH_3$, C_2H_5, C_3H_7-i, or C_6H_5; $R^2 = R^4 = H$
$R^1 = CH_3$, $R^3 = C_2H_5$; $R^2 = R^4 = H$
$R^1 = R^2 = R^4 = CH_3$; $R^3 = H$
$R^1 = R^2 = R^3 = R^4 = CH_3$

The compounds in Table 8 are prepared by the following methods:

Method I: Reaction of Na[Fp] in THF with RX (X = F, Cl, Br [16, 28, 36, 44, 48, 57, 59, 60, 69, 71, 111, 118, 125, 128, 142], X = $OSO_2C_6H_4CH_3$-4 [98]) usually for several hours at room temperature [59, 71, 98, 118, 125, 128]; at 0 °C then warmed to room temperature [57]; at 0 °C, then at 50 °C (~1 h) [28], or at −80 °C then warmed to room temperature [16, 69, 111]. After solvent removal the residue is worked up by chromatography, generally on Al_2O_3, in some cases on Florisil (Nos. 15, 100, 147, 148, 173) or SiO_2 (Nos. 82, 83, 123 to 127, 196). Some more references containing special details on the reaction procedure and workup are given under "Further information".

Method II: Deprotonation of the corresponding $[Fp(\eta^2$-cycloalkene)]$^+$ cations by treatment with $N(C_2H_5)_3$ in CH_2Cl_2 at room temperature for ~30 min [69], see also [37, 48, 60]. A workup procedure is given only for the cyclobutenyl complex No. 13, which results from the reaction of $[Fp(\eta^2$-cyclobutene)]BF_4 with $N(C_3H_7$-i)$_2C_2H_5$ in CH_2Cl_2 for 2 h. Thus, after solvent removal the residue is extract with ether and the extracts are filtered, concentrated, and chromatographed on Al_2O_3 with ether/petroleum ether (1:5) as eluent [69]. The deprotonation reaction is highly stereospecific, proceeding by loss of an allylic proton trans to the metal olefin bond [60, 69].

Method III: Photochemical decarbonylation of FpCOR to give FpR in CD_3COCD_3 at 0 °C for 2 to 3 h (to give Nos. 3, 4, 151 [116], 142 [136]) or at 10 °C for 70 min (No. 1 [116]), in pentane at room temperature for 4 to 15 min (Nos. 152 to 155 [141], 198 [68]), or in petroleum ether for ~45 min (No. 109 [98]). Before workup by chromatography on Al_2O_3 with hexane or pentane as eluent, the reaction mixture is either concentrated [141], evaporated [98], or added to Al_2O_3 [136], followed by solvent removal in vacuum [116]. The reaction proceeds with retention of configuration [141].

Method IVa: Addition of $(NC)_2C=C(CN)_2$ to $FpCH_2CR^1=CR^2R^3$, $FpC(=CH_2)CH=CH_2$, $FpCH=CHCR^1=CH_2$, $FpC\equiv CC_6H_5$, or $FpCH_2C\equiv CCH_3$, generally in CH_2Cl_2 and/or THF, at room temperature for 5 to 30 min affords the cyclic compounds No. 60 to 77 [29, 30, 40, 52, 61, 67, 69, 78, 89, 90], 134 to 136 [138], 21 [148], and 94 [30, 52]. Compound No. 92 is obtained from $FpCH=C=CH_2$ [43], see also [54].

In most cases the products remaining after solvent removal are further purified by recrystallization from CH_2Cl_2/hexane. Compounds No. 65 and 77, which are prepared at 0 °C to room temperature, are isolated from the reaction mixture by addition of hexane, followed by chromatography on Al_2O_3 and recrystallization from CH_2Cl_2/hexane (No. 77) [78]. According to [69], any geometric isomerism associated with a substituent at C-3 in $FpCH_2CR^1=CR^2R^3$ is preserved in the product by the relationship between this substituent and the adjacent Fp group.

Method IV b: Similarly, bicyclic products are obtained from the corresponding cyclic derivatives and $(NC)_2C=C(CN)_2$: No. 149 from No. 13, No. 156 from No. 78, No. 157 from No. 79, No. 158 from No. 81, No. 168 from No. 101, No. 177 from No. 119, and No. 190 from No. 141 [48, 69], see also [60]. These cycloaddition reactions have been shown to occur stereospecifically by a suprafacial addition of the electrophile trans to the Fp group [69].

Method V a: Reaction of $FpCH_2CR^1=CHR^2$ ($R^1=R^2=H$; $R^1=OCH_3$, $R^2=H$; $R^1=H$, $R^2=OCH_3$) with $R^3R^4C=CR^5R^6$ or $R^3C≡CR^4$ in CH_2Cl_2 or DMF at room temperature for 1 to 90 h gives the cycloadducts II and III, respectively. After solvent removal under reduced pressure, the residue is taken up in benzene and chromatographed on Al_2O_3 or Florisil, generally with mixtures of ether/light petroleum as eluent [91, 114, 115, 120], a procedure by which mixtures of diastereomers, if formed, could also be separated [120]. The yields depend on the solvents used.

II   III

Method V b: Cycloadducts of type IV and V are obtained from the reactions of FpC_5H_5 with $R^1R^2C=CR^3R^4$ or $R^1C≡CR^2$, respectively, in CH_2Cl_2 at 25 °C for 0.5 to 72 h and workup by solvent removal under reduced-pressure and medium pressure chromatography on Al_2O_3 with hexane/ethyl acetate (Nos. 163, 164, 171) or benzene (Nos. 159 to 162) as eluent. However, with dimethyl maleate or methyl acrylate the reactions take place only in the presence of $Al(C_2H_5)_2Cl$ (25% solution in toluene at 0 °C). After 30 min at 25 °C the reaction is quenched with water and the aqueous layer extracted with CH_2Cl_2. The single stereoisomer (No. 163), thus formed with dimethyl maleate (73% yield), is the same as that obtained with dimethyl fumarate in the absence of $Al(C_2H_5)_2Cl$ (90% yield) [130]. Whereas only one isomer (No. 165) results from the reaction of FpC_5H_5 with fumaronitrile in CH_2Cl_2 at 0 °C to room temperature (1 h) and chromatography on Al_2O_3 with hexane/CH_2Cl_2 as eluent, (Z)-NCCH=CHCN gives two isomers (Nos. 166, 167) under similar conditions [132]. In each case two isomeric cycloadducts are produced in the reactions of FpC_5H_5 with 2-chloroacrylonitrile (No. 164), acrylonitrile (Nos. 161, 162), and methyl acrylate (Nos. 159, 160). No. 164 could not be separated by medium-pressure chromatography [130], in contrast to Nos. 159 to 162.

References on pp. 268/72

Method VIa: Addition of FpC≡CCH$_3$, FpCH$_2$CH=CH$_2$, or FpCH$_2$C≡CCH$_3$ to cyclohexenone and freshly sublimed AlBr$_3$ in CH$_2$Cl$_2$ at −78 °C, warming the mixture to 0 °C over a period of 1 to 2.5 h, then quenching it with water lead to formation of the bicyclic compounds No. 178, 187, and 191, respectively. After solvent removal from the organic layer dried with MgSO$_4$ the residue is taken up with ether or CH$_2$Cl$_2$ and chromatographed on Al$_2$O$_3$ (No. 187) or SiO$_2$ (No. 191) with ether/petroleum ether as eluent. In the case of No. 178, a portion of the residue is chromatographed by thin layer chromatography on SiO$_2$ with ether/petroleum ether (3:2). The yield of No. 178 decreases from 54 to 24%, when FpCH=C=CH$_2$ is used in place of FpC≡CCH$_3$ [121].

Method VIb: Similar cycloaddition products result from the reactions of FpCH$_2$CH=CH$_2$ (Nos. 175, 189, 197) or FpCH$_2$C(CH$_3$)=CH$_2$ (No. 176) with the corresponding 2-ethoxycarbonylcycloalkenones in refluxing CH$_2$Cl$_2$ overnight and workup of the reaction residue taken up in ether, by chromatography on Al$_2$O$_3$ eluting with ether/petroleum ether and recrystallization from the same solvent mixture [121].

Method VII: Addition of the lithium enolates LiCH(COOR^1)$_2$, LiCH(COCH$_3$)COOR^1, or LiCH(CN)COOR^1 (R^1=CH$_3$, C$_2$H$_5$) (generated by treatment of the activated methylene compounds with Li[N(Si(CH$_3$)$_3$)$_2$] in THF at −78 °C) to [Fp(η^2-C$_5$H$_7$R^1-cyclo)]$^+$ (R^1=H, CH(COOR)$_2$) or to [Fp(η^2-C$_6$H$_{10}$-cyclo)]$^+$ at −78 °C, followed by warming to room temperature over a period of 2 to 3 h, removal of solvent under reduced pressure and chromatography of the residue on Al$_2$O$_3$ with ether or CH$_2$Cl$_2$ as eluent give the products (Nos. 26 to 28, 32, 33, 115 to 117). These are assigned trans stereochemistry in conformity with the generally observed stereochemical course of the additions of nucleophiles to coordinated olefins [46, 78].

The relatively large magnetic anisotropy associated with the Fp group is reflected in the ^{1}H NMR spectra of the malonate adducts No. 26, 27, and 116, which clearly show the presence of diastereotopic ester groups through the doubling of signals for methyl and methylene ester protons. For Nos. 28 and 115 the presence of equal proportions of diastereomers is likewise evidenced by the multiplicity and relative intensity of these signals. The small coupling constant of the substituent methine proton in the cyanoacetic ester adduct No. 117 with the adjacent ring proton (J = 2 Hz) is consistent with an equatorial conformation for this group, as would be anticipated for trans addition of the nucleophile to the iron olefin bond (see Method VII) [78].

The mass spectra of the perhalogenated compounds No. 15, 100, and 137 exhibit the molecular ion and the fragments [M−nCO]$^+$ and [M−F−nCO]$^+$ (n = 0, 1, or 2). The CO-free ions resulting from the cyclobutene (No. 15) and cyclohexene complex (No. 137) may then lose FeF or FeF$_2$ with concomitant transfer of the C$_5$H$_5$ group to the fluorocarbon residue, while the cyclopentene complex (No. 100) may fragment by loss of FeF$_2$ or FeFCl. An alternative process is elimination of neutral C$_5$H$_5$FeF or C$_5$H$_5$FeCl. Ions containing no fluorine, i.e., $[C_5H_5Fe(CO)_2]^+$, $[C_5H_5Fe(CO)]^+$, $[C_5H_5Fe]^+$, and $[C_3H_3Fe]^+$, have also been found. Fragmentation schemes giving suggested routes for most of the ions observed as well as the mass spectra are represented [14].

Treatment of compounds No. 31, 36, 39 to 41, 84 [115], and 60 [90] in CO-saturated R^1OH ($R^1 = CH_3$, C_2H_5) with $[NH_4]_2[Ce(NO_3)_6]$ at room temperature overnight in an atmosphere of CO leads to substitution of the Fp moiety by a $COOR^1$ group to give the corresponding esters. On the basis of the ^1H NMR spectra, it is concluded that these products are obtained as a mixture of isomers [115]. However, according to [120] the ^1H NMR spectra of the crude products of the reactions of compounds No. 46 to 51 and 85 with $[NH_4]_2[Ce(NO_3)_6]$ under similar conditions indicate that the carboxylation proceeds with retention of configuration. Isomerization takes place only on workup by chromatography. Stereospecific replacement of the Fp moiety by a $COOCH_3$ group with retention of configuration also occurs by oxidation of Nos. 159 to 162 [130], 163, 164 [126, 130] (at 20 °C), 165 to 167 [132], and 171 [126, 130] (at 0 °C to room temperature) with $[NH_4]_2[Ce(NO_3)_6]$ in CO saturated CH_3OH for 1 to 4 h, see also [127]. Oxidative carboxylation is further observed when No. 189 (at room temperature for 0.5 h) [121] or No. 200 (at 0 °C, then at 24 °C for 15 min) [138] is allowed to react with $[NH_4]_2[Ce(NO_3)_6]$ in pure CH_3OH. In contrast, the Fp moiety is replaced by an OR^1 group in the reaction of No. 44 ($R^2 = H$, $R^3 = COOC_2H_5$) [114] or No. 62 ($R^2 = R^3 = CN$) [90] with $[NH_4]_2[Ce(NO_3)_6]$ in CO-saturated R^1OH ($R^1 = CH_3$, C_2H_5) at room temperature overnight yielding the acetal VI and a small amount of the olefin VII.

The mechanism of the reaction of compound No. 1 with $(NC)_2C=C(CN)_2$ to give the cyclopentyl complex No. 59 is assumed to involve the cleavage of the C(1)–C(2) bond to produce a dipolar intermediate, which then collapses to the product by cycloaddition [31, 48], see also [36]. For the reactions of Nos. 13, 78, 79, 81, 101, 119, and 141 with $(NC)_2C=C(CN)_2$ see Method IV b.

The reaction of Nos. 82, 83, 123, 125, and 126 with $[O(C_2H_5)_3]BF_4$ in CH_2Cl_2 at 20 °C gives the carbene salts VIII ($R^1 = H$ or CH_3) and IX ($R^2 = R^3 = H$; $R^2 = H$, $R^3 = C_6H_5$; $R^2 = R^3 = CH_3$), respectively [111].

Treatment of Nos. 31, 36, and 39 with N-bromopyridinium bromide (1:1 mole ratio) in CH_2Cl_2 at −70 °C for 1 h, then warming to −20 °C over a period of 3 h followed by addition of a further equivalent of $C_5H_5NBr_2$ and warming to room temperature afford a mixture of X and XI ($R^1 = COOCH_3$, $COOC_2H_5$, or CN; $R^2 = H$, $COOCH_3$, or $COOC_2H_5$). In contrast, 3,4,5-tribromocyclopentene was the product isolated from the corresponding reactions (20 h

                                   References on pp. 268/72

at room temperature) of compounds No. 40 and 41 with $C_5H_5NBr_2$. This unusual product is thought to result from the bromination of the C_5H_5 group of the Fp moiety [115].

$$
\begin{array}{ccc}
\text{Br}\underset{\underset{R^2}{\overset{\displaystyle COOC_2H_5}{\underset{H}{--R^1}}}}{\wedge} & \underset{\underset{R^2}{\overset{\displaystyle COOC_2H_5}{\underset{H}{--R^1}}}}{\bigcirc} & \text{or} \quad \underset{\underset{R^2}{\overset{\displaystyle COOC_2H_5}{\underset{H}{--R^1}}}}{\bigcirc} \\
X & XI &
\end{array}
$$

UV irradiation of compound No. 16 and $P(C_6H_5)_3$ in benzene [35], of No. 146 and $P(C_6H_5)_3$ or $P(C_4H_9\text{-}n)_3$ in THF [125], or of Nos. 145 or 202 and $P(C_4H_9\text{-}n)_3$ in THF [117] for several hours leads to substitution of one CO group by the ^2D ligand to give $C_5H_5Fe(CO)(^2D)R$. The same type of product is obtained, although only in 3.5% yield, by heating No. 16 with an ~80% excess of $P(C_6H_5)_3$ in boiling methylcyclohexane for 25 h [35]. However, treatment of compound No. 108 with $P(C_6H_5)_3$ in $(CH_3)_2SO$ at room temperature [50], of Nos. 198 and 199 with $P(C_6H_5)_3$ in refluxing benzene for 0.5 h, or of No. 198 with $P(CH_3)_3$ at $-196\,°C$, then at reflux for 1 h [68] causes formation of the acyl compounds $C_5H_5Fe(CO)(^2D)COR$. For the reaction of Nos. 101 and 110 with phosphines see "Further information".

Condensation of SO_2 onto compounds No. 61, 76, or 94 and maintaining the resulting solution at reflux for 12 to 16 h afford, after evaporation of the solvent, only the starting materials [52].

Table 8
$C_5H_5Fe(CO)_2R$ Compounds with R = Other Carbocycle.
Further information on compounds with numbers preceded by an asterisk is given at the end of the table. For abbreviations and dimensions, see p. X.

No. compound method of preparation (yield)	properties and remarks
*1 Fp—◁ I [31, 36] III [122], (75%) [116]	amber oil [31], orange–brown solid [116] ^1H NMR: -0.12 (m, 2H) and 0.52 (m, 3H), assignable to protons cis and trans, respectively, in relation to the Fe–C bond, in CS_2 [31]; -0.05 (m, 2H), 0.58 (m, 3H), 4.90 (s, C_5H_5) in CD_3COCD_3 [116] $^{13}C\{^1H\}$ NMR (CD_3COCD_3): -9.8 (CH), 8.6 (CH_2), 86.9 (C_5H_5), 218.1 (CO) [116] IR: 1961, 2014 (both CO) [31], the complete spectrum (neat) 465 to 3128 given [116]
*2 Fp ⋈ D	—
3 Fp ⟨2,1,3 triangle⟩ C_2H_5 / C_2H_5 III (75%) [116]	brownish oil [116] ^1H NMR: 0.2 (m, 1H), 1.10 (m, 12H), 4.85 (s, C_5H_5) [116] $^{13}C\{^1H\}$ NMR (CD_3COCD_3): 7.5 (C-1), 14.1, 15.1 (C-2,3), 27.4 (CH_3), 30.1, 30.3 (both CH_2), 86.4 (C_5H_5), 218.2, 218.5 (both CO) [116] IR (neat): complete spectrum from 560 to 2960 given [116]

References on pp. 268/72

Table 8 (continued)

No.	compound method of preparation (yield)	properties and remarks
*4	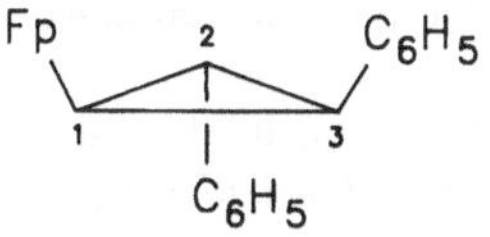 III (97%) [116]	orange solid, air-stable [116] ^{1}H NMR (CDCl$_3$): 1.5 (dd, H-1), 2.1 (dd, H-2), 2.45 (dd, H-3, ^{3}J(H-1,3) = 5.2, ^{3}J(H-2,3) = 2.3), 4.4 (s, C$_5$H$_5$), 7.12 (m, 10H, C$_6$H$_5$) [116] ^{13}C{^{1}H} NMR (CD$_3$COCD$_3$): 15.8 (C-1), 31.2, 34.9 (C-2,3), 85.1 (C$_5$H$_5$), 124.9, 125.2, 125.8, 127.7, 128.1, 129.2, 143.6, 145.8 (all C$_6$H$_5$), 215.7, 216.4 (both CO) [116] IR (CDCl$_3$): complete spectrum 530 to 3100 given [116]
*5	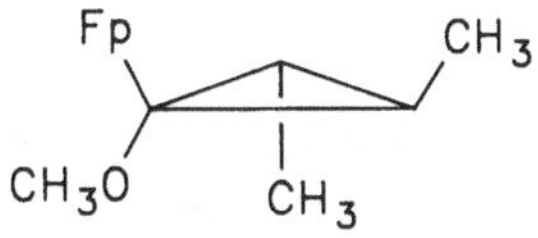	—
*6	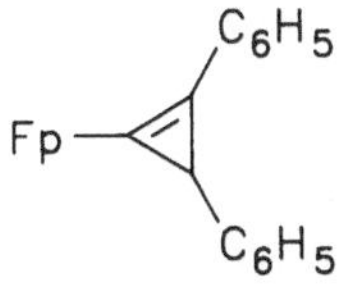	yellow crystals, dec. above 60°, stable, when excluded from light and air [143] ^{1}H NMR (CDCl$_3$): 2.47 (s, 1H), 4.86 (s, 5H), 6.87 to 7.45 (m, 10H) [143] IR (KBr): 1958, 2020 (both CO), 1753 (C=C) [143]
*7	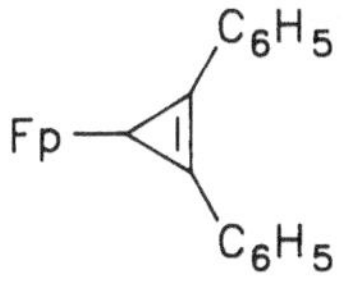	yellow crystals, m.p. 66 to 67° (dec.) [86] ^{1}H NMR (CS$_2$): 2.63 (s, 1H), 4.59 (s, 5H), 7.15 to 7.5 (m, 6H), 7.55 to 7.8 (m, 4H) [86] IR (CS$_2$): 1937, 1995 (both CO), 1782 (C=C), 2840, 3055, 3063 (all CH) [86]
*8	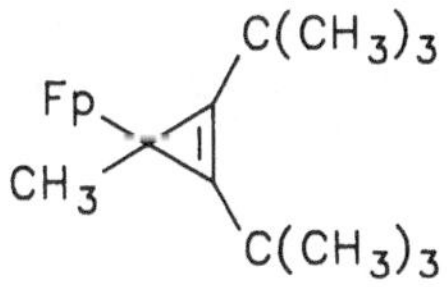	yellow crystals [86] ^{1}H NMR (CS$_2$): 1.18 (s, 18H), 1.49 (s, 3H), 4.67 (s, 5H) [86]
*9	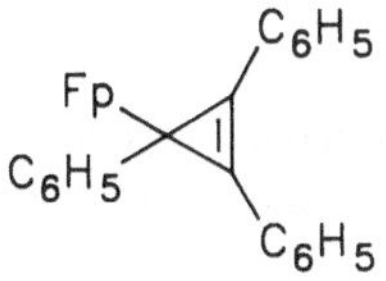	golden-yellow crystals [97] ^{1}H NMR (CDCl$_3$): 4.65 (s, 5H), 7.0 to 7.6 (m, 11H), 7.7 to 8.0 (m, 4H) [97] IR: 1922, 1985 (both CO), 830, 1412 (both C$_5$H$_5$), 3050, 3075 (both CH) in CS$_2$, 1779 (C=C) in KBr [97]
*10	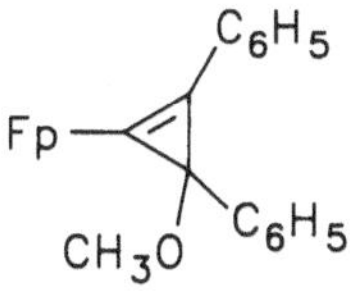	yellow crystals, m.p. 95 to 97° (dec.), stable, when excluded from light and air [143] ^{1}H NMR (CDCl$_3$): 3.29 (s, 3H), 4.98 (s, 5H), 7.25 (m, 10H) [143] IR (KBr): 1958, 2020 (both CO), 1729 (C=C) [143]

References on pp. 268/72

Table 8 (continued)

No.	compound method of preparation (yield)	properties and remarks

*11

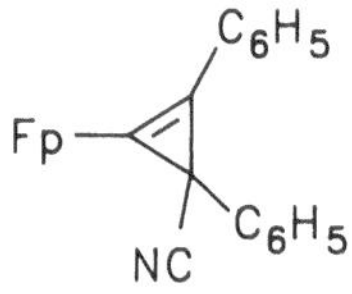

yellow crystals, m.p. 160 to 162° (dec.), stable, when excluded from light and air [143]
^{1}H NMR (CDCl$_3$): 5.09 (s, 5H), 7.0 to 7.5 (m, 10H) [143]
IR (KBr): 1984, 2030 (both CO), 1787 (C=C), 2205 (CN) [143]

*12

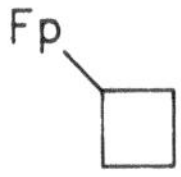

I (58%) [69]

^{1}H NMR (CS$_2$): 1.74 to 2.40 (m, 6H, CH$_2$), 3.35 (m, FeCH), 4.60 (s, C$_5$H$_5$) [69]
IR (neat): 1910, 2007 (both CO) [69]

*13

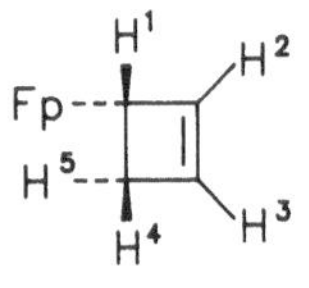

II [48, 60], (59%) [69]

^{1}H NMR: 2.34 (m, H–5), 3.04 (m, H–4, J(H–4,5) = 14.7), 3.88 (m, H–1, J(H–1,4) = 3.7), 4.75 (s, C$_5$H$_5$), 5.77 (m, H–3), 6.31 (m, H–2, J(H–2,3) = 2.3) in CS$_2$ [72], 2.34 (d, H–5, J = 14), 3.00 (dd, H–4, J = 14.0, 3.0), 3.86 (br d, H–1, J = 3.0), 4.74 (s, C$_5$H$_5$), 5.79 (m, H–3), 6.31 (d, H–2, J = 2.5) [69]
IR (CH$_2$Cl$_2$): 1947, 1999 (both CO) [69]

*14

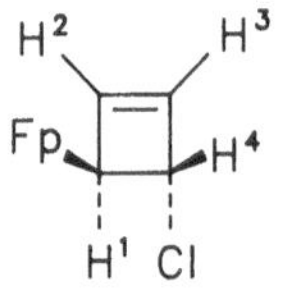

I [76], (17%) [65], (18%) [72]

dark yellow crystals, m.p. 65 to 67° [65, 72]
^{1}H NMR (CS$_2$): 3.83 (m, H–1), 4.41 (m, H–4), 4.82 (s, C$_5$H$_5$), 5.63 (m, H–3), 6.28 (m, H–2) [53, 65, 72]
IR (Nujol): 1950, 2000 (both CO) [53, 65, 72]

*15

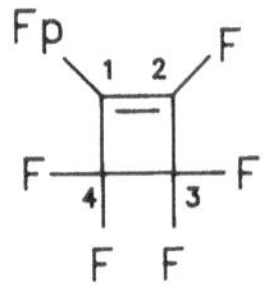

I (35%) [8]

orange crystals, m.p. 50 to 51° [8], 49 to 51° [9]
^{19}F NMR (CS$_2$): 119.1 (C^{4}F$_2$), 123.4 (C^{3}F$_2$), 124.1 (C^{2}F) [8]
IR: 2008, 2053 (both CO) in C$_6$H$_{12}$, bands from 634 to 1632 in CS$_2$ [8]

*16

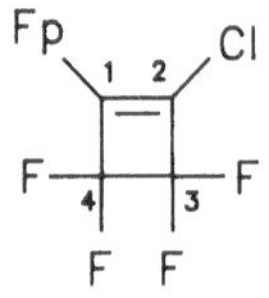

I (60%) [35]

yellow-orange, m.p. 73 to 74° [35]
^{1}H NMR (CDCl$_3$): 5.09 (s, C$_5$H$_5$) [35]
^{19}F NMR (CH$_2$Cl$_2$): 110.9 (s, 2F, C^{3}F$_2$), 115.0 (s, 2F, C^{4}F$_2$) [35]
IR: 1997, 2007, 2056 in C$_6$H$_{12}$, 1982, 1996, 2037 (all CO), 1078, 1106, 1131, 1234, 1237, 1312 (all CF), 3128 (CH) in KBr, bands from 587 to 1435 in KBr [35]

*17

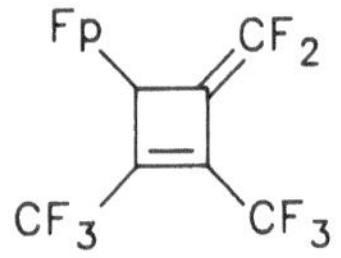

—

References on pp. 268/72

Table 8 (continued)

No.	compound method of preparation (yield)	properties and remarks
*18	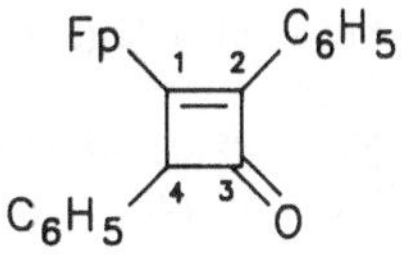 (structure: Fp, C_6H_5, C_6H_5, O cyclobutenone; positions 1,2,3,4)	yellow crystals, m.p. 148 to 149° [109], gold crystals, m.p. 147 to 148° [93] IR (KBr): 1980, 2025 (both CO), 1705 (C=O) [93]
*19	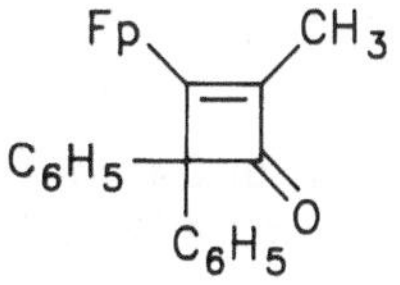 (structure: Fp, CH_3, C_6H_5, C_6H_5, O)	yellow, m.p. 178 to 179°, air-stable in the solid state [109] ^{1}H NMR (CDCl$_3$): 2.0 (CH$_3$), 2.72? (C$_6$H$_5$), 4.58 (C$_5$H$_5$) [109] IR (KBr): 1983, 2035 (both CO), 1710 (C=O) [109]
*20	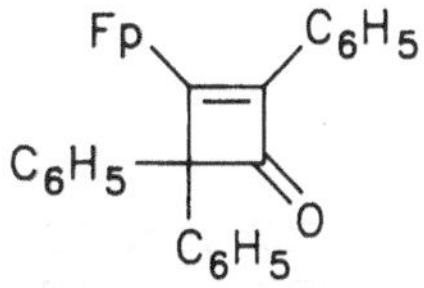 (structure: Fp, C_6H_5, C_6H_5, C_6H_5, O)	yellow crystals, m.p. 189 to 190°, air-stable in the solid state [109] ^{1}H NMR (CDCl$_3$): 4.46 (C$_5$H$_5$), 7.1 to 7.95 (C$_6$H$_5$) [109] IR (KBr): 1982, 2035 (both CO), 1711 (C=O) [109]
*21	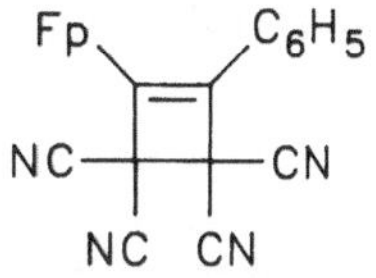 (structure: Fp, C_6H_5, NC, CN, NC, CN cyclobutene) IVa (82%) [92], (50%) [148]	yellow, m.p. 78 to 79° [148], yellow-brown crystals [92] unstable in the solid state, even under N$_2$ [148] ^{1}H NMR: 5.18 (s, C$_5$H$_5$), 7.55 (m, C$_6$H$_5$) in CDCl$_3$ [148], 5.13 (s, C$_5$H$_5$) [92] ^{13}C NMR (CD$_3$COCD$_3$/CH$_3$COCH$_3$ at −73°): 87.0 (s, C$_5$H$_5$), 111.4, 112.3 (both: s, CN), 130.0 (m, C$_6$H$_5$), 212.5 (s, CO) [148] IR: 2001, 2045 (both CO), 2229 (CN) in CH$_2$Cl$_2$, 1555, 1582, 1599 (all C=C), other bands from 692 to 1447 in Nujol [148], 2005, 2050 (both CO) [92] mass spectrum (fast atom bomdardment): [M+H]$^+$ (18), [M]$^+$ (20), [M−CN]$^+$ (8), [M−2CO]$^+$ (66), [M−2CO−CN]$^+$ (46), [M+H−C$_2$(CN)$_4$]$^+$ (100), [M+H−2CO−C$_6$H$_5$]$^+$ (26), 268 (17), [FeC$_2$(C$_6$H$_5$)C$_2$(CN)$_3$]$^+$ (11) [148]
22	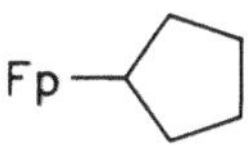 (structure: Fp–cyclopentyl)	^{13}C NMR (CD$_3$NO$_2$): 23.69 (C-2,5), 41.36 (C-3,4) [123]
*23	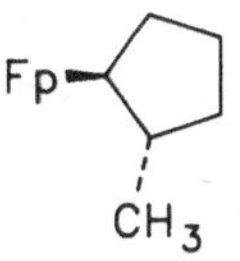 (structure: Fp–cyclopentyl with CH$_3$)	oil [78] ^{1}H NMR (CS$_2$): 0.97 (CH$_3$), 4.7 (s, C$_5$H$_5$) [78]
*24	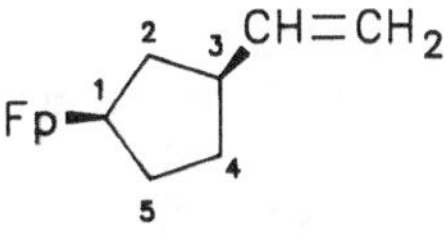 (structure: Fp–cyclopentyl, CH=CH$_2$; positions 1,2,3,4,5) I (24%) [99]	^{1}H NMR (C$_6$D$_6$): 1.3 to 2.5 (m, 8H at C-1 to C-5), 4.10 (s, C$_5$H$_5$), 5.01 (m, 2H, CH$_2$=), 5.83 (m, 1H, –CH=) [99] ^{13}C NMR (CD$_3$NO$_2$): 23.37 (C-1), 40.59, 49.91 (C-2,5), 33.47, 46.28 (C-3,4), 86.61 (C$_5$H$_5$), 112.05 (CH$_2$=), 145.06 (–CH=), 218.85 (CO) [99] IR (neat): 1942, 2000 (both CO), 1639 (C=C) [99]

References on pp. 268/72

Table 8 (continued)

No.	compound method of preparation (yield)	properties and remarks

*25

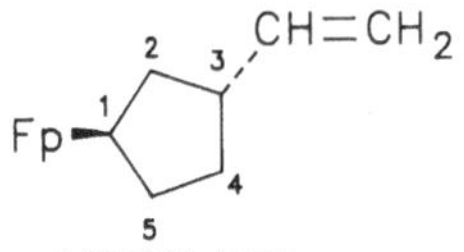

Fp, positions 1,2,3,4,5, CH=CH₂

I (46%) [99]

^{1}H NMR (C$_6$D$_6$): 1.3 to 2.7 (m, 8H at C-1 to C-5), 4.12 (s, C$_5$H$_5$), 5.00 (m, 2H, CH$_2$=), 5.85 (m, 1H, CH=) [99]
^{13}C NMR (CD$_3$NO$_2$): 22.07 (C-1), 35.08 (C-4), 42.66, 47.38 (C-2,5), 44.86 (C-3), 86.55 (C$_5$H$_5$), 111.59 (CH$_2$=), 145.71 (CH=), 218.79 (CO) [99]
IR (neat): 1950, 2000 (both CO), 1637 (C=C) [99]

*26

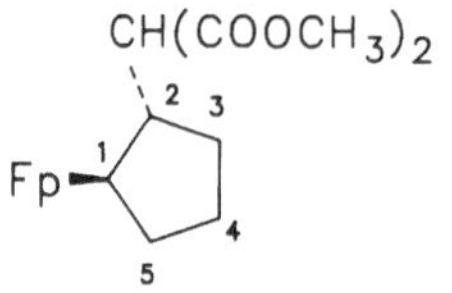

CH(COOCH$_3$)$_2$

Fp

VII (93%) [78]

oil, air-stable [78]
^{1}H NMR (CS$_2$): 1.3 to 2.5 (m, 8H at C-1 to C-5), 3.47 (d, 1H, OCCHCO, J=5.5), 3.56, 3.58 (2s, 6H, OCH$_3$), 4.72 (s, C$_5$H$_5$) [78]
IR (neat): 1942, 2008 (both CO) [78]

*27

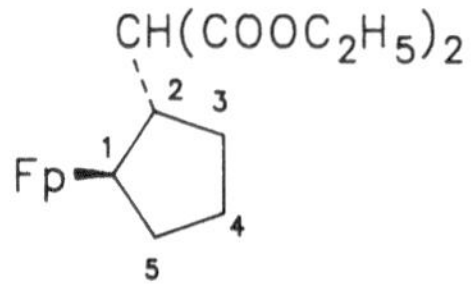

CH(COOC$_2$H$_5$)$_2$

Fp

VII (89%) [78], (82%) [46]

oil, air-stable [78]
^{1}H NMR (CS$_2$): 1.19, 1.23 (2t, 6H, CH$_3$, J=7.0), 1.2 to 2.2 (m, 8H at C-1 to C-5), 3.8 (d, 1H, OCCHCO, J=4), 4.0, 4.08 (2q, 4H, OCH$_2$, J=7.0), 4.71 (s, C$_5$H$_5$) [78]
IR (neat): 1940, 2000 (both CO) [78]

*28   CH$_3$CO—CH—COOCH$_3$

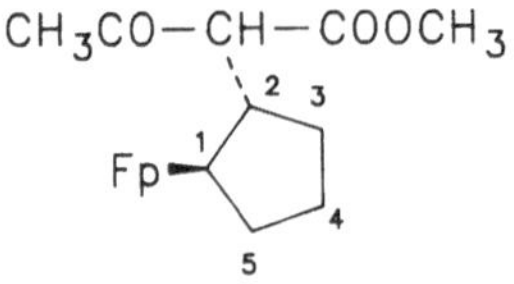

Fp

VII (61%) [78]

oil, air-stable [78]
^{1}H NMR (CS$_2$): 1.55 (br s, 8H at C-1 to C-5), 2.08, 2.1 (2s, 3H, CH$_3$CO), 3.34, 3.57 (2d, 1H, OCCHCO), 3.64, 3.67 (2s, 3H, OCH$_3$), 4.75 (s, C$_5$H$_5$) [78]
IR (neat): 1938, 2008 (both CO) [78]

*29

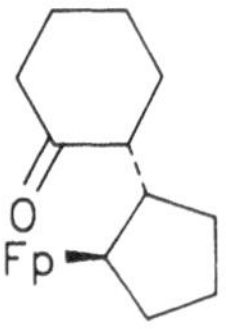

VII [78]

oil [78]
^{1}H NMR (CS$_2$): 1.2 to 2.5 (m, 17H), 4.70 (s, C$_5$H$_5$) [78]
IR (neat): 1934, 2000 (both CO) [78]

30

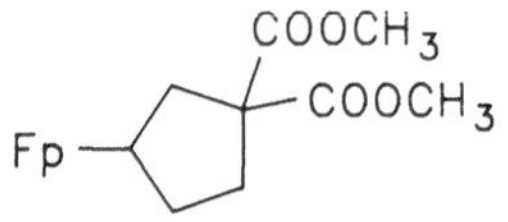

Fp, COOCH$_3$, COOCH$_3$

Va (64%) [115], (66%) [69]

yellow crystals, m.p. 82 to 82.5 (from ether/hexane) [69], air-stable [115]
^{1}H NMR (CDCl$_3$): 1.5 to 2.85 (m, 7H, CH, CH$_2$), 3.70 (s, 6H, CH$_3$), 4.79 (s, C$_5$H$_5$) [69]
IR (KBr): 1940, 2020 (both CO), 1730 (C=O) [69]

*31

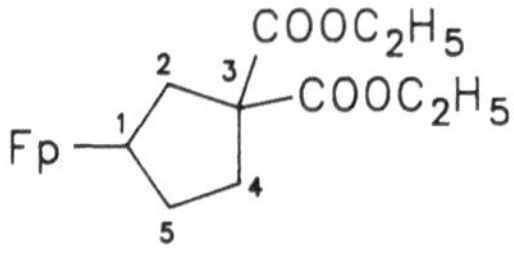

Fp, COOC$_2$H$_5$, COOC$_2$H$_5$

Va (70%) [91, 115]

m.p. 70° (from hexane/ether), air-stable [115]
^{1}H NMR (CDCl$_3$): 1.24 (t, 6H, CH$_3$), 1.73 to 2.78 (m, 7H at C-1,2,4,5), 4.18 (q, 4H, OCH$_2$), 4.80 (s, C$_5$H$_5$) [115]
IR (CHCl$_3$): 1945, 2000 (both CO), 1720 (C=O) [115]
mass spectrum: 334 (44), 293 (48), 122 (36), 121 (100), 95 (43), 67 (70), 66 (37), no parent peak [115]

References on pp. 268/72

Table 8 (continued)

No.	compound method of preparation (yield)	properties and remarks
*32	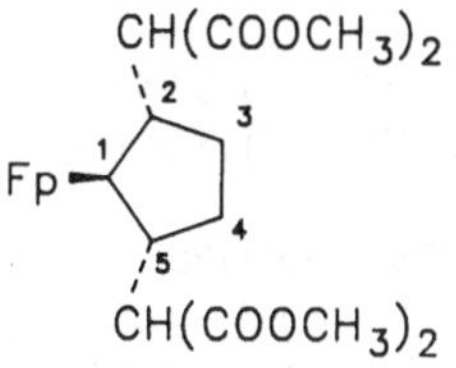 CH(COOCH$_3$)$_2$ Fp CH(COOCH$_3$)$_2$ VII (23%) [78]	oil, air–stable [78] ^{1}H NMR (CS$_2$): 1.2 to 1.9 (m, CH$_2$CH$_2$), 2.39 (br s, 3H, FeCH, FeCCH), 3.40 (d, 2H, OCCHCO, J=6), 3.62 (s, 12H, OCH$_3$), 4.80 (s, C$_5$H$_5$) [78] ^{13}C NMR (CD$_3$NO$_2$): 26.53 (C-1), 29.58 (C-3), 52.80, 53.06, 53.58, 56.18 (C-2, acyclic CH, OCH$_3$), 87.89 (C$_5$H$_5$), 170.67, 171.32 (COO), 218.98 (CO) [78] IR (KBr): 1921, 2004 (both CO) [78]
33	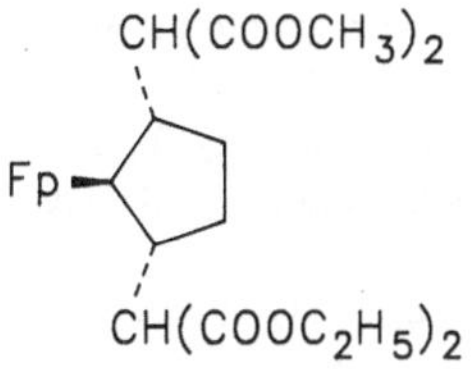 CH(COOCH$_3$)$_2$ Fp CH(COOC$_2$H$_5$)$_2$ VII (41%) [78]	oil, air–stable [78] ^{1}H NMR (CS$_2$): 1.20 (t, 6H, CH$_3$, J=7), 3.6 (s, 6H, OCH$_3$), 4.08 (q, 4H, OCH$_2$, J=7), 4.72 (s, C$_5$H$_5$) [78] IR (neat): 1949, 2004 (both CO) [78]
*34	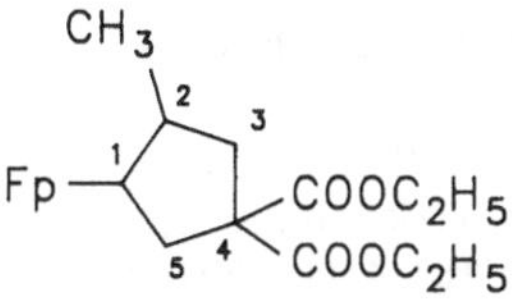 CH$_3$ Fp COOC$_2$H$_5$ COOC$_2$H$_5$	yellow oil [121] ^{1}H NMR (CDCl$_3$): 1.05 (d, 3H, C^{2}CH$_3$, J=6), 1.23 (t, 6H, CH$_3$, J=7), 1.5 to 2.8 (m, 6H at C-1,2,3,5), 4.20 (q, 4H, OCH$_2$, J=7), 4.76, 4.80 (2s, 5H, C$_5$H$_5$) [121] IR (CH$_2$Cl$_2$): 1950, 2000 (both CO), 1720 (C=O) [121]
*35	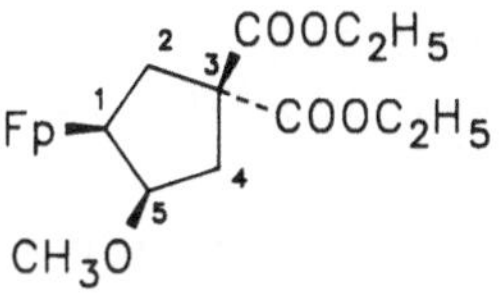 COOC$_2$H$_5$ Fp COOC$_2$H$_5$ CH$_3$O Va (25%) [120]	m.p. 65° [120] ^{1}H NMR (CDCl$_3$): 1.23 (t, 6H, CH$_3$-C-O), 1.95 to 2.90 (m, 5H at C-1,2,4), 3.20 (s, 3H, OCH$_3$), 3.57 (m, 1H at C-5), 4.20 (q, 4H, OCH$_2$), 4.80 (s, C$_5$H$_5$) [120] IR (CHCl$_3$): 1952, 2010 (both CO), 1722 (C=O) [120] mass spectrum: [M−2CO]$^+$ (24), 300 (15), [M−2CO−C$_5$H$_5$]$^+$ (100), 269 (31), 122 (17), 121 (37), no parent peak [120]
*36	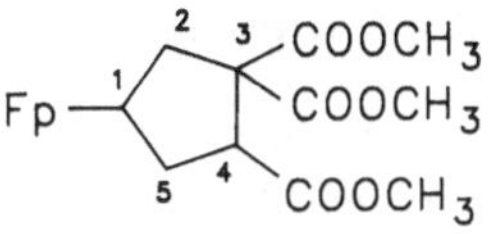 COOCH$_3$ Fp COOCH$_3$ COOCH$_3$ Va (80%) [91], (50%) [115]	m.p. 105° (from hexane/ether), air-stable [115] ^{1}H NMR (CDCl$_3$): 1.60 to 3.06 (m, 5H at C-1,2,5), 3.67, 3.70, 3.75, 3.97 (each s, 9H, COOCH$_3$), ~3.70 (m, 1H, H-4), 4.83 (s, C$_5$H$_5$) [115] IR (CHCl$_3$): 1950, 2000 (both CO), 1730 (C=O) [115] mass spectrum: 364 (53), 270 (100), 212 (35), 151 (39), 121 (80), 93 (39), 65 (40), no parent peak [115]
*37	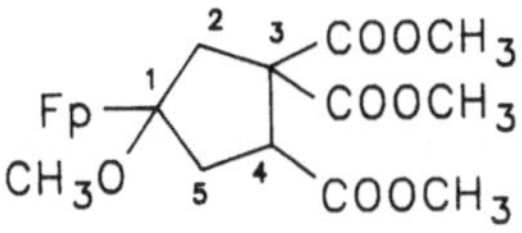 COOCH$_3$ Fp COOCH$_3$ CH$_3$O COOCH$_3$ Va [91], (22%) [114]	oil [114] ^{1}H NMR (CDCl$_3$): 1.90 to 3.00 (br m, 4H, 2CH$_2$), 3.04 (s, 3H, OCH$_3$), 3.68 (s, 6H, COOCH$_3$), 3.79 (s, 3H, COOCH$_3$), ~3.80 (m, 1H, CH), 4.90 (s, C$_5$H$_5$) [114] IR (CHCl$_3$): 1950, 2000 (both CO), 1735 (C=O) [114] mass spectrum: 362 (100), 302 (56), 213 (88), 181 (66), 153 (92), 121 (74), 69 (54), no parent peak [114]

References on pp. 268/72

Table 8 (continued)

No.	compound method of preparation (yield)	properties and remarks

*38

Fp, COOCH$_3$ COOCH$_3$ COOCH$_3$ CH$_3$O
Va (16%) [120]

m.p. 125° [120]
^{1}H NMR (CDCl$_3$): 2.20 to 2.90 (m, 3H at C-1, 2), 3.34 (s, 3H, OCH$_3$), 3.69, 3.72 ("each: s, 3H, COOCH$_3$"), 3.96 (m, 2H at C-4, 5), 4.78 (s, C$_5$H$_5$) [120]
IR (CHCl$_3$): 1955, 2005 (both CO), 1735 (C=O) [120]
mass spectrum: [M−2CO]$^+$ (11), 330 (15), [M−2CO−C$_5$H$_5$]$^+$ (100), 299 (22), 151 (29), 121 (32), 71 (24), 65 (15), 59 (20), no parent peak [120]

*39

Fp, COOC$_2$H$_5$ CN COOC$_2$H$_5$
Va (81%) [115]

amber oil, mixture of diastereomers, air-stable [115]
^{1}H NMR (CDCl$_3$): 1.29, 1.36 (each: t, 3H, CH$_3$), 1.78 to 3.06 (m, 5H at C-1, 2, 5), 3.18 to 3.56 (m, 0.5H, H-4), 3.86 (dd, 0.5H, H-4, J=11, 7.5), 4.23, 4.30 (each: q, 2H, OCH$_2$), 4.83, 4.86 (both: s, 2.5H, C$_5$H$_5$) [115]
IR (CH$_2$Cl$_2$): 1952, 2010 (both CO), 1740 (C=O), 2245 (CN) [115]
mass spectrum: 359 (19), 293 (57), 221 (23), 149 (24), 148 (20), 121 (100), 93 (27), 92 (21), no parent peak [115]

40

Fp, COOC$_2$H$_5$ CN CN
Va (94%) [115]

m.p. 116° (from ether), mixture of the 2 diastereomers, air-stable [115]
^{1}H NMR (CDCl$_3$): 1.36 (t, 3H, CH$_3$), 1.75 to 3.00 (m, 5H at C-1, 2, 5), 3.43 (br, dd, 0.25H, H-4, J~9.5, 6), 3.61 (dd, 0.75H, H-4, J=10, 5.5), 4.33 (q, 2H, OCH$_2$), 4.83, 4.89 (each: s, C$_5$H$_5$, intensity ratio 5:1) [115]
IR (CHCl$_3$): 1955, 2000 (both CO), 1740 (C=O), 2240 (CN) [115]
mass spectrum: 312 (10), 246 (23), 121 (100), 93 (43), 92 (31), 66 (38), 65 (42), no parent peak [115]

41

Fp, CN CN COOC$_2$H$_5$
Va (67%) [91, 115]

m.p. 88° (from hexane/ether) [115], mixture of stereoisomers [91], air-stable [115]
^{1}H NMR (CDCl$_3$): 1.32, 1.33, 1.35 (each t, 3H, CH$_3$), 1.76 to 2.95 (m, 5H at C-1, 2, 5), 3.32 (dd, 0.3H, H-4, J~11, 7), 3.41 (dd, 0.7H, H-4, J=12, 6), 4.24, 4.25, 4.28 (each q, 2H, OCH$_2$), 4.88 (s, C$_5$H$_5$) [115]
IR (CHCl$_3$): 1955, 2008 (both CO), 1737 (C=O), 2248 (CN) [115]
mass spectrum: 312 (81), 285 (52), 270 (48), 246 (82), 219 (100), 149 (34), 147 (40), 121 (68), no parent peak [115]

*42

Fp, CN COOC$_2$H$_5$ CH$_3$O COOC$_2$H$_5$

^{1}H NMR (CDCl$_3$): 1.26 (br t, 6H, CCH$_3$), 3.20 (s, 3H, OCH$_3$), 4.14, 4.19 (each: q, 2H, OCH$_2$), 4.89 (s, C$_5$H$_5$) [114]

*43

Fp, COOC$_2$H$_5$ CN CN CH$_3$O
Va (71) [114]

yellow crystals, m.p. 148 to 149° (from hexane/ether), mixture of diastereomers [114]
^{1}H NMR (CDCl$_3$): 1.36 (t, 3H, CCH$_3$), 2.35 (t, 1H, H-5, J=12), 2.64 (d, 1H, H-2, J=14), 2.78 to 3.04 (m, 1H, H-5), 3.06 (br, d, 1H, H-2, J=14), 3.18 (s, 3H, OCH$_3$), 3.66 (dd, 1H,

References on pp. 268/72

Table 8 (continued)

No.	compound method of preparation (yield)	properties and remarks

H-4, J = 12, 6), 4.36 (q, 2H, OCH₂), 3.91 (s, C₅H₅) [114]

IR (CHCl₃): 1960, 2000 (both CO), 1747 (C=O), 2240 (CN) [114]

mass spectrum: 186 (17), 147 (38), 122 (34), 121 (100), 95 (19), 56 (37), 42 (19), no parent peak [114]

*44

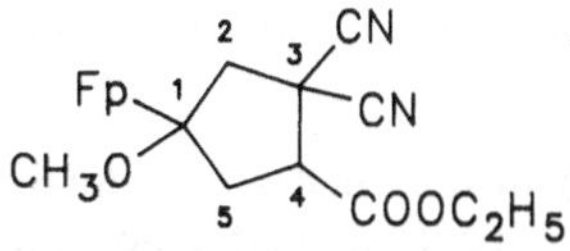

Fp–CH₃O ... CN, CN, COOC₂H₅ (positions 1,2,3,4,5)

Va (58%) [91, 114]

yellow crystals, m.p. 96 to 98° (from hexane/ether) [114], mixture of diastereomers [91]

^{1}H NMR (CDCl₃): 1.35 (t, 3H, CCH₃), 2.13 (dd, 1H, H-5, J = 14, 12), 2.63 (d, 1H, H-2, J = 14), 2.81 (ddd, 1H, H-5, J = 14, 6.5, 3), 3.16 (s, 3H, OCH₃), 3.26 (dd, 1H, H-2, J = 14, 3), 3.65 (dd, 1H, H-4, J = 12, 6.5), 4.30, 4.31 (each: q, OCH₂, intensity ratio 1:1), 4.92 (s, C₅H₅) [114]

IR (CHCl₃): 1960, 2000 (both CO), 1735 (C=O), 2242 (CN) [114]

mass spectrum: M⁺ (0.9), 276 (27), 221 (47), 194 (50), 166 (48), 147 (23), 122 (36), 121 (100), 56 (27) [114]

*45

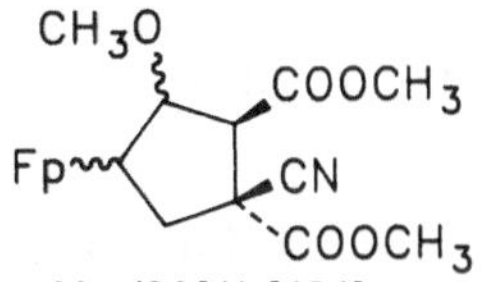

CH₃O, COOCH₃, Fp, CN, COOCH₃

Va (86%) [131]

mixture of diastereomers [131]

*46

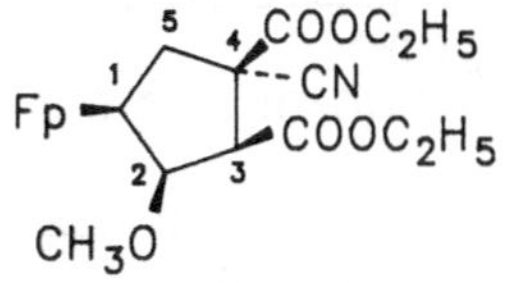

Fp–CH₃O ... COOC₂H₅, CN, COOC₂H₅ (positions 1,2,3,4,5)

Va (56%) [120]

m.p. 103° (from hexane/ether) [120]

^{1}H NMR (CDCl₃): 1.30, 1.34 (both: t, 3H, CCH₃), 2.35 to 2.84 (m, 3H at C-1,5), 3.47 (s, 3H, OCH₃), 3.60 (d, 1H, H-3, J = 3.5), 4.02 (br t, 1H, H-2, J = 3.5), 4.23, 4.30 (both: q, 2H, OCH₂), 4.82 (s, C₅H₅) [120]

IR (CH₂Cl₂): 1960, 2010 (both CO), 1740 (C=O) [120]

mass spectrum: [M − 2CO]⁺ (21), 325 (13), 324 (100), 294 (21), 293 (15), 121 (22), no parent peak [120]

*47

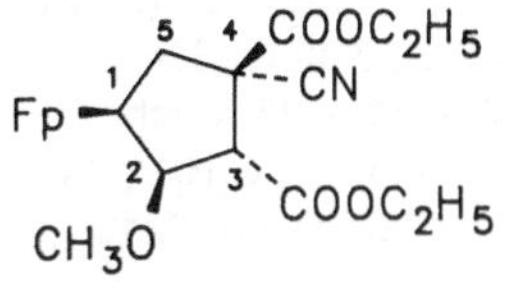

Fp–CH₃O ... COOC₂H₅, CN, COOC₂H₅ (positions 1,2,3,4,5)

Va (33%) [120]

m.p. 99° (from hexane/ether) [120]

^{1}H NMR (CDCl₃): 1.30, 1.35 (both: t, 3H, CCH₃), 2.10 to 2.50 (m, 2H at C-5), 3.01 (ddd, 1H, FeCH, J = 12, 6), 3.18 (s, 3H, OCH₃), 3.71 (d, 1H at C-3, J = 2.5), 3.98 (dd, 1H at C-2, J = 6, 2.5), 4.08 to 4.42 (m, 4H, OCH₂), 4.82 (s, C₅H₅) [120]

IR (CH₂Cl₂): 1960, 2010 (both CO), 1740 (C=O) [120]

mass spectrum: [M − 2CO]⁺ (24), [M − 2CO − C₅H₅]⁺ (100), 294 (28), 122 (29), 121 (44), 93 (24), 92 (20), 40 (30), no parent peak [120]

48

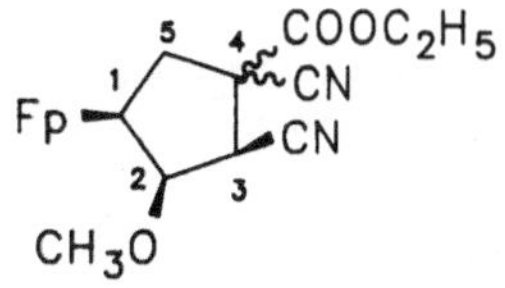

Fp–CH₃O ... COOC₂H₅, CN, CN (positions 1,2,3,4,5)

Va (42%) [120]

m.p. 163° (from hexane/ether), mixture of the 2 diastereomers, eluted with CH₂Cl₂/acetone (Method Va in CH₂Cl₂) [120]

^{1}H NMR (CD₃COCD₃): 1.33 (t, 3H, CCH₃), 2.18 to 2.82 (m, 3H at C-1,5), 3.65 (s, 3H, OCH₃), 3.74 (d, 1H at C-3, J = 3),

References on pp. 268/72

Table 8 (continued)

No. compound method of preparation (yield)	properties and remarks
48 (continued)	4.07 (br t, 1H at C-2, J=3), 4.32 (q, 2H, OCH$_2$), 5.06 (s, C$_5$H$_5$) [120] IR (CHCl$_3$): 1967, 2010 (both CO), 1740 (C=O), 2242 (CN) [120] mass spectrum: [M−2CO]$^+$ (18), 271 (32), 247 (30), 246 (57), 186 (40), 152 (47), 122 (61), 121 (100), 117 (51), no parent peak [120]

49 Fp — (cyclopentane ring, positions 5,4,1,2,3) COOC$_2$H$_5$, CN, CN, CH$_3$O

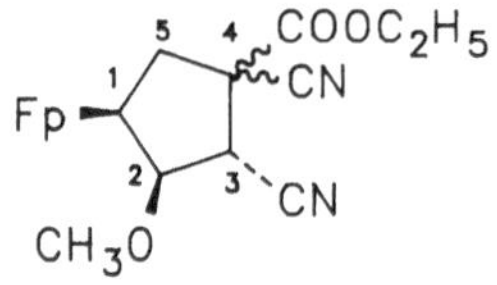

Va (44%) [120]

m.p. 136° (from hexane/ether), mixture of the 2 diastereomers, eluted with ether/light petroleum (Method V a in CH$_2$Cl$_2$) [120]
^{1}H NMR (CDCl$_3$): 1.35 (t, 3H, CCH$_3$), 2.12 to 3.09 (m, 3H at C-1,5), 3.34, 3.37 (each s, 3H, OCH$_3$), 3.75 (d, 1H, H-3, J=1.5), 3.90 (dd, 1H, H-2, J=5.5, 1.5), 4.31 (q, 2H, OCH$_2$), 4.84 (s, C$_5$H$_5$) [120]
IR (CHCl$_3$): 1970, 2010 (both CO), 1750 (C=O), 2242 (CN) [120]
mass spectrum: [M]$^+$ (0.1), [M−2CO]$^+$ (10), 246 (36), 152 (40), 149 (38), 122 (95), 121 (100), 117 (31) [120]

*50 Fp — (cyclopentane ring, positions 5,4,1,2,3) CN, CN, COOC$_2$H$_5$, CH$_3$O

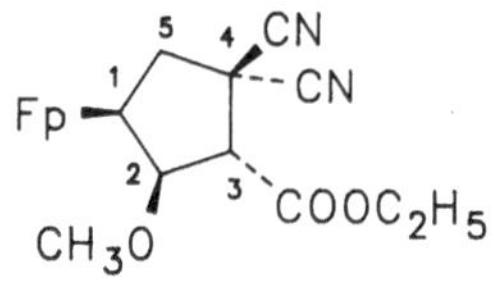

Va [91], (49%) [120]

m.p. 89° (from hexane/ether) [120]
^{1}H NMR (CDCl$_3$): 1.37 (t, 3H, CCH$_3$), 2.30 to 3.02 (m, 3H at C-1,5), 3.29 (s, 3H, OCH$_3$), 3.45 (d, 1H, H-3, J=2.5), 3.95 (dd, 1H, H-2, J=5.5, 2.5), 4.32, 4.35 (each q, 2H, OCH$_2$, intensity ratio 1:1), 4.83 (s, C$_5$H$_5$) [120]
IR (CH$_2$Cl$_2$): 1960, 2005 (both CO), 1735 (C=O), 2240 (CN) [120]
mass spectrum: [M−2CO]$^+$ (27), 246 (22), 219 (18), 152 (100), 149 (19), 122 (90), 121 (73), no parent peak [120]

*51 Fp — (cyclopentane ring, positions 5,4,1,2,3) CN, CN, COOC$_2$H$_5$, CH$_3$O

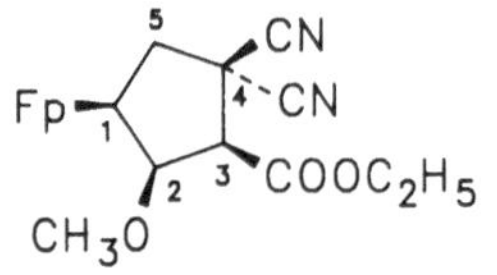

Va [91], (36%) [120]

m.p. 140° (from hexane/ether) [120]
^{1}H NMR (CDCl$_3$): 1.36 (t, 3H, CCH$_3$), 2.38 to 2.92 (m, 3H at C-1,5), 3.34 (d, 1H, H-3, J=3.5), 3.43 (s, 3H, OCH$_3$), 4.04 (t, 1H, H-2, J=3.5), 4.31, 4.35 (each q, 2H, OCH$_2$, intensity ratio 1:1), 4.85 (s, C$_5$H$_5$) [120]
IR (CH$_2$Cl$_2$): 1960, 2010 (both CO), 1737 (C=O), 2242 (CN) [120]
mass spectrum: [M−2CO]$^+$ (30), 246 (22), 219 (24), 152 (100), 149 (21), 122 (94), 121 (96), 91 (68), no parent peak [120]

*52 Fp — (cyclopentane ring) CN, CN, with o-Cl-C$_6$H$_4$ substituent

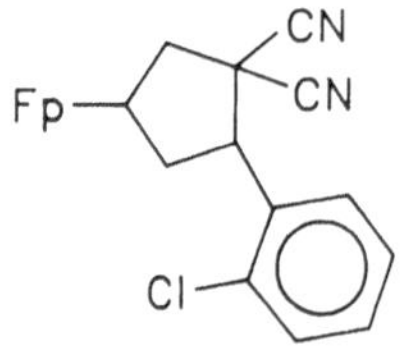

yellow, m.p. 125 to 128° [69]
^{1}H NMR (CDCl$_3$): 2.2 to 3.1 (m, 5H, FeCH, CH$_2$), 4.3 to 4.65 (m, 1H, CH-aryl), 4.85 (s, C$_5$H$_5$), 7.1 to 7.7 (m, 4H, C$_6$H$_4$) [69]
IR (KBr): 1930, 2000 (both CO) [69]

References on pp. 268/72

Table 8 (continued)

No.	compound method of preparation (yield)	properties and remarks

*53

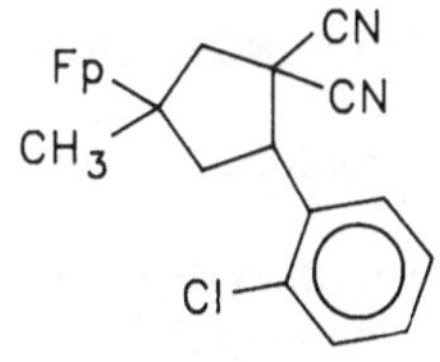

m.p. 123 to 124° [69], mixture of isomers
^{1}H NMR (CDCl$_3$): 1.59, 1.80 (both: s, CH$_3$), 2.4 (m, 2H, CH$_2$), 2.94 (m, 2H, CH$_2$), 4.67 (m, 1H, CH), 4.82 (s, C$_5$H$_5$), 7.34 (m, 4H, C$_6$H$_4$) [69]
IR (KBr): 1960, 2020 (both CO) [69]

*54

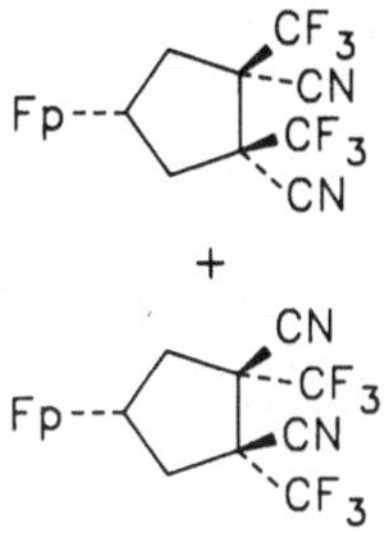

^{19}F NMR (CDCl$_3$): 67.46, 67.75 (both: s, CF$_3$; no assignment to the isomers) [83]

*55

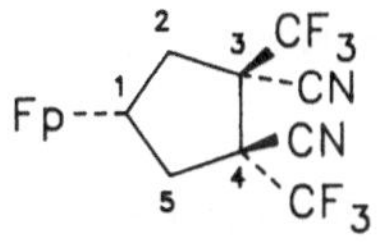

yellow crystals, m.p. 115 to 115.7° [83]
^{1}H NMR (CDCl$_3$): 2.5 (br, m, FeCH), 4.88 (s, C$_5$H$_5$) [83]
^{13}C NMR (CDCl$_3$): 12.48 (C–1), 49.96 (C–5), 50.92 (C–2), 53.8, 55.9 (C–3,4, J(C,F)=30), 85.54 (C$_5$H$_5$), 113.30, 113.98 (both CN), 122.35, 123.14 (both CF$_3$, J(C,F)= 285), 215.25 (CO) [83]
^{19}F NMR (CDCl$_3$): 68.33 (q), 69.09 (q, J(F,F)=2.8) [83]
IR: 1962, 2016 (both CO) in CH$_2$Cl$_2$, 2250 (CN) in KBr [83]

*56

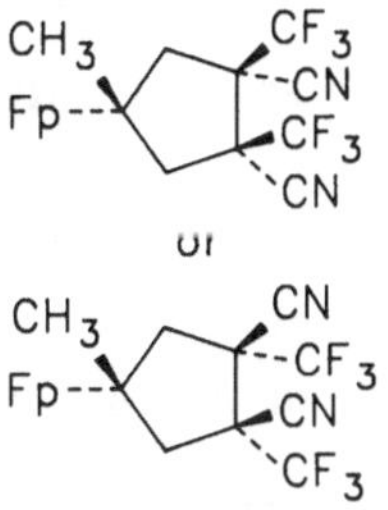

yellow, m.p. 149 to 150°, one of the isomers, but structural assignment uncertain [83]
^{1}H NMR (CDCl$_3$): 1.78 (s, CH$_3$), 2.87 (s, CH$_2$), 4.86 (s, C$_5$H$_5$) [83]
^{19}F NMR (CDCl$_3$): 67.08 (s) [83]
IR (CH$_2$Cl$_2$): 1962, 2016 (both CO) [83]

*57

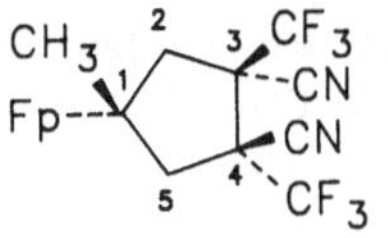

yellow, m.p. 139 to 139.5° [83]
^{1}H NMR (CDCl$_3$): 1.63 (s, CH$_3$), 2.89 (s, C^{2}H$_2$), 2.94 (s, C^{5}H$_2$), 4.87 (s, C$_5$H$_5$) [83]
^{13}C NMR (CDCl$_3$): 30.74 (C–1), 39.23 (CH$_3$), 53.53, 56.20 (C–3,4, J(C,F)=30), 56.20 (C–5), 57.88 (C–2), 86.33 (C$_5$H$_5$), 113.64, 114.15 (both CN), 122.85, 122.91 (both CF$_3$, J(C,F)=285), 215.93, 216.04 (both CO) [83]
^{19}F NMR (CDCl$_3$): 68 (m) [83]
IR: 1960, 2014 (both CO) in CH$_2$Cl$_2$, 2250 (CN) in KBr [83]

References on pp. 268/72

Table 8 (continued)

No.	compound method of preparation (yield)	properties and remarks
58	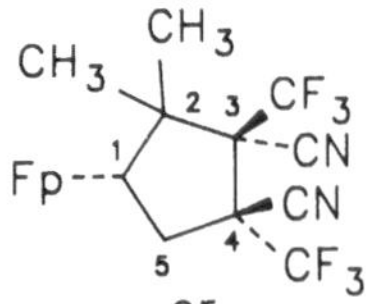 or 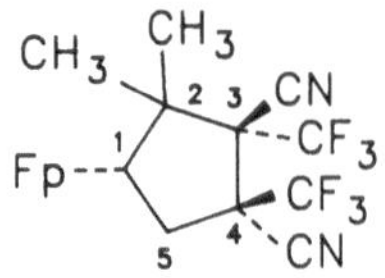	from FpCH$_2$CH=C(CH$_3$)$_2$ and (E)-CF$_3$(NC)C=C(CN)CF$_3$ in benzene for 71 h [83] yellow, m.p. 113 to 113.7°, structural assignment uncertain [83] ^{1}H NMR (CDCl$_3$): 1.19 (q, J(H,H)=2), 1.49 (s, CH$_3$), 2.65 (s, FeCH), 2.70 (s, CH$_2$), 4.88 (s, C$_5$H$_5$) [83] ^{13}C NMR (CDCl$_3$): 22.40 (CH$_3$, J(C,F)=4), 26.70 (CH$_3$), 29.17 (FeC), ~52 (C-3,4), 49.29 (CH$_2$), 57.38 (C-2), 86.21 (C$_5$H$_5$), 113.5, 114.3 (both CN), 122.6 (CF$_3$, J(C,F)=289), 123.5 (CF$_3$, J(C, F)=285), 215.76, 216.77 (both CO) [83] ^{19}F NMR (CDCl$_3$): 61.58 (m, J(H,F)=2.5), 62.33 (m), 68.60 (q, J(F,F)=2.9), 69.48 (q, J(F,F)=5.2) [83] IR: 1962, 2016 (both CO) in CH$_2$Cl$_2$, 2254 (CN) in KBr [83]
59	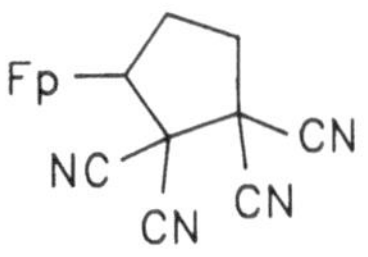	obtained from FpC$_3$H$_5$–cyclo and (NC)$_2$C=C(CN)$_2$ [31, 48] m.p. 119 to 121° [31] ^{1}H NMR (CD$_3$NO$_2$): 2.0 to 3.2 (m, 5H, CH, CH$_2$), 5.34 (s, C$_5$H$_5$) [31] IR (KBr): 1965, 2021 (both CO), 2250 (CN) [31]
*60	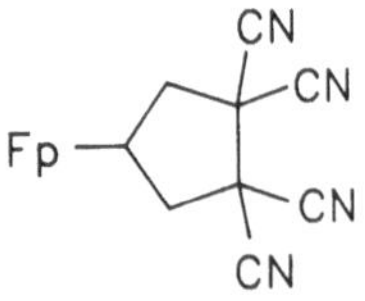 IVa ("good yield") [29, 69]	m.p. 193 to 196° (dec.) [31], dec. 200 to 204° [69] ^{1}H NMR (CD$_3$NO$_2$): 2.26 to 3.42 (m, 5H, CH, CH$_2$), 5.07 (s, C$_5$H$_5$) [69] IR (KBr): 1940, 2000 (both CO) [29, 69], 2250 (CN) [29]
*61	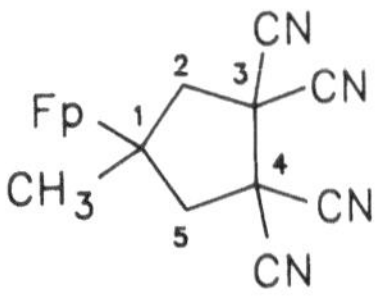 IVa [30, 40, 67, 69], (95%) [29, 52]	yellow crystals [40], amber, dec. 140° [52], dec. 147 to 148° [69] ^{1}H NMR: 1.68 (s, CH$_3$), 3.02, 3.78 (q, CH$_2$, J=14.5), 5.14 (s, C$_5$H$_5$) in CD$_3$COCD$_3$ [30, 52], see also [29], spectrum also depicted [52]; 1.64 (s, CH$_3$), 2.98 (d, 2H, cis-FeCCH$_2$), 3.36 (d, 2H, trans-FeCCH$_2$, J=14.5), 5.06 (s, C$_5$H$_5$) in CD$_3$NO$_2$ [69] ^{13}C NMR (CD$_3$COCD$_3$): 33.68 (FeC), 37.20 (CH$_3$), 45.32 (C-3,4), 61.38 (CH$_2$), 87.25 (C$_5$H$_5$), 113.06, 113.77 (both CN), 216.67 (CO) [83] IR: 1968, 2018 in CH$_2$Cl$_2$ [52], 1958, 2012 in THF [79], 1950, 2010 [69], 1940, 2000 (all CO), 2250 (CN) in KBr [29], 852 (C$_5$H$_5$), 2250 (CN) in Nujol [52]

References on pp. 268/72

Table 8 (continued)

No.	compound method of preparation (yield)	properties and remarks

62

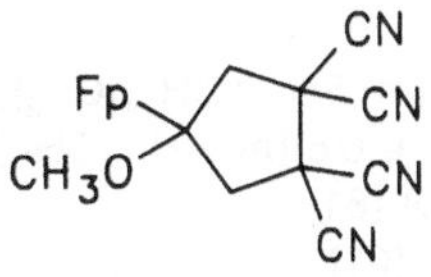

IVa (98%) [89, 90]

yellow-green crystals, m.p. 158 to 159° (dec.) [89, 90], air-stable [89]
^{1}H NMR (CD$_3$COCD$_3$): 2.84, 3.81 (both: br, q, CH$_2$, J = 13), 3.25 (s, OCH$_3$), 5.29 (s, C$_5$H$_5$) [89, 90]
IR (CH$_3$COCH$_3$): 1956, 2008 (both CO) [89]

*63

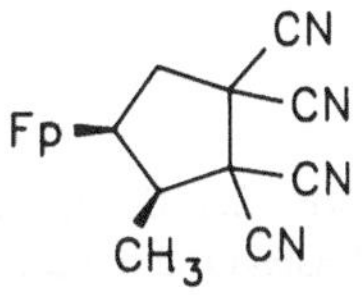

IVa (77%) [61]

m.p. 149 to 151° [61]
^{1}H NMR (CD$_3$NO$_2$): 1.39, 1.52 (2d, 3H, CH$_3$, J=7), 2.2 to 3.3 (m, 4H, CH, CH$_2$), 5.06, 5.10 (2s, C$_5$H$_5$) [61], 1.39 (cis-CH$_3$), 1.49 (trans-CH$_3$), cf. No. 64 [78]
IR (KBr): 1960, 2020 (both CO) [61]

*64

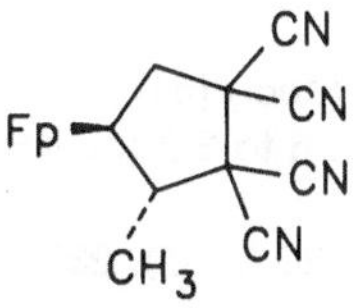

IVa [69]

dec. 162° [69]
^{1}H NMR (CD$_3$NO$_2$): 1.49 (d, CH$_3$, J=6.0), 2.14 to 3.26 (m, 4H, CH, CH$_2$), 5.12 (s, C$_5$H$_5$) [69], see also No. 63
IR (KBr): 1945, 2005 (both CO) [69]

65

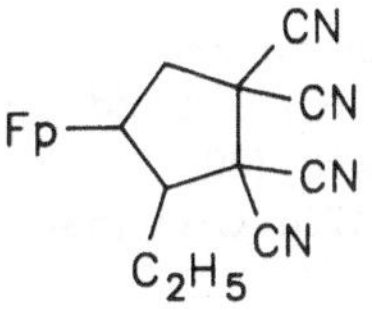

IVa (86%) [78]

m.p. 117.5 to 118.5° [78]
IR: 1980, 2024 (both CO), 2257 (CN) [78]

66

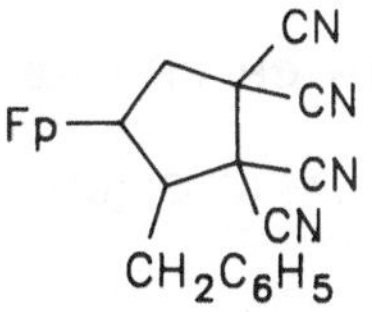

IVa (79%) [78]

m.p. 200° (dec.) [78]

67

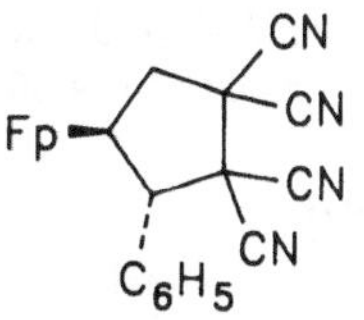

IVa [69]

dec. 134° [69]
^{1}H NMR (CD$_3$NO$_2$): 2.70 to 4.00 (m, 4H, CH, CH$_2$), 4.95 (s, C$_5$H$_5$), 7.54 (s, C$_6$H$_5$) [69]
IR (KBr): 1960, 2010 (both CO) [69]

References on pp. 268/72

Table 8 (continued)

No.	compound method of preparation (yield)	properties and remarks

68

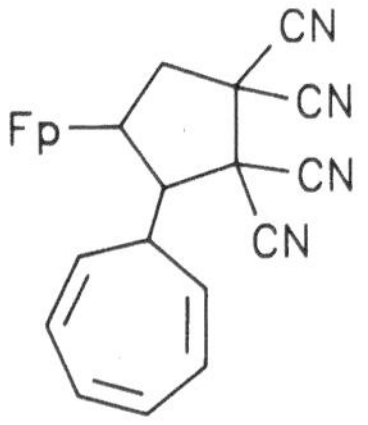

IVa (69%) [61]

m.p. 143 to 144° (dec.) [61]
^{1}H NMR (CD$_3$NO$_2$): 1.9 to 3.3 (m, 5H, >CH–, CH$_2$), 5.03 (s, C$_5$H$_5$), 5.05 (m, 2H, =CH–), 6.5 (m, 4H, =CH–) [61]
IR (KBr): 1970, 2040 (both CO) [61]
mass spectrum (70 eV): [M]$^+$ [61]

69

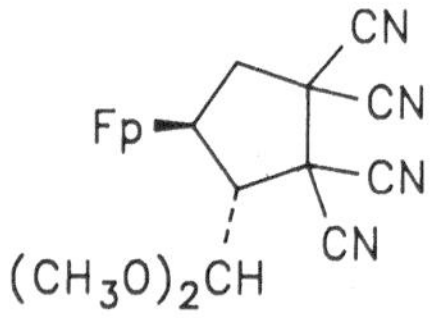

(CH$_3$O)$_2$CH
IVa (88%) [61]

m.p. 135 to 137° (dec.) [61]
^{1}H NMR (CD$_3$NO$_2$): 2.2 to 3.4 (m, 4H, CH, CH$_2$), 3.59, 3.61 (2s, 6H, OCH$_3$), 4.75 (d, O–CH–O, J=3.5), 5.08 (s, C$_5$H$_5$) [61]
IR (KBr): 1990, 2040 (both CO) [61]

70

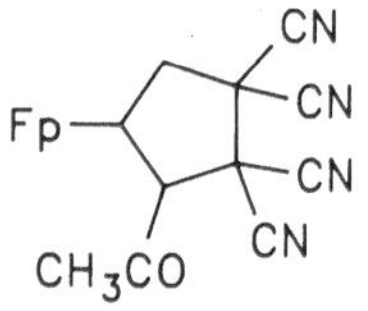

CH$_3$CO
IVa (70%) [61]

yellow needles, m.p. 150 to 151° (dec.) [61]
^{1}H NMR (CD$_3$NO$_2$): 2.55 (s, CH$_3$), 2.6 to 3.9 (m, 4H, CH, CH$_2$), 5.1 (s, C$_5$H$_5$) [61]
IR (KBr): 1950, 2020 (both CO), 1720 (C=O), 2375 (CN) [61]

71

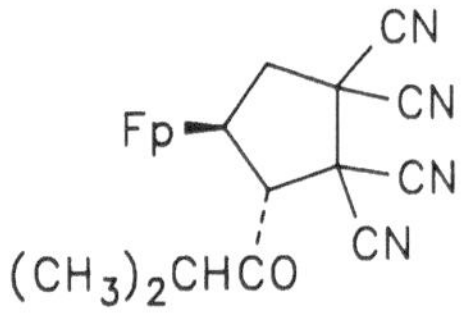

(CH$_3$)$_2$CHCO
IVa [69]

^{1}H NMR (CD$_3$NO$_2$): 1.22, 1.32 (both: d, 3H, CH$_3$, J=5.0), 2.46 to 3.40 (m, 4H, CH, CH$_2$), 3.92 (dd, 1H, FeCCH, J=8.5, 1.5), 5.06 (s, C$_5$H$_5$) [69]
IR (KBr): 1975, 2030 (both CO), 1720 (C=O) [69]

72

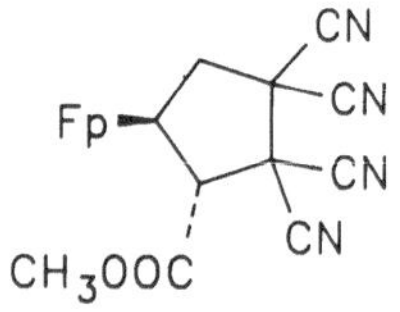

CH$_3$OOC
IVa (83%) [61]

m.p. 131 to 132° [61]
^{1}H NMR (CDCl$_3$): 2.4 to 3.5 (m, 4H, CH, CH$_2$), 3.9 (s, OCH$_3$), 4.92 (s, C$_5$H$_5$) [61]
IR (KBr): 1980, 2040 (both CO), 1745 (C=O) [61]

73

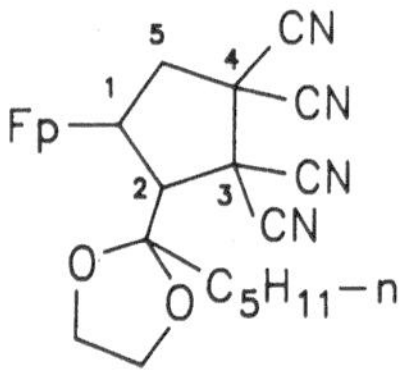

IVa (71%) [61]

m.p. 138 to 139° (dec.) [61]
^{1}H NMR (CDCl$_3$): 0.9 to 2.2 (m, 11H, C$_5$H$_{11}$), 2.4 to 3.2 (m, 4H at C-1,2,5), 4.12 (br, s, 4H, OCH$_2$), 4.90 (s, C$_5$H$_5$) [61]
IR (KBr): 1970, 2030 (both CO) [61]

References on pp. 268/72

Table 8 (continued)

No.	compound method of preparation (yield)	properties and remarks
74	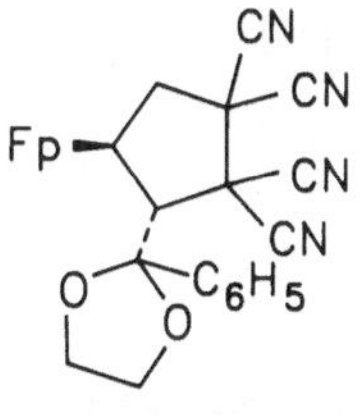 IVa (71%) [61]	m.p. 168 to 170° (dec.) [61] ^{1}H NMR (CD$_3$NO$_2$): 2.4 to 4 (m, 4H, CH, CH$_2$), 4.35 (m, 4H, OCH$_2$), 4.45 (s, C$_5$H$_5$), 7.65 (m, C$_6$H$_5$) [61] IR (KBr): 1960, 2020 (both CO), 2270 (CN) [61]
75	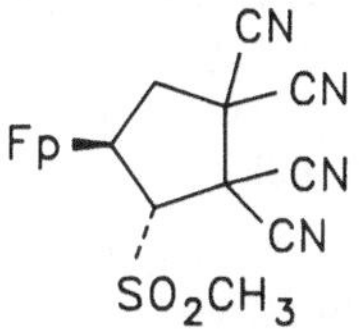 IVa (80%) [61]	dec. 142° [61] ^{1}H NMR (CD$_3$NO$_2$): 2.5 to 3.4 (m, 4H, CH, CH$_2$), 3.30 (s, CH$_3$), 5.10 (s, C$_5$H$_5$) [61] IR (KBr): 1970, 2030 (both CO), 1140, 1320 (both SO$_2$) [61]
76	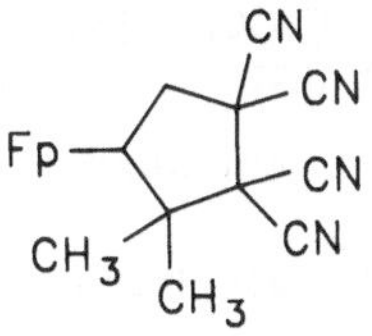 IVa [30, 69], (95%) [52]	bright yellow, dec. 150° [52], dec. 155° [69] ^{1}H NMR: 0.83, 1.18 (2s, CH$_3$), 2.34 (complex, CH$_2$CH), 4.82 (s, C$_5$H$_5$) [52], 1.32 (cis-CH$_3$), 1.54 (trans-CH$_3$) [78] in CD$_3$COCD$_3$, 1.32 (s, 3H, CH$_3$), 1.54 (s, 3H, CH$_3$), 2.75 to 3.35 (m, 3H, CH, CH$_2$), 5.13 (s, C$_5$H$_5$) in CD$_3$NO$_2$ [69] ^{13}C NMR (CD$_3$COCD$_3$): 23.30, 26.14 (both CH$_3$), FeC obscured by CD$_3$COCD$_3$, 43.9, 56.3 (C bonded to CN), 53.19 (CH$_2$), 60.01 (C(CH$_3$)$_2$), 87.18 (C$_5$H$_5$), 111.82, 114.09 (CN), 216.3, 217.2 (CO) [83] IR: 1968, 2023 in CH$_2$Cl$_2$, 2245 (CN), 854 (C$_5$H$_5$) in Nujol [52], 1950, 2010 in KBr [69] same stability and solubility as No. 61 [52] reacts with gaseous HCl in THF to give FpCl [52]
*77	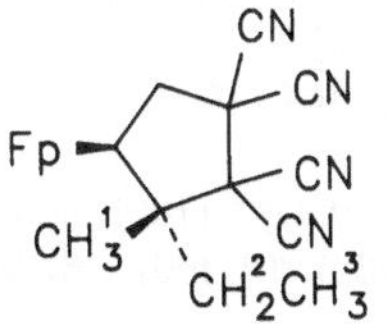 IVa (78%) [78]	m.p. 122 to 123° [78] ^{1}H NMR (CD$_3$COCD$_3$): 1.18 (t, 3H, CH$_3^3$, J=7), 1.30 (s, 3H, CH$_3^1$), 1.75 (m, 1H, CH$_2^2$), 2.26 (m, 1H, CH$_2^2$), 2.73 (dd, 1H, FeCH, J=13,8), 3.04 (dd, 1H, FeCCH$_2$, J=13,13), 3.35 (dd, 1H, FeCCH$_2$, J=13,8), 5.23 (s, C$_5$H$_5$) [78] IR (KBr): 1961, 2028 (both CO), 2278 (CN) [78]
*78	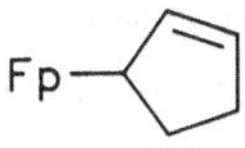 I [48], (60%) [69] II [48, 60], (100%) [37]	^{1}H NMR (CDCl$_3$): 2.1 (m, 4H, CH$_2$), 3.78 (m, FeCH), 4.74 (s, C$_5$H$_5$), 5.52 (m, CH=), 6.08 (m, CH=) [37], 1.58 to 2.52 (m, 4H, CH$_2$), 3.60 to 3.92 (m, 1H, FeCH), 4.70 (s, C$_5$H$_5$), 5.37 to 5.60 (m, 1H, FeCC=CH), 5.94 to 6.20 (m, 1H, FeCCH=) [69] IR (CH$_2$Cl$_2$): 1945, 1997 (both CO) [69]

References on pp. 268/72

Table 8 (continued)

No.	compound method of preparation (yield)	properties and remarks
79	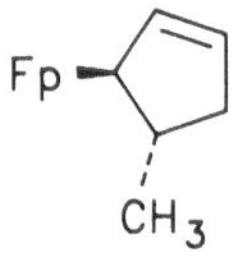 II [48, 60, 69]	IR (CH$_2$Cl$_2$): 1950, 2002 (both CO) [69]
80	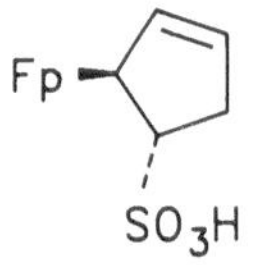 II [37], (50%) [69]	^{1}H NMR (CDCl$_3$): 2.34 (m, 3H, CH$_2$, CHS), 3.84 (m, FeCH), 4.84 (s, C$_5$H$_5$), 5.64 (m, CH=), 6.16 (m, CH=) [37], 2.0 to 2.5 (m, 3H, CH$_2$, CHS), 3.84 (m, 1H, FeCH), 4.8 (s, C$_5$H$_5$), 5.4 to 5.6 (m, 1H, FeCC=CH), 5.9 to 6.2 (m, 1H, FeCCH=) [69] IR (CH$_2$Cl$_2$): 1930, 1988 (both CO) [69]
81	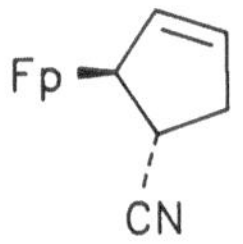 II [37, 48, 60], (70%) [69]	^{1}H NMR (CDCl$_3$): 2.82 (m, 3H, CH$_2$, CHCN), 3.90 (m, FeCH), 4.92 (s, C$_5$H$_5$), 5.45 (m, CH=), 6.0 (m, CH=) [37], 2.56 to 3.12 (m, 3H, CH, CH$_2$), 3.76 to 3.96 (m, 1H, FeCH), 4.86 (s, C$_5$H$_5$), 5.28 to 5.52 (m, 1H, FeCC=CH), 5.86 to 6.10 (m, 1H, FeCCH=) [69] IR (CH$_2$Cl$_2$): 1957, 2005 (both CO) [69]
82	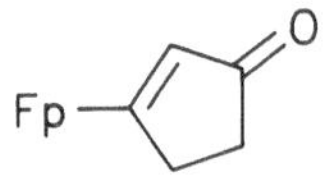 I (51%) [111]	m.p. 94 to 95°, air-stable as solid, somewhat light-sensitive [111] IR (CH$_2$Cl$_2$): 1976, 2024 (both CO), 1516, 1665 [111]
83	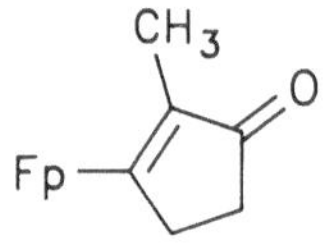 I (59%) [111]	m.p. 129 to 130°, air-stable as solid, somewhat light-sensitive [111] IR (CH$_2$Cl$_2$): 1971, 2022 (both CO), 1557, 1668 [111]
*84	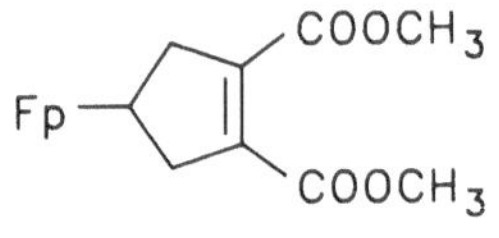 Va [91], (42%) [115]	yellow, m.p. 94 to 96° [115] ^{1}H NMR (CDCl$_3$): 2.40 to 3.12 (m, 5H, CH, CH$_2$), 3.77 (s, 6H, OCH$_3$), 4.81 (s, C$_5$H$_5$) [115] IR (CHCl$_3$): 1945, 1999 (both CO), 1642 (C=C), 1720 (C=O) [115] mass spectrum: [M]$^+$ (1.3), [M$-$2CO]$^+$ (100), 244(33), 210(70), 180(72), 152(71), 122(64), 121(60), 56(32) [115]
85	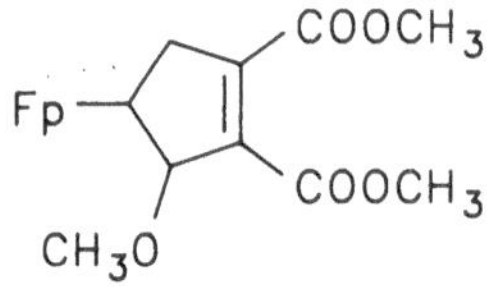 Va (77%) [91, 120]	yellow, m.p. 94 to 96° [120] ^{1}H NMR (CDCl$_3$): 2.58 to 3.10 (m, 3H, CH$_2$, FeCH), 3.38 (s, 3H, OCH$_3$), 3.79, 3.81 (s's, each 3H, COOCH$_3$), 4.26 (br, dd, 1H, CHO, J~4,1.5), 4.85 (s, C$_5$H$_5$) [120] IR (CHCl$_3$): 1955, 2000 (both CO), 1632 (C=C), 1715 (C=O) [120] mass spectrum: 237(55), 122(43), 121(54), 119(100), 95(35), 92(29), 77(25), no parent peak [120]

References on pp. 268/72

Table 8 (continued)

No.	compound method of preparation (yield)	properties and remarks
*86	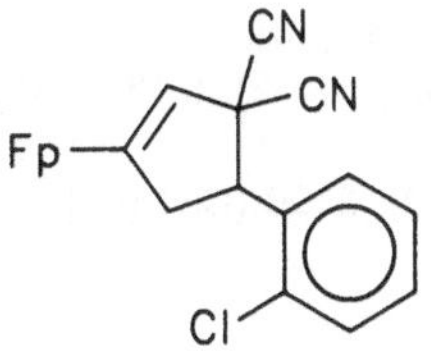	m.p. 114 to 115° [43] ^{1}H NMR (CDCl$_3$): 3.00 (2m, CH$_2$, J=7.5), 4.72 (t, >CH–, J=7.5), 4.92 (s, C$_5$H$_5$), 5.40 (t, =CH, J=1.5), 6.35 (m, C$_6$H$_4$) [43]
*87	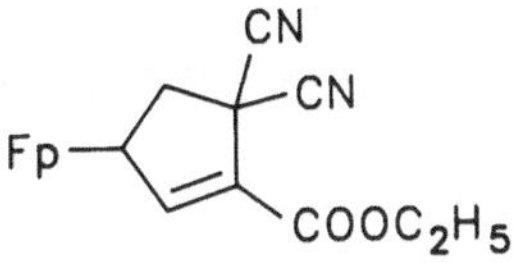	^{1}H NMR (CDCl$_3$): 1.37 (t, 3H, CH$_3$), 2.64 (br, dd, 1H, CH$_2$), 3.12 to 3.4 (m, 1H, CH$_2$), 3.56 (m, 1H, FeCH), 4.34 (br, q, 2H, OCH$_2$), 4.9 (s, C$_5$H$_5$), 7.39 (br, d, 1H, –CH=, J=2.5) [120] IR (CH$_2$Cl$_2$): 1970, 2010 (both CO), 1595 (C=C), 1700 (C=O), 2242 (CN) [120]
*88	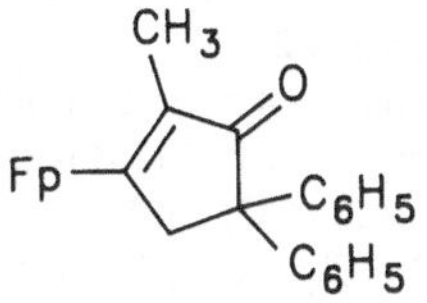	yellow crystals, m.p. 188 to 192° (dec.) [80] ^{1}H NMR (CDCl$_3$): 1.93 (br, CH$_3$), 3.76 (br, CH$_2$), 4.89 (s, C$_5$H$_5$) [80] IR (CHCl$_3$): 1980, 2042 (both CO), 1670 (C=O) [80]
*89	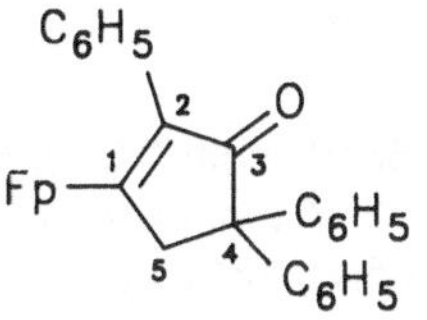	yellow, m.p. 165 to 167° [80] ^{1}H NMR (CDCl$_3$): 3.87 (s, CH$_2$), 4.69 (s, C$_5$H$_5$) [80] ^{13}C NMR (CDCl$_3$): 62.66 (C–4), 66.43 (C–5), 85.76 (C$_5$H$_5$), 126.23, 126.90, 128.08, 128.25, 130.28, 138.65, 144.78 (all C$_6$H$_5$), 156.69 (C–2), 199.85 (C–1), 206.04 (C=O), 213.68 (CO) [80] IR (CHCl$_3$): 1978, 2029 (both CO), 1673 (C=O) [80]
*90	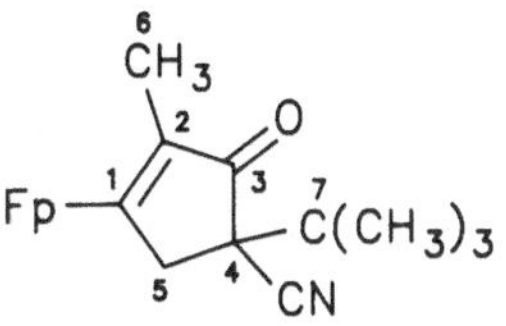	yellow, m.p. 145 to 146° (dec.) [80] ^{1}H NMR (CD$_3$COCD$_3$): 1.05 (s, C$_4$H$_9$), 1.90 (t, CH$_3$, J=1.5), 3.30 (q, CH$_2$, J=1.5), 5.28 (s, C$_5$H$_5$) [80] ^{13}C NMR (CDCl$_3$): 13.86 (C–6), 25.68 [(CH$_3$)$_3$], 36.90 (C–7), 54.28 (C–4), 55.79 (C–5), 85.57 (C$_5$H$_5$), 120.96 (CN), 149.88 (C–2), 193.88 (C–1), 203.85 (C–3), 213.55 (CO) [80] IR (KBr): 1962, 2023 (both CO), 1666 (C=O), 2229 (CN) [80]
*91	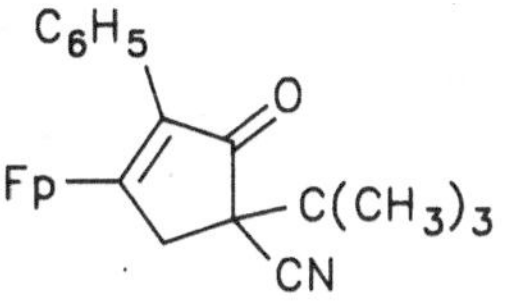	yellow, m.p. 187 to 188° (dec.) [80] ^{1}H NMR (CDCl$_3$): 1.13 (s, C$_4$H$_9$), 3.30, 3.46 (q, CH$_2$, J=10), 4.73 (s, C$_5$H$_5$) [80] IR: 1978, 2030 (both CO), 1681 (C=O) in CH$_2$Cl$_2$, 2234 (CN) in Nujol [80]

References on pp. 268/72

Table 8 (continued)

No.	compound method of preparation (yield)	properties and remarks
92	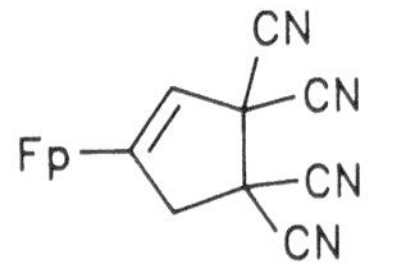 IVa [54], (42%) [43]	yellow crystals, m.p. 172 to 173° [43] ^{1}H NMR (CD$_3$NO$_2$): 3.65 (d, CH$_2$, J=2), 5.13 (s, C$_5$H$_5$), 5.60 (t, =CH–, J=2) [43] IR (KBr): 1980, 2040 (both CO), 2190 (CN) [43]
*93	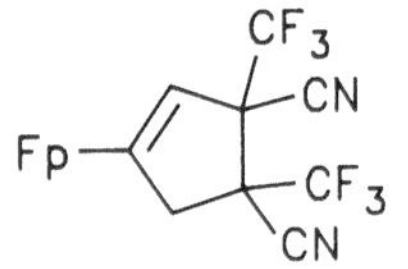	tan, m.p. 94 to 96° [83] ^{19}F NMR (CDCl$_3$): 68.22 (q), 70.55 (q, J(F,F)=10.3), 71.82 (s), 72.11 (s) [83] IR (CH$_2$Cl$_2$): 1988, 2037 [83]
94	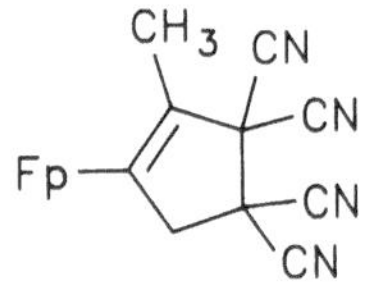 IVa [30], (75%) [52]	amber, dec. 175° [52] ^{1}H NMR (CD$_3$COCD$_3$): 1.70 (t, CH$_3$, J=2.0), 3.23 (q, CH$_2$, J=2.0), 4.74 (s, C$_5$H$_5$) [52] ^{13}C NMR (CD$_3$COCD$_3$): 15.5 (CH$_3$), 44.5, 54.2 (both >C<), 57.4 (CH$_2$), 86.5 (C$_5$H$_5$), 111.2, 112.8 (both CN), 128.0 (=C<), 156.2 (FeC), 214.3 (CO) [83] IR: 1990, 2034 (both CO) in CH$_2$Cl$_2$, 852 (C$_5$H$_5$), 2234 (CN) in Nujol [52] same stability and solubility as No. 61 [52]
95	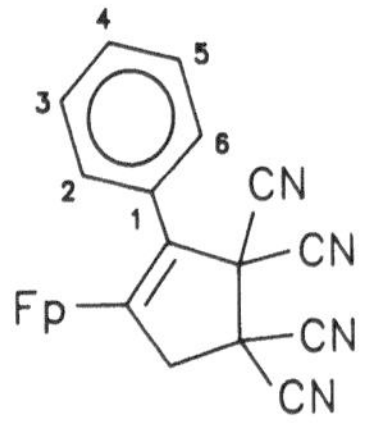	^{13}C NMR (CD$_3$COCD$_3$): 45.54, 55.01 (both >C<), 57.61 (CH$_2$), 86.73 (C$_5$H$_5$), 111.4, 112.74 (both CN), 129.4 (C–4), 129.24, 130.74 (C–2,3,5,6), 134.05 (C–1), 135.48 (=C<), 164.4 (FeC), 213.55 (CO) [83]
*96	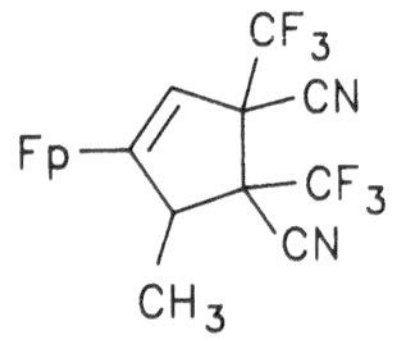	brown oil [83] ^{1}H NMR (CDCl$_3$): 1.5 (br), 3.5 (s) (CH, CH$_3$), 4.9 (br, C$_5$H$_5$) [83] ^{19}F NMR (CDCl$_3$): 66.23 (br, s), 73.14 (br, s) [83] IR (CH$_2$Cl$_2$): 1981, 2031 (both CO) [83]
*97	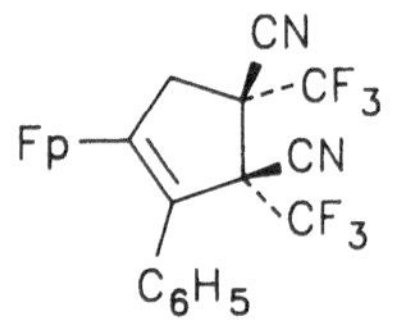	^{1}H NMR (CDCl$_3$): 3.28, 3.38 (CH$_2$, J(H,H)=17), 4.66 (s, C$_5$H$_5$), 7.36 (br, C$_6$H$_5$) [83] ^{19}F NMR (CDCl$_3$): 67.67 (q), 68.26 (q, J(F,F)=9.8) [62, 83]

References on pp. 268/72

Table 8 (continued)

No.	compound method of preparation (yield)	properties and remarks

*98

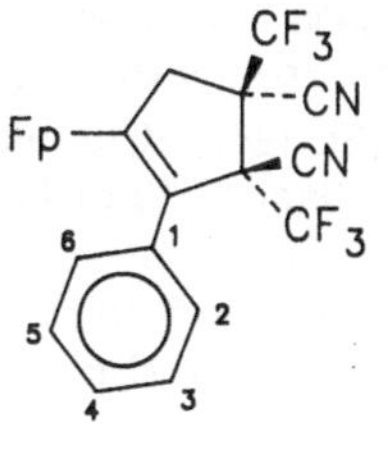

yellow, m.p. 144.7 to 145.3° [83]
^{1}H NMR (CDCl$_3$): 3.47 (s, CH$_2$), 4.66 (s, C$_5$H$_5$), 7.36 (br, C$_6$H$_5$) [83]
^{13}C NMR (CDCl$_3$): 53.32 (J(C,F)=29), 62.5 (possibly q, J(C,F)=30), both >C<, 56.54 (CH$_2$), 85.45 (C$_5$H$_5$), 112.20, 113.52 (both CN), 122.71 (J(C,F)=287), 123.19 (J(C,F)=285), both CF$_3$, 128.55 (C-4), 128.33, 131.58 (C-2,3,5,6), 136.37 (C-1), 135.54 (=C<), 160.21 (FeC=), 212.68, 213.45 (both CO) [83]
^{19}F NMR (CDCl$_3$): 68.62 (s), 71.06 (d, J(H,F)=1.4) [62, 83]
IR: 1983, 2033 (both CO) in CH$_2$Cl$_2$ [62, 83], 2254 (CN) in KBr [83]

*99

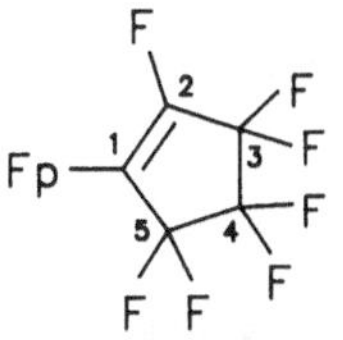

I (14%) [28]

m.p. 87° [28]
^{19}F NMR (CH$_3$COCH$_3$, referred to CF$_3$COOH): 22.0 (m, F-5), 40.2 (m, F-3), 47.0 (m, F-2), 51.8 (m, F-4) [28]
IR (C$_5$H$_{12}$): 1972, 1998, 2006, 2050 (all CO), other bands 680 to 1629 [28]

*100

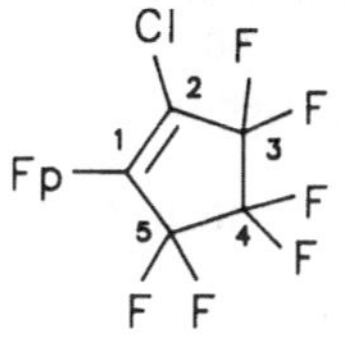

I (10%) [9]

large orange plates, m.p. 99 to 101° [9]
^{19}F NMR (THF): 99.6 (sept, F-5), 112.6 (nonet, F-3), 130.0 (quint, F-4), J(3,4)=J(4,5)=6.82, J(3,5)=3.65 [9]
IR: 2008, 2050 (both CO) in C$_6$H$_{12}$, 663 (δCF), 780, 858 (both νCCl), 840 (C$_5$H$_5$), 985, 1550 (both νC-C), 1000, 1350 (both δCH of C$_5$H$_5$), 1079, 1096, 1181, 1198, 1254, 1268, 1323 (all νCl'), 1675 (νC-C), other bands at 740, 1125 in CS$_2$ [9]

*101

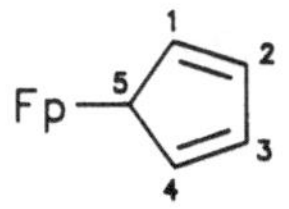

deep red crystals [1], orange-red plates [3], m.p. 46° [1, 3, 4], orange crystals, m.p. 47 to 48° [13]
^{1}H NMR: 4.09 (H-5), 5.13 (C$_5$H$_5$), 6.18 (H-2), 6.53 (H-1), J(C,H-1)=164, J(C,H-2)=162 in CH$_3$COCH$_3$ at −75° [25], 0.6 (σ-C$_5$H$_5$), 2.1 (π-C$_5$H$_5$) in CS$_2$ referred to C$_6$H$_6$ [4]; spectrum depicted at various temperatures in CD$_3$COCD$_3$ [17], CS$_2$ [10], at −85° in ether [25]
^{13}C NMR: 87.1 (π-C$_5$H$_5$), 113.1 (br, σ-C$_5$H$_5$) in CH$_3$COCH$_3$ [34], 80.7±0.3 (br, σ-C$_5$H$_5$, J(^{13}C,H)=161.2±0.6), 106.7±0.3 (π-C$_5$H$_5$, J(^{13}C,H)=180.8±0.5) in acetone referred to CS$_2$ [39], −23.7 (CO), 47.5 (C-1), 71.2 (C-2), 106.9 (π-C$_5$H$_5$), 164.6 (C-5) in CS$_2$/toluene (3:1) at −78° referred to CS$_2$ [32]

References on pp. 268/72

Table 8 (continued)

No.	compound · method of preparation (yield)	properties and remarks
*101 (continued)		IR: 1965, 2013 (both CO), 2907, 3080 (both CH), 3920 (overtone CO) in CCl_4 [4], 1966, 1973, 2016, 2022 in C_6H_{12} [21], 1954, 2031 in KBr [3], 1965, 2015 (all CO), bands 740 to 1785 in Nujol [1], bands 710 to 1110 in CS_2 and 1360 to 1780 in CCl_4 [4] UV: 211 (log ε = 4.36), 320 (log ε = 3.9), 596 (log ε = 2.99) in C_2H_5OH (95%) [3], 318 (log ε = 3.9) in C_6H_{12} [4]
*102	$FpC_5H_4CH_3$	fluxional rearrangement of $C_5H_4CH_3$ ^{1}H NMR (C_6D_6): C_5H_5 region (2 broad peaks, ratio 1:1, AA′BB′ system) depicted in [112] ^{13}C NMR ($CDCl_3$): 15.6 (q, CH_3, $^1J(C,H) = 123$ in gated-decoupled spectrum), 85.6 (d of quint, C_5H_5, $^1J(C,H) = 179$, $^2J(C,H) = ^3J(C,H) = 6$), 215.8 (CO) [112]
*103	$FpC_5H_4CH_2C_6H_5$	probably one of the possible isomers or fluxional rear- rangement dark red oil [3] IR (neat): 1941, 1992 [3]
104	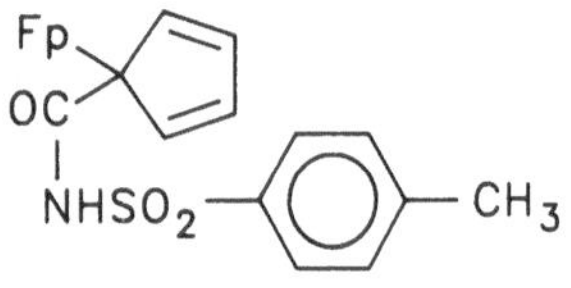	obtained from FpC_5H_5 and $4\text{-}CH_3C_6H_4SO_2NCO$ [69] m.p. 153° [69] ^{1}H NMR (CD_3NO_2): 2.0 to 2.5 (m, NH), 2.45 (s, CH_3), 4.90 (s, C_5H_5), 5.74 (m, 2H, –CH=), 6.58 (m, 2H, –CH=), 7.37 (d, 2H, C_6H_4, J = 8.5), 7.98 (d, 2H, C_6H_4, J = 8.5) [69] IR (CD_3COCD_3): 1960, 2025 (both CO), 1670 (C=O) [69]
*105	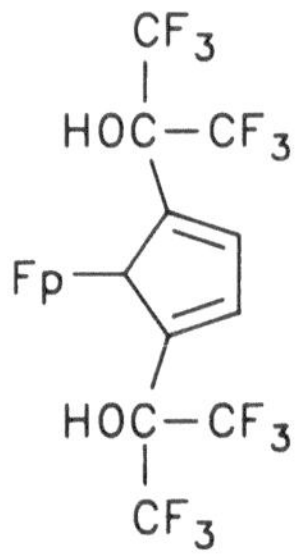	orange crystals, m.p. 97 to 99° [75] ^{1}H NMR ($CDCl_3$): 3.25 (s, 2H), 4.38 (s, C_5H_5), 5.50 (s, 2H), 6.96 (s, 1H) [75] ^{19}F NMR (CH_2Cl_2): 75.30 (q, 6F, CF_3, J(F,F) = 9.8), 76.89 (q, 6F, CF_3, J(F,F) = 9.8) [75] IR (CCl_4): 1973, 2026 (both CO), 1150, 1214, 1235, 1270 (all CF), 3551 (OH) [75]
*106	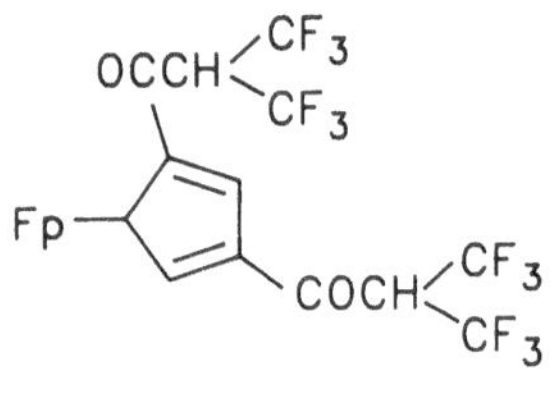	light purple crystals [139] ^{1}H NMR (CD_2Cl_2): 4.72 (hept, 2H, $CH(CF_3)_2$, J(F,H) = 7.6), 5.09 (s, C_5H_5), 6.21 (s, 2H, C_5H_3), 7.56 (s, 1H, C_5H_3) [139] ^{13}C NMR (CD_2Cl_2): 85.73 (C_5H_5), 177.85 (C=O), 210.62 (CO) [139] IR (CH_2Cl_2): 2017, 2057 (both CO), 1636 (C=O) [139]

References on pp. 268/72

Table 8 (continued)

No.	compound method of preparation (yield)	properties and remarks
*107	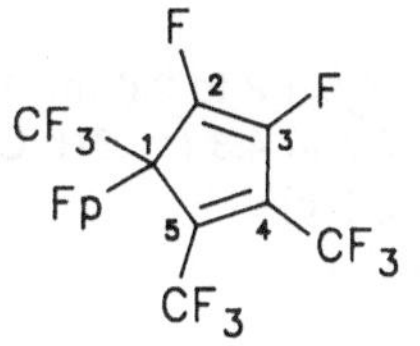	yellow, waxy solid, m.p. 40 to 42° [24] ^{1}H NMR: 4.8 (C_5H_5) [24] ^{19}F NMR: 60.5 (m, CF_3 at C-5), 62.9 (m, CF_3 at C-1), 63.9 (m, CF_3 at C-4) (intensity ratio 1:1:1), 84.3 (m, F-3), 87.5 (m, F-2), J(2,3)=33.5, J(3,4)=9.4, J(4,5)=5.7, J(1,5)=3.2, J(1,2)=7.0 [24] IR (mull): bands from 510 to 3140 given, intense bands at 1650 and 1780 indicate conjugated C=C bonds [24]
108	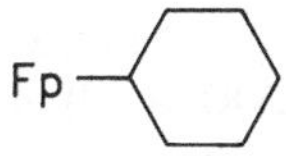 Fp—	treatment with $CuCl_2 \cdot 2H_2O$ (1:3 mole ratio) in R^1OH ($R^1=CH_3$, C_2H_5) at 0° for 1 h leads to rapid precipitation of CuCl and formation of the corresponding esters cyclo-$C_6H_{11}COOR^1$, yields are improved in the presence of CO [42]
*109	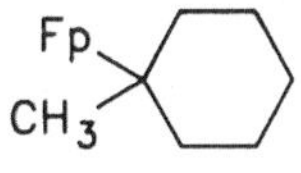	oil [98] ^{1}H NMR ($CDCl_3$): 1.52 (s, CH_3), 1.0 to 2.1 (m, CH_2), 4.63 (s, C_5H_5) [98] IR (petroleum ether): 1940, 1995 (both CO) [98] III (40%) [98]
*110	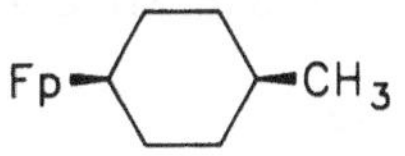	yellow oil [42, 105], thermally unstable [105] ^{1}H NMR: 1.0 (d, CH_3, J=7), 1.3 to 2.1 (m, 9H, CH, CH_2), 2.6 (m, FeCH), 4.6 (s, C_5H_5) in CS_2 [42], 1.00 (d, CH_3, J=7), 1.20 to 2.10 (m, CH, CH_2), 2.63 (m, FeCH), 4.67 (s, C_5H_5) in $CDCl_3$ [105] I (26%) [42], (22%) [105]
*111	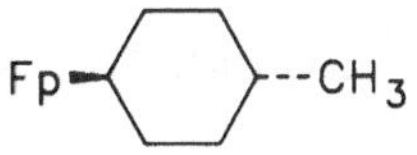	^{1}H NMR (CS_2): 0.8 (d, CH_3, J=5), 0.9 to 2.7 (m, 10H, CH, CH_2), 4.6 (s, C_5H_5) [42] I [42]
*112	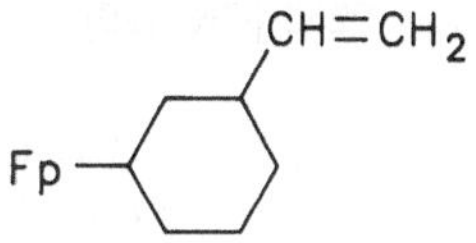	—
*113	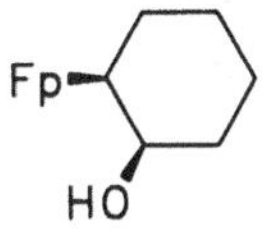	yellow solid [38] ^{1}H NMR: 3.10 (br m, CHO) [38]
*114	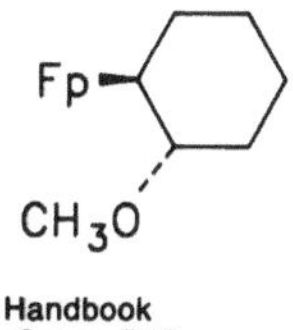	—

References on pp. 268/72

Table 8 (continued)

No.	compound method of preparation (yield)	properties and remarks
115	Fp — cyclohexene with CH bearing C₂H₅OOC and COCH₃ substituent C_2H_5OOC $COCH_3$ VII (76%) [46, 78]	oil, air-stable [78] ^{1}H NMR (CS$_2$): 1.20, 1.25 (2t, CH$_3$), 1.2 to 3.0 (m, 10H, CH, CH$_2$), 2.05 (2s, COCH$_3$), 3.7 to 4.3 (m, 3H, OCH$_2$, CHCOO), 4.69, 4.72 (2s, C$_5$H$_5$) [78] IR (neat): 1938, 1996 (both CO) [78]
116	Fp — cyclohexene with CH bearing C₂H₅OOC and COOC₂H₅ substituent C_2H_5OOC $COOC_2H_5$ VII (65%) [46, 78]	m.p. 69 to 71° [46, 78], air-stable [78] ^{1}H NMR (CS$_2$): 1.17, 1.23 (2t, 6H, CH$_3$), 1.2 to 2.3 (m, 10H, CH, CH$_2$), 3.8 (d, 1H, CHCOO, J=4), 4.0, 4.08 (2q, 4H, OCH$_2$), 4.72 (s, C$_5$H$_5$) [78] IR (KBr): 1942, 2000 (both CO) [78] reacts with [C(C$_6$H$_5$)$_3$]$^+$ at 0° to give the unsubstituted complex [Fp(η^2-C$_6$H$_{10}$-cyclo)]$^+$ [78]
117	Fp — cyclohexene with CH bearing NC and COOC₂H₅ substituent NC $COOC_2H_5$ VII (94%) [78]	m.p. 132 to 133°, air-stable [78] ^{1}H NMR (CS$_2$): 1.30 (t, CH$_3$), 1.2 to 2.5 (m, 10H, CH, CH$_2$), 4.26 (q, 2H, OCH$_2$, J=7), 4.26 (d, 1H, CHCN, J=2), 4.77 (s, C$_5$H$_5$) [78] IR (KBr): 1930, 1992 (both CO) [78]
118	Fp — cyclohexane with CN, CN, CN, CN substituents	obtained from FpCH$_2$C$_3$H$_5$-cyclo and (NC)$_2$C=C(CN)$_2$ [29] ^{1}H NMR (CD$_3$NO$_2$): 2.0 to 3.0 (m, 7H, CH, CH$_2$), 5.03 (s, C$_5$H$_5$) [29] IR (KBr): 1945, 2009 (both CO), 2250 (CN) [29]
*119	Fp — cyclohexene (positions 1-6 labeled) I [48], (91%) [69] II [48, 60], (100%) [37]	^{1}H NMR (CDCl$_3$): 1.34 to 2.34 (m, 6H, CH$_2$), 3.14 to 3.50 (m, 1H, FpCH), 4.68 (s, C$_5$H$_5$), 5.20 to 5.54 (m, 1H, FpCC=CH), 5.68 to 6.04 (m, 1H, FpCCH=) [69] IR (CH$_2$Cl$_2$): 1947, 1998 (both CO) [69]
*120	Fp — cyclohexene with CH₃O substituent CH_3O	—

Table 8 (continued)

No.	compound method of preparation (yield)	properties and remarks
*121	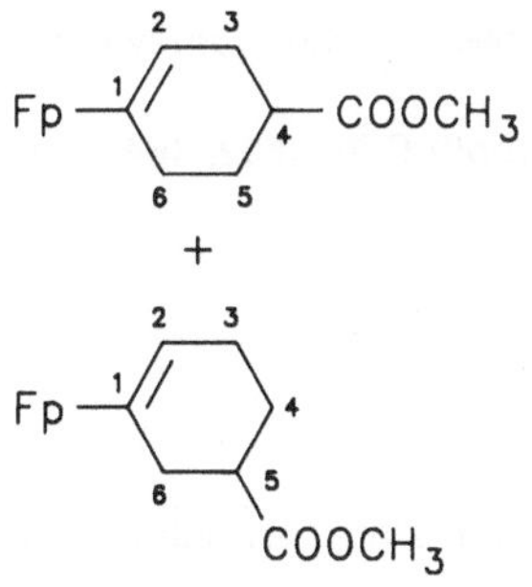	brown oil, equimolar mixture of the two isomers and Fp$_2$ [138] ^{1}H NMR (CS$_2$): 2.30 (H-3,4,5), 3.49, 3.50 (both OCH$_3$), 5.40 (H-2), 5.60, 5.61 (both C$_5$H$_5$) [138]
*122	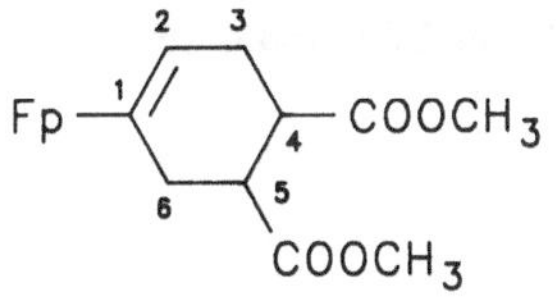	yellow, m.p. 104° [138] ^{1}H NMR (CDCl$_3$): 2.40 (H-3,6), 2.75 (H-4,5), 3.65, 3.67 (both OCH$_3$), 4.72 (C$_5$H$_5$), 5.50 (H-2) [138] IR (KBr): 1935, 2005 (both CO), 1720, 1740 (both C=O) [138]
123	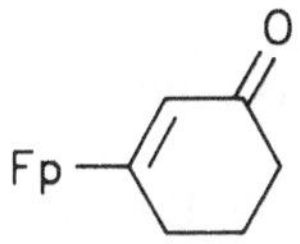 I (61%) [111]	m.p. 92 to 94°, air-stable as solid, somewhat light-sensitive [111] IR (CH$_2$Cl$_2$): 1970, 2017 (both CO), 1538, 1644 [111]
124	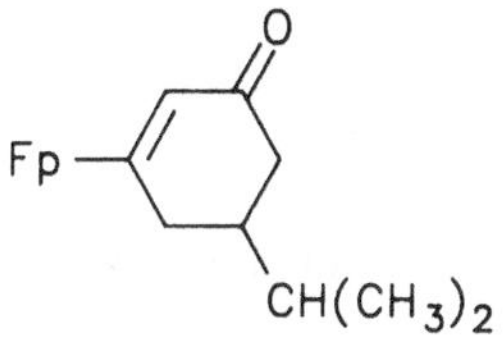 I (40%) [111]	m.p. 100 to 101°, air-stable as solid, somewhat light-sensitive [111] IR (CH$_2$Cl$_2$): 1970, 2024 (both CO), 1543, 1638 [111]
125	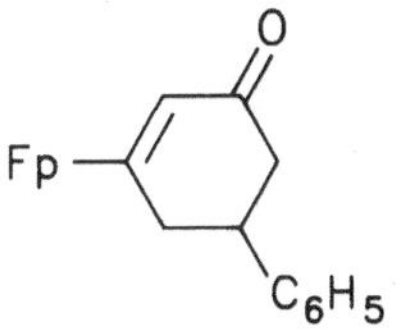 I (66%) [111]	m.p. 118 to 119°, air-stable as solid, somewhat light-sensitive [111] IR (CH$_2$Cl$_2$): 1974, 2025 (both CO), 1545, 1641 [111]
126	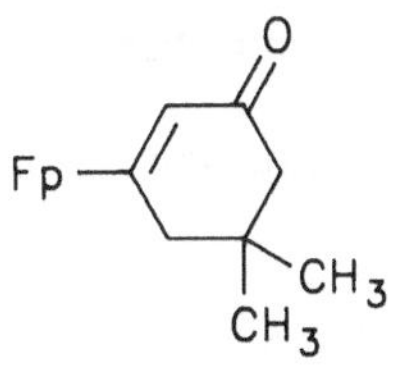 I (76%) [111]	m.p. 115°, air-stable as solid, somewhat light-sensitive [111] IR (CH$_2$Cl$_2$): 1968, 2021 (both CO), 1542, 1643 [111]

     References on pp. 268/72

Table 8 (continued)

No.	compound method of preparation (yield)	properties and remarks
127	CH₃ structure	m.p. 115 to 117°, air-stable as solid, somewhat light-sensitive [111] IR (CH₂Cl₂): 1964, 2013 (both CO), 1552, 1640 [111]
	I (55%) [111]	
*128	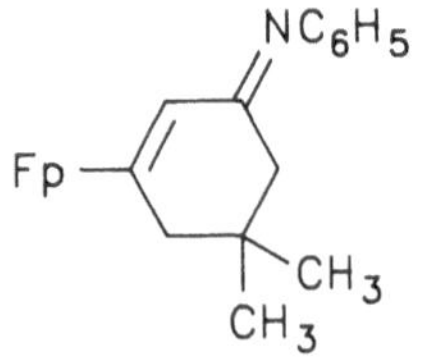 NC₆H₅ structure	yellow, m.p. 115 to 116°, air-stable as solid, somewhat light-sensitive [111] IR (CH₂Cl₂): 1963, 2011 (both CO), 1555, 1591 [111]
*129	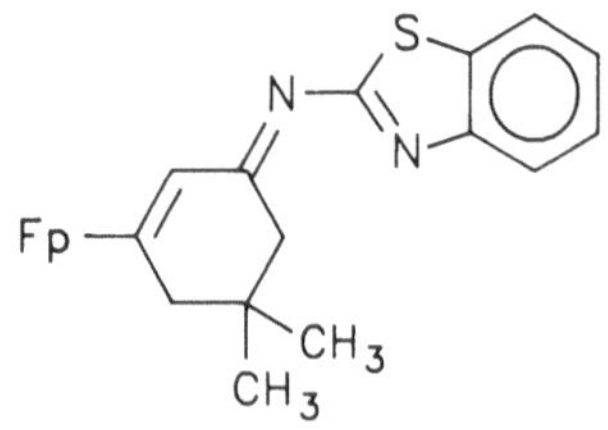 benzothiazole structure	yellow, m.p. 146 to 147°, air-stable as solid, somewhat light-sensitive [111] IR (CH₂Cl₂): 1971, 2019 (both CO), 1480, 1540, 1580, 1601 [111]
*130	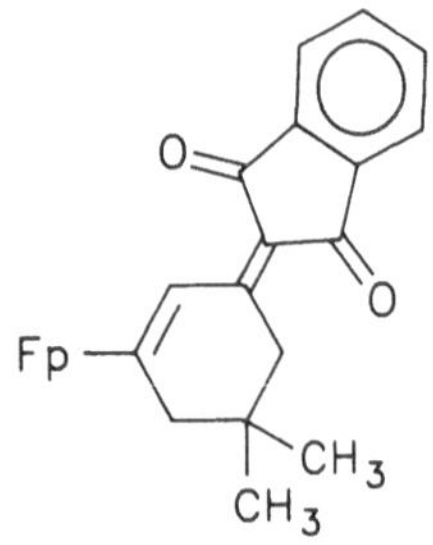 indanedione structure	m.p. 190 to 191°, air-stable as solid, somewhat light-sensitive [111] IR (CH₂Cl₂): 1971, 2018 (both CO), 1510, 1663 [111]
*131	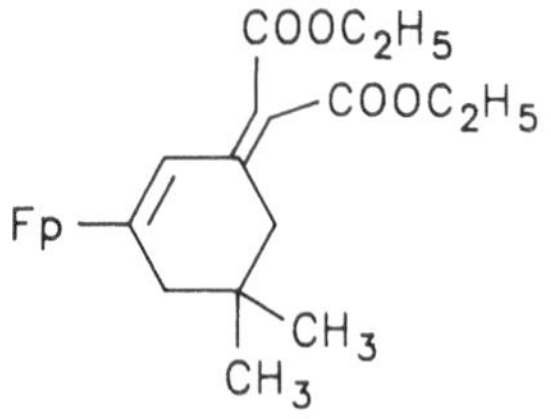 COOC₂H₅ structure	m.p. 76 to 77°, air-stable as solid, somewhat light-sensitive [111] IR (CH₂Cl₂): 1968, 2020 (both CO), 1220, 1586, 1707 [111]

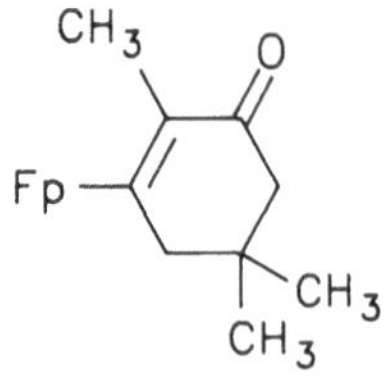

References on pp. 268/72

Table 8 (continued)

No.	compound method of preparation (yield)	properties and remarks

*132

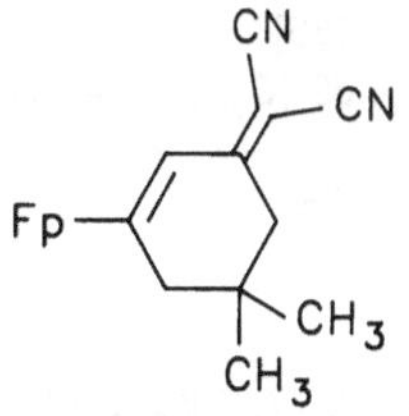

m.p. 171 to 172°, air-stable as solid, somewhat
   light-sensitive [111]
IR (CH$_2$Cl$_2$): 1976, 2022 (both CO), 1526, 2218 [111]

*133

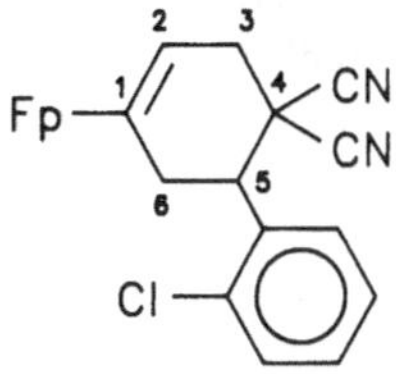

yellow, m.p. 145 to 146° [138]
^{1}H NMR (CDCl$_3$): 2.80 (H-3,6), 4.05 (H-5), 4.85 (C$_5$H$_5$),
   5.57 (H-2), 7.30 (C$_6$H$_4$) [138]
IR (KBr): 1955, 2007 (both CO), 2253, 2300 (both CN)
   [138]

134

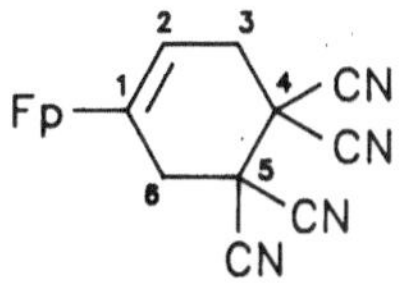

IVa (93%) [138]

yellow, m.p. 228 to 229° [138]
^{1}H NMR (CD$_3$COCD$_3$): 3.25 (H-6), 3.50 (H-3), 5.17
   (C$_5$H$_5$), 5.72 (-CH=) [138]
IR (KBr): 1970, 2020 (both CO), 2320, 2365 (both CN)
   [138]

135

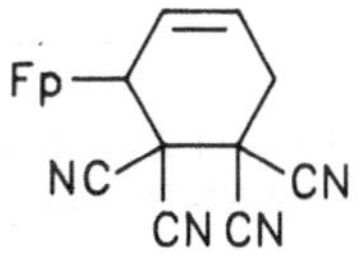

IVa [138]

green crystals contaminated with (NC)$_2$C=C(CN)$_2$ [138]
^{1}H NMR (CD$_3$COCD$_3$): 3.12 (CH$_2$), 3.67 (FeCH), 5.35
   (C$_5$H$_5$), 6.04, 6.20 (both -CH=) [138]
IR (KBr): 1960, 2010 (both CO) [138]

136

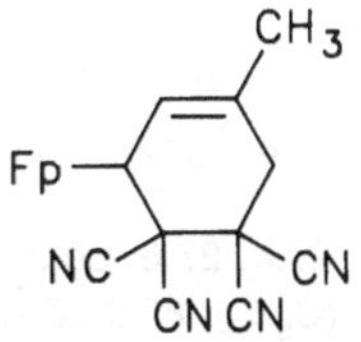

IVa (79%) [138]

yellow powder [138]
^{1}H NMR (CD$_3$COCD$_3$): 1.76 (CH$_3$), 2.95 (CH$_2$), 3.65
   (FeCH), 5.15 (C$_5$H$_5$), 5.75 (-CH=) [138]
IR (KBr): 1960, 2005 (both CO) [138]

*137

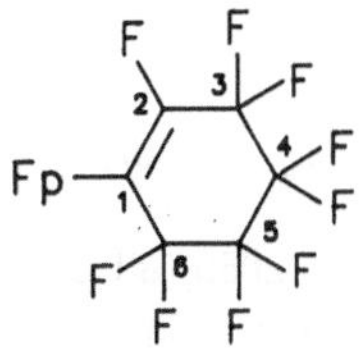

orange crystals, m.p. 97 to 98° [8]
^{19}F NMR (CS$_2$): 99.0 (F-6), 104.5 (F-2), 124.1 (F-3),
   141.0 (F-5), 143.5 (F-4) [8]
IR: 2012, 2055 (both CO) in C$_6$H$_{12}$, bands from 750 to
   1635 in CS$_2$ [8]

References on pp. 268/72

Table 8 (continued)

No.	compound method of preparation (yield)	properties and remarks
*138	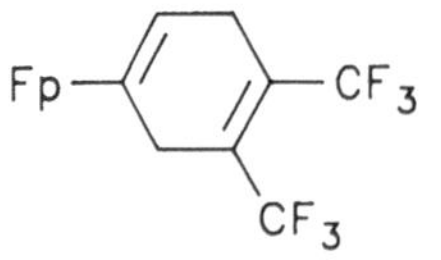 Fp—CF$_3$ / CF$_3$	bright yellow crystals, m.p. 65° [138] ^{1}H NMR (CS$_2$): 2.95 (4H, CH$_2$), 4.69 (C$_5$H$_5$), 5.40 (–CH=) [138] IR (KBr): 1945, 2010 (both CO) [138]
*139	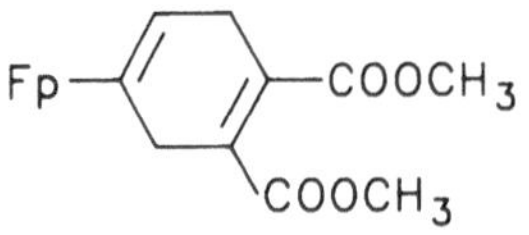 Fp—COOCH$_3$ / COOCH$_3$	yellow, m.p. 106° [138] ^{1}H NMR (CDCl$_3$): 3.10 (4H, CH$_2$), 3.77, 3.83 (both OCH$_3$), 4.80 (C$_5$H$_5$), 5.52 (–CH=) [138] IR (KBr): 1935, 2015 (both CO) [138]
*140	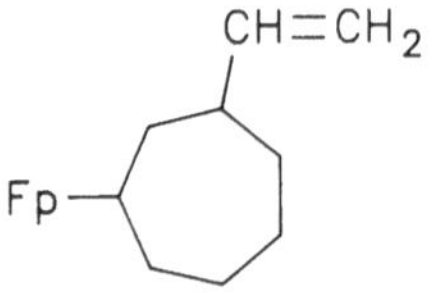 CH=CH$_2$ / Fp—	—
*141	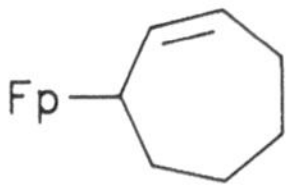 Fp— I [48, 60]	—
*142	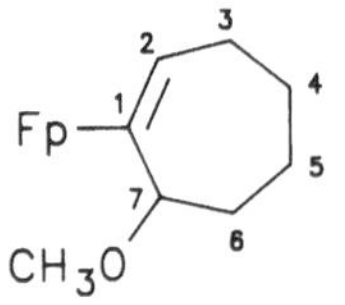 Fp—(ring, positions 1–7) CH$_3$O III (53%) [136]	brown oil, air-sensitive [136] ^{1}H NMR (CDCl$_3$): 0.80 to 2.50 (m, 8H, CH$_2$), 3.30 (s, 3H, OCH$_3$), 3.8 (m, 1H, OCH<), 4.75 (s, C$_5$H$_5$), 5.80 (t, 1H, –CH=) [136] ^{13}C{^{1}H} NMR (CDCl$_3$): 27.2, 27.5, 29.6, 31.7 (C–3 to C–6), 56.1 (OCH$_3$), 85.4 (C$_5$H$_5$), 90.3 (C–7), 140.7 (C–2), 147.8 (C–1), 216.7 (CO) [136] IR (neat): complete spectrum from 575 to 3110 [136]
*143	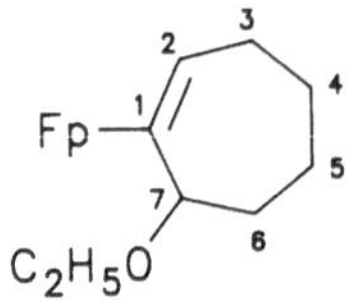 Fp—(ring, positions 1–7) C$_2$H$_5$O	brown oil [136] ^{1}H NMR (CDCl$_3$): 1.55 (m, CH$_2$, CH$_3$), 3.45 (q, 2H, OCH$_2$), 4.00 (m, 1H, OCH<), 4.80 (s, C$_5$H$_5$), 5.81 (t, 1H, –CH=) [136] ^{13}C{^{1}H} NMR (CDCl$_3$): 15.7 (CH$_3$), 27.4, 27.6, 30.3, 31.7 (C–3 to C–6), 63.4 (OCH$_2$), 85.3 (C$_5$H$_5$), 87.9 (C–7), 140 (C–2), 148.9 (C–1), 216.8 (CO) [136] IR (neat): complete spectrum from 570 to 3100 [136]
*144	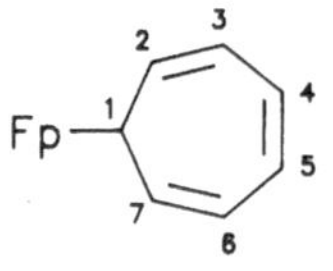 Fp—(ring, positions 1–7)	see also No. 145 red crystals, m.p. 98 to 99° [22] ^{1}H NMR (C$_6$D$_5$CD$_3$, 30°): 4.1 (s, C$_5$H$_5$), 4.8 (br, ~18 Hz at half-height, C$_7$H$_7$) [22] IR: 1930 br in CHCl$_3$, 1920 br, 2010, 2055 (all CO) in CCl$_4$, bands from 525 to 3020 in CS$_2$ [22]

References on pp. 268/72

Table 8 (continued)

No.	compound method of preparation (yield)	properties and remarks

*145

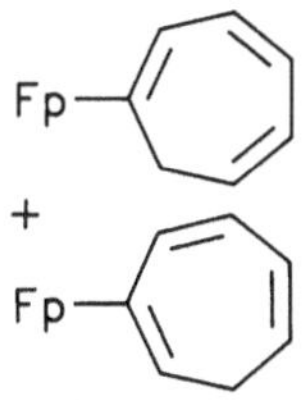

yellow oil, air-sensitive, mixture of isomers, contains also No. 144 [119]
^{1}H NMR (CDCl$_3$): 2.15, 2.25, 2.62 (2t, d, 2H, saturated CH), 4.65, 4.79, 4.84 (3s, C$_5$H$_5$), 4.98 to 5.50, 5.78 to 6.83 (2br m, 5H, =CH–) [85, 119]
IR (neat): 1950, 2020 (both CO) [85], bands at 725, 830, 1015, 1430, 1495, 3020 [119]

*146

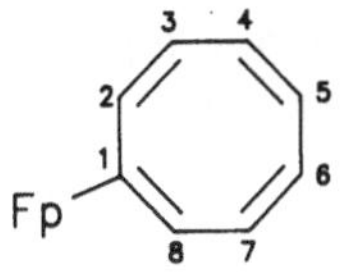

I [125]

yellow crystals, m.p. ~30° [58], yellow oil [125]
^{1}H NMR: 4.16 (s, C$_5$H$_5$), 5.33 (d, 1H, J=10), 5.6 to 6.1 (m, 5H), 6.23 (d, 1H, J=11) in C$_6$D$_6$ [58], 4.81 (s, C$_5$H$_5$), 5.22 (dd, H–7, J(7,8)=11.1, J(6,7)=3.4), 5.51 (dd, H–5, J(5,6)=10.9, J(4,5)=3.2), 5.68 (d, H–2, J(2,3)=2.8), 5.75 (dd, H–6), 5.77 to 5.85 (m, H–3,4), 6.16 (d, H–8) in CD$_2$Cl$_2$ [125]
^{13}C NMR (CD$_3$COCD$_3$): 87.2 (C$_5$H$_5$), 119.3, 129.2, 132.6 (C–2), 136.7, 139.3, 147.7, 151.7 (C–1), 217.2, 217.7 (both CO) [125]
IR: 1970, 1976 sh, 2021, 2024 sh in C$_6$H$_{14}$ [58], 1925, 1990 (all CO), other bands at 702, 714, 827, 909, 1005, 1400, 1560, 2960 (neat) [125]

*147

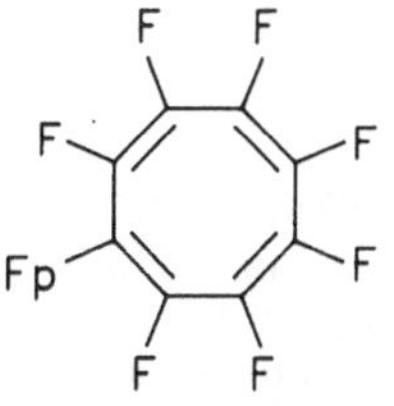

I [142], (40%) [128]

yellow crystals, m.p. 92°, sublimable [128]
^{13}C{^{19}F} NMR (CDCl$_3$): 85.51 (d, C$_5$H$_5$, J(C,H)=181), 126.07 (FeC=), 127.60, 136.30, 137.73, 138.84, 140.24, 149.81, 152.87 (all =CF), 212.45, 212.53 (both CO) [128]
^{19}F NMR (CDCl$_3$): 74.55, 93.82, 119.03, 124.84, 128.97, 130.83, 140.29 (all m of equal intensity) [128]
IR (C$_6$H$_{14}$): 1997, 2041 (both CO) [128]

*148

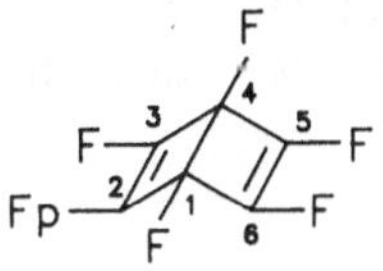

I (62%) [16], (56%) [57]

yellow, m.p. 58 to 60° (from C$_5$H$_{12}$ at −20° [57]) [16], 56 to 57° [15]
^{19}F NMR: 107.5 (F–3), 123.0 (F–6), 129.2 (F–5), 174.9 (F–1), 187.9 (F–4) in CHCl$_3$ [16], 29.8 (F–3), 45.2 (F–6), 46.0 (F–5), 97.6 (F–1), 109.0 (F–4), referred to CF$_3$COOH [15], signals of equal intensity [15, 16]
IR: 1999, 2041 (both CO) in C$_6$H$_{12}$, 1603 (FeC=C), 1740 (FC=CF), bands from 652 to 1500 in CHCl$_3$ [16], 1984, 2030 (both CO), 1600 (FeC=C), 1740 (FC=CF) [15]

149

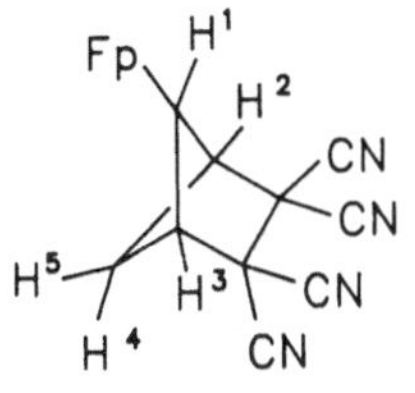

IVb [48, 69]

dec. 178° [69]
^{1}H NMR (CD$_3$NO$_2$): 1.66 (t, 1H, H–4, J=10.0), 2.46 (d, 1H, H–1, J=10.0), 2.92 (dt, 1H, H–5, J=2.5,10), 3.22 (d, 2H, H–2,3, J=2.5), 5.07 (s, C$_5$H$_5$) [69]
IR (KBr): 1941, 2001 (both CO) [69]

References on pp. 268/72

Table 8 (continued)

No.	compound method of preparation (yield)	properties and remarks
*150	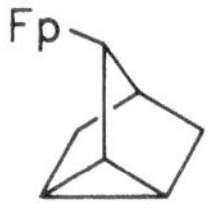 Fp	—
*151	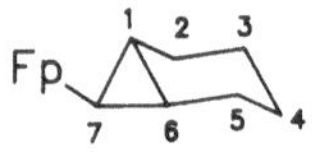 Fp I [31], III (52%) [116]	orange-brown liquid, air-stable [116] ^1H NMR (CDCl$_3$): 0.56, 1.2, 1.7 (3m), 4.70 (s, C$_5$H$_5$) [116] ^{13}C{^{1}H} NMR (CD$_3$COCD$_3$): 4.6 (C-7), 21.4, 22.1, 26.2 (C-1 to C-6), 86.3 (C$_5$H$_5$), 205.8 (CO) [116] IR (neat): spectrum from 720 to 2930 [116]
*152	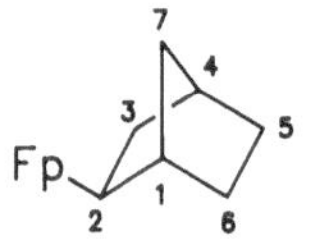 Fp III (44%) [141]	amber oil [141] ^1H NMR (CDCl$_3$): 1.21 (m, 3H), 1.61 (m, 5H), 2.07 (br s, >CH-, bridgehead), 2.21 (br s, >CH-, bridgehead), 2.55 (m, FeCH), 4.78 (s, C$_5$H$_5$) [141] ^{13}C{^{1}H} NMR (20% CD$_2$Cl$_2$/CH$_2$Cl$_2$): 27.4 (C-2), 29.0 (C-6), 35.0 (C-5), 37.2 (C-7), 39.6 (C-4), 47.9 (C-3), 49.1 (C-1), 86.9 (C$_5$H$_5$), 218.6, 218.9 (both CO) [141] IR: 1939, 1997 (both CO) [141] mass spectrum: [M]$^+$, [M−CO]$^+$, [M−2CO]$^+$, [M− Fp]$^+$ [141]
*153	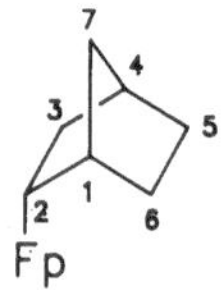 Fp III (83%) [141]	amber oil [141] ^1H NMR (CDCl$_3$): 1.45 (m, 8H), 2.03 (br s, 2H, >CH-, bridgehead), 2.83 (m, FeCH), 4.67 (s, C$_5$H$_5$) [141] ^{13}C{^{1}H} NMR (20% CD$_2$Cl$_2$/CH$_2$Cl$_2$): 27.0 (C-2), 29.0 (C-6), 30.1 (C-5), 37.0 (C-3), 40.6 (C-4), 43.9 (C-7), 47.9 (C-6), 86.0 (C$_5$H$_5$), 218.5, 218.9 (both CO) [141] IR (CHCl$_3$): 1940, 1998 (both CO) [141] mass spectrum: [M]$^+$, [M−CO]$^+$, [M−2CO]$^+$, [M− Fp]$^+$ [141]
*154	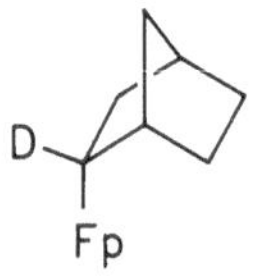 D Fp III (77%) [141]	—
*155	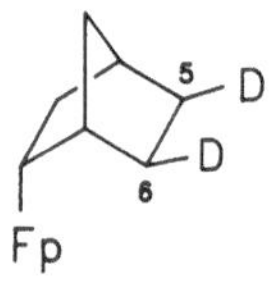 D, D Fp III (89%) [141]	—

References on pp. 268/72

Table 8 (continued)

No.	compound method of preparation (yield)	properties and remarks

156

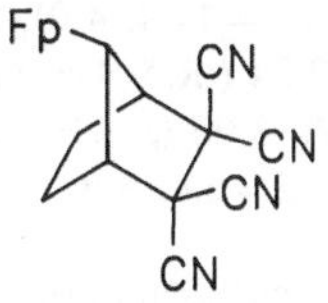

IVb [48, 69]

dec. 137° [69]
^{1}H NMR (CD$_3$NO$_2$): 1.51 to 2.35 (m, 4H, CH$_2$), 2.91 (br s, 1H, FeCH), 3.13 (br s, 2H, CCHC), 5.09 (s, C$_5$H$_5$) [69]
IR (KBr): 1945, 2010 (both CO) [69]

157

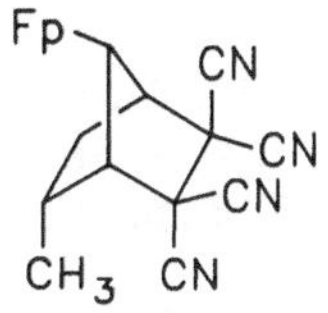

IVb [48, 69]

dec. 55° [69]
^{1}H NMR (CD$_3$NO$_2$): 1.38 (d, 3H, CH$_3$, J = 6.5), 1.7 to 2.5 (m, 2H, CH$_2$), 2.50 to 3.18 (m, 4H, CH), 5.07 (s, C$_5$H$_5$) [69]
IR (KBr): 1945, 2005 (both CO) [69]

158

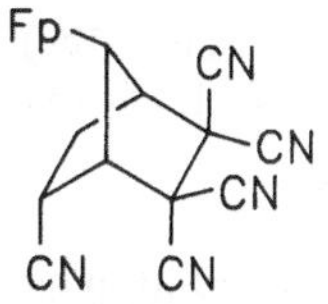

IVb [48, 60, 69]

obtained as CH$_2$Cl$_2$ monosolvate, dec. 122° [69]
^{1}H NMR (CD$_3$NO$_2$): 1.8 to 2.9 (m, 2H, CH$_2$), 2.95 to 3.07 (m, 1H, FeCH), 3.07 to 3.43 (m, 2H, CH), 3.71 (ddd, 1H, CHCN, J = 12.0, 7.0, 3.5), 5.14 (s, C$_5$H$_5$), 5.43 (s, 2H, CH$_2$Cl$_2$) [69]
IR (KBr): 1950, 2010 (both CO) [69]

159

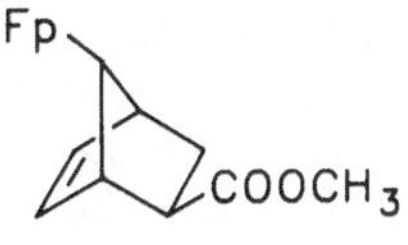

Vb (43%) [130]

^{1}H NMR (CDCl$_3$): 1.48 (dd, 1H, CH, J = 10.1, 11.0), 1.83 (ddd, 1H, CH, J = 4.0, 3.8, 11), 2.29 (dd, 1H, CH, J = 4.4, 10.0), 2.72 (s, 1H, CH), 2.77 (s, 1H, CH), 2.91 (s, 1H, CH), 3.69 (s, 3H, COOCH$_3$), 4.71 (s, C$_5$H$_5$), 5.99 (m, 1H, HC=), 6.03 (m, 1H, HC=) [130]
IR (CH$_2$Cl$_2$): 1944, 2003 (both CO) [130]

160

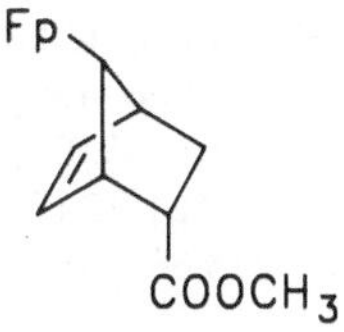

Vb (42%) [130]

^{1}H NMR (CDCl$_3$): 1.50 (dd, 1H, J = 4.1, 11.5), 1.87 (ddd, 1H, CH, J = 3.7, 8.8, 11.5), 2.51 (s, 1H, CH), 2.77 (s, 1H, CH), 2.90 (ddd, 1H, CH, J = 4.0, 3.9, 8.8), 3.06 (m, 1H, CH), 3.61 (s, 3H, COOCH$_3$), 4.68 (s, C$_5$H$_5$), 5.79 (dd, 1H, HC=, J = 2.8, 5.6), 6.10 (dd, 1H, HC=, J = 3.1, 5.6) [130]

*161

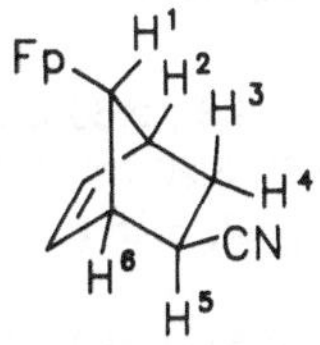

Vb [127], (42%) [130]

^{1}H NMR (CDCl$_3$): 1.59 (t, 1H, H-4, J = 11.3), 1.90 (dd, 1H, H-3, J = 11.3, 3.8), 2.23 (dd, 1H, H-5, J = 9.2, 4.1), 2.70 (s, 1H, H-1), 2.88 (s, 1H, H-2), 3.05 (s, 1H, H-6), 4.73 (s, C$_5$H$_5$), 5.92 (dd, 1H, HC=, J = 5.5, 3.0), 6.05 (dd, 1H, HC=, J = 5.5, 3.1) [130]

References on pp. 268/72

Table 8 (continued)

No.	compound method of preparation (yield)	properties and remarks
*162	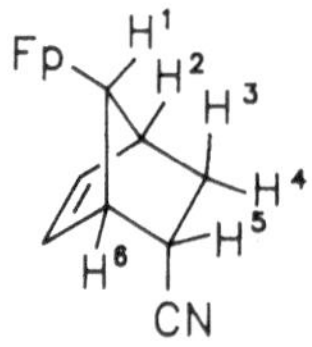 CN Vb [127], (40%) [130]	^{1}H NMR (CDCl$_3$): 1.38 (dd, 1H, H-4, J = 11.4, 4.0), 2.03 (m, 1H, H-3), 2.25 (s, 1H, H-1), 2.74 (ddd, 1H, H-5, J = 9.2, 5.3, 3.9), 2.84 (s, 1H, H-2), 3.06 (s, 1H, H-6), 4.67 (s, C$_5$H$_5$), 6.07 (dd, 1H, HC=, J = 5.4, 2.8), 6.21 (dd, 1H, HC=, J = 5.5, 2.3) [130] IR (CH$_2$Cl$_2$): 1949, 2005 (both CO), 2230 (CN) [130]
163	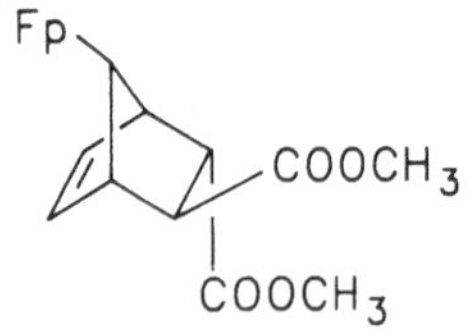 COOCH$_3$ COOCH$_3$ Vb [139], (90%) [126, 130], (73%) [130]	yellow solid [126] ^{1}H NMR (CDCl$_3$): 2.74 (s, 1H, CH), 2.78, 3.01, 3.12, 3.26 (all m, 4H, CH), 3.62 (s, 3H, CH$_3$), 3.71 (s, 3H, CH$_3$), 4.71 (s, C$_5$H$_5$), 5.86 (m, 1H, CH=), 6.18 (m, 1H, CH=) [126, 130] ^{13}C NMR (CDCl$_3$): 45.21, 47.81, 48.78 (all CH), 51.46, 51.69 (both CH$_3$), 55.96, 57.98 (both CH), 85.61 (C$_5$H$_5$), 134.80, 137.47 (both CH=), 173.22, 175.40 (both C=O), 217.05, 217.20 (both CO) [126, 130] IR (CH$_2$Cl$_2$): 1949, 2005 (both CO), 1728 [126, 130]
164	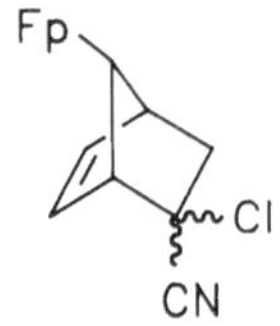 CN Vb (87%) [126, 130]	yellow oil, mixture of epimers [130] major isomer: ^{1}H NMR (CDCl$_3$): 1.74 (d, 1H, CH$_2$, J = 12.8), 2.68 (dd, 1H, CH$_2$, J = 3.7, 12.8), 2.83 (s, 1H, CH), 2.88 (m, 1H, CH), 3.28 (m, 1H, CH), 4.74 (s, C$_5$H$_5$), 6.01 (m, 1H, CH=), 6.29 (m, 1H, CH=) [126, 130] ^{13}C NMR (CDCl$_3$): 44.56 (CH), 47.23 (CH$_2$), 53.07 (CH), 55.11 (CCl), 65.75 (CH), 85.69 (C$_5$H$_5$), 123.90 (CN), 131.96, 139.44 (both CH=), 216.51, 216.66 (both CO) [126, 130] minor isomer: ^{1}H NMR (CDCl$_3$): 2.29 (d, 2H, J = 2.6), 2.88 (s, 1H, CH), 3.05 (s, 1H, CH), 3.12 (m, 1H, CH), 4.74 (s, C$_5$H$_5$), 6.11 (m, 1H, CH=), 6.31 (m, 1H, CH=) [126, 130] ^{13}C NMR (CDCl$_3$): 42.37 (CH), 47.97 (CH$_2$), 53.07 (CH), 55.11 (CCl), 66.48 (CH), 85.69 (C$_5$H$_5$), 121.22 (CN), 132.78, 142.20 (both CH=), 216.51, 216.66 (both CO) [126, 130] both isomers: IR (CH$_2$Cl$_2$): 1960, 2009 (both CO) [126, 130]
*165	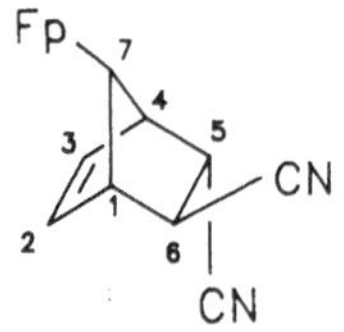 CN CN Vb (85%) [132]	yellow, m.p. 124 to 124.5° [84] ^{1}H NMR (CDCl$_3$): 2.55 (d, 1H, H-6, J = 3.8), 2.60 (m, 1H, H-7), 3.04 (dd, 1H, H-5, J = 3.9, 3.8), 3.20 (m, 1H, H-1), 3.25 (m, 1H, H-4), 4.74 (s, C$_5$H$_5$), 6.25 (m, 2H, H-2,3) [132], see also [84] IR: 1958, 2013 (both CO) in CH$_2$Cl$_2$, 1658 (C=C), 2234 (CN) in KBr [84]

References on pp. 268/72

Table 8 (continued)

No.	compound method of preparation (yield)	properties and remarks

*166 Fp

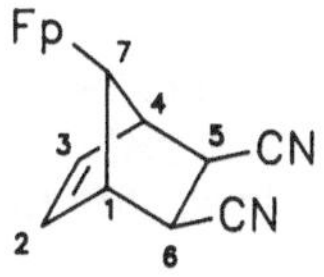

Vb [134], (49%) [132]

yellow solid [132]
^{1}H NMR (CDCl$_3$): 2.69 (s, 2H, H-5,6), 2.88 (m, 1H, H-7), 3.20 (m, 2H, H-1,4), 4.77 (s, C$_5$H$_5$), 6.08 (m, 2H, H-2,3) [132]
IR (KBr): 1943, 2003 (both CO), 2234, 2240 (both CN) [132]
UV (C$_6$H$_6$): 351 (ε=866) [132]

*167 Fp

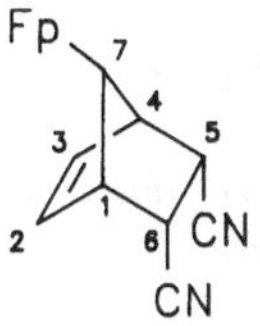

Vb [134], (45%) [132]

yellow solid [132]
^{1}H NMR (CDCl$_3$): 2.18 (m, 1H, H-7), 3.15 (m, 2H, H-5,6), 3.22 (m, 2H, H-1,4), 4.70 (s, C$_5$H$_5$), 6.37 (m, 2H, H-2,3) [132]
IR (KBr): 1947, 2004 (both CO), 2236, 2241 (both CN) [132]
UV (C$_6$H$_6$): 351 (ε=866) [132]

168 Fp

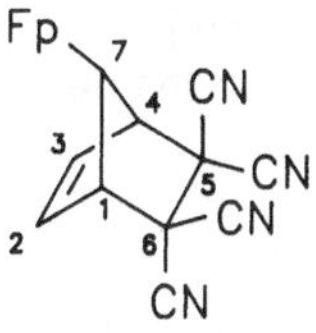

IVb [69]

m.p. 108° (dec.) [69]
^{1}H NMR (CD$_2$Cl$_2$): 2.87 (m, 1H, H-7), 3.88 (m, 2H, H-1,4), 4.90 (s, C$_5$H$_5$), 6.60 (m, 2H, H-2,3) [69]
IR (KBr): 1960, 2015 (both CO) [69]

*169 Fp

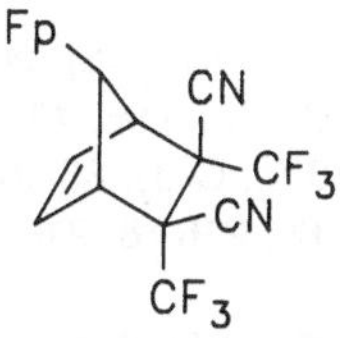

yellow, m.p. 89 to 91° [84]
^{1}H NMR (CDCl$_3$): 2.72 to 4.13 (m, FeCH, CH), 4.81 (s, C$_5$H$_5$), 6.22 to 6.72 (m, CH=CH) [84]
^{19}F NMR: 65.11 (s), 66.44 (s) [84]
IR: 1965, 2016 (both CO) in CH$_2$Cl$_2$, 1585 (C=C), 2234, 2247 (both CN) [84]

*170 Fp

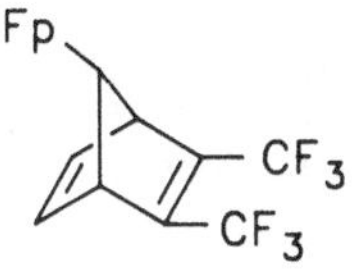

yellow crystals, m.p. 51.5 to 53°, subl. at 40°/0.1 Torr [84]
^{1}H NMR (CDCl$_3$): 3.72 (br, CH), 4.00 (t, FeCH, J(H,H) = 0.5), 4.69 (s, C$_5$H$_5$), 6.78 (t, d, CH=CH, J(H,H)=2, 0.5) [84]
^{19}F NMR: 63.07 (s) [84]
IR: 1967, 2017 (both CO) in CH$_2$Cl$_2$, 1563, 1674 (C=C) in KBr [84]

171 Fp

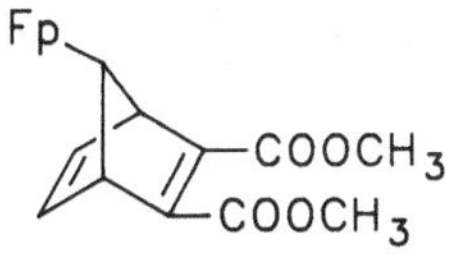

Vb (85%) [126, 130]

^{1}H NMR (CDCl$_3$): 3.76 (s, 6H, CH$_3$), 3.77 (m, 3H, CH), 4.70 (s, C$_5$H$_5$), 6.78 (t, 2H, CH=CH, J=2.0) [126, 130]
^{13}C NMR (CDCl$_3$): 51.78 (CH$_3$), 64.62 (CH), 85.17 (CH), 85.70 (C$_5$H$_5$), 141.66 (-CH=), 155.45 (=C<), 165.96 (C=O), 216.89 (CO) [126, 130]
IR (CH$_2$Cl$_2$): 1951, 2007 (both CO), 1709, 1729 [126]

References on pp. 268/72

Table 8 (continued)

No.	compound method of preparation (yield)	properties and remarks
*172	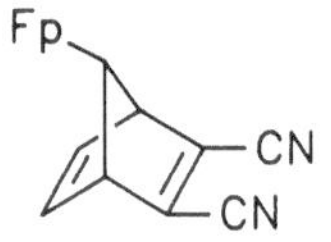 	yellow crystals, m.p. 165° (dec.) [84] ^{1}H NMR (CDCl$_3$): 3.85 (br, CCHC), 4.18 (br, FeCH), 4.76 (s, C$_5$H$_5$), 6.82 (t, CH=CH, J(H,H)=2.2) [84] IR: 1960, 2014 (both CO) in CH$_2$Cl$_2$, 1554, 1579 (both C=C), 2206 (CN) [84]
173	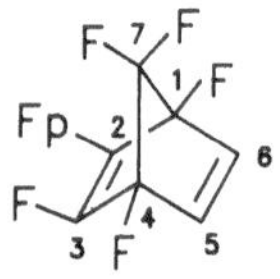 I (13%) [118]	yellow crystals [118] ^{1}H NMR (CDCl$_3$): 5.0 (s, C$_5$H$_5$), 6.6 (m, 2H, CH=CH) [118] ^{19}F NMR (CDCl$_3$, referred to CF$_3$COOH): 54 (F-3), 56 (F-7), 104 (F-1), 125 (F-4) [118] IR: 1975, 2038 (both CO) in CH$_2$Cl$_2$, 1614 (FeC=C) in Nujol and hexachlorobutadiene [118]
174	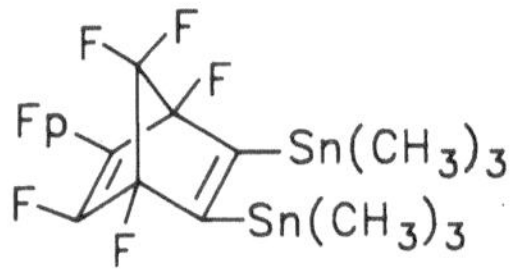 I (15%) [118]	yellow crystals [118] IR: 2000(?), 2050 (both CO) in CH$_2$Cl$_2$, 1615 (FeC=C) in Nujol and hexachlorobutadiene [118] ^{1}H NMR, ^{19}F NMR: solubility too low for the determination [118]
*175	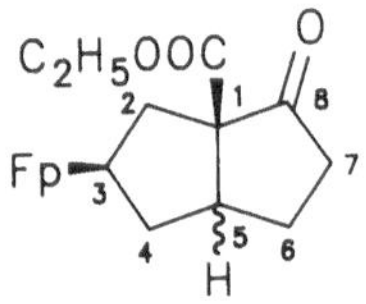 VIb (50%) [121]	yellow crystals, m.p. 96 to 98° [121] ^{1}H NMR (CDCl$_3$): 1.12 (t, 3H, CH$_3$, J=7), 1.40 to 3.10 (m, 10H, CH$_2$, CH), 4.13 (q, 2H, OCH$_2$, J=7), 4.75 (s, C$_5$H$_5$) [121] ^{13}C NMR (CDCl$_3$): 14.10 (q, CH$_3$), 18.64 (d, C-3), 26.63 (t, C-6), 39.82 (t, C-7), 47.49 (d, C-5), 48.66, 48.79 (2t, C-2,4), 61.13 (t, OCH$_2$), 67.36 (s, C-1), 85.36 (d, C$_5$H$_5$), 171.76 (s, COOR), 216.65, 216.78, 216.91 (3s, CO, C=O?) [121] IR (CHCl$_3$): 1950, 2000 (both CO), 1715, 1740 (both C=O) [121]
176	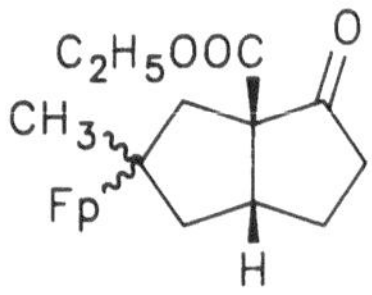 VIb (45%) [121]	yellow crystals, m.p. 84 to 87° [121] ^{1}H NMR (CDCl$_3$): 1.22 (t, 3H, CH$_3$, J=7), 1.25 (s, 3H, CH$_3$), 1.50 to 3.50 (m, 9H, CH$_2$, CH), 4.15 (q, 2H, OCH$_2$, J=7), 4.76 (s, C$_5$H$_5$) [121] IR (KBr): 1935, 1995 (both CO), 1720, 1745 (both C=O) [121]
177	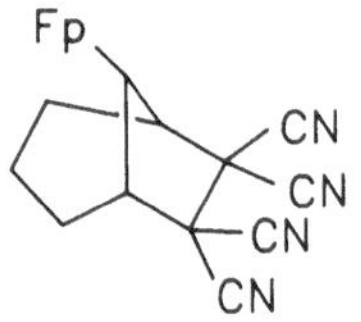 IVb [48, 69]	m.p. 125° (dec.) [69] ^{1}H NMR (CD$_3$NO$_2$): 1.35 to 2.30 (m, 6H, CH$_2$), 3.12 (3H, br s, CH), 5.10 (s, C$_5$H$_5$) [69] IR (KBr): 1950, 2010 (both CO) [69]

References on pp. 268/72

Table 8 (continued)

No.	compound method of preparation (yield)	properties and remarks

178

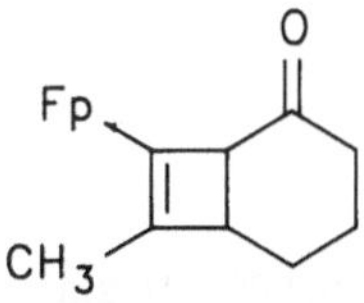

VIa (24, 54%) [121]

amber oil [121]
^{1}H NMR (CDCl$_3$): 1.3 to 3.6 (d, m, 9H, CH$_2$, CH$_3$, J =
1.5), 3.17 (m, 1H, CH), 3.42 (m, 1H, CH), 4.92
(s, C$_5$H$_5$) [121]
^{13}C NMR (CDCl$_3$): 14.55 (q, CH$_3$), 17.60 (t, CH$_2$), 24.88
(t, CH$_2$), 40.08 (t, CH$_2$CO), 46.25 (d, CH), 63.92
(d, CH), 84.84 (d, C$_5$H$_5$), 140.12 (s, >C=), 165.72 (s,
>C=), 213.85, 214.57 (2s, C=O, CO) [121]
IR (CH$_2$Cl$_2$): 1955, 2010 (both CO), 1673 [121]

*179

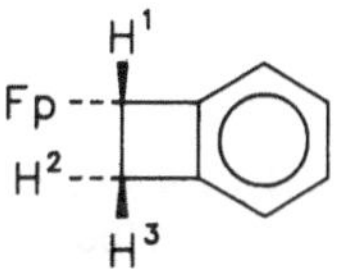

I [76], (50%) [44, 71],
(60%) [147]

m.p. 66 to 68° (from petroleum ether at −78°) [71]
^{1}H NMR (CS$_2$): 2.69 (m, H-2), 3.44 (m, H-3, J(2,3) =
13.5), 4.05 (m, H-1, J(1,2) = 2.1, J(1,3) = 4.9), 4.63
(s, C$_5$H$_5$), 6.82 (m, C$_6$H$_4$) [44, 71], 2.74 (q, H-2,
J(1,2) = 2.2, J(2,3) = 14.4), 3.44 (q, H-3, J(1,3) = 4.8),
4.07 (q, H-1), 4.59 (s, C$_5$H$_5$), 6.85 (m, C$_6$H$_4$) [147]
IR: 1942, 2003 (neat) [44, 71], 1940, 2005 (all CO), 634,
743, 836, 845, 1041, 1410, 1445, 3055, 3125 (mulls) [147]

*180

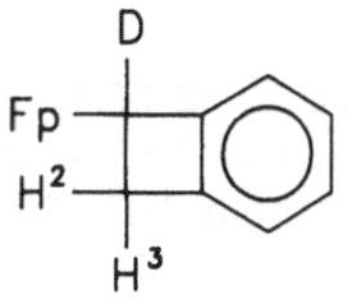

^{1}H NMR (CS$_2$): 2.72 (d, H-2), 3.47 (d, H-3, J(2,3) = 14.7),
4.63 (s, C$_5$H$_5$), 6.87 (m, C$_6$H$_4$) [71], see also [44]

*181

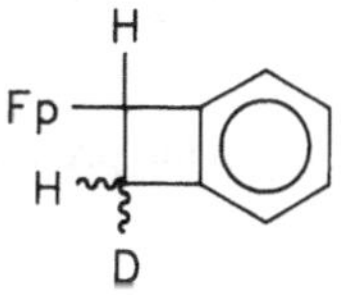

yellow crystals, 1:1 mixture of cis- and trans-
isomers [66]
^{1}H NMR (CS$_2$): 2.66 (s, br, 0.5H, CHD), 3.40 (s, br, 0.5H,
CHD, cis to H), 4.05 (d, br, 1H), 4.55 (s, C$_5$H$_5$), 6.80
(m, C$_6$H$_4$) [66]

*182

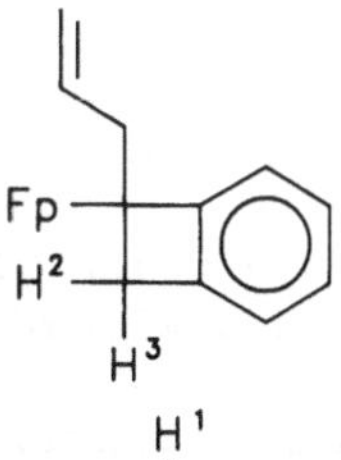

m.p. 59 to 62° [71]
^{1}H NMR (CS$_2$): 2.64 (m, H-2,3), 3.07 (d, CH$_2$, J = 4.5), 4.69
(s, C$_5$H$_5$), 4.90 (m, =CH$_2$), 5.65 (m, =CH-), 6.96
(m, C$_6$H$_4$) [71]
IR (neat): 1920, 1985 (both CO) [71]

*183

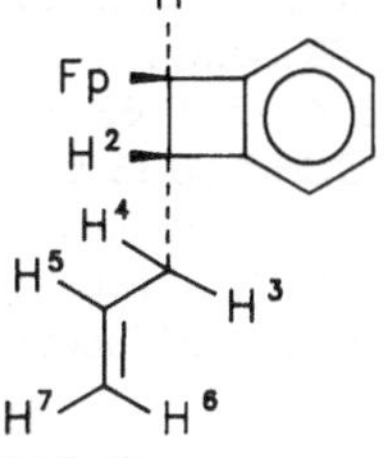

orange oil [71]
^{1}H NMR (CS$_2$): 2.32 (m, H-3,4), 2.97 (m, H-2), 3.62
(d, H-1, J(1,2) = 1.8), 4.67 (s, C$_5$H$_5$), 4.74 (m, H-7), 5.02
(m, H-6), 5.82 (m, H-5, J(5,6) = 4.5), 6.82 (m, C$_6$H$_4$)
[71], see also [55]
IR (neat): 1930, 1995 (both CO) [55, 71]

          References on pp. 268/72

Table 8 (continued)

No.	compound method of preparation (yield)	properties and remarks

*184

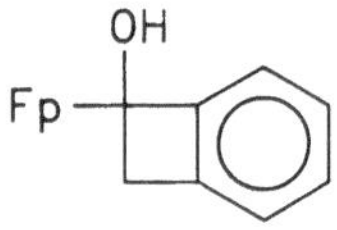

Fp—[structure with OH]

—

*185

[structure with OCH₃, Fp, H², H³]

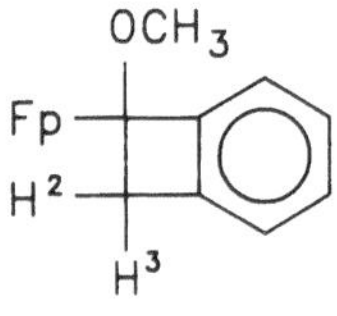

yellow–orange crystals, m.p. 88.5 to 90° [71], stable [45]
^{1}H NMR (CS$_2$): 3.22 (d, H–2), 3.24 (s, CH$_3$), 3.57 (d, H–3,
J(2,3) = 13.5), 4.75 (s, C$_5$H$_5$), 6.99 (m, C$_6$H$_4$) [44, 71]
IR (Nujol): 1916, 1989 (both CO) [44, 71]

*186

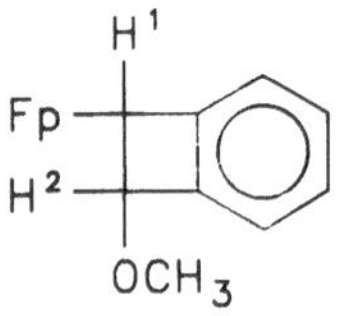

[structure with H¹, Fp, H², OCH₃]

m.p. 86 to 88° [71]
^{1}H NMR (CS$_2$): 3.39 (s, CH$_3$), 4.00 (d, H–2), 4.41 (d, H–1,
J(1,2) = 1.5), 4.77 (s, C$_5$H$_5$), 7.04 (m, C$_6$H$_4$) [55, 71]
IR: 1940, 1995 (both CO) in Nujol [71], same in KBr [55]

*187

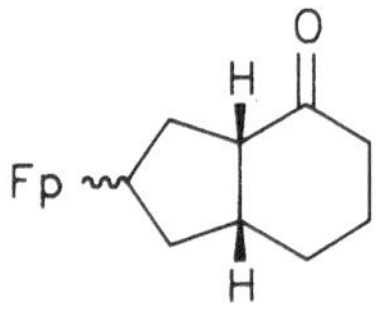

[structure with O, H, Fp, H]

VIa (45%) [121]

orange oil, mixture of the two stereoisomers [121]
^{1}H NMR (CDCl$_3$): 1.3 to 3.0 (m, 13H, CH$_2$, CH), 4.78
(s, C$_5$H$_5$) [121]
^{13}C NMR (CDCl$_3$): 17.86, 20.53, 23.00, 23.52, 27.41, 29.17,
39.24, 39.63, 41.77, 42.61, 42.87, 43.00, 46.38, 47.36,
53.46, 85.29 (C$_5$H$_5$), 214.11, 214.37 (both C=O), 217.23
(CO) [121]
IR (CH$_2$Cl$_2$): 1935, 1995 (both CO), 1700 (C=O) [121]

*188

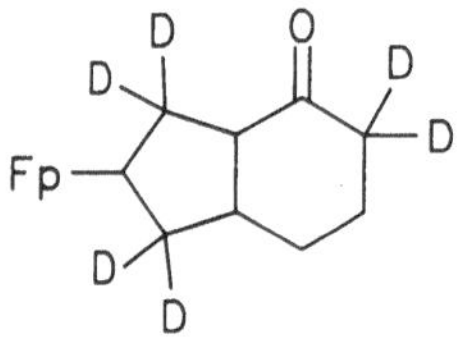

[structure with D, O, D, D, Fp, D, D]

prepared from FpCD$_2$CH=CD$_2$ and
6,6-D$_2$-cyclohex-2-enone in CH$_2$Cl$_2$ in the presence
of AlBr$_3$ [121]

*189

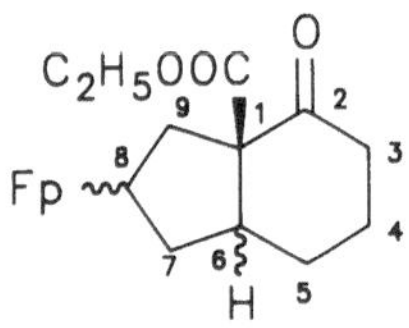

[structure with C₂H₅OOC, O, numbered positions 9,8,1,2,7,6,5,3,4, Fp, H]

VIb (63%) [121]

yellow crystals, m.p. 87 to 89°, mixture of two diastereomers [121]
^{1}H NMR (CDCl$_3$): 1.20 (t, 3H, CH$_3$, J = 7), 1.50 to 3.00
(m, 12H, CH$_2$, CH), 4.16 (q, 2H, OCH$_2$, J = 7), 4.76
(s, C$_5$H$_5$) [121]
^{13}C NMR (CDCl$_3$): 14.10 (q, CH$_3$), 18.25, 18.71 (both:
d, C–8), 22.22 (t, C–4), 26.83, 27.09 (both: t, C–5), 39.28,
40.28 (both: t, C–3), 44.37, 46.45, 46.97, 47.62 (all:
C–7,9), 48.72 (d, C–6), 61.13 (t, OCH$_2$), 66.13, 66.85
(both: s, C–1), 85.36 (d, C$_5$H$_5$), 172.21, 173.25 (both
COOR), 208.46, 208.85 (both C=O), 216.97 (CO) [121]
IR (CHCl$_3$): 1950, 2000 (both CO), 1700, 1720 (both C=O)
[121]

References on pp. 268/72

Table 8 (continued)

No.	compound method of preparation (yield)	properties and remarks
190	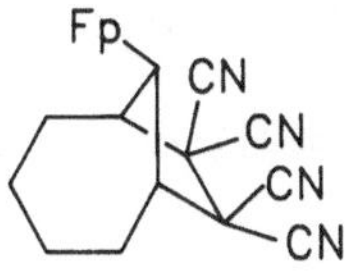 IVb [48, 69]	dec. 91° [69] ^{1}H NMR (CD$_3$NO$_2$): 1.35 to 2.45 (m, 8H, CH$_2$), 3.35 (br s, 3H, CH), 5.09 (s, C$_5$H$_5$) [69] IR (KBr): 1950, 2000 (both CO) [69]
*191	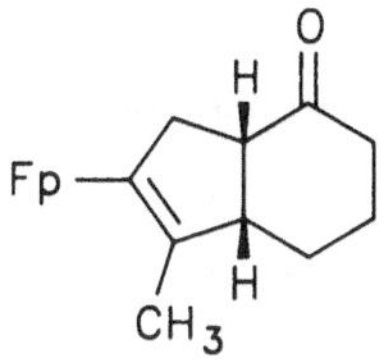 VIa (20%) [121]	yellow oil [121] ^{1}H NMR (CDCl$_3$): 1.77 (d, m, 7H, CH, CH$_2$, CH$_3$, J=2), 2.31 (m, 2H, CH$_2$), 2.65 (m, 2H, CH$_2$), 2.87 (m, 2H, CH$_2$), 4.80 (s, C$_5$H$_5$) [121] ^{13}C NMR (CDCl$_3$): 17.21 (q, CH$_3$), 22.35, 27.61 (both: t, CH$_2$), 40.02 (t, CH$_2$CO), 51.32 (t, CH$_2$), 51.84, 52.68 (both: d, CH), 85.66 (d, C$_5$H$_5$), 132.52, 144.6 (both: s, >C=), 215.02, 215.61, 215.74 (all: s, C=O, CO) [121] IR (CH$_2$Cl$_2$): 1950, 2005 (both CO), 1690 [121]
192	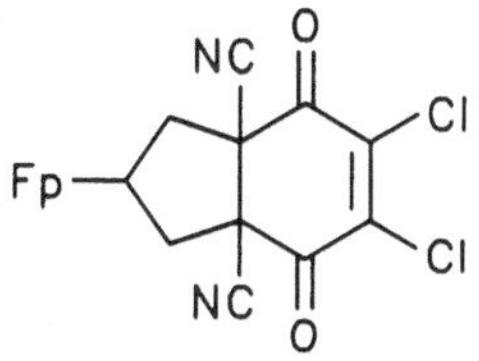	prepared from FpCH$_2$CH=CH$_2$ and 2,3–dichloro–5,6–dicyanoquinone [29, 69]
*193	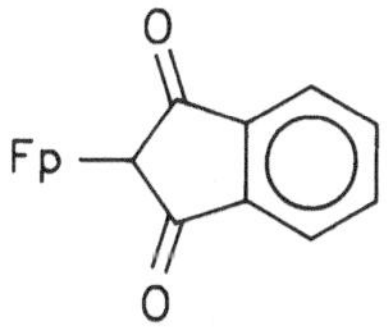	yellow–brown prisms, m.p. 129 to 131° (dec.) [106] ^{1}H NMR (CDCl$_3$): 3.32 (s, CH), 5.00 (s, C$_5$H$_5$), 6.8 to 7.6 (m, C$_6$H$_4$) [106] IR: 1946, 1974, 1987, 2017, 2022 sh (all CO), 1666 sh, 1672, 1700 (all C–O) in KBr, 1990, 2029 (both CO), 1676, 1702 br (both C=O) in CH$_2$Cl$_2$ [106]
*194	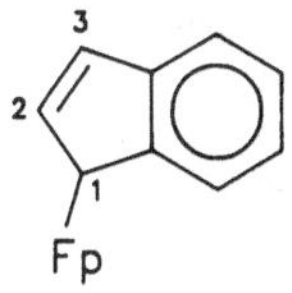	red crystals, m.p. 62 to 64° [11] IR (C$_7$H$_{16}$): 1955, 2010 (both CO) [11]
*195	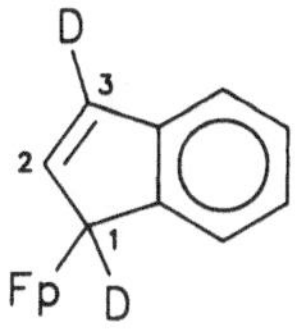	^{1}H NMR spectrum is depicted in [11]

References on pp. 268/72

Table 8 (continued)

No.	compound method of preparation (yield)	properties and remarks
196	 C_6H_5 Fp I (52%) [111]	m.p. 230 to 231°, air-stable as solid, somewhat light-sensitive [111] IR (CH_2Cl_2): 1981, 2028 (both CO), 1690 [111]
197	 C_2H_5OOC Fp∿ H VIb (9%) [121]	yellow crystals, m.p. 117 to 120°, stereochemistry not given [121] ^{1}H NMR (CDCl$_3$): 1.20 (t, 3H, CH$_3$, J = 7), 1.00 to 3.00 (m, 14H, CH$_2$, CH), 4.12 (q, 2H, OCH$_2$, J = 7), 4.67 (s, C$_5$H$_5$) [121] IR (KBr): 1935, 1995 (both CO), 1700, 1725 (both C=O) [121]
*198	 Fp III (39%) [68]	yellow needles, m.p. 93 to 94° [68] IR: 1945, 1999 (both CO) [68]
*199	 Fp I (11%) [68]	yellow, m.p. 107 to 108°, moderately air-stable [68] IR: 1942, 1998 (both CO) [68]
*200	 O Fp O	yellow, m.p. 101° [138] ^{1}H NMR (CS$_2$): 2.50 (>CH–CH<), 3.02 (CH$_2$), 4.75 (C$_5$H$_5$), 5.37 (FeC=CH), 6.41 (CCH=CHC) [138] IR (KBr): 1960, 2002 (both CO) [138]
*201	 F F Fp F CH$_3$ F F CH$_3$	yellow, m.p. 165 to 167° [59] IR (CHCl$_3$): 1992, 2042 (both CO), 1587 (FeC=C), 1712 (CC=CC) [59]

Table 8 (continued)

No.	compound method of preparation (yield)	properties and remarks
*202	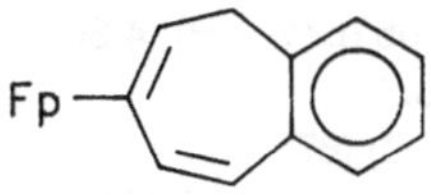	yellow-brown, m.p. 114.5 to 115.5° (dec.) [102] ^{1}H NMR (CDCl$_3$): 3.0 (d, 2H, CH$_2$, J=7), 4.6 (s, C$_5$H$_5$), 5.6 (t, 1H, FeC=CH, J=7), 6.62 (s, 2H, FeCCH=CH), 7.2 (m, 4H, C$_6$H$_4$) [102] IR (KBr): 1935, 1960, 2010 (all CO) [102]
*203	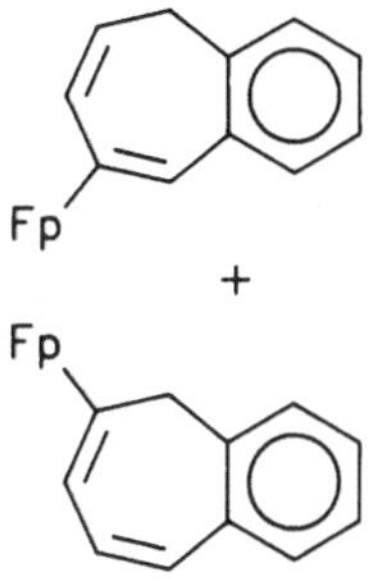	yellow-brown oil, air-sensitive, mixture of isomers [102, 119] ^{1}H NMR (CDCl$_3$): 2.9, 3.45 (d, s, 2H, CH$_2$), 4.9 (s, C$_5$H$_5$), 5.35, 6.2, 6.7, 7.2 (8H, all: m, CH) [102, 119] IR (neat): 1950, 2020 (both CO) [102], 730, 830, 910, 1020, 1430, 1480, 1525, 1775, 3020 [119]
*204	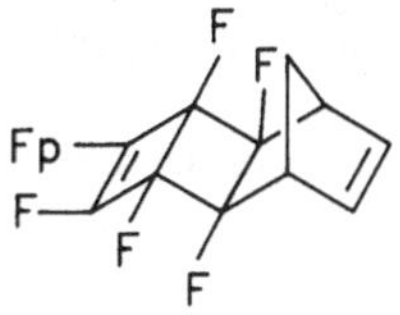 I (8%) [59]	pale yellow crystals, m.p. 123 to 124° [59] ^{1}H NMR: 1.75 (CH$_2$), 3.16 (>CH–), 4.88 (C$_5$H$_5$), 6.53 (=CH–), 4 broad signals, intensity ratio 2:2:5:2 [59]
*205	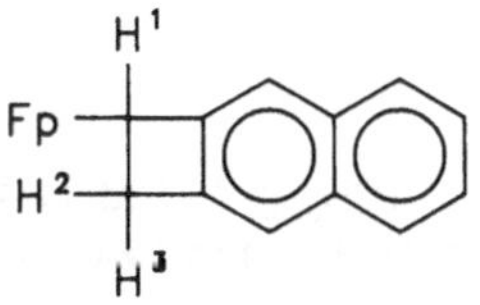 I [76], (67%) [71]	m.p. 157 to 158° (from petroleum ether) [71] ^{1}H NMR (CS$_2$): 2.81 (m, H-2), 3.58 (m, H-3, J(2,3)=15.0), 4.22 (m, H-1, J(1,3)=6.0, J(1,2)=3.75), 4.77 (s, C$_5$H$_5$), 7.37 (m, 6H, aryl) [71] IR (Nujol): 1926, 2000 (both CO) [71]
*206	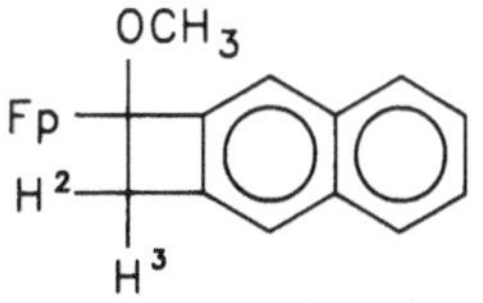	m.p. 98 to 99° [71], stable [45] ^{1}H NMR (CS$_2$): 3.29 (s, CH$_3$), 3.37 (d, H-2), 3.39 (d, H-3, J(1,3)=15.0), 4.71 (s, C$_5$H$_5$), 7.37 (m, 6H, aryl) [71] IR (KBr): 1922, 1995 (both CO) [71]
*207	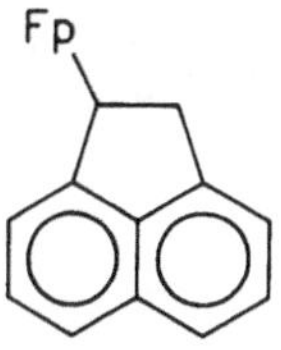	yellow-orange crystals, m.p. 71 to 73° [87] ^{1}H NMR (CS$_2$): 3.30, 3.65 (both: m, 1H), 3.95 (br s, 1H), 4.28 (s, C$_5$H$_5$), 7.25 (m, 6H, aryl) [87] IR: 1960, 1985 in CH$_2$Cl$_2$ [87], 1953, 2008 (all CO) in C$_6$H$_6$ [110] UV (C$_6$H$_{14}$): depicted from 200 to 750 [94] mass spectrum: [M]$^+$, [M−CO]$^+$, [M−2CO]$^+$ [87]

References on pp. 268/72

Table 8 (continued)

No.	compound method of preparation (yield)	properties and remarks

*208

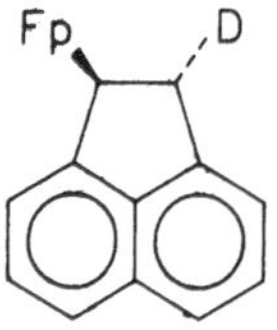

^{1}H NMR (CS$_2$): 3.30 (m, 1H), 4.05 (br s, 1H), 4.30 (s, C$_5$H$_5$), 7.3 (m, 6H, aryl) [87]
IR (CH$_2$Cl$_2$): 1960, 1985 (both CO) [87]
mass spectrum: [M]$^+$, [M−CO]$^+$, [M−2CO]$^+$ [87]

*209   C$_5$D$_5$(CO)$_2$Fe

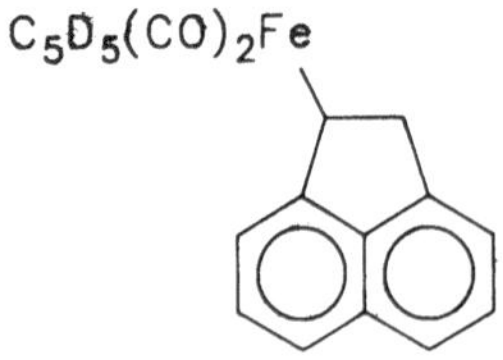

—

*210

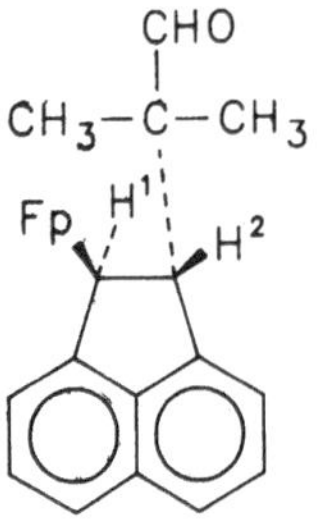

yellow oil [63]
^{1}H NMR (CS$_2$): 0.8, 0.9 (both: s, CH$_3$), 3.70 (br s, H-2), 3.90 (br s, H-1), 4.15 (s, C$_5$H$_5$), 7.0 to 7.7 (m, aromatic), 9.50 (s, CHO) [63]
IR (CHCl$_3$): 1960, 1990 (both CO), 1725 (C=O) [63]

*211

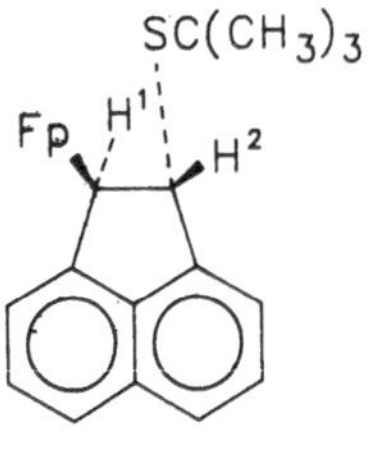

yellow oil [63]
^{1}H NMR (CD$_3$COCD$_3$): 1.50 (s, C(CH$_3$)$_3$), 4.15 (d, H-1, J=1), 4.60 (s, C$_5$H$_5$), 4.75 (d, H-2), 7.1 to 7.7 (m, aromatic) [63]
IR (CH$_3$CN): 1960, 1995 (both CO) [63]

*212

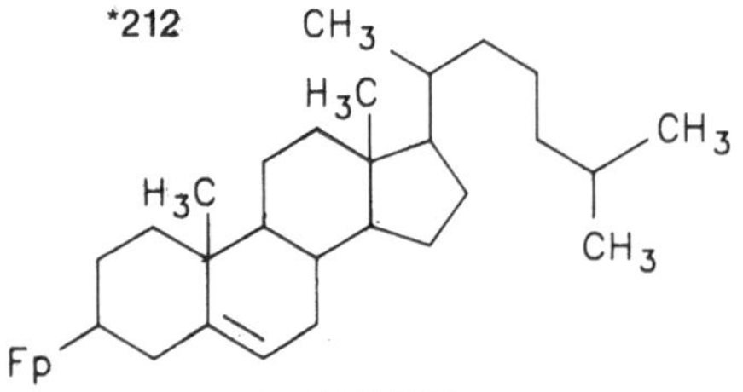

I (40%) [98]

glass [98]
^{1}H NMR (CDCl$_3$): 0.7 to 2.3 (m, CH, CH$_2$, CH$_3$), 3.51 (m, FpCH), 4.68 (s, C$_5$H$_5$), 5.1 to 5.35 (m, =CH−) [98]
IR (petroleum ether): 1952, 2005 (both CO) [98]

References on pp. 268/72

Table 8 (continued)

No.	compound method of preparation (yield)	properties and remarks

213

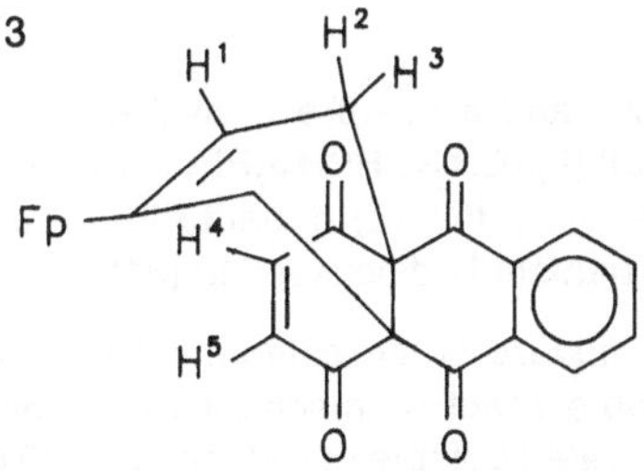

prepared from FpC(=CH$_2$)CH=CH$_2$ and quinazarinquin-one [138]

m.p. 184° [138]

^{1}H NMR (CS$_2$): 2.61 (H-2,3), 4.85 (C$_5$H$_5$), 5.50 (H-1), 6.68 (H-4,5), 7.80 (aryl) [138]

IR (KBr): 1940, 2010 (both CO), 1692, 1715 (both C=O) [138]

supplement

*214

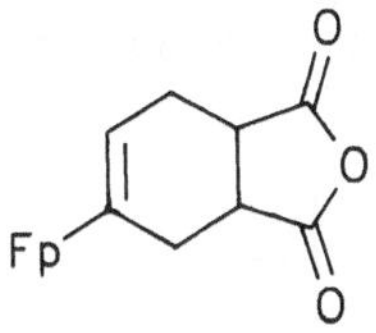

yellow, m.p. 136° [138]

^{1}H NMR (CDCl$_3$): 2.50 (>CH-CH<), 3.28 (CH$_2$), 4.79 (C$_5$H$_5$), 5.87 (FeC=CH) [138]

IR (KBr): 1945, 2002 (both CO), 1769, 1861 (both O=C-O-C=O) [138]

*215

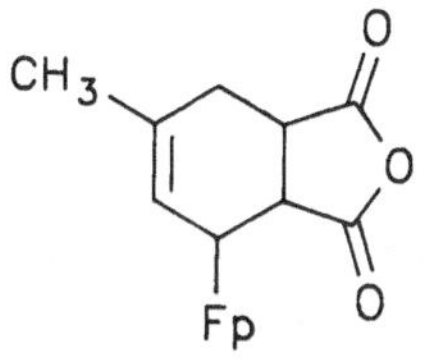

brown oil [138]

^{1}H NMR (CS$_2$): 1.70 (CH$_3$), 2.30 (CH$_2$), 2.65 (FeCH), 3.22 (>CH-CH<), 4.65 (C$_5$H$_5$), 5.65 (=CH-) [138]

IR (neat): 1940, 2000 (both CO), 1765 (C=O) [138]

*216

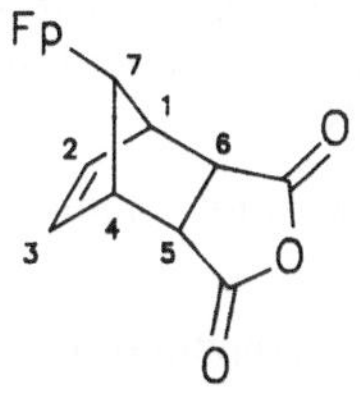

^{1}H NMR (CDCl$_3$): 2.59 (t, H-7, J=0.8), 3.34 (m, H-1,4), 3.54 (m, H-5,6), 4.73 (s, C$_5$H$_5$), 6.20 (m, 2H, CH=CH) [126, 130]

^{13}C NMR (CDCl$_3$): 48.25 (C-5,6), 50.77 (C-7), 56.83 (C-1,4), 85.59 (C$_5$H$_5$), 135.47 (C-2,3), 170.97 (C=O), 216.51 (CO) [126, 130]

IR (CH$_2$Cl$_2$): 1955, 2011 (both CO), 1776 (C=O) [126, 130]

*217

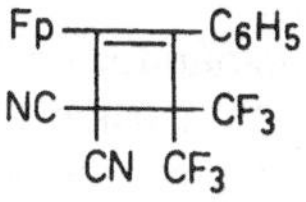

pale yellow crystals, m.p. 96 to 97° (from CH$_2$Cl$_2$/petro-leum ether) [150]

^{1}H NMR: 4.95 (s, C$_5$H$_5$), 7.3 (m, C$_6$H$_5$) [150]

IR: 1995, 2044 (both CO), 2243 (CN) in CH$_2$Cl$_2$, 1110, 1155, 1190 sh, 1200, 1212 sh, 1222 sh, 1275 (all CF), 1580, 1590, 1615 (all C=C), further bands from 620 to 1440 in Nujol [150]

mass spectrum (fast atom bombardment): [M+H]$^+$ (18), [M]$^+$ (11), [M−CN]$^+$ (7), [M−2CO]$^+$ (100), [M−2CO−F]$^+$ (21), [M−2CO−CN]$^+$ (23), [M−2CO−2F]$^+$ (7), [M−2CO−CF$_3$]$^+$ (8), [C$_2$C$_6$H$_5$(NC)$_2$C$_2$(CF$_3$)$_2$−F+H]$^+$ (17), [M−H−(NC)$_2$C$_2$(CF$_3$)$_2$]$^+$ (37), [C$_2$C$_6$H$_5$(NC)C$_2$(CF$_3$)$_2$−F]$^+$ (45), [C$_5$H$_5$FeC$_2$C$_6$H$_5$]$^+$ (6), [Fe(C$_5$H$_5$)$_2$]$^+$ (27), [Fp]$^+$ (17), also containing ions m/e >500 [150]

References on pp. 268/72

*Further information:

C₅H₅Fe(CO)₂C₃H₅ and **C₅H₅Fe(CO)₂C₃H₄D** (Table 8, Nos. 1 and 2). Lower yields of No. 1 are obtained when Method III is carried out in C_6D_6 (52%) or petroleum ether (16%) [116]. Compound No. 1 has also been formed by alkylation of FpBr with cyclopropyl lithium [31].

FpC₃H₅ is air-stable; it may be distilled at room temperature under high vacuum without decomposition [31].

Compound No. 1 reacts with gaseous HCl to give FpCl and a mixture of cyclopropane and propene, whereas the reaction with HBF₄ yields the salt [Fp(CH₂=CHCH₃)]BF₄ [31]. Treatment of the deuterio derivative with HBF₄ affords exclusively the (Z)-1-deuteriopropene salt [Fp(CHD=CHCH₃)]BF₄ [31, 48], see also [88]. The mechanism is discussed in [31].

The reaction of FpC₃H₅ with SO₂ in benzene gives a mixture of stereoisomeric sultines XIIa and a small amount of FpSO₂CH₂CH=CH₂. A possible reaction mechanism is given [31], see also [48]. The corresponding reaction of the deuterated complex produces XIIb [31].

On treatment of No. 1 with [C(C₆H₅)₃]⁺ the expected β-hydride abstraction does not take place. Instead, only the addition product [Fp(CH₂=CHCH₂C(C₆H₅)₃)]⁺ is formed [31, 49].

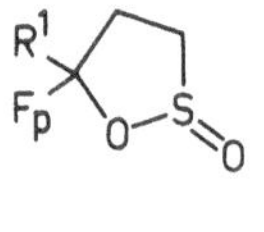

XII a: R¹=H
XII b: R¹=D

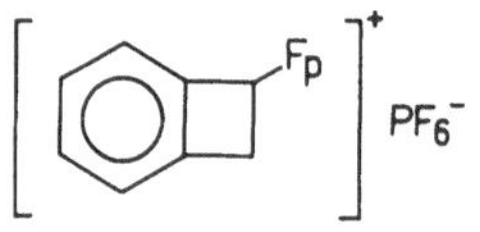

XIII

The reaction of FpC₃H₅ with XIII in CH₂Cl₂ at room temperature for 4 d affords an orange-yellow precipitate of [Fp(CH₂=C=CH₂)]PF₆ admixed with varying amounts of the acidolysis product [Fp(CH₂=CHCH₃)]PF₆ and [C₅H₅Fe(CO)₃]PF₆; compound No. 179 could be isolated from the solution. The formation of the allene salt is explained in terms of α-hydride abstraction by XIII concomitant with the opening of the three-membered ring [49].

C₅H₅Fe(CO)₂C₃H₃(C₆H₅)₂ (Table 8, No. 4). Lower yields are obtained when Method III is carried out in C_6D_6 (64%) or petroleum ether (51%) [116].

C₅H₅Fe(CO)₂C₃H₂(CH₃)₂(OCH₃) (Table 8, No. 5). The formation of the compound by photolysis of the corresponding acyl complex is given as an equation in [137].

Photolysis of the compound in benzene results in formation of the ring-expanded carbene complex XIV. The proposed mechanism involves migration of the alkyl from carbon to 16-electron coordinately unsaturated iron, thus generating the carbene complex XV, which subsequently ring expands to give XIV [137].

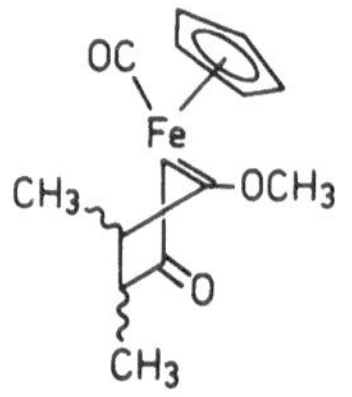

XIV

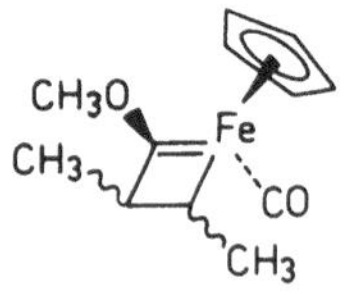

XV

References on pp. 268/72

C₅H₅Fe(CO)₂C₃H(C₆H₅)₂ $C_5H_5Fe(CO)_2C_3H(C_6H_5)_2$ (Table **8**, No. 6) has been prepared by stirring the cyclopropenylium salt [FpC₃(C₆H₅)₂]BF₄ and NaBH₃CN (1:1 mole ratio) in THF for several hours, evaporation of the solvent, chromatography of the residue on SiO_2 with toluene as eluent, and recrystallization from ether/hexane at −40 °C, 44% yield [143].

C₅H₅Fe(CO)₂C₃H(C₆H₅)₂ $C_5H_5Fe(CO)_2C_3H(C_6H_5)_2$ (Table **8**, No. 7) has been prepared by addition of Na[Fp] in THF to a suspension of the cyclopropenylium salt XVIII (X=ClO₄) in the same solvent at −78 °C. After warming to −10 °C the solvent is removed (high vacuum), the residue is taken up in ether/dioxane (2:1), the solution is then cooled to −10 °C and filtered. The filtrate is evaporated at −10 °C (high vacuum), the residue is taken up in ether and the solution stirred at −40 to −70 °C for several hours. The crystals formed are washed with ether/ hexane and dried in vacuum, 46% yield [86].

The compound decomposes rapidly at room temperature [143].

The reactions of the compound differ substantially from those of No. 9. With the electrophiles studied [C(C₆H₅)₃]BF₄, [HO(C₂H₅)₂]BF₄, HCl, Br₂, I₂, SbCl₅, and [C(OCH₃)₃]BF₄, the cyclopropenylium salts XX (X=BF₄, Cl, Br, or I) are formed almost exclusively [86].

C₅H₅Fe(CO)₂C₃(C₄H₉-t)₂CH₃ $C_5H_5Fe(CO)_2C_3(C_4H_9\text{-}t)_2CH_3$ (Table **8**, No. 8) has been obtained from Na[Fp] and the cyclopropenylium salt [C₃(C₄H₉-t)₂CH₃]ClO₄ in THF at −70 °C [86].

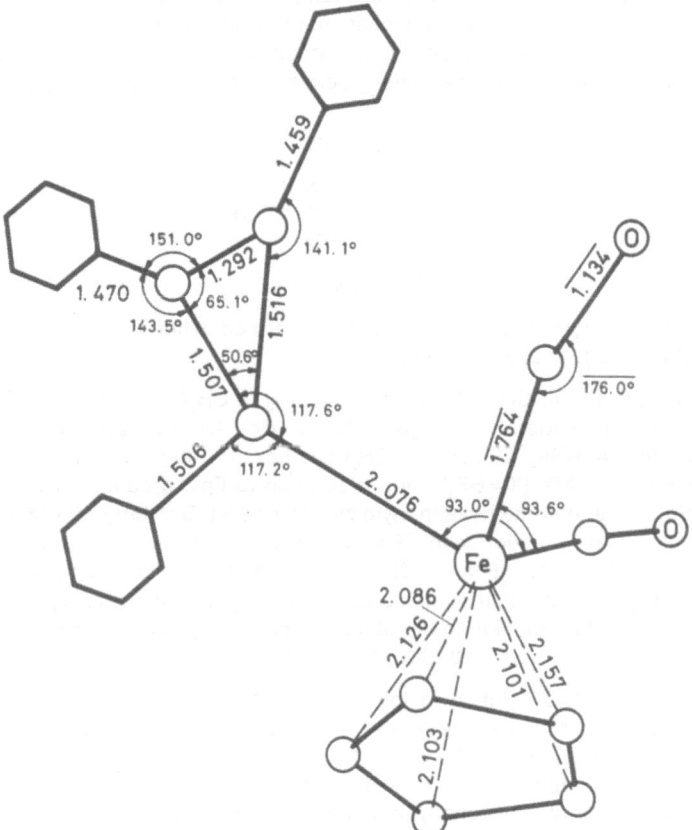

Fig. 6. Molecular structure of $C_5H_5Fe(CO)_2C_3(C_6H_5)_3$ (No. 9) with selected bond lengths (in Å) and angles.

References on pp. 268/72

The compound is very unstable. Analoguously to No. 9, it reacts with I_2 in CH_2Cl_2 at $-70\,°C$ to give FpI and the cyclopropenyliumpentaiodide $[C_3(C_4H_9\text{-}t)_2CH_3]I_5$ [86].

$C_5H_5Fe(CO)_2C_3(C_6H_5)_3$ (Table **8**, No. **9**) has been prepared by addition of Na[Fp] in THF to a suspension of XIX ($X=BF_4$) in THF at $-70\,°C$, removal of the solvent at $0\,°C$ (high vacuum), dissolution of the residue in toluene and concentration of the filtered solution at -10 to $0\,°C$ under high vacuum. The crystals formed are washed with toluene/ether and dried in vacuum, 69% yield. If XIX ($X=Br$) is used in place of XIX ($X=BF_4$), the yield is 61% [97].

The compound crystallizes in the orthorhombic system, space group Pbca $-D_{2h}^{15}$ (No. 61), $a=16.697(1)$, $b=17.264(6)$, $c=15.006(5)$ Å; $Z=8$, $d_m=1.02$ g/cm³. The molecular structure is shown in **Fig. 6** (see p. 235) [97].

The compound is very slightly soluble in alkanes, ether, and CCl_4, and very soluble in aromatic hydrocarbons, $CHCl_3$, CH_2Cl_2, THF, and CS_2. It is very sensitive towards air, light, and heat. However, below $-20\,°C$, when light and air are excluded, the solid is indefinitely stable. Solutions decompose slowly, even at lower temperatures, to give Fp_2 and hexaphenylbicyclopropenyl [97].

On heating, the crystals decompose without melting. Thermal decomposition (at room temperature after three weeks in the solid state, after 6 days in solution or after 10 min in refluxing benzene) affords dark, iron-containing products, almost insoluble in all usual solvents. The only identifiable substance was triphenylcyclobutenone. In the mass spectrum only the fragments ferrocene and hexaphenylbenzene could be observed [97].

XVI

A mixture of triphenylcyclobutenone and XVI is formed, when the compound is irradiated in ether at $-40\,°C$ for 16 h under argon [97].

No. 9 reacts with Br_2 in CCl_4 at $-40\,°C$ or with I_2 in CH_2Cl_2/hexane at $-70\,°C$ followed by warming to room temperature to give XIX ($X=Br_3$, I_5) and FpBr or FpI, respectively. Also, the reaction with $HBF_4 \cdot O(C_2H_5)_2$ in CH_2Cl_2/ether at $-40\,°C$ and warmed to room temperature yields mainly XIX ($X=BF_4$) and, in addition to Fp_2, a cationic complex, assumed to be [Fp-H-Fp]BF_4, and some hexaphenylbicyclopropenyl. Similarly, XIX ($X=CF_3COO$) has been obtained from the reaction with CF_3COOH in CH_2Cl_2/ether at -70 to $0\,°C$. Treatment of the compound with CH_3OSO_2F in CH_2Cl_2 at room temperature for 2 d gives Fp_2, hexaphenylbicyclopropenyl and also small amounts of the corresponding cyclopropenylium salt. The latter is mainly formed in addition to the other two products, when the compound is reacted with [$O(C_2H_5)_3$]BF_4 in CH_2Cl_2 at $-20\,°C$ for two weeks [97].

Passing HCl through a solution in toluene at $-70\,°C$ for 30 min and workup at room temperature afford 1,2-diphenylindene and FpCl. Similarly, the reaction with LiBr in THF or [$N(C_4H_9\text{-}n)_4$]I in CH_2Cl_2 and $HBF_4 \cdot O(C_2H_5)_2$ in CH_2Cl_2 at $-70\,°C$ or at room temperature, respectively, gives 1,2-diphenylindene and FpX ($X=Br$, I). Possible reaction mechanisms leading to the formation of 1,2-diphenylindene are discussed [97].

Addition to R^1COCl ($R^1=CH_3$, C_6H_5) and $AlCl_3$ in CH_2Cl_2 at $-70\,°C$ and stirring the mixture at room temperature for 24 h followed by treatment with [$N(C_4H_9\text{-}n)_4$]I or 1 N HCl, respectively, give 3-acyl-1,2-diphenylindene and FpX (X=I, Cl). In the reaction with $CH_3COCl/AlCl_3$ small amounts of 2-methyl-3,4,5-triphenylfuran are also formed [97].

Treatment with SO_2 in CH_2Cl_2 at $-70\,°C$ and warmed to room temperature over 12 h affords small amounts of $FpSO_2Fp$ and a brown solid, which is assumed to be XIX (X=$FpSO_2$) in equilibrium in solution with complex XVII from its spectroscopic data and solubility in weakly polar solvents [97].

XVII

XVIII : R^1 = H
XIX : R^1 = C_6H_5
XX   : R^1 = Fp
XXI : R^1 = OCH_3

The reaction with $HgCl_2$ (1:1 mole ratio) in benzene at room temperature for 15 h forms hexaphenylbicyclopropenyl as the main product along with FpHgCl and FpCl [97].

$C_5H_5Fe(CO)_2C_3(C_6H_5)_2OCH_3$ (Table 8, No. 10) has been prepared by stirring XX (X=BF_4) and $NaOCH_3$ (1:2 mole ratio) in THF for several hours, evaporation of the solvent, dissolution of the residue in CH_2Cl_2, addition of hexane, and concentration of the filtered solution. The crystals formed are washed with hexane and dried in high vacuum, 54% yield. The formation is reversible. Addition of $HBF_4 \cdot O(C_2H_5)_2$ regenerates XX (X=BF_4). The compound also forms when Na[Fp] is allowed to react with XXI (X=SO_3F) [143].

$C_5H_5Fe(CO)_2C_3(C_6H_5)_2CN$ (Table 8, No. 11) has been prepared by stirring XX (X=BF_4) and KCN (1:5 mole ratio) in CH_3CN for several hours followed by evaporation of the solvent, chromatographic purification of the residue on SiO_2 with $CHCl_3/O(C_3H_7-i)_2$ (1:1) as eluent, and recrystallization from CH_2Cl_2/hexane, 70% yield [143].

$C_5H_5Fe(CO)_2C_4H_7$ (Table 8, No. 12). Addition to $[C(C_6H_5)_3]BF_4$ in CH_2Cl_2 at 0 °C and further reaction at room temperature for 1 h afford the salt $[Fp(\eta^2-C_4H_6-cyclo)]BF_4$ [69].

$C_5H_5Fe(CO)_2C_4H_5$ (Table 8, No. 13) has also been obtained by addition of $N(C_2H_5)_3$ to XXII in CH_2Cl_2 at $-78\,°C$. After warming to 24 °C and solvent removal, the residue is taken up in petroleum ether and the filtrate evaporated, giving the compound in 58% yield [72].

XXII

$C_5H_5Fe(CO)_2C_4H_4Cl$ (Table 8, No. 14) prepared from Na[Fp] and cis-3,4-dichlorocyclobutene in THF at -78 to 0 °C [53, 65] or -78 to $+24\,°C$ [72] (Method I) is isolated from the reaction mixture by solvent removal, extraction with petroleum ether and fractional crystallization at $-78\,°C$.

The trans stereochemistry has been established by the observation of a small vicinal coupling (~ 2 Hz) between the allylic protons of the four-membered ring [72].

References on pp. 268/72

The compound undergoes quantitative thermal isomerization to the (E,E)-butadiene complex FpCH=CHCH=CHCl by refluxing in toluene for 1 h [72], see also [65].

Hydrolysis to the (E)-butenal complex FpCH$_2$CH=CHCHO occurs when a benzene solution of the compound is chromatographed on neutral Al$_2$O$_3$. Possible reaction mechanisms are discussed [72], see also [65, 76].

Addition of a CH$_2$Cl$_2$ solution of AgPF$_6$ to the compound in the same solvent at −78 °C leads to formation of XXIII. The inverse addition did not yield XXIII, rather it gave an incompletely characterized substance. Although the compound failed to react with cyclopentadiene in CH$_2$Cl$_2$ at 25 °C over a 4 h period, a rapid reaction ensued when the compound in CH$_2$Cl$_2$ was added to a mixture of AgPF$_6$ and excess cyclopentadiene in CH$_2$Cl$_2$ at −78 °C, giving XXIV. Addition of the compound to a mixture of AgPF$_6$ and 1,3-diphenylisobenzofuran in CH$_2$Cl$_2$ at −78 °C yields the Diels–Alder adduct XXV in low yield. This product was not obtained when the reaction was carried out in the absence of AgPF$_6$. The compound failed to give a Diels–Alder adduct on treatment with dimethyl fumarate in the presence of AgPF$_6$ in CH$_2$Cl$_2$. Based on these results, a mechanism for the formation of XXIII, XXIV, and XXV is proposed, according to which Ag$^+$ generates the highly reactive cyclobutadiene complex XXVI by β-chloride abstraction from FpC$_4$H$_4$Cl, which is then trapped by the dienes [65, 73], see also [76].

XXIII   XXIV   XXV   XXVI

C$_5$H$_5$Fe(CO)$_2$C$_4$F$_5$ (Table 8, No. 15). The reaction mixture, obtained from Na[Fp] and perfluorocyclobutene in THF at 0 °C to room temperature (Method I), is worked up by solvent removal at 0 °C/30 Torr followed by sublimation of the residue at 40 °C/0.05 Torr onto a probe cooled by solid CO$_2$ [8]. The compound is formed in 8% yield by addition of Na[Fp] to 1,2-dichlorohexafluorocyclobutane in THF at 0 °C and leaving the mixture at room temperature overnight. Workup is carried out by solvent removal, chromatography on Florisil with benzene/light petroleum as eluent, sublimation, and crystallization from light petroleum [9].

The compound appears to be indefinitely stable on exposure to air. It decomposes in solution, but is not decomposed by water [8].

C$_5$H$_5$Fe(CO)$_2$C$_4$F$_4$Cl (Table 8, No. 16) has been prepared from Na[Fp] and either ClFC=CFCF=CFCl or the cyclobutenes XXVII or XXVIII in THF at −78 °C to room temperature (Method I) and workup by chromatography on Florisil with CH$_2$Cl$_2$/hexane (1:10) as eluent, recrystallization from pentane and sublimation at 50 °C/0.05 Torr [35].

XXVII   XXVIII

The presence of only two ^{19}F NMR resonances requires an effective plane of symmetry bisecting all four carbon atoms of the cyclobutene ring. This implies free rotation about the Fe–C σ-bond to the C_4F_4Cl group. This rotation could not be slowed down to the NMR time scale by cooling to −80 °C, as indicated by the essential equivalence of the ^{19}F NMR spectrum at −80 °C and at ambient temperature (~30 °C) [35].

The presence of pyrolysis products such as $[(C_5H_5)_2Fe]^+$ and $[C_5H_5FeC_5H_4C_4F_4Cl]^+$ complicated the mass spectrum, even when the sample inlet temperature was held at 85 °C. At more elevated temperature (~100 °C) an ion at m/e = 467 corresponding to $[C_4F_4C_5H_4FeC_5H_4C_4F_4Cl]^+$ was even observed. Other possible processes include apparent losses of CO groups, fluorine atoms and hydrogen fluoride. The fragments observed (sample temperature 85 °C) and their relative intensities are given [35].

$C_5H_5Fe(CO)_2C_4H(=CF_2)(CF_3)_2$ (Table 8, No. 17) is mentioned in [20] to form from the 1:1 insertion product of $FpSn(CH_3)_3$ and $CF_3C≡CCF_3$ by elimination of $(CH_3)_3SnF$ (reaction conditions are not given). A later publication by the same authors [24] reports that UV irradiation of $FpSn(CH_3)_3$ and $CF_3C≡CCF_3$ at 76 °C affords compound No. 107 as main product in addition to (E)–$FpC(CF_3)=CHCF_3$ and $(CH_3)_3SnF$. Compound XXIX is only assumed as an intermediate, which readily rearranges to No. 107.

$C_5H_5Fe(CO)_2C_4H(C_6H_5)_2O$ (Table 8, No. 18) has been prepared by addition of $C_6H_5CH_2COCl$ in benzene to $FpC≡CC_6H_5$ and $N(C_2H_5)_3$ in the same solvent at 0 °C. After stirring at room temperature for 5 h, the $[N(C_2H_5)_4]Cl$ is filtered off, the filtrate evaporated and the residue chromatographed on Al_2O_3 with benzene/ethyl acetate (10:1) as eluent, 87% yield [109]. The compound has also been obtained by the reaction of XXX with $N(C_2H_5)_3$ in THF at 20 °C [93].

XXIX   XXX

The compound crystallizes in the monoclinic space group $P2_1/c – C_{2h}^5$ (No. 14), a = 9.2371(9), b = 26.4502(9), c = 8.2450(7) Å, β = 115.502(6)°; Z = 4. The molecular structure is shown in **Fig. 7** (see p. 240). The four-membered cycle is slightly bent around the C(1)–C(3) line (cf. No. 18 in Table 8) with a dihedral angle of 7°. The atoms C(1), C(2), C(3) have planar trigonal (sp^2) coordination and the atom C(4) tetrahedral (sp^3) coordination. The Fe–C bond length of 1.935(6) Å is somewhat shorter than most of the known values in related σ-vinyl complexes [103].

$C_5H_5Fe(CO)_2C_4(CH_3)(C_6H_5)_2O$ (Table 8, No. 19) has been obtained in 95% yield from $FpC≡CCH_3$ and $(C_6H_5)_2C=C=O$ by the procedure described for compound No. 20 [109].

$C_5H_5Fe(CO)_2C_4(C_6H_5)_3O$ (Table 8, No. 20) has been prepared from $FpC≡CC_6H_5$ and $(C_6H_5)_2C=C=O$ (1:1 mole ratio) in benzene at room temperature for 1 d. After solvent removal the residue is chromatographed on Al_2O_3 with benzene/ethyl acetate (10:1) as eluent and the product is recrystallized from benzene/hexane, 80% yield [109].

$C_5H_5Fe(CO)_2C_4(C_6H_5)(CN)_4$ (Table 8, No. 21). On addition of $(NC)_2C=C(CN)_2$ to a solution of $FpC≡CC_6H_5$ in ether, a dark green solution forms. This color is said to result from a diradical like $FpC^·=C(C_6H_5)C(CN)_2C^·(CN)_2$. After 15 min the solution lightens to pale yellow and No. 21 crystallizes out [148]. In [92] the same color changes and crystallization were

References on pp. 268/72

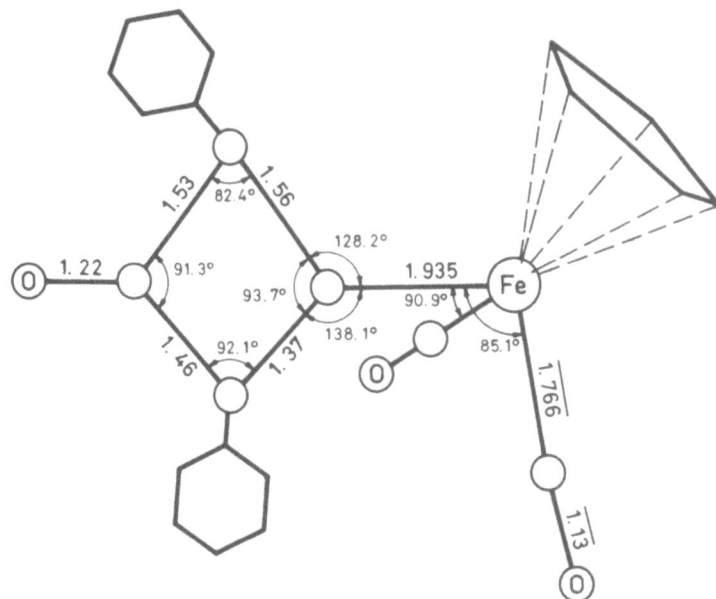

Fig. 7. Molecular structure of $C_5H_5Fe(CO)_2C_4H(C_6H_5)_2O$ (No. 18) with selected bond lengths (in Å) and angles.

reported, but they are explained in a different way and other structures, including charge-transfer complexes, are suggested. The reaction in CH_2Cl_2 is believed to give No. 21 [92], but the product is probably mostly $C_5H_5Fe(CO)_2C[=C(CN)_2]C(C_6H_5)=C(CN)_2$ [148] (see Section 1.5.2.3.16.1.11, Table 5, No. 179).

$C_5H_5Fe(CO)_2C_5H_8CH_3$ (Table **8**, No. **23**) has been obtained in 10% yield along with the main product, Fp_2, by addition of $LiCu(CH_3)_2$ in ether to $[Fp(\eta^2-C_5H_8-cyclo)]^+$ in THF at temperatures between -78 to $-20\,°C$. After 1 h the solvent is removed and the product is taken up in petroleum ether and chromatographed on Al_2O_3 [78].

$C_5H_5Fe(CO)_2C_5H_8(CH=CH_2)$ (Table **8**, Nos. **24** and **25**). The cis–isomer has been prepared from Na[Fp] and trans–3–vinylcyclopentyl bromide in THF at 0 °C to room temperature (Method I). After solvent removal, the residue is extracted with Skelly–B and filtered through Al_2O_3. The concentrated filtrate is chromatographed on Al_2O_3 with Skelly–B as eluent. The trans–isomer has been prepared from Na[Fp] and cis–3–vinylcyclopentyl tosylate in THF at room temperature followed by workup as described for the cis–isomer. The ^{13}C NMR spectra of the isomers obtained by Method I show each to be contaminated by 20 to 30% of the other isomer [99]. The trans–derivative has also been isolated in 28% yield from the reaction of XXXI with $N(C_2H_5)_3$ in CH_3NO_2 for 4 h and after solvent removal with NaI in acetone at room temperature for 2 h [99], see also [46]. A 1:1 mixture of No. 24 or No. 25 and XXXII (40% total yield) is formed when $FpCH_2CH=CH_2$ is allowed to react with $[Fp(CH_2=CHCH=CH_2)]BF_4$ in CH_3NO_2 at room temperature for 3 h and then with NaI in acetone

$$\left[Fp\diagdown\!\!/\!\!\diagup\!\!\!\diagdown\!\!\diagup\!\!\!\diagdown\!\!Fp \right]^{2+}$$

<u>XXXI</u>   <u>XXXII</u>

for 2 h. In both cases workup is performed by chromatography on Al_2O_3 [99], see also [46, 47, 54].

$C_5H_5Fe(CO)_2C_5H_8CH(COOCH_3)_2$ and $C_5H_5Fe(CO)_2C_5H_8CH(COOC_2H_5)_2$ (Table 8, Nos. 26 and 27). The reaction with $[C(C_6H_5)_3]^+$ at 0 °C leads to hydride abstraction giving XXXIII [78], see also [46]; cf. different behavior of the corresponding cyclohexane complex No. 116.

$C_5H_5Fe(CO)_2C_5H_8CH(COCH_3)COOCH_3$ (Table 8, No. 28) reacts with $[C(C_6H_5)_3]^+$ at 0 °C to give the unsubstituted alkene complex $[Fp(\eta^2\text{-}C_5H_8\text{-cyclo})]^+$ [78].

$R^1 = CH_3$ or C_2H_5

XXXIII   XXXIV

$C_5H_5Fe(CO)_2C_5H_8C_6H_9O$ (Table 8, No. 29) has been prepared by treatment of $[Fp(\eta^2\text{-}C_5H_8\text{-}$cyclo)]^+$ in CH_3CN with one equivalent of 1-pyrrolidino-cyclohexene at 0 °C for 3 h and hydrolysis of the iminium salt formed (XXXIV) by heating the mixture on a steam bath for 5 to 20 min with one equivalent of NaOH in aqueous ethanol. Extraction into ether, workup and chromatography on neutral Al_2O_3 give the product in 79% yield. It is assumed that a mixture of diastereomeric adducts is formed in this reaction [78], see also [46, 54].

$C_5H_5Fe(CO)_2C_5H_7(COOC_2H_5)_2$ (Table 8, No. 31). Treatment in CH_2Cl_2 with gaseous HCl at 0 °C for 1 h and keeping the mixture at room temperature for 40 h afford $CH_2=$ $CHCH_2CH_2CH(COOC_2H_5)_2$ as sole product [115]. The reaction with $[C(C_6H_5)_3]BF_4$ in CH_2Cl_2 at 45 °C for 2 h gives the salt $[Fp\{CH_2=CHCH_2CH_2C(OOCC_2H_5)=C(OC_2H_5)OC(C_6H_5)_3\}]BF_4$ [115].

$C_5H_5Fe(CO)_2C_5H_7(CH(COOCH_3)_2)_2$ (Table 8, No. 32). The single metal carbonyl resonance and the two closely spaced ester carbonyl signals observed in the ^{13}C NMR spectrum are in accord with a structure having a plane of symmetry passing through the Fp group [78].

$C_5H_5Fe(CO)_2C_5H_6(COOC_2H_5)_2CH_3$ (Table 8, No. 34) has been prepared by stirring $FpCH_2CH=CHCH_3$ ((E),(Z)-mixture) and diethyl methylenemalonate (1:1 mole ratio) in ether overnight and purifying by chromatography on Al_2O_3 with pentane as eluent, 64% yield [121].

$C_5H_5Fe(CO)_2C_5H_6(COOC_2H_5)_2OCH_3$ (Table 8, No. 35) prepared by Method Va in dimethylformamide is isolated by chromatography on Florisil and then on Al_2O_3 [120]. According to [91] only a 5% yield has been obtained after chromatography on Al_2O_3 alone.

$C_5H_5Fe(CO)_2C_5H_6(COOCH_3)_3$ (Table 8, No. 36), prepared by Method Va in dimethylformamide [91, 115], reacts with gaseous HCl under the conditions described for compound No. 31 to give an inseparable mixture of $CH_2=CHCH_2CH(COOCH_3)CH(COOCH_3)_2$ and the corresponding cyclopentane derivative (ratio 1:1), which results from Fe-C cleavage [115].

$C_5H_5Fe(CO)_2C_5H_5(COOCH_3)_3OCH_3$ (Table 8, No. 37). In the reaction of $FpCH_2C(OCH_3)=CH_2$ with $(CH_3OOC)_2C=CHCOOCH_3$ (Method Va) No. 37 is formed together with linear products. The highest yields are obtained in CH_2Cl_2: 22% No. 37, 28% $FpCH_2C(OCH_3)=$ $CHCH(COOCH_3)CH(COOCH_3)_2$, and 5% $FpCH_2COCH_2CH(COOCH_3)CH(COOCH_3)_2$ (from hydrolysis during chromatography). In THF and benzene $FpCH_2C(OCH_3)=CHCH(COOCH_3)CH\text{-}$ $(COOCH_3)_2$ is formed almost exclusively [91, 114].

References on pp. 268/72

$C_5H_5Fe(CO)_2C_5H_5(COOCH_3)_3OCH_3$ (Table **8**, No. **38**) prepared by Method Va in dimethyl-formamide is isolated by chromatography on Florisil and then on Al_2O_3 [120]. According to [91] only a 4% yield has been obtained after chromatography on Al_2O_3 alone.

$C_5H_5Fe(CO)_2C_5H_6(COOC_2H_5)_2CN$ (Table **8**, No. **39**) reacts with gaseous HCl under the conditions described for compound No. 31 to give an inseparable mixture of $CH_2=CHCH_2CH(COOC_2H_5)CH(CN)COOC_2H_5$ and the corresponding cyclopentane derivative (ratio 1:2) resulting from cleavage of the Fe–C bond [115].

$C_5H_5Fe(CO)_2C_5H_5(OCH_3)(COOC_2H_5)_2CN$ (Table **8**, No. **42**) is formed in only 2% yield by Method Va (in CH_2Cl_2) in addition to the polymeric adducts $FpCH_2C(OCH_3)=CH[CH(COOC_2H_5)C(CN)COOC_2H_5]_nH$ (n = 2, 3, etc.) [114].

$C_5H_5Fe(CO)_2C_5H_5(OCH_3)(COOC_2H_5)(CN)_2$ (Table **8**, No. **43**) is formed as a mixture of diastereomers by Method Va in CH_2Cl_2 as indicated by the presence of a number of methoxy resonances in the ^1H NMR spectrum of the crude product: $\delta = 3.02$, 3.18, and 3.28 ppm (each s, OCH_3, ratio ~3:4:1) [114].

$C_5H_5Fe(CO)_2C_5H_5(OCH_3)(COOC_2H_5)(CN)_2$ (Table **8**, No. **44**) decomposed during chromatography (unstable on Al_2O_3, SiO_2, and Florisil) [91]. Thus, the low yield obtained by Method Va is probably due to decomposition during chromatography, as the ^1H NMR spectrum of the crude product indicated a quantitative conversion to the title compound [114].

$C_5H_5Fe(CO)_2C_5H_5(OCH_3)(COOCH_3)_2CN$ (Table **8**, No. **45**) is prepared by Method Va in CH_2Cl_2. Exposure of the compound to anhydrous HCl in CH_2Cl_2 followed by irradiation with visible light yields the cyclopentene XXXV [131].

XXXV : R^1 = CH_3
XXXVI : R^1 = C_2H_5
                                 XXXVII

$C_5H_5Fe(CO)_2C_5H_5(OCH_3)(COOC_2H_5)_2CN$ (Table **8**, Nos. **46** and **47**). Treatment of compound No. 46 or 47 (both prepared by Method Va in CH_2Cl_2) with gaseous HCl under the conditions described for No. 50 results in exclusive formation of compound XXXVI. The reaction of No. 47 with $[C(C_6H_5)_3]BF_4$ in CH_2Cl_2 at 45 °C for 2 h gives the salt XXXVII [120].

$C_5H_5Fe(CO)_2C_5H_5(OCH_3)(COOC_2H_5)(CN)_2$ (Table **8**, Nos. **50** and **51**) are prepared by Method Va in CH_2Cl_2. Chromatography of the compounds on Al_2O_3 causes elimination of CH_3OH, yielding complex No. 87. In general, No. 50 is isolated in low yield (<15%) from the column, whereas No. 51 always gives the product of elimination [120].

Bubbling HCl through a solution of No. 50 in CH_2Cl_2 at 0 °C for 1 h and setting aside the mixture for 20 to 24 h leads to formation of 3-ethoxycarbonyl-4,4-dicyanocyclopent-1-ene [120].

$C_5H_5Fe(CO)_2C_5H_6(C_6H_4Cl-2)(CN)_2$ (Table **8**, No. **52**) has been obtained in 40% yield by the reaction of $FpCH_2CH=CH_2$ with $2-ClC_6H_4CH=C(CN)_2$ (1:1 mole ratio) in CH_2Cl_2 at room temperature for 3 h and workup by addition of hexane to the concentrated solution and recrystallization of the solid formed from ether/hexane [69].

$C_5H_5Fe(CO)_2C_5H_5(CH_3)(C_6H_4Cl-2)(CN)_2$ (Table **8**, No. **53**) has been prepared by refluxing $FpCH_2C(CH_3)=CH_2$ and $2-ClC_6H_4CH=C(CN)_2$ (1.1:1 mole ratio) in CH_2Cl_2 for 2 h followed by addition of hexane to the concentrated solution and recrystallization of the solid from ether/hexane, 85% yield [69].

$C_5H_5Fe(CO)_2C_5H_5(CF_3)_2(CN)_2$ (Table **8**, Nos. **54** and **55**). The three diastereomers obtained in a 94% total yield from the reaction of $FpCH_2CH=CH_2$ with (E)-$CF_3(NC)C=C(CN)CF_3$ in benzene for 0.5 h could be separated into No. 55 (97.2%) and an isomeric mixture (No. 54) (2.8%) by repeated chromatography on Florisil, eluting with pentane/CH_2Cl_2 and fractional crystallization from pentane/benzene. They are characterized by their ^{19}F NMR spectra [83].

$C_5H_5Fe(CO)_2C_5H_4(CH_3)(CF_3)_2(CN)_2$ (Table **8**, Nos. **56** and **57**). From the reaction of $FpCH_2C(CH_3)=CH_2$ with (E)-$CF_3(NC)C=C(CN)CF_3$ in benzene for 6 min, only two diastereomers could be separated chromatographically (79% total yield) as indicated by the ^{19}F NMR spectra. These are No. 57 (98.3%) and one of the isomers of No. 56 (1.7%). The corresponding reaction with (Z)-$CF_3(NC)C=C(CN)CF_3$ gives a total yield of 92% with 73% of No. 57 and 27% of No. 56 [83].

$C_5H_5Fe(CO)_2C_5H_5(CN)_4$ (Table **8**, No. **60**). Electron impact mass spectra are given, applying the usual technique or laser induced evaporation conditions. In the former case, the molecular ion is absent, the highest molecular value (relative intensities in parentheses) is m/e $= 305$ (7) from $[Fp(NC)_2C=C(CN)_2]^+$, the base peak is $[C_3H_3]^+$. Under laser induced evaporation conditions a weak molar peak appears (1.5), the more abundant ions are [M − 2CO]$^+$ (46), [M − $C_5H_5(CN)_4$]$^+$ (22), $[C_5H_4Fe(CN)_2]^+$ (79), $[C_5H_5FeCO]^+$ (35), $[C_5H_5FeCN]^+$ (37), and $[C_5H_5Fe]^+$ (100) [149].

$C_5H_5Fe(CO)_2C_5H_4(CN)_4CH_3$ (Table **8**, No. **61**) crystallizes in the triclinic space group P$\bar{1}$ − C$_i^1$ (No. 2), a $= 13.1986(28)$, b $= 10.1648(16)$, c $= 13.8088(43)$ Å, $\alpha = 115.55(2)°$, $\beta = 103.46(2)°$, $\gamma = 87.06(1)°$; Z $= 4$, $d_c = 1.474$, $d_m = 1.45(2)$ g/cm^3. The crystal contains two crystallographically independent molecules with bond distances that agree within the limits of experimental error. Their molecular geometries differ only in the rotational orientation of their $C_5H_4(CN)_4CH_3$ groups about the Fe–C(1) bonds, see **Fig. 8**. Within each molecule the cyclo-

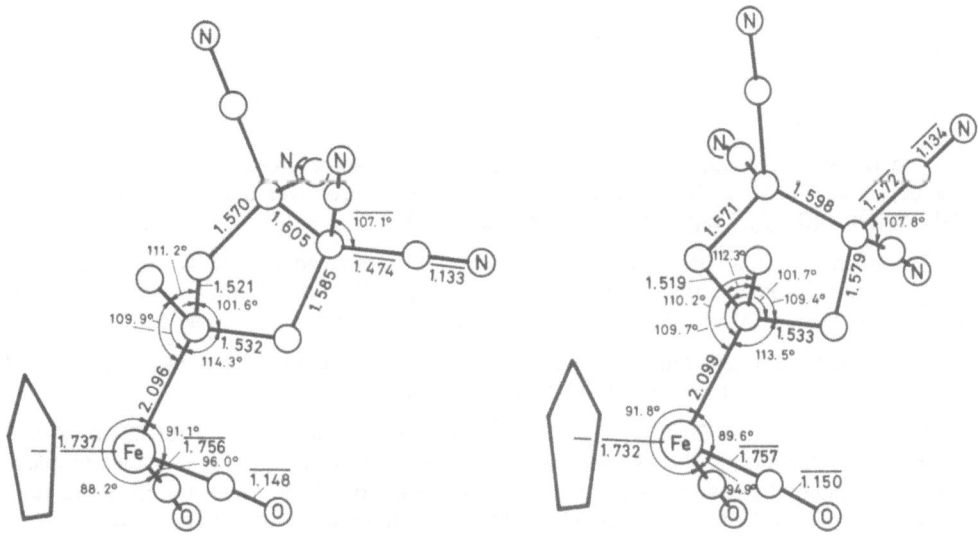

Fig. 8. Molecular structure of $C_5H_5Fe(CO)_2C_5H_4(CN)_4CH_3$ (No. 61) with selected bond lengths (in Å) and angles. Views showing the different rotational conformations of the molecules about their Fe–C(1) bonds.

pentane ring is markedly nonplanar. It is assumed that the observed conformation minimizes repulsive interactions between substituents on adjacent carbon atoms [40].

The solid is stable to air at room temperature, but solutions are less stable. The compound is soluble in acetone, moderately soluble in $CHCl_3$ or CH_2Cl_2, slightly soluble in benzene, and insoluble in saturated hydrocarbons [52].

$C_5H_5Fe(CO)_2C_5H_4(CN)_4CH_3$ (Table 8, Nos. **63** and **64**). Apparently (see ^{1}H NMR data) the product from the reaction of (E)-$FpCH_2CH=CHCH_3$ with $(NC)_2C=C(CN)_2$ (Method IVa) is a mixture of diastereomers (No. 63 and No. 64) [61]. However, in a later publication [69] it is reported that the $(NC)_2C=C(CN)_2$ adducts derived from (Z)- and (E)-$FpCH_2CH=CHCH_3$ correspond to only one isomer that has the CH_3 group in the cis-position (No. 63) (only $\delta(CH_3) = 1.39$ ppm reported [78]) or trans-position (No. 64), respectively, to the Fp moiety.

$C_5H_5Fe(CO)_2C_5H_3(CN)_4(CH_3)C_2H_5$ (Table 8, No. **77**). A comparison of the methyl proton chemical shifts with those observed for the $(NC)_2C=C(CN)_2$ adducts of methyl–substituted $(\eta^1$-allyl)Fp complexes of defined stereochemistry provides strong evidence for the cis-relationship of the methyl and Fp substituents on the cyclopentane ring [78].

$C_5H_5Fe(CO)_2C_5H_7$ (Table 8, No. **78**) forms in small amount by deprotonation of $[Fp(\eta^2-C_5H_8$-cyclo)$]^+$ with 1-pyrrolidinocyclohexene at 0 °C [78].

UV irradiation of the compound and the cyclic phosphite $P(OCH_2)_3CCH_3$ in petroleum ether gives only the dinuclear complex $(C_5H_5)_2Fe_2(CO)_3P(OCH_2)_3CCH_3$ and not the expected substitution product [104].

Addition of a solution of Br_2 in ether at 0 °C causes immediate precipitation of the bromide salt of the cation XXXVIII ($R^1 = Br$) [69], see also [60].

XXXVIII　　　　　　　　　　XXXIX

The reaction with $[O(CH_3)_3]PF_6$ in CH_3NO_2 at room temperature for 1 h [69], see also [48, 60], or with $[C(C_6H_5)_3]PF_6$ in CH_2Cl_2 at room temperature gives the PF_6^- salts of XXXVIII ($R^1 = CH_3$ or $C(C_6H_5)_3$, respectively) [69], see also [60].

Treatment with a solution of $CH_3OOCN=SO_2 \cdot THF$ (prepared at -78 °C from methyl N-(chlorosulfonyl)urethane and NaH (56% oil dispersion) in THF) at 0 °C, then at room temperature for 1 h affords XXXVIII, $R^1 = SO_3^-$, by loss of methyl cyanate [69], see also [37].

The compound reacts with $ClSO_2NCO$ in CH_2Cl_2 at room temperature [69] or below [37] to give XXXVIII, $R^1 = CN$, by loss of $ClSO_3^-$, see also [48, 60].

The reaction with 4-$CH_3C_6H_4SO_2NCO$ in CH_2Cl_2 at room temperature for 30 min yields the bicyclic γ-lactam XXXIX [69], see also [37, 48, 60]. This cycloaddition is shown to occur stereospecifically by a suprafacial addition of the electrophile trans to the Fp group [69]. In contrast, no well-defined product could be isolated from the reaction with CH_3O-SO_2NCO [37].

The compound condenses with $[Fp(CH_2=CH_2)]^+$ in acetone [46] or CH_3NO_2 [99] at room temperature to give XXXVIII, $R^1 = (CH_2)_2Fp$ (see also [47]), but fails to react with $[Fp(CH_3CH=$

CH$_2$)]$^+$ [99]. FpC$_5$H$_7$ is transformed to the dinuclear complex XL on brief heating in CH$_2$ClCH$_2$Cl at 70 °C in the presence of excess [Fp(CH$_2$=C(CH$_3$)$_2$)]BF$_4$ [113].

XL

C$_5$H$_5$Fe(CO)$_2$C$_5$H$_5$(COOCH$_3$)$_2$ (Table **8**, No. **84**). In the reaction of FpCH$_2$CH=CH$_2$ with CH$_3$OOCC≡CCOOCH$_3$ (Method Va in DMF) the compound is formed in a 42% yield together with the linear products FpCH$_2$CH=CHC(COOCH$_3$)=CHCOOCH$_3$ (3%) and FpC(COOCH$_3$)= C(COOCH$_3$)CH$_2$CH=CH$_2$ (69%) [91, 115].

Bubbling HCl through a solution of the compound in CH$_2$Cl$_2$ at 0 °C for 1 h and setting the mixture aside for 40 h at room temperature leads to cleavage of the Fe–C bond and formation of dimethyl cyclopent-1-ene-1,2-dicarboxylate [115].

C$_5$H$_5$Fe(CO)$_2$C$_5$H$_4$(CN)$_2$C$_6$H$_4$Cl-2 (Table **8**, No. **86**) has been obtained by a cycloaddition reaction of FpCH=C=CH$_2$ with 2-ClC$_6$H$_4$CH=C(CN)$_2$ [43], see also [54]. No further details are given.

C$_5$H$_5$Fe(CO)$_2$C$_5$H$_4$(CN)$_2$COOC$_2$H$_5$ (Table **8**, No. **87**) is obtained when the crude product resulting from the reaction of (Z)-FpCH$_2$CH=CHOCH$_3$ with (NC)$_2$C=CHCOOC$_2$H$_5$ in CH$_2$Cl$_2$ (Method Va) is chromatographed on Al$_2$O$_3$ with ether as eluent, 32% yield. In addition, 13% of compound No. 50 can be eluted with ether/light petroleum (1:1) [120].

C$_5$H$_5$Fe(CO)$_2$C$_5$H$_2$O(C$_6$H$_5$)$_2$CH$_3$ and **C$_5$H$_5$Fe(CO)$_2$C$_5$H$_2$O(C$_6$H$_5$)$_3$** (Table **8**, Nos. **88** and **89**) have been prepared by addition of (C$_6$H$_5$)$_2$C=C=O in benzene to FpCH$_2$C≡CR^1 (R^1 = CH$_3$, C$_6$H$_5$) (1.4:1 mole ratio) in CH$_2$Cl$_2$ at −40 °C, stirring at room temperature for 5 h, and chromatography of the concentrated solution on Al$_2$O$_3$ with pentane/CH$_2$Cl$_2$ (1:3) as eluent. After evaporation of the solvent, the residue is crystallized from CH$_2$Cl$_2$/pentane. The yields are 71% (No. 88) and 68% (No. 89) [80].

The air-stable compounds are soluble in organic solvents. The proposed structures are based on the spectroscopic data [80].

C$_5$H$_5$Fe(CO)$_2$C$_5$H$_2$O(CH$_3$)(C$_4$H$_9$-t)CN and **C$_5$H$_5$Fe(CO)$_2$C$_5$H$_2$O(C$_6$H$_5$)(C$_4$H$_9$-t)CN** (Table **8**, Nos. **90** and **91**) have been obtained in 45 and 40% yields, respectively, by addition of t-C$_4$H$_9$(NC)C=C=O in benzene to FpCH$_2$C≡CR^1 (R^1 = CH$_3$, C$_6$H$_5$) in the same solvent and stirring the mixture at room temperature for 3 h. Solvent removal affords an oil that is chromatographed on Al$_2$O$_3$ with CH$_2$Cl$_2$ as eluent. Treatment of the semisolid remaining after solvent removal with pentane yields the product as a solid [80].

The air-stable compounds are soluble in organic solvents. The proposed structures are based on the spectroscopic data [80].

C$_5$H$_5$Fe(CO)$_2$C$_5$H$_3$(CF$_3$)$_2$(CN)$_2$ (Table **8**, No. **93**). Two apparently isomeric solids have been isolated from the reaction of FpCH=C=CH$_2$ with (E)-CF$_3$(NC)C=C(CN)CF$_3$ by chromatography on Al$_2$O$_3$ eluting with pentane/benzene, then with benzene. The first solid was obtained in insufficient amount for satisfactory characterization (m.p. 72 to 76 °C, ν(CO) = 1983, 2032 cm^{-1}), and the second solid appeared from spectroscopic data to be a mixture of two diastereomers with the CF$_3$ groups both cis and trans [83].

$C_5H_5Fe(CO)_2C_5H_2(CF_3)_2(CN)_2CH_3$ (Table 8, No. **96**). The product resulting from the reaction of FpCH=C=CHCH$_3$ with (E)-CF$_3$(NC)C=C(CN)CF$_3$ could not be separated into the expected isomers either by chromatography on Al$_2$O$_3$ or by crystallization from 2:1 pentane/benzene and 4:1 pentane/CH$_2$Cl$_2$. However, the spectra indicate the predominance of one of the possible four diastereomers with the CN groups trans to the CF$_3$ groups [83].

$C_5H_5Fe(CO)_2C_5H_2(CF_3)_2(CN)_2C_6H_5$ (Table 8, Nos. **97** and **98**) have been obtained (5% cis-isomer, 49% trans–isomer) from the reaction of FpCH$_2$C≡CC$_6$H$_5$ with (E)-CF$_3$(NC)C=C(CN)CF$_3$ in benzene for 5 min. A similar proportion results from the corresponding reaction with (Z)-CF$_3$(NC)C=C(CN)CF$_3$, presumably due to isomerization of the (Z)-reactant to the trans-isomer in the course of the reaction [62, 83], see also [74]. Separation of the diastereomers has been accomplished by a combination of chromatography on Al$_2$O$_3$, elution with 2:1 pentane/benzene and fractional crystallization from the same solvents [83].

The diastereomers can be distinguished by their ^{19}F NMR spectra, each of which shows two ^{19}F resonances for the nonequivalent CF$_3$ groups (for a graphical representation see [62]). However, the J(CF$_3$, CF$_3$) values for nonequivalent cis-CF$_3$ groups are much larger (~10 Hz) than those for nonequivalent trans-CF$_3$ groups, which appear to be zero, because of through-space F–F coupling. Thus, the spectrum of the cis-isomer consists of two 1:3:3:1 equal intensity quartets, whereas the spectrum of the trans-isomer comprises two equal-height singlets. Moreover, the chemical shifts of the trans-isomer occur at higher fields than those of the cis-isomer. An additional spectroscopic difference is observed in the methylene region of the ^{1}H NMR spectra. Whereas the cis-isomer exhibits an AB pattern with J(H$_A$, H$_B$) = 17 Hz, expected for magnetically nonequivalent geminal protons, the trans-isomer shows the corresponding resonance as a singlet [83].

The air-stable compounds are only slightly soluble in pentane but increasingly soluble in benzene, CH$_2$Cl$_2$, and acetone. They have been found to display configurational stability toward prolonged storage in CHCl$_3$ or benzene [83].

$C_5H_5Fe(CO)_2C_5F_7$ (Table 8, No. **99**). The mass spectrum shows the molecular ion (relative intensity 18%) and fragments resulting from successive loss of the CO groups (20 and 26%). Other ions observed are [C$_5$H$_5$FeC$_5$F$_2$H]$^+$ (100), [C$_5$H$_5$FeC$_2$F$_3$]$^+$ (11), [C$_5$F$_5$]$^+$ (72), [C$_3$HF$_6$]$^+$ (20), [C$_5$H$_5$FeF]$^+$ (97), [C$_3$HF$_2$]$^+$ (24), [C$_5$H$_5$]$^+$ (21), Fe$^+$ (17), and [C$_3$H$_3$]$^+$ (25) [28].

$C_5H_5Fe(CO)_2C_5F_6Cl$ (Table 8, No. **100**). The reaction residue (Method I) is worked up by sublimation at 100 °C/0.1 Torr giving a mixture of orange and purple solids (color from Fp$_2$), followed by chromatography on Florisil with benzene/light petroleum as eluent. Removal of solvent from the first eluate, sublimation at 80 °C/10^{-2} Torr and crystallization from light petroleum afford the product [9].

$C_5H_5Fe(CO)_2C_5H_5$ (Table 8, No. **101**) has been obtained in 77% yield by treatment of [Fp(η^2-C$_5$H$_8$-cyclo)]BF$_4$ in CH$_3$NO$_2$ with NaBH$_3$CN (1:1.25 mole ratio) at 0 °C for several minutes, evaporation of the solvent, extraction of the residue with petroleum ether, and evaporation of the extracts [70]. FpC$_5$H$_5$ has been prepared in 40% yield by addition of an excess of NaC$_5$H$_5$ in THF to [Fp$_2$I]BF$_4$ in THF, solvent removal after 1.5 to 2 h, and chromatography of the residue on Al$_2$O$_3$, eluting with benzene/petroleum ether. The amounts of Fp$_2$ and ferrocene formed increase with increasing reaction time [51]. The same three products also result from the reaction of FpX (X=Cl, Br, I) with NaC$_5$H$_5$ or LiC$_5$H$_5$ in THF at room temperature for 1 h and similar workup of the reaction as aforementioned, 15% yield [1, 3, 4]. The compound has been purified by crystallization from ligroin (b.p. 40 to 60 °C) at −70 °C (orange-red plates) [3] or from pentane at −78 °C followed by a slow vacuum sublimation for 10 h at 25 °C and 10^{-2} Torr(?) onto a water-cooled probe (small red prisms) [10]. FpC$_5$H$_5$ forms in 25% yield in addition to Fp$_2$ when FpBr is allowed to react with $NaOCH_3$ in anhydrous CH_3OH at 0 °C for 30 min and at 20 °C for 30 min further. The two products are separated by chromatography on Al_2O_3. For purification, the title compound is recrystallized from pentane at −70 °C [13]. The analogous reaction of FpCl with $NaOCH(CF_3)_2$ in THF for 1 h gives also Fp_2 and a small amount of a product, which due to its IR spectrum, is suggested to be FpC_5H_5 [26]. The compound has been obtained along with ferrocene and $FpCH_2CH=CHCH_2CH=CH_2$ by treatment of the dinuclear complex $[Fp(CH_2=CHCH_2CH_2CH=CH_2)Fp]^{2+}$ with $N(C_4H_9\text{-}n)_3$ in CH_3NO_2 and then with NaI in acetone [99]. The formation of FpC_5H_5 by deprotonation of $[Fp(\eta^2\text{-}C_5H_6\text{-cyclo})]^+$ with either isobutyral-dehyde or 1-pyrrolidinocyclohexene or lithium methyl acetoacetate at −70 °C is briefly mentioned in [78].

FpC_5H_5 has been found by ^1H NMR spectroscopy to be a "fluxional" molecule in solution characterized by a dynamic nature of bonding between iron and one C_5H_5 ligand at room temperature and a static structure at low temperatures. Thus, the ^1H NMR spectrum in CS_2 at 30 °C exhibits two resonances, one at $\delta = 4.40$ ppm $(\pi\text{-}C_5H_5)$ and another of essentially the same integrated intensity, but broader, at $\delta = 5.70$ ppm $(\sigma\text{-}C_5H_5)$. However, as the temperature is lowered, the latter signal broadens, and the signal at $\delta = 4.40$ ppm remains unchanged. At −60 °C three bands are observed at $\delta = 3.5$, 4.4, and ~6 ppm with relative intensities 1:5:4. The signal at 3.5 ppm is fairly broad even at −100 °C and is assigned to the proton on the carbon bonded to iron (H-5) [10]. A study of the fine structure of this resonance in ether at −85 °C showed it to be a triplet with $J(H\text{-}1,5) = J(H\text{-}4,5) = 1.2$ Hz [25]. The band assigned to the two remaining pairs of protons of the monohapto ring $(\delta \sim 6$ ppm) begins as a single broad, asymmetric band at ~−40 °C (multiplet; H-1 to H-4) and separates into two bands, each of relative intensity 2 below −60 °C, indicating that the static structure of the monohapto ring is attained. From a consideration of the fine structure of the A_2B_2 multiplet, particularly of the asymmetric variation in its envelope as a function of temperature, it is concluded that in solution an intramolecular reorientation process occurs by repeated 1,2-shifts, probably passing through a polar metal–alkene transition state [10]. However, in [17] it is demonstrated that this transition state is essentially nonpolar and it is suggested that the 1,2-shifts must be regarded as degenerate [1.5]sigmatropic rearrangements. A consideration of the HOMO also predicts a [1.5]sigmatropic shift occurring in a suprafacial manner [100]. The assignment of the low-field side of the asymmetric band to H-1,4, which proves 1,2-shifts, is evidenced by computer simulated spectra, obtained after iterative refinement of coupling constants [21], and by the observation of the corresponding satellite lines due to coupling to ^{13}C by data accumulation at low temperature [25]. The ^1H NMR spectrum shows the same behavior in ether as in CS_2, indicating that there is no specific solvent dependence of the rearrangement process [10]. Exchange rates (in the range 10 to 10^3 s^{-1}) have been calculated from the width of line, due to the σ-bonded ring observed between −35 to 0 °C, and values of the activation parameters have been obtained from an Arrhenius plot: $E_a = 8.5 \pm 0.8$ kcal/mol [25].

A solid state (wide line) ^1H NMR study revealed that the monohapto ring reorientates not only in solution but also in the solid state. Thus, from a comparison of the observed and calculated second moment of the ^1H NMR curve at various temperatures, it is concluded that at 120 K, motion of only the pentahapto ring is responsible for the change in the spectrum, whereas between room temperature and the melting point, the monohapto ring is also dynamic [33], see also [146].

The fluxional behavior of FpC_5H_5 is also briefly mentioned in [4, 7, 56]. A correlation between the chemical shift of the C_5H_5 protons and the bonding character of the C_5H_5 ring (σ or π), as well as the reactivity of the compound towards maleic anhydride is pointed out in [5] for FpC_5H_5 and related compounds.

References on pp. 268/72

Fourier transform ^{13}C NMR spectra have been recorded over the temperature range +52 to -88 °C, showing a behavior similar to the ^{1}H NMR spectra. Thus, at +27 °C the proton-decoupled ^{13}C NMR spectrum consists of a sharp resonance due to the pentahapto ring, a broad resonance due to the monohapto ring, which is markedly narrowed at +60 °C [34, 39], and a resonance due to the two CO groups. As a solution of FpC$_5$H$_5$ in CS$_2$/toluene (3:1) is cooled to -78 °C, the broad signal collapses completely and is supplanted by three signals in the intensity ratio 2:2:1. Assignment of the latter resonances has been made by comparison with the chemical shifts of the allyl group in FpCH$_2$CH=CH$_2$. The fact that the signal at $\delta = 47.5$ ppm sharpens more slowly with decreasing temperature than that at 71.2 ppm is seen as evidence for 1,2-shifts being the predominant mode of site exchange. Kinetic data have been extracted from the ^{13}C NMR spectrum over the range from +52 to -69 °C (corresponding to exchange rates of 10^5 to 10 s^{-1}). Line shape analysis based on a one spin–three site model and the corresponding Arrhenius plots give E$_a$ = 10.7 ± 0.5 kcal/mol. E$_a$ is a little higher than values obtained from ^{1}H NMR spectra (8.5 ± 0.8 kcal/mol [25] and 9.8 ± 0.1 kcal/mol [21]) [32]. A solvent effect on the energy barrier for the fluxional process has been found in [139]: E$_a$ = 11.1 ± 0.2 kcal/mol in CS$_2$/C$_6$D$_5$CD$_3$ (3:1) and E$_a$ = 10.3 ± 0.2 kcal/mol in CD$_3$COCD$_3$.

As criteria for distinguishing between π- and σ-bonded C$_5$H$_5$ groups, the ^{1}H and ^{13}C NMR chemical shifts and the J(H,H) and J(^{13}C,H) coupling constants are discussed for a series of metal cyclopentadienyls in [39].

For an assignment of the complete IR spectrum see [145]. According to [3] a rather weak absorption band observable in the C–H stretching region below 3000 cm^{-1} and attributed to the "aliphatic" C–H bonds also indicates a structure in which one C$_5$H$_5$ ring is σ-bonded to iron.

The He(I) photoelectron spectrum of FpC$_5$H$_5$ is represented graphically in [101]. The measured vertical ionization potentials and empirical assignments are given below:

IP (in eV)	assignment (No. of IP's)
7.58	Fe 3d (2)
8.18	Fe 3d (1) + Fe–C σ (1)
8.76	π C$_5$H$_5$ (1)
10.20	Fe–π C$_5$H$_5$ (3)
11.00	π C$_5$H$_5$ (1)

A comparison with the spectrum of FpCH$_3$ shows that the lowest lying ion states resulting from ionization of molecular orbitals with large Fe 3d character move to lower energy in FpC$_5$H$_5$. Extended Hückel-type calculations revealed that the orbitals in FpC$_5$H$_5$, which are primarily olefinic in character, have substantial amounts of metal d character. A good correlation between the cyclic voltammetric oxidation potentials (for FpC$_5$H$_5$, E$_{1/2}$ = 0.75 eV, measured in THF/0.05 M [N(C$_4$H$_9$)$_4$-n]BF$_4$ vs. SCE) and the energies of the lowest ion states (from UV photoelectron spectra) of FpCH$_3$, FpCH$_2$CH=CH$_2$, and FpC$_5$H$_5$ has been noted [101].

FpC$_5$H$_5$ crystallizes in the monoclinic system, space group P2$_1$/c – C$_{2h}^5$ (No. 14), a = 12.53 ± 0.03, b = 7.50 ± 0.02, c = 11.29 ± 0.02 Å, β = 98.0 ± 0.5°; Z = 4, d$_c$ = 1.54 and d$_m$ = 1.5 g/cm^3. The molecular structure (see **Fig. 9**) shows that one C$_5$H$_5$ ring is present as a normal σ-bonded 2,4-cyclopentadienyl group. The Fe–CO distances, averaging 1.70 Å, are shorter than those found in related compounds, thus indicating considerable π-character in the Fe–C bonds [10]. The structure is discussed in relation to current theories concerning the bonding in compounds of the ferrocene type [3].

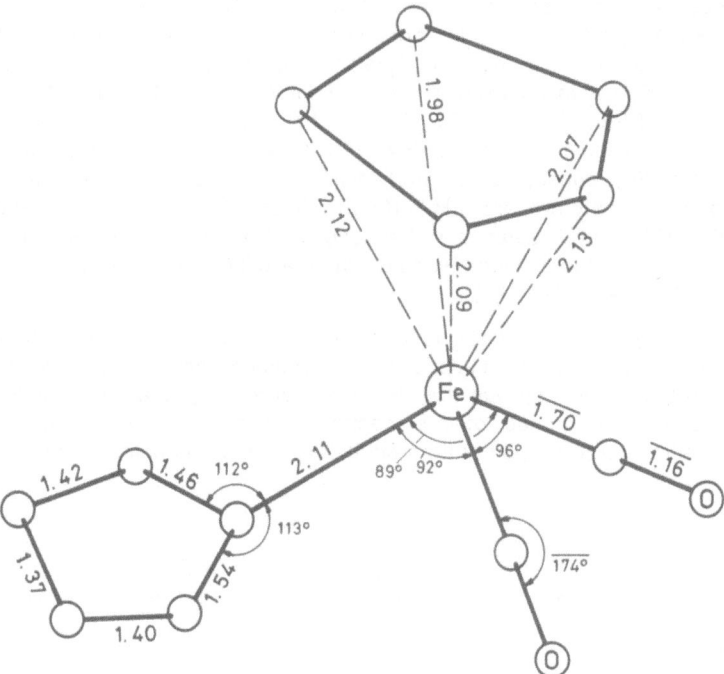

Fig. 9. Molecular structure of $C_5H_5Fe(CO)_2C_5H_5$ (No. 101) with selected bond lengths (in Å) and angles.

After several weeks, samples protected from light have decomposed almost completely to ferrocene together with only a small amount of brown insoluble material [3], see also [10]. At 90 °C rapid decomposition ensued, but only a trace of ferrocene sublimed. Fp_2 was readily extracted from the residue by acetone [3].

Polarographic reduction of FpR (R=e.g. C_5H_5) in $CH_3CN/0.1$ M $[N(C_2H_5)_4]ClO_4$ at 25 °C takes place in an irreversible two-electron step giving the anions $[Fp]^-$ and R^-. For FpC_5H_5 a half-wave potential $E_{1/2} = -1.32$ V vs. SCE has been measured. A good linear correlation between the half-wave potentials of a series of FpR compounds including FpC_6H_6 and the pK_a values of the respective hydrocarbons RH has been found [12, 41], see also [27].

FpC_5H_5 is readily oxidized by quinone in the presence of picric acid or aqueous mineral acids. The cation $[(C_5H_5)_2Fe(CO)_2]^+$ is assumed to be formed, because the aqueous solutions obtained give an instantaneous precipitate with Reinecke salt [1].

Stirring FpC_5H_5 in ligroin with deaerated boiling HCl under N_2 for 40 min leads to cleavage of the Fe–C bond and formation of FpCl [3], whereas treatment of the compound in petroleum ether with gaseous HCl causes immediate precipitation of the salt $[Fp(\eta^2-C_5H_6-cyclo)]Cl$ [6].

For the following reactions with phosphines and phosphites, a radical chain mechanism is proposed [95, 129]. From the reaction with $P(C_6H_5)_2CH_3$ in acetone, Fp_2 was isolated as the only identifiable product, in addition to a material that appeared to be an Fe^{II} salt [81]. However, when the same reaction is carried out in $CHCl_3/CH_3COCH_3$ (1:4) for 1 h, $[C_5H_5Fe(CO)(P(C_6H_5)_2CH_3)_2]^+Cl^-$ is formed. In pure $CHCl_3$ this reaction proceeds only very slowly [95, 129]. In contrast, FpC_5H_5 reacts rapidly (15 min) with $P(CH_3)_3$ in benzene at room temperature to give a yellow precipitate of $[C_5H_5Fe(CO)(P(CH_3)_3)_2]^+[C_5H_5]^-$. No reaction

References on pp. 268/72

has been observed between FpC_5H_5 and $P(C_6H_5)_3$ in $CHCl_3/CH_3COCH_3$. UV irradiation of FpC_5H_5 and $P(C_6H_5)_3$ in benzene in the presence of Fp_2 affords only ferrocene [129]. Formation of the CO substitution products $C_5H_5Fe(CO)(^2D)C_5H_5$ occurs when FpC_5H_5 is allowed to react with $P(OC_6H_5)_3$ in benzene in the dark for 4 to 8 h [81, 129] or briefly irradiated in the presence of $P(OC_6H_5)_3$ or $P(OC_3H_7-i)_3$ in benzene. On longer irradiation (>5 min), conversion to ferrocene takes place. However, if the irradiated solution also contains a small amount of Fp_2, substitution is markedly enhanced [95, 129]. The catalytic effect of Fp_2 has also been used to prepare $C_5H_5Fe(CO)(PF_2N(CH_3)_2)C_5H_5$ from FpC_5H_5 and a threefold excess of $PF_2N(CH_3)_2$ by irradiation in benzene for 1 h [112]. Rapid reaction occurs with $P(OR^1)_3$ ($R^1 = CH_3$, C_2H_5) in benzene leading, via an Arbuzov-like rearrangement, to C_5H_5Fe-$(CO)(P(OR^1)_3)(P(O)(OR^1)_2)$ [81, 129].

FpC_5H_5 reacts with $4-CH_3C_6H_4SO_2NCO$ to give the bicyclic γ-lactam XLI. The stereochemistry assigned to this product corresponds to that resulting from sterically preferred cycloaddition trans to the bulky Fp group [37]. However, according to a later publication by the same authors [69], the product of this reaction is compound No. 104, assumed to form through proton transfer in a dipolar intermediate followed by sigmatropic shift of the C–metal bond.

XLI

Reactions of FpC_5H_5 with electrophilic alkenes and alkynes afford 1:1 adducts of the reactants containing bicyclic norbornenyl and norbornadienyl rings. For details, see Preparation Method Vb on p. 193 and further information on compounds No. 165, 166, 167, 169, 170, and 172. The nature of the stereochemical results obtained for the products precluded distinguishing between a dipolar cycloaddition mechanism and a concerted Diels–Alder one [84]. As reported in Method Vb, a mixture of cycloadducts (Nos. 166 and 167) isomeric with compound No. 165 results from the reaction of FpC_5H_5 with (Z)-NCCH=CHCN. Kinetic measurements in various solvents studied under pseudo-first-order conditions (using a large excess of (Z)-NCCH=CHCN) revealed that these reactions obey first-order kinetics for at least 3 half-lives. The observed rates (in s^{-1}) are as follows: benzene, 3.8×10^{-4}; CH_2Cl_2, 7.3×10^{-4}; CH_3OH, 16×10^{-4}. Although the relative rates of reaction are comparable, a solvent dependence of the exo (No. 166): endo (No. 167) product ratio has been found: 6:1 in C_6H_6, 1:1 in CH_2Cl_2, 1:2 in CH_3OH. Mechanistic considerations support a concerted [4+2]cycloaddition for these reactions [132], see also [134]. Competitive reaction of FpC_5H_5 with (E)- and (Z)-NCCH=CHCN in $CDCl_3$ revealed the two alkenes to be of comparable reactivity [132] in contrast to the results with dimethyl fumarate and dimethyl maleate (see Method Vb) [126, 130]. Competitive reaction of FpC_5H_5 and $(CH_3)_5C_5Fe(CO)_2C_5H_5$ with dimethyl fumarate in $CDCl_3$ showed FpC_5H_5 to be ~5 times less reactive than the methyl substituted derivative [139].

FpC_5H_5 does not react with methyl acrylate or dimethyl maleate at 25 °C even after several days, but see Method Vb on p. 193. Slight heating of the solution containing FpC_5H_5 and methyl acrylate to 50 °C only causes formation of ferrocene [126].

No. 216 has been obtained by treatment of FpC_5H_5 with maleic anhydride in CH_2Cl_2 at 0 °C [126]. However, according to [4] the corresponding reaction in benzene gives a

References on pp. 268/72

product which is not sufficiently stable to allow isolation and characterization. The reactivity of FpC$_5$H$_5$ with maleic anhydride is also mentioned in [2, 5] without details on the reaction product.

UV irradiation of FpC$_5$H$_5$ with an excess of F$_2$C=CF$_2$ in hexane at room temperature for 50 h causes unusual carbon–carbon double bond cleavage of F$_2$C=CF$_2$ and formation of small quantities of compound XLII, in addition to ferrocene [64, 75], see also [77].

XLII

XLIII: R^1 = CF$_3$
XLIV: R^1 = COOCH$_3$

Photolytic reaction of FpC$_5$H$_5$ with an excess of CF$_3$C≡CCF$_3$ in hexane at room temperature for 30 h gives the 1:1 adduct XLIII corresponding to a formal [4+2]cycloaddition to an uncoordinated substituted cyclopentadiene, followed by a CO insertion reaction [64, 75], see also [77]. A similar reaction (20 h irradiation) with CH$_3$OOCC≡CCOOCH$_3$ affords compound XLIV [75], see also [77].

Compound No. 105 is obtained when an excess of CF$_3$COCF$_3$ is condensed ($-196\,^\circ$C) onto a solution of FpC$_5$H$_5$ in hexane, followed by warming to room temperature [75], see also [77]. Compound No. 106 results from the reaction with (CF$_3$)$_2$C=C=O in CH$_2$Cl$_2$ at 0 $^\circ$C [139].

For the reaction of FpC$_5$H$_5$ with (NC)$_2$C=C(CN)$_2$, see p. 195.

C$_5$H$_5$Fe(CO)$_2$C$_5$H$_4$CH$_3$ (Table **8**, No. **102**). In the ^{13}C NMR spectrum the η^1-ring carbons give a signal which is poorly resolved at room temperature, but at 55 $^\circ$C three separate (but still broad) peaks can be distinguished at $\delta=98.9$, 111.6, and 117.6 ppm; only slight decomposition is observed at this temperature [112].

The compound reacts with P(OC$_6$H$_5$)$_3$ in benzene to give C$_5$H$_5$Fe(CO)(P(OC$_6$H$_5$)$_3$)C$_5$H$_4$CH$_3$. Substitution of one CO group by a ^{2}D ligand also takes place when a solution in benzene containing a threefold excess of PF$_2$N(CH$_3$)$_2$ is irradiated in the presence of catalytic amounts of Fp$_2$ for 1 h [112].

C$_5$H$_5$Fe(CO)$_2$C$_5$H$_4$CH$_2$C$_6$H$_5$ (Table **8**, No. **103**) has been obtained by addition of FpBr to a suspension of LiC$_5$H$_4$CH$_2$C$_6$H$_5$ prepared from LiC$_6$H$_5$ and benzylcyclopentadiene in ether. A mixture of ether/THF (3:1) is then added dropwise to the suspension during 1 h and after stirring for 2 h further the solvent is removed, the residue dissolved in benzene and chromatographed on Al$_2$O$_3$ with benzene as eluent. Evaporation of the eluate and rechromatography of the residue in ligroin afford the product [3].

The compound freezes to orange crystals in CO$_2$/acetone. It decomposes slowly in air. At room temperature it loses CO, producing benzylferrocene [3].

The reaction with 35% aqueous HBr in ligroin at room temperature for 1 h gives almost quantitatively FpBr [3].

C$_5$H$_5$Fe(CO)$_2$C$_5$H$_3$(C(CF$_3$)$_2$OH)$_2$ (Table **8**, No. **105**) has been obtained in 46% yield by condensation of an excess of CF$_3$COCF$_3$ onto a solution of FpC$_5$H$_5$ in hexane and subsequent warming to room temperature. After removal of the volatile material in vacuum the residue is recrystallized from CH$_2$Cl$_2$/hexane [75].

References on pp. 268/72

UV irradiation in ether at room temperature for 6 h leads to loss of CO and formation of the ferrocene derivative XLV [75].

$C_5H_5Fe(CO)_2C_5H_3(C(=O)CH(CF_3)_2)_2$ (Table **8**, No. **106**) has been prepared by bubbling $(CF_3)_2C=C=O$ into a solution of FpC_5H_5 in CH_2Cl_2 at 0 °C for 20 min. After addition of toluene and concentration, the mixture is set aside at −25 °C for 24 h. The precipitated crystals are washed with cold toluene and dried under reduced pressure, 70% yield [139].

XLV : $R^1 = C(CF_3)_2OH$
XLVI : $R^1 = COCH(CF_3)_2$

The single heptet at $\delta = 4.72$ ppm for the two $CH(CF_3)_2$ methine protons requires that the iron group rapidly migrates partially or entirely around the η^1-ring. A ^1H NMR spectrum at 203 K did not show any significant broadening of the heptet or the η^1-ring proton resonances, suggesting the energy barrier for the fluxional process to be very small [139].

From the IR data (high $\nu(CO)$ and low $\nu(C=O)$ frequencies) it is concluded that the Fe–C bond has considerable ionic character [139].

The compound converts to the ferrocene derivative XLVI by heating in toluene at 90 °C for 20 min [139].

$C_5H_5Fe(CO)_2C_5F_2(CF_3)_3$ (Table **8**, No. **107**) has been obtained by UV irradiation of $FpSn(CH_3)_3$ and $CF_3C{\equiv}CCF_3$ (1:2.7 mole ratio) in hexane at 76 °C for 40 h in a sealed tube. After removal of the reaction volatiles and extraction of the residue with hexane, the compound is separated from (E)-$FpC(CF_3)=CHCF_3$ and $(CH_3)_3SnF$, also formed, by repeated chromatography on SiO_2 with hexane/benzene as eluent. A reaction mechanism leading to the formation of the title compound is proposed [24].

$C_5H_5Fe(CO)_2C_6H_{10}CH_3$ (Table **8**, No. **109**). The electron–impact mass spectrum (70 eV) is characterized by three competing decomposition pathways of FpR commencing with loss of CO, R or R−H. The following ions (relative intensities) have been observed: $[M]^+$ (2), $[M-CO]^+$ (5), $[M-2CO]^+$ (<1), $[M-2CO-H_2]^+$ (35), $[M-2CO-2H_2]^+$ (7), $[M-R]^+$ (31), $[M-R-CO]^+$ (24), $[M-R-2CO]^+$ (56), $[M-C_7H_{12}]^+$ (31), $[M-C_7H_{12}-CO]^+$ (17), $[M-C_7H_{12}-2CO]^+$ (24), $[Fe]^+$ (64), $[M-RH-CO]^+$ (100), $[R]^+$ (60) [107].

Attempted hydrogen abstraction by treatment with $[C(C_6H_5)_3]BF_4$ in CH_2Cl_2 at 0° C has given only traces of the alkene complex XLVII, perhaps because of steric hindrance, although the reactants were consumed [98].

XLVII

$C_5H_5Fe(CO)_2C_6H_{10}CH_3$ (Table **8**, Nos. **110** and **111**). The preparation from Na[Fp] and cis- or trans-$CH_3C_6H_8X$-4 (X = $OSO_2C_6H_5$ [42] or $OSO_2C_6H_4CH_3$-4 [96, 105]) at −78 to +10 °C [105] (Method I) proceeds with inversion of configuration. After chromatography on Florisil and elution with hexane, the cis-complex No. 110 is obtained in 26% [42] or 22% yield

[105], whereas the trans–isomer No. 111 could be isolated only in very small amounts [42]. According to [96] the cis– and trans–complexes are formed in ~40% yields.

Treatment of the isomers with DX (X=Cl, CF$_3$COO) in CH$_2$Cl$_2$ proceeds with retention of configuration at the α–carbon atom to give FpX and 4–CH$_3$C$_6$H$_{10}$D. A mechanism is proposed that involves protonation of the iron followed by reductive elimination of alkane. The relative rates of this cleavage (trans>cis) are assumed to be the result of steric effects [96, 108]. Retention of configuration has also been observed in the reaction with CuCl$_2$ · 2H$_2$O (1:3 mole ratio) in ethanol at 0 °C, which causes rapid precipitation of CuCl and formation of the respective esters 4–CH$_3$C$_6$H$_{10}$COOC$_2$H$_5$ (cis and trans), the yields are improved in the presence of CO [42]. The reaction of the cis–isomer with P(C$_6$H$_5$)$_3$ (1:1 mole ratio) in CH$_3$CN at 39 °C for 18 h gives the cis–derivative of C$_5$H$_5$Fe(CO)(P(C$_6$H$_5$)$_3$)COC$_6$H$_{10}$CH$_3$–4 [105]. In contrast, inversion of configuration occurs when the cis–complex is allowed to react with a large excess of liquid SO$_2$ at room temperature over 18 h in an ampule, affording the trans S–sulfinate insertion product FpSO$_2$C$_6$H$_{10}$CH$_3$–4 [105].

C$_5$H$_5$Fe(CO)$_2$C$_6$H$_{10}$(CH=CH$_2$) (Table **8**, No. **112**) (cis or trans) is formed in 6% yield in addition to the main product XLVIII by deprotonation of the 1,7–octadiene complex [Fp(CH$_2$= CH(CH$_2$)$_4$CH=CH$_2$)Fp]$^{2+}$ with N(C$_4$H$_9$-n)$_3$ in CH$_3$NO$_2$ at 0 °C to [Fp(CH$_2$=CH(CH$_2$)$_3$CH= CHCH$_2$Fp)]$^+$ and further reaction with NaI in acetone. Protonation of the (E, E)– or (Z, Z)–diene complex FpCH$_2$CH=CH(CH$_2$)$_2$CH=CHCH$_2$Fp with HBF$_4$ · O(C$_2$H$_5$)$_2$ in CH$_2$Cl$_2$ at 0 °C followed by demetalation of the monocation formed by brief exposure to NaI in acetone gives 5 and 3%, respectively [123].

XLVIII

C$_5$H$_5$Fe(CO)$_2$C$_6$H$_{10}$OH (Table **8**, No. **113**) has been obtained as a labile and air-sensitive solid by treatment of cyclohexene epoxide with Na[Fp] in THF and quenching the solution of the alkoxide formed (LI) with water [38].

The observed ^{1}H NMR chemical shift at δ=3.10 ppm (CHO) is consistent with an axial conformation for this proton, resulting from trans opening of the epoxide ring [38].

C$_5$H$_5$Fe(CO)$_2$C$_6$H$_{10}$OCH$_3$ (Table **8**, No. **114**) has been obtained together with the carbene complex IL (each in 40% yield) by treatment of [Fp]$^-$ with cyclohexene oxide and quenching the anion formed (L) with [O(CH$_3$)$_3$]BF$_4$ at 0 °C. The formation of the intermediate as L rather than its acyclic tautomer LI (cf. formation of No. 113 via an alkoxide [38]) is based on spectroscopic data [82].

IL   L   LI

C$_5$H$_5$Fe(CO)$_2$C$_6$H$_9$ (Table **8**, No. **119**) is also formed by the reaction of FpH with cyclohexa–1,3–diene (2:1 mole ratio) in C$_6$D$_6$. Chemically Induced Dynamic Nuclear Polarization (CIDNP) studies indicate that this process involves a radical pair mechanism. The sites

References on pp. 268/72

of observed emissions are the positions 1 and 3; at all other positions (see Table 8) emissions are not observed because of overlap [144].

The compound decomposes rapidly in CH_2ClCH_2Cl at 50 °C ($t_{1/2} \sim 2$ min) [113]. The reaction with $4\text{-}CH_3C_6H_4SO_2NCO$ (1:1 mole ratio) in CH_2Cl_2 at room temperature for 30 min gives the bicyclic γ-lactam LII [69], see also [37, 48, 60]. A similar product LIII is obtained from FpC_6H_9 and CH_3OSO_2NCO in CH_2Cl_2 at 0 °C after 30 min. These cycloaddition reactions are shown to occur stereospecifically by a suprafacial addition of the electrophile trans to the Fp group [69], see also [37]. No well-defined product could be isolated from the reaction with $ClSO_2NCO$ [37].

LII : $R^1 = SO_2C_6H_4CH_3-4$
LIII : $R^1 = SO_2(OCH_3)$

$C_5H_5Fe(CO)_2C_6H_8OCH_3$ (Table 8, No. 120). Methoxy abstraction with $(CH_3)_3SiOSO_2CF_3$ did not afford the corresponding carbocyclic allene complex as in the case of compound No. 142; rather rearranged products formed from carbocation intermediates [122].

$C_5H_5Fe(CO)_2C_6H_8COOCH_3$ (Table 8, No. 121) has been obtained as an equimolar mixture of two positional isomers together with Fp_2 by refluxing $FpC(=CH_2)CH=CH_2$, $CH_2=CHCOOCH_3$, and hydroquinone (1:3.1:0.02 mole ratio) in CH_2ClCH_2Cl for 120 h, evaporation of the solvent and chromatography on Al_2O_3 with benzene as eluent. Although the three components of the resulting oil could not be separated, the ^1H NMR spectrum is given as evidence for the presence of the two isomers as [4+2]cycloadducts [138].

$C_5H_5Fe(CO)_2C_6H_7(COOCH_3)_2$ (Table 8, No. 122) has been prepared by refluxing $FpC(=CH_2)CH=CH_2$ and dimethyl fumarate (1:1 mole ratio) in CH_2Cl_2 for 96 h, evaporation of the solvent, trituration of the remaining oil, and addition of the resulting solid to hot ligroin, 70% yield [138].

$C_5H_5Fe(CO)_2C_6H_5(CH_3)_2NC_6H_5$ (Table 8, No. 128) has been prepared from the carbene salt LIV and $C_6H_5NH_2$ in CH_2Cl_2 at 20 °C, followed by addition of NaOH, 49% yield [111].

LIV

$C_5H_5Fe(CO)_2C_6H_5(CH_3)_2=NC_7H_4SN$ (Table 8, No. 129) has been prepared by heating the carbene salt LIV and 2-amino-benzothiazole in THF in the presence of $N(C_2H_5)_3$, 84% yield [111].

$C_5H_5Fe(CO)_2C_6H_5(CH_3)_2=C_9H_4O_2$ (Table 8, No. 130) has been prepared from the carbene salt LIV and sodium indandione, 66% yield. No further details are given [111].

$C_5H_5Fe(CO)_2C_6H_5(CH_3)_2=C(COOC_2H_5)_2$ (Table 8, No. 131) has been prepared from the carbene salt LIV and sodium diethyl malonate, 49% yield. No further details are given [111].

$C_5H_5Fe(CO)_2C_6H_5(CH_3)_2=C(CN)_2$ (Table 8, No. 132) has been prepared from the carbene salt LIV and sodium malononitrile, 61% yield. No further details are given [111].

References on pp. 268/72

C$_5$H$_5$Fe(CO)$_2$C$_6$H$_6$(CN)$_2$C$_6$H$_4$Cl-2 (Table **8**, No. **133**) has been prepared by refluxing FpC(=CH$_2$)CH=CH$_2$ and 2-ClC$_6$H$_4$CH=C(CN)$_2$ in CH$_2$ClCH$_2$Cl for 2 h, evaporation of the solvent, dissolution of the remaining oil in CH$_2$Cl$_2$, and addition of petroleum ether, 83% yield [138].

C$_5$H$_5$Fe(CO)$_2$C$_6$F$_9$ (Table **8**, No. **137**). The reaction mixture obtained from Na[Fp] and perfluorocyclohexene in THF at 0 °C to room temperature (Method I) is worked up by solvent removal at 0 °C/30 Torr and sublimation of the residue at 40 °C/0.05 Torr onto a probe cooled by dry ice [8].

The compound is mentioned in connection with the mass spectra of various FpR compounds, all of which exhibit the ion [(C$_5$H$_5$)$_2$Fe]$^+$, suggested to arise from ferrocene produced by thermal decomposition of the compounds during their introduction into the mass spectrometer [19].

The compound appears to be indefinitely stable on exposure to air. It decomposes in solution, but is not decomposed by water [8].

C$_5$H$_5$Fe(CO)$_2$C$_6$H$_5$(CF$_3$)$_2$ (Table **8**, No. **138**) has been prepared by sealing FpC(=CH$_2$)CH=CH$_2$ and CF$_3$C≡CCF$_3$ in an ampule with CH$_2$Cl$_2$ at −78 °C and allowing the mixture to stand at 24 °C for 3 h. After removal of the volatile components in vacuum the remaining oil is dissolved in petroleum ether, chromatographed on Al$_2$O$_3$ with petroleum ether as eluent and crystallized from the same solvent, 85% yield [138].

C$_5$H$_5$Fe(CO)$_2$C$_6$H$_5$(COOCH$_3$)$_2$ (Table **8**, No. **139**) has been prepared by refluxing FpC(=CH$_2$)CH=CH$_2$ and CH$_3$OOCC≡CCOOCH$_3$ (1:1 mole ratio) in CH$_2$ClCH$_2$Cl for 1.5 h, evaporation of the solvent, trituration of the remaining oil with petroleum ether, and crystallization from cyclohexane, 90% yield [138].

Addition of I$_2$ to the compound in CH$_2$Cl$_2$ at −78 °C and stirring the solution at 24 °C for 20 min leads to cleavage of the Fe–C bond to afford 4-iodocyclohexa-1,4-dien-1,2-dicarboxylic acid dimethyl ester which could not be obtained free of FpI [138].

C$_5$H$_5$Fe(CO)$_2$C$_7$H$_{12}$CH=CH$_2$ (Table **8**, No. **140**) (cis or trans) is suggested to be formed in addition to the cyclohexane complex LV by treatment of [Fp(CH$_2$=CH(CH$_2$)$_4$CH=CHCH$_2$Fp)]$^+$ (obtained either by deprotonation of [Fp(CH$_2$=CH(CH$_2$)$_5$CH=CH$_2$)Fp]$^{2+}$ with N(C$_4$H$_9$-n)$_3$ or by protonation of (E,E)-FpCH$_2$CH=CH(CH$_2$)$_3$CH=CHCH$_2$Fp with HBF$_4$ · O(C$_2$H$_5$)$_2$) with NaI in acetone [123].

IV   LVI   LVII

C$_5$H$_5$Fe(CO)$_2$C$_7$H$_{11}$ (Table **8**, No. **141**). Directly after the metalation reaction (Method I), the relatively unstable complex has not been isolated but was derivatized with (NC)$_2$C=C(CN)$_2$, yielding No. 190 [69].

C$_5$H$_5$Fe(CO)$_2$C$_7$H$_{10}$OCH$_3$ (Table **8**, No. **142**). The reaction with HBF$_4$ · O(C$_2$H$_5$)$_2$ at 0 °C gives the salt LVI (X=BF$_4$, yield 40%) and a substance assumed to be LVII (38% yield). Clean methoxy abstraction affording LVI (X=OSO$_2$CF$_3$ or BF$_4$, respectively) takes place either by addition of (CH$_3$)$_3$SiOSO$_2$CF$_3$ to the compound in CH$_2$Cl$_2$ at −78 °C (see also [122]) or by the reaction with [O(CH$_3$)$_3$]BF$_4$ in CH$_2$Cl$_2$ at 0 °C [136].

C₅H₅Fe(CO)₂C₇H₁₀(OC₂H₅) (Table **8**, No. **143**) has been prepared by reaction of the allene salt LVI with Na_2CO_3 and ethanol at room temperature for 2 h, solvent removal in vacuum, chromatography on Al_2O_3, and elution with hexane. The yield is 77% when the borate (LVI, X=BF_4) is used as starting material, whereas the triflate (LVI, X=OSO_2CF_3) gives only 54% [136], see also [122].

C₅H₅Fe(CO)₂C₇H₇ (Table **8**, No. **144**) is formed in 4% yield along with Fp₂ and ditropyl, when [Fp]⁻ is allowed to react with tropylium tetrafluoroborate in THF at $-70\,°C$. The product is purified by chromatography on Al_2O_3 [22].

The temperature dependence of the ¹H NMR spectrum is typical of fluxional molecules. While the C_5H_5 signal remains unchanged on lowering the temperature, the C_7H_7 peak centered at $\delta=4.8$ ppm splits into four broad resonances at about $-15\,°C$, which further convert into four well-defined complex signals (relative areas 1:2:2:2) at $-50\,°C$: $\delta=2.44$ (t, H-1, J=7 Hz), 4.32 (t, H-2,7, J=7 Hz), 5.20 (m, AA'XX' set, H-4,5, J(4,5)=7.5, J(3,4)=11.2, J(3,5)=1.1 Hz), and 6.05 (m, H-3,6) ppm. The assignments result from low-temperature double resonance experiments [22].

The compound is moderately air-stable in the crystalline state and can be kept for prolonged periods in degassed solutions at $0\,°C$. It is, however, rapidly decomposed in hydroxylic media and when its otherwise stable organic solutions are exposed to air [22].

The mass spectrum shows the molecular ion and fragments resulting from successive loss of the CO groups. Other intense peaks are $[C_7H_7Fe]^+$, $[C_5H_5Fe]^+$, $[C_7H_7]^+$, and Fe^+ [22].

C₅H₅Fe(CO)₂C₇H₇ (Table **8**, No. **145**) has been obtained as isomeric mixture by addition of 1-, 2-, and 3-lithiocycloheptatrienes, prepared from 1-, 2-, and 3-bromocycloheptatrienes and n-C_4H_9Li in THF at $-78\,°C$, to FpI in the same solvent at $-78\,°C$. After stirring for 45 min the mixture is warmed to room temperature and stirred for 20 min further, followed by addition of a small amount of Al_2O_3. Solvent removal and chromatography of the products (coated on Al_2O_3) on Al_2O_3 with pentane as eluent afford the mixture of the three isomers in 29% yield [119], see also [117]. Attempts to separate individual isomers have been unsuccessful [85].

Treatment of the isomers with $[C(C_6H_5)_3]PF_6$ in CH_2Cl_2 at $-78\,°C$ (30 min) to room temperature (30 min) gives the carbene salt [Fp=C_7H_6-cyclo]PF_6 [85, 119].

C₅H₅Fe(CO)₂C₈H₇ (Table **8**, No. **146**) is formed in 10% yield by the reaction of cyclooctatetraenyl lithium, prepared from bromocyclooctatetraene in ether and n-C_4H_9Li in hexane at $-80\,°C$, with FpCl at $-80\,°C$. After 30 min at room temperature, hydrolysis with saturated NH_4Cl solution and chromatography on Al_2O_3 in hexane give the product [58], see also [125].

The isodynamic interconversions of the cyclooctatetraene moiety have been observed. Exchange of diastereotopic CO ligand signals (¹³C NMR) allowed calculation of ΔG^+ (ring inversion) $=16.6$ kcal/mol; broadening and coalescence of the cyclooctatetraene ¹H NMR signals due to bond shift processes were consistent with ΔG^+ (bond shift) ~18 kcal/mol [125].

The IR spectrum shows a shoulder on each of the two strong CO bands (2021 and 1970 cm^{-1}), suggesting the presence of two conformations in solution possibly due to the known C_8-ring inversion [58].

The compound is surprisingly robust, it is recovered unchanged after heating under reflux in methylcyclohexane (2 h). No evidence has been obtained for loss of CO with concomitant coordination to Fe of one of the double bonds of the C_8H_7 ring [58]. However, according to [125], FpC_8H_7 is an air-sensitive oil which is best stored refrigerated in a solvent other than $CHCl_3$.

Protonation of the compound at $-40\,°C$ with either excess dry HCl or two equivalents of FSO_3H or CF_3COOH in dry CH_2Cl_2 produces the homoaromatic carbene complex LVIII. Addition of 1.1 equivalent of $[C(C_6H_5)_3]PF_6$ followed by dilution with cold, dry ether gives LVIII as PF_6^- salt [125].

LVIII   LIX

FpC_8H_7 reacts with $Fe_2(CO)_9$ in heptane to give the dinuclear complex LIX [58].

$C_5H_5Fe(CO)_2C_8F_7$ (Table **8**, No. **147**). In the $^{13}C\{^1H\}$ NMR spectrum the carbonyl carbons appear as multiplets owing to coupling with ^{19}F. ^{19}F and $^{13}C\{^{19}F\}$ NMR studies reveal that the cyclooctatetraene ring does not undergo rapid ring inversion or bond shifting on the NMR time scale. The ^{19}F NMR spectrum remains unchanged at $+60\,°C$ [128].

The mass spectrum shows the molecular ion and the fragments $[M-CO]^+$ and $[M-2CO]^+$ [128].

FpC_8F_7 reacts with $[Fp]^-$ to give the corresponding 1,5-disubstituted cyclooctatetraene complex [133, 142].

The reactivity towards a variety of ^2D donor ligands under photochemical conditions, giving a mixture of diastereomers of $C_5H_5Fe(CO)(^2D)C_8F_7$, is briefly mentioned in [142].

$C_5H_5Fe(CO)_2C_6F_5$ (Table **8**, No. **148**) is stable at ambient temperature as solid [16, 57] and for several weeks in solution [57], but isomerizes to the corresponding aromatic isomer FpC_6F_5 (see Section 1.5.2.3.16.2, Table 7, No. 20) on heating [16].

The Diels–Alder adduct No. 204 is obtained when a solution of the compound and cyclopentadiene in CH_2Cl_2 is shaken at room temperature for 12 h in a tube sealed under vacuum. Similar treatment of FpC_6F_5 in furan for three days or in 2,3-dimethylbuta-1,3-diene for one month in the absence of light affords No. 55 in Table 1 in "Organoiron Compounds" B 14, 1988, and No. 201 in Table 8, respectively. In contrast, heating of the compound and pyrrole in $CHCl_3$ at $60\,°C$ for four days in a sealed tube gives only FeC_6F_5, presumably formed by rearrangement of the starting material [59].

$C_5H_5Fe(CO)_2C_7H_9$ (Table **8**, No. **150**). Protonation yields the exo–norbornene complex LX. No further details are given [54].

LX

$C_5H_5Fe(CO)_2C_7H_{11}$ (Table **8**, No. **151**) has been obtained from Na[Fp] and a mixture of stereoisomeric norcarane bromides (Method I) [31].

The Fp group is assigned as exo, since models show that the alternative endo-form would be prohibitively crowded [31].

The compound reacts with SO_2 to give $FpSO_2C_7H_{11}(C_7H_{11}=2$-methylencyclohexyl). The reaction mechanism is discussed [31].

$C_5H_5Fe(CO)_2C_7H_{11}$ (Table **8**, Nos. **152** and **153**) are oxidatively, photolytically, and thermally labile oils. As monitored by low temperature ^{13}C NMR, the endo-compound is stable for one week at $-20\,°C$ under N_2, while the exo-derivative is significantly decomposed after two days. This may explain the lower yield of the exo-compound derived by photolysis of the corresponding acyl complex [75].

The endo-norbornyl complex reacts with $[C(C_6H_5)_3]BF_4$ in CH_2Cl_2 at $0\,°C$ for 15 h under N_2 or Ar to give a mixture of Fp (endo-norbornylcarbonyl) and $[Fp(\eta^2$-exo-norbornene)]BF_4. Under similar conditions, but at $25\,°C$ (6 h), only the π-complex is formed, whereas at $-20\,°C$ (24 h) the acyl complex is obtained along with traces of the π-complex. The acyl derivative is also the main product when the reaction is carried out at $25\,°C$ for 5 h under excess CO. Addition of triphenylmethyl radical to the reaction mixtures inhibits the acyl formation. It is assumed that the CO insertion follows a catalytic radical chain process, initiated by a reversible one-electron transfer from FpR to the cation $[C(C_6H_5)_3]^+$. This assumption could be confirmed by EPR studies and electrochemical measurements. Mechanistic studies indicate the resulting exo-Fp-norbornene salt (no endo π-complex observed) is formed by β-elimination/σ-π rearrangement. The exo-compound is less reactive. Thus, under Ar or N_2 Fp(exo-norbornylcarbonyl) and $[Fp(\eta^2$-exo-norbornene)]BF_4 are formed at $+25\,°C$ (7 h), whereas at $-20\,°C$ (24 h) no reaction takes place. However, when Ar is replaced by CO in the latter reaction and the mixture is stirred for an additional 2 h, the acyl complex is formed as the sole product [141], see also [135].

$C_5H_5Fe(CO)_2C_7H_{10}D$ (Table **8**, No. **154**). In the $^{13}C\{^1H\}$ NMR spectrum all signals except the one at $\delta = 27.8$ ppm, which is split and shifted slightly upfield by deuterium, coincide with the corresponding resonances in the spectrum of the nondeuterated compound [141].

Treatment of the compound with $[C(C_6H_5)_3]BF_4$ in CD_2Cl_2/CH_2Cl_2 at $33\,°C$ for 10 h under N_2 causes formation of the salt LXI [141].

$C_5H_5Fe(CO)_2C_7H_9D_2$ (Table **8**, No. **155**). The $^{13}C\{^1H\}$ NMR spectrum is identical with that of the undeuterated compound, except for the splitting and slight upfield shift of the singlets at $\delta = 30.1$ and 29.0 ppm caused by deuterium at C-5 and C-6 [141].

The salt LXII has been formed exclusively in the reaction with $[C(C_6H_5)_3]BF_4$ in CH_2Cl_2 at $25\,°C$ for 36 h under N_2. No deuterium-scrambled π-complex could be detected by $^{13}C\{^1H\}$ NMR [141].

$C_5H_5Fe(CO)_2C_7H_8(CN)$ (Table **8**, Nos. **161** and **162**). The highest yields have been obtained when the reaction of FpC_5H_5 with acrylonitrile (Method Vb) was carried out in the presence of $Al(C_2H_5)_2Cl$ [130].

References on pp. 268/72

$C_5H_5Fe(CO)_2C_7H_7(CN)_2$ (Table **8**, Nos. **165** to **167**). A cycloadduct of undetermined stereo-chemistry has been obtained by the reaction of FpC_5H_5 with (E)-NCCH=CHCN in benzene for 1 h, solvent removal, dissolution of the residue in CH_2Cl_2, and addition of pentane (cf. Method Vb) [84]. However, according to [132] this product is identical with No. 165. The yields resulting from the reaction of FpC_5H_5 with (Z)-NCCH=CHCN (Method Vb) depend on the solvent used: No. 166, 77% (C_6H_6), 49% (CH_2Cl_2), 33% (CH_3OH) and No. 167, 13% (C_6H_6), 45% (CH_2Cl_2), 60% (CH_3OH) [132].

The structural assignments of the three cycloadducts are based on their ^1H NMR spectra and supported by the properties of their oxidation products with $[NH_4]_2[Ce(NO_3)_6]$, see p. 195 [132].

Each of the pure isomers is rapidly isomerized with $(CH_3)_3COK$ in $(CH_3)_3COH$ at 30 °C to an equilibrium at which the ratio of No. 165: No. 166: No. 167 is 10:1:3 [132].

$C_5H_5Fe(CO)_2C_7H_5(CF_3)_2(CN)_2$ (Table **8**, No. **169**) has been obtained by stirring a solution of FpC_5H_5 and (Z)- or (E)-$CF_3(NC)C=C(CN)CF_3$ (~1:1 mole ratio) in pentane at 25 °C for 10 min and subsequent concentration in a stream of nitrogen. The precipitated product is collected by filtration [84].

The observation of two singlets of equal intensity in the ^{19}F NMR spectrum is thought to be consistent with the presence of the diastereomer with trans-CF_3 groups or, less likely, equal amounts of the two meso compounds. It is noteworthy that the stereochemistry does not change when (E)-$CF_3(NC)C=C(CN)CF_3$ is replaced by the (Z)-isomer in the reaction with FpC_5H_5 [84].

$C_5H_5Fe(CO)_2C_7H_5(CF_3)_2$ (Table **8**, No. **170**) has been prepared by addition of excess $CF_3C\equiv CCF_3$ to FpC_5H_5 in CH_2Cl_2 under a dry ice condenser (cf. Method Vb). After 10 min the solvent is removed, the residue dissolved in pentane, and the filtered solution evaporated to dryness, 92% yield. The compound has also been obtained in addition to LXIII, ferrocene, (E)-$FpC(CF_3)=CHCF_3$, and other unidentified products by the reaction of $FpCH_2CH=CHCH_3$ with $CF_3C\equiv CCF_3$ in $CHCl=CCl_2$ at 60 to 65 °C for 12 h and chromatographic workup on Al_2O_3 [84].

Storage of the compound in $CDCl_3$ for several days leads to a color change from yellow to orange and formation of complex LXIV. Photolysis in cyclohexane for 15 min gives compound LXV by loss of one CO molecule [84].

LXIII   LXIV   LXV

$C_5H_5Fe(CO)_2C_7H_5(CN)_2$ (Table **8**, No. **172**) has been prepared by addition of NCC≡CCN in pentane/CH_2Cl_2 (1:1) to FpC_5H_5 in pentane (cf. Method Vb). After solvent removal and extraction of the residue with pentane, the pentane-insoluble material is dissolved in CH_2Cl_2 and the solution is passed through Al_2O_3. Addition of pentane to the filtrate and partial removal of the solvent afford the product [84].

References on pp. 268/72

$C_5H_5Fe(CO)_2C_8H_{10}OCOOC_2H_5$ (Table **8**, No. **175**). The ^1H and ^{13}C NMR spectra show that the compound is a single cis-diastereomer [121].

$C_5H_5Fe(CO)_2C_8H_7$ (Table **8**, No. **179**) has been obtained in 45 and 21% yield, respectively, by reduction of $[Fp(\eta^2-benzocyclobutadiene)]PF_6$ with $NaBH_4$ or $LiBH_4$ in THF (several min) and chromatographic workup on Al_2O_3, eluting with petroleum ether. Evaporation of the eluate gives a yellow oil that exhibits a ^1H NMR spectrum identical to that of an authentic sample of the title compound. A similar procedure with FpH as reducing agent gives a 40% yield of the compound. Reduction of the benzocyclobutadiene salt with Na[Fp] in THF at $-78\,°C$ and workup at $24\,°C$ as aforementioned affords 16% of the compound and the dinuclear complex LXVI [71]. The compound has also been obtained as yellow oil by the reduction of $[Fp(\eta^1-benzocyclobutenylidene)]PF_6$ with $LiAlH_4$ in ether. After 1 h, the solvent is evaporated, the residue taken up in petroleum ether, percolated through Al_2O_3 and eluted with petroleum ether and ether. Evaporation of the eluate gives the product in 60% yield [71], see also [44]. The reaction of FpC_3H_5-cyclo with $[Fp(\eta^1-benzocyclobutenylidene)]PF_6$ in CH_2Cl_2 at room temperature for 4 d affords an orange-yellow precipitate of $[Fp(CH_2=C=CH_2)]PF_6$ admixed with varying amounts of $[Fp(CH_2=CHCH_3)]PF_6$ and $[C_5H_5Fe(CO)_3]PF_6$; the solution contains the title compound, which is isolated by chromatography in better than 80% yield. The compound forms in addition to $[Fp(CH_2=CHCH=CH_2)]PF_6$, when $FpCH_2C_3H_5$-cyclo is allowed to react with the benzocyclobutenylidene salt [49]. A 5 to 9% yield of the compound is obtained in addition to the dinuclear complex LXVI by the reaction of Na[Fp] with trans-1,2-dibromo-, cis-1,2-diiodo-, or trans-1,2-diiodobenzocyclobutene in THF at $-78\,°C$ to room temperature. The yields increase when the reaction with the dibromo derivative is carried out in the presence of t-butanol (12%), methanol (37%), or excess cyclopentadiene (51%); the presence of water leads to less than 1% yield. The addition of styrene to the methanol reaction mixture reduces the yield to 28%. On the basis of these results and some other experiments a mechanism for the formation of the title compound and complex LXVI is proposed which involves initial formation of benzocyclobutadiene, addition of an Fp radical to benzocyclobutadiene and subsequent addition of a second Fp radical to form LXVI, or hydrogen abstraction from FpH to form No. 179 [66].

Addition of the compound to $[C(C_6H_5)_3]PF_6$ in CH_2Cl_2 at $-78\,°C$ and further reaction at $24\,°C$ afford the cationic carbene complex LXVII. The assumption that this product may be formed by α-hydride abstraction was supported by the corresponding reaction of the deuterated derivative No. 180, which gives the same product without incorporation of deuterium [44, 71], see also [45, 76].

For the cyclic voltammogram see [124].

$C_5H_5Fe(CO)_2C_8H_6D$ (Table **8**, No. **180**) has been prepared by reduction of $[Fp(\eta^1-benzocyclobutenylidene)]PF_6$ with $LiAlD_4$ in ether overnight, evaporation of the solvent, extraction of the residue with petroleum ether and passing the extracts through Al_2O_3 with petroleum ether and ether as eluents. Evaporation of the eluate gives the product in 57% yield [71], see also [44]. Also, treatment of $FpCH(CD_3)_2$ with the benzocyclobutenylidene salt leads to formation of the compound [49].

References on pp. 268/72

The observation of a pair of doublets with a large coupling constant $(J(2,3) = 14.7$ Hz) and the absence of a resonance at $\delta = 4.05$ ppm indicates that the protons on the four-membered ring are geminal and that the deuterium is attached to the ligating carbon of the benzocyclobutenyl ligand [44, 71].

For the reaction with $[C(C_6H_5)_3]^+$ see further information on No. 179.

$C_5H_5Fe(CO)_2C_8H_6D$ (Table 8, No. 181). The trans-isomer is formed in 23% yield along with less than 5% of the cis-isomer by the reaction of Na[Fp] with trans-1,2-dibromobenzo-cyclobutene in THF at $-78\,°C$ in the presence of a limited amount of D_2O. After warming to room temperature and addition of petroleum ether, the mixture is filtered through Celite, the solvents are evaporated and the residue is extracted with benzene and chromatographed on Al_2O_3. A 1:1 mixture of the cis- and trans-isomers is obtained when Na[Fp] is treated with trans-1-bromo-2-deuterobenzocyclobutene in THF at $-78\,°C$ to room temperature followed by a workup similar to that described before [66].

$C_5H_5Fe(CO)_2C_8H_6CH_2CH=CH_2$ (Table 8, No. 182) has been prepared by reductive demetalation of the dinuclear complex LXVIII with Na[Fp] in THF, addition of petroleum ether, filtration through Celite, evaporation of the solvent and chromatography on Al_2O_3 with petroleum ether and ether as eluents [71].

LXVIII   LXIX

$C_5H_5Fe(CO)_2C_8H_6CH_2CH=CH_2$ (Table 8, No. 183) has been prepared by treatment of the dinuclear complex LXIX with Na[Fp] in THF, evaporation of the solvent, extraction of the residue with petroleum ether, and passing the solution through Al_2O_3. Elution with petroleum ether and evaporation of the eluate give 89% yield of the trans-isomer [71]. The compound forms in 30% yield when LXIX is allowed to react with NaI in acetone [55].

A small coupling (1.8 Hz) between H-1 and H-2 confirms the trans-orientation of the iron and propenyl groups [71].

$C_5H_5Fe(CO)_2C_8H_6OH$ (Table 8, No. 184) is formed by treatment of $[Fp(\eta^1$-benzocyclobuten-ylidene)]PF_6$ with aqueous NaHCO_3 in CH_2Cl_2 [44, 71], see also [45].

The compound is unstable and readily decomposes during workup of the reaction mixture to benzocyclobutenone and Fp_2 [44, 45, 71].

$C_5H_5Fe(CO)_2C_8H_6OCH_3$ (Table 8, No. 185) has been prepared by treatment of $[Fp(\eta^1$-benzocyclobutenylidene)]PF_6$ with a mixture of CH_3OH and NaHCO_3 overnight, evaporation of the solvent, extraction of the residue with petroleum ether, and evaporation of the filtered solution, 79% yield [71], see also [44, 45].

$C_5H_5Fe(CO)_2C_8H_6OCH_3$ (Table 8, No. 186) has been prepared by stirring $[Fp(\eta^2$-benzocyclobutadiene)]PF_6$ and NaHCO_3 in CH_3OH overnight, followed by evaporation of the solvent, extraction of the residue with petroleum ether, passing the solution through Al_2O_3, and solvent removal, 82% yield [71], see also [55].

For the cyclic voltammogram see [124].

$C_5H_5Fe(CO)_2C_9H_{13}O$ and $C_5H_5Fe(CO)_2C_9H_7D_6O$ (Table **8**, Nos. **187** and **188**). Compound No. 187 is formed in 10% yield when $FpCH_2CH=CH_2$ is allowed to react with $[Fp(\eta^1\text{-cyclohex-enone})]BF_4$ in refluxing CH_2Cl_2 [121].

Evidence that No. 187 is a mixture of two stereoisomers is seen on examination of its ^{1}H NMR spectrum, which shows two C_5H_5 proton resonances at $\delta = 5.10$ and 5.15 ppm of equal intensity after addition of $Eu(fod)_3$ [tris(heptafluoro-2,2-dimethyloctan-3,5-diona-to)europium(III)]. Examination of the ^{1}H NMR spectrum of the hexadeuterio complex No. 188 in the presence of $Eu(fod)_3$ exhibits a signal at $\delta = 3.35$ ppm, assignable to H-3, as a doublet with $J(3,8) = 8$ Hz, consistent with the assignment of a cis-hydrindane structure LXX, LXXI. The C=O resonance observed in the ^{13}C NMR spectrum at $\delta = 214$ ppm supports this assignment [121].

LXX   LXXI   LXXII

Compound No. 187 reacts with HCl in CH_2Cl_2 to give the ring-opened complex LXXII [121].

$C_5H_5Fe(CO)_2C_9H_{12}O(COOC_2H_5)$ (Table **8**, No. **189**). The ^{1}H NMR spectrum, even in the presence of $Eu(fod)_3$, failed to give evidence for the existence of more than one isomer, but the ^{13}C NMR spectrum reveals that the compound is a mixture of two diastereomers. These show ^{13}C carbonyl resonances at $\delta = 208$ ppm and are assigned trans-hydrindanone structures [121].

For the reaction with $[NH_4]_2[Ce(NO_3)_6]$ see p. 195.

$C_5H_5Fe(CO)_2C_9H_{10}OCH_3$ (Table **8**, No. **191**). The ^{1}H and ^{13}C NMR spectra show the product to be a single stereoisomer, this is assigned a cis-hydrindenone structure ($\delta(CO) = 215$ ppm) [121].

$C_5H_5Fe(CO)_2C_9H_5O_2$ (Table **8**, No. **193**) has been obtained in 18% yield by treatment of Fp_2 with 2-diazoindan-1,3-dione in THF at $-110\,^\circ$C followed by UV irradiation at $-80\,^\circ$C for 5 h. After the photolysis, the mixture is stirred at room temperature for 12 h. Workup involves evaporation, chromatography of the oily residue on SiO_2, elution with benzene/ether (1:1), concentration of the eluate, and recrystallization of the solid formed from CH_2Cl_2/ether at $-35\,^\circ$C [106].

$C_5H_5Fe(CO)_2C_9H_7$ (Table **8**, No. **194**) has been prepared by treatment of indenylsodium with FpI (1:1 mole ratio) in THF at 25 °C for 5 h, solvent removal, extraction of the residue with benzene, and chromatography of the filtered and concentrated solution on Al_2O_3 with benzene/petroleum ether (1:3) as eluent. Rechromatography of the orange band, eluting with petroleum ether and solvent removal afford an oily, sticky residue. After drying under vacuum for 1 h and standing at 25 °C for several hours the product solidifies. It is recrystallized from petroleum ether, 20% yield [11]. The formation of the compound by deprotonation of the indenyl complex $[Fp(\eta^2\text{-}C_9H_8)]^+$ with either isobutyraldehyde or 1-pyrrolidinocyclohex-ene is briefly mentioned in [78].

No. 194 has been found to be a rigid (nonfluxional) molecule up to 70 °C, displaying a ^{1}H NMR spectrum with an ABX pattern. By a combination of a spin-decoupling experiment and study of the 1,3-dideuterio compound (No. 195) the assignment of the spectrum (in

CDCl$_3$) has been established (all three spectra are depicted). Thus, the two doublets at $\delta = 6.72$ and 6.53 ppm, each of relative intensity 1, are due to the nonequivalent protons (A, B) on carbon 2 and 3, while the peak of relative intensity 1 at $\delta = 3.97$ ppm is due to the proton (X) on carbon atom 1, the coupling constants being $J_{BX} \ll J_{AX} \approx 2$ Hz. The complex signal of total relative intensity 4 lying between $\delta = 6.95$ and 7.65 ppm results from the four aromatic protons. From these results a 1,2-shift mechanism is concluded [11]. A 1,3-shift is excluded because the 2-indenyl structure is so unfavorable, as shown by Hückel LCAO-MO calculations, that it does not serve as a suitable transition state or intermediate for the interconversion of the 1-indenyl to the 3-indenyl complex at a rate sufficient to give an averaged spectrum at room temperature [11, 23]. A parallel between the 1-indenyl system and the σ-C$_5$H$_5$ system of FpC$_5$H$_5$ which rearranges by a series of 1,2-shifts is pointed out [11].

FpC$_9$H$_7$ undergoes rapid decomposition above 70 °C [23].

C$_5$H$_5$Fe(CO)$_2$C$_9$H$_5$D$_2$ (Table **8**, No. **195**). The 1,3-dideuterio compound has been prepared from FpI and 1,1,3-trideuterioindene; the deuterioindene was made by successively hydrolizing the corresponding monolithium indenide in ether with D$_2$O, repreparing the lithium indenide, etc., through three cycles [11].

C$_5$H$_5$Fe(CO)$_2$C$_{10}$H$_{15}$ (Table **8**, No. **198**), prepared according to Method III, is separated from the coproduct Fp$_2$ ($\sim$1:1 mixture) by dissolving the residue remaining after solvent removal with pentane/ether (10:1) and chromatographing the filtrate on Al$_2$O$_3$ with pentane/ether (4:1) as eluent. Solvent removal results in deposition of crystals which are further purified by sublimation at 80 °C/10^{-3} Torr. The compound has also been obtained by decarbonylation of FpCOC$_{10}$H$_{15}$ with Rh(P(C$_6$H$_5$)$_3$)$_3$Cl (1:2 mole ratio) in benzene at room temperature for 1 h followed by solvent removal from the filtrate, digestion of the residue with hexane/benzene (3:2), chromatography on Al$_2$O$_3$, elution with hexane/benzene (3:2), and purification as aforementioned, 15% yield [68].

The compound is stable to sublimation and undergoes only superficial oxidation in air over a period of two years, ferrocene is one product of this decomposition [68].

C$_5$H$_5$Fe(CO)$_2$C$_{10}$H$_{15}$ (Table **8**, No. **199**). The residue, obtained from the reaction of Na[Fp] with 2-bromoadamantane in THF (room temperature, 2 d) after solvent removal (Method I), is dissolved in hot cyclohexane. The cooled extracts are then chromatographed on Al$_2$O$_3$ with cyclohexane/benzene (4:1) as eluent. The product is further purified by sublimation [68].

The compound is readily converted to FpCOC$_{10}$H$_{15}$ by heating under reflux in heptane and bubbling CO through the solution [68].

C$_5$H$_5$Fe(CO)$_2$C$_{10}$H$_9$O$_2$ (Table **8**, No. **200**) has been prepared by refluxing FpC(=CH$_2$)CH=CH$_2$ and p-quinone ($\sim$1:1 mole ratio) in CH$_2$ClCH$_2$Cl for 1.5 h. The black solid, formed on cooling, is filtered off and the filtrate is concentrated. Dissolution of the resulting oil in hot petroleum ether (b.p. 30 to 60 °C) and cooling of the filtered solution to -78 °C give the product in 85% yield [138].

Treatment of the compound with HPF$_6$ in ether at 0 °C leads to immediate precipitation of the salt LXXIII [138].

LXXIII

References on pp. 268/72

$C_5H_5Fe(CO)_2C_{10}H_4(CH_3)_2F_5$ (Table **8**, No. **201**) is formed in 5% yield, when a solution of compound No. 148 in 2,3-dimethylbuta-1,3-diene is kept in a sealed tube at room temperature for one month in the absence of light. Workup consists of removal of the volatile materials, chromatography of the remaining solid on Al_2O_3 with $CHCl_3$ as eluent and recrystallization from $CHCl_3$/light petroleum (b.p. 40 to 60 °C) at −20 °C [59].

The mass spectrum shows the molecular ion and fragments resulting from stepwise loss of CO, accompanied by cleavage of CH_3 groups and/or loss of fluorines. The retro-Diels-Alder cleavage of the parent ion is another major fragmentation pathway and gives rise to the peaks $[C_5H_5Fe(CO)_2C_6F_5]^+$, $[C_5H_5Fe(CO)C_6F_5]^+$, $[C_5H_5FeC_6F_5]^+$, and $[C_6H_{10}]^+$ [59].

$C_5H_5Fe(CO)_2C_{11}H_9$ (Table **8**, No. **202**) has been obtained by addition of Fpl in THF to a cold (−78 °C) solution of 4,5-benzo-1-lithiocycloheptatriene (prepared from the corresponding bromo derivative in THF and $n-C_4H_9Li$ in hexane) and further reaction at −78 °C for 16 h, then at room temperature for 1 h [117]. The product is eluted with a pentane/benzene mixture from Al_2O_3, 26% yield [102].

The reaction of the compound with $[C(C_6H_5)_3]PF_6$ to give the carbene salt $[Fp(\eta^1\text{-4,5-}$ benzocycloheptatrienylidene)]PF_6 is briefly mentioned in [102].

$C_5H_5Fe(CO)_2C_{11}H_9$ (Table **8**, No. **203**) has been obtained as a mixture of two isomers in 37% yield by addition of a cold (−78 °C) solution of 6- and 8-lithio-5H-benzocycloheptatriene (prepared from the respective bromo derivatives in THF and $n-C_4H_9Li$ in hexane) to Fpl in THF. After stirring for 45 min, the mixture is warmed to room temperature and Al_2O_3 is added. The reaction products coated on the Al_2O_3 are chromatographed on Al_2O_3 after solvent removal, eluting with pentane and then with benzene/pentane [119], see also [102].

Treatment of the isomers with $[C(C_6H_5)_3]PF_6$ in CH_2Cl_2 at −78 °C for 1 h, then warming to room temperature afford the carbene salt $[Fp(\eta^1\text{-3,4-benzocycloheptatrienylidene}]PF_6$ [119], see also [102].

$C_5H_5Fe(CO)_2C_{11}H_6F_5$ (Table **8**, No. **204**) has been prepared by shaking compound No. 148 and cyclopentadiene in CH_2Cl_2 at room temperature for 12 h in a sealed tube under vacuum, followed by removal of the volatiles and chromatography of the residue on Al_2O_3 with $CHCl_3$ as eluent, 65% yield [59].

Under low resolution the ^{19}F NMR spectrum shows five band groups of approximately equal intensity. The similarity between the chemical-shift value of the peak at lowest field (24.8 ppm) for the vinylic fluorine atom adjacent to the metal atom and that observed for the same fluorine atom in compound No. 148 (29.8 ppm referred to CF_3COOH [15]) suggests that Diels-Alder addition has occurred across the CF=CF bond in No. 148 [59].

The mass spectrum shows the molecular ion and fragments corresponding to stepwise loss of two CO molecules. The major fragmentation pattern is a retro-Diels-Alder reaction of the parent ion to give cyclopentadiene and $C_5H_5Fe(CO)_2C_6F_5$ [59].

The IR spectrum (in C_6H_{12}) shows six bands in the ν(CO) region. It is assumed that they are due to rotamers arising from restricted rotation about the Fe-C bond [59].

UV irradiation of the compound and $P(OC_6H_5)_3$ in hexane at room temperature for 2 h gives a trace of a yellow solid, which is tentatively formulated as $C_5H_5Fe(CO)(P(OC_6H_5)_3)R$ [59].

$C_5H_5Fe(CO)_2C_{12}H_9$ (Table **8**, No. **205**) has been obtained by addition of 1,1- and 1,2-dibromonaphthocyclobutenes (in place of 1-bromonaphthocyclobutene, Method I) in THF to a solution of Na[Fp] in the same solvent containing some methanol at $-78\,°C$. After warming to room temperature and filtration through Celite, the solvents are evaporated and the residue is chromatographed on Al_2O_3 with petroleum ether/benzene (1:1) as eluent, 44% yield [66].

Treatment of the compound with $[C(C_6H_5)_3]PF_6$ in CH_2Cl_2 yields the carbene salt $[Fp(\eta^1-$naphthocyclobutenylidene)]PF_6 via α-hydride abstraction [71], see also [45, 76].

$C_5H_5Fe(CO)_2C_{12}H_8(OCH_3)$ (Table 8, No. 206) has been prepared by stirring $[Fp(\eta^1-$naphthocyclobutenylidene)]PF_6 and $NaHCO_3$ in CH_3OH overnight, evaporation of the solvent, extraction of the residue with petroleum ether, and filtration through Celite. The residue remaining after solvent removal is then dissolved in ether and chromatographed on Al_2O_3 with ether as eluent, 59% yield [71], see also [45].

$C_5H_5Fe(CO)_2C_{12}H_9$ (Table 8, No. 207) has been prepared by treatment of $[Fp(\eta^2-$acenaphthylene)]BF_4 with $NaBH_3CN$ (1:1.25 mole ratio) in CH_3CN at $0\,°C$ for a few minutes, evaporation of the solvent, extraction of the residue with petroleum ether, and evaporation of the extracts, 81% yield [70]. Purification of the oily product by dissolution in petroleum ether and chromatography on Al_2O_3, eluting with petroleum ether into a flask cooled at $-78\,°C$, gives the crystalline compound in 77% yield [87]. The preparation of No. 207 (=FpR) from Na[Fp] and RCl is briefly mentioned in [94].

On heating in vacuum at $150\,°C$ for 30 min the compound decomposes to give acenaphthene, acenaphthylene (2:1), and Fp_2. When this thermolysis is carried out under 2 atm CO for 1 h, acenaphthene and acenaphthylene are obtained in a significantly increased ratio (4.5:1), but lower combined yield. In addition to the aforementioned three products, biacenaphthenyl is formed by refluxing the compound in toluene for 30 min under N_2. Thermolysis in toluene-d_8 produces deuterium-enriched acenaphthene. These observations, together with those resulting from Nos. 208 and 209, are explained by a mechanistic scheme involving competing β-elimination, iron-carbon bond homolysis to produce the acenaphthenyl radical, and C_5H_4 abstraction by an undetermined pathway [87].

Photolysis of No. 207 (=FpR) leads to cleavage of the Fe–C bond. Whereas in argon the R ligand forms acenaphthene by reaction with the solvent, in air the R ligand is oxidized to acenaphthenol and acenaphthenone. In both cases the formation of Fp_2 is also observed. The quantum yields of the photodecomposition in C_6H_{14}, CH_3OH, and THF are practically independent of the solvent and the atmosphere (argon or air) under which the photolysis is carried out [94, 110]. A proposed mechanism uses a coordinatively unsaturated species $C_5H_5Fe(CO)R$ (indicated by only one ν(CO) band in the IR spectrum) which undergoes σ-π rearrangement of the R ligand, followed by transformation of the intermediate into the final products by CO addition and dimerization [110].

$C_5H_5Fe(CO)_2C_{12}H_8D$ (Table 8, No. 208). The trans-deuteroacenaphthene complex has been obtained in 75% yield by the same procedure as described for compound No. 207 but with $NaBD_3CN$ [87], see also [70].

Thermal decomposition proceeds in a way similar to that of the undeuterated complex (see No. 207), producing deuterium enriched acenaphthene in vacuum and in toluene [87].

$C_5D_5Fe(CO)_2C_{12}H_9$ (Table 8, No. 209) has been prepared in 46% yield from $[C_5D_5Fe-(CO)_2(\eta^2-$acenaphthylene)]BF_4 and $NaBH_3CN$ in the previously described manner, see No. 207 [87].

The compound gives the same types of decomposition products as No. 207 on thermolysis, except that the acenaphthene formed contains deuterium [87].

References on pp. 268/72

$C_5H_5Fe(CO)_2C_{12}H_8C(CH_3)_2CHO$ (Table **8**, No. **210**) has been prepared by addition of N-(2-methylprop-1-enyl)pyrrolidine (enamine from isobutyraldehyde and pyrrolidine) in CH_3CN to [Fp(η^2-acenaphthylene)]BF_4 ($\sim$1:1 mole ratio) in the same solvent. After 15 min the reaction mixture is poured into water and extracted with ether. The dried and concentrated extracts are chromatographed on Al_2O_3 with 1:9 ether/hexane as eluent. Evaporation of the solvent gives the product as an oil in 83% yield [63]. The compound is converted to the corresponding 2,4-dinitrophenylhydrazone in the standard fashion [63]. For spectroscopic data see No. 211.

$C_5H_5Fe(CO)_2C_{12}H_8SC(CH_3)_3$ (Table **8**, No. **211**) has been prepared by stirring a solution of [Fp(η^2-acenaphthalene)]BF_4 in CH_3CN containing $(CH_3)_3CSH$ and $NaHCO_3$ at room temperature for 15 h. After evaporation of the solvent and excess thiol, the residue is extracted with ether. Chromatography of the concentrated extracts on Al_2O_3 with 1:9 ether/pentane as eluent affords the thiol adduct (76% yield), which is further purified by rechromatography [63].

The most important feature of the ^1H NMR data of Nos. 210 and 211 is the appearance of the H-1 and H-2 resonances as broad singlets (width at half height, 2.5 Hz, No. 210) or weakly coupled doublets (J = 1 Hz, No. 211). Compared with published coupling constants for 1,2-substituted acenaphthenes, these results indicate a trans relationship between H-1 and H-2. None of the isomeric cis-complexes were detected chromatographically or spectroscopically, suggesting a high degree of stereoselectivity for these addition processes. Mechanistic aspects of the reactions are discussed [63].

$C_5H_5Fe(CO)_2C_{17}H_{22}(CH_3)_2CH(CH_3)(CH_2)_3CH(CH_3)_2$ (Table **8**, No. **212**) undergoes rapid decomposition in the mass spectrometer. Treatment of the compound in CH_2Cl_2 at 0 °C with [$C(C_6H_5)_3$]BF_4 in CH_2Cl_2 and completion of the reaction at room temperature (monitored by IR) results in abstraction of a β-hydrogen atom, giving the cationic complex LXXIV [98].

LXXIV

$C_5H_5Fe(CO)_2C_8H_7O_3$ (Table **8**, No. **214**) has been prepared by refluxing $FpC(=CH_2)CH=CH_2$ and maleic anhydride (1:1 mole ratio) in CH_2ClCH_2Cl for 1.5 h followed by addition of petroleum ether at 20 °C and cooling to −78 °C. The product is recrystallized from CH_2Cl_2/petroleum ether, 80% yield [138].

$C_5H_5Fe(CO)_2C_8H_6O_3(CH_3)$ (Table **8**, No. **215**) has been obtained in 89% yield by refluxing (E)-$FpCH=CHC(CH_3)=CH_2$ and maleic anhydride (1:1 mole ratio) in CH_2ClCH_2Cl for 15 min, evaporating the solvent in a stream of N_2 and extracting the residue with petroleum ether and then CH_2Cl_2. Evaporation of the CH_2Cl_2 extract gives the product as an oil. Trace impurities, visible in the ^1H NMR spectrum, could not be removed because of decomposition when column chromatography was attempted and the failure of this material to crystallize at low temperature [138].

References on pp. 268/72

$C_5H_5Fe(CO)_2C_9H_7O_3$ (Table **8**, No. **216**) has been obtained as single stereoisomer in 73% yield from FpC_5H_5 and maleic anhydride (1:1 mole ratio) in CH_2Cl_2 at 0 °C (30 min). After evaporation of the solvent under reduced pressure the residue is recrystallized from $CHCl_3$/ ether [126] or from ether at −25 °C [130].

The compound crystallizes in the monoclinic system, space group $P2_1/c - C_{2h}^5$ (No. 14), a = 15.626(3), b = 6.596(3), c = 14.239(5) Å, β = 99.05(2)°; Z = 4. The molecular structure is shown in **Fig. 10** [126].

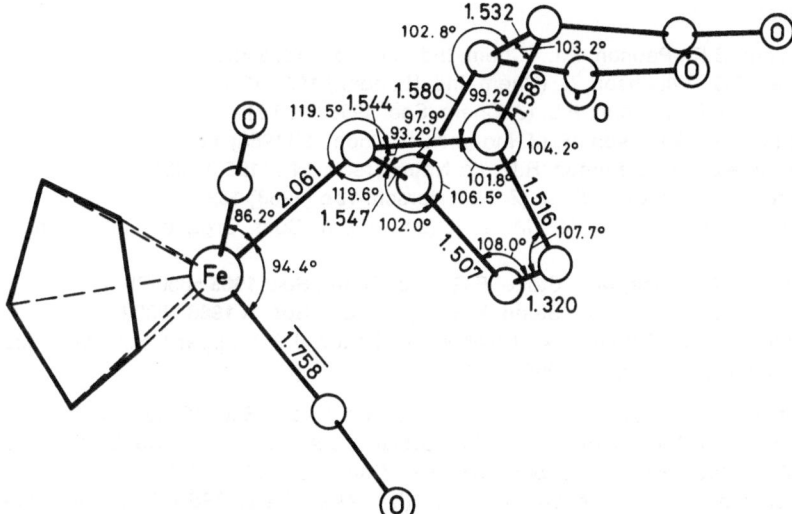

Fig. 10. Molecular structure of $C_5H_5Fe(CO)_2C_9H_7O_3$ (No. 216) with selected bond lengths (in Å) and angles.

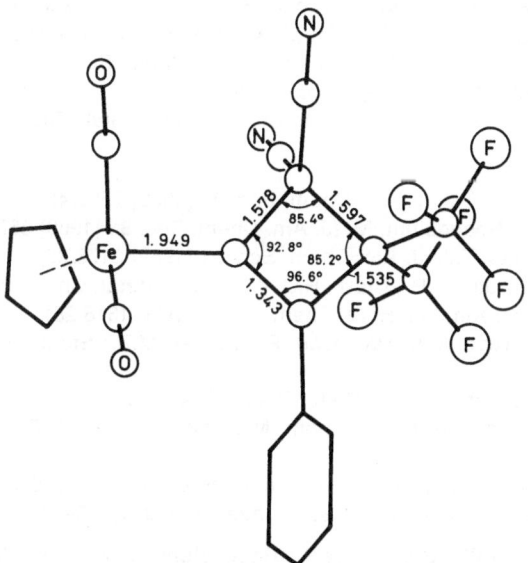

Fig. 11. Molecular structure of $C_5H_5Fe(CO)_2C_4(CN)_2(CF_3)_2C_6H_5$ (No. 217) with selected bond lengths (in Å) and angles.

                          References on pp. 268/72

$C_5H_5Fe(CO)_2C_4(CN)_2(CF_3)_2C_6H_5$ (Table **8**, No. **217**) has been prepared from $FpC\equiv CC_6H_5$ and $(CF_3)_2C=C(CN)_2$, molar ratio 3:5, in the dark in ether at room temperature, 98% [150].

The compound crystallizes in the monoclinic space group $P2_1/a(P2_1/c) - C_{2h}^5$ (No. 14), $a = 11.430(3)$, $b = 15.201(3)$, $c = 11.887(2)$ Å, $\beta = 102.50(2)°$; $Z = 4$, $d_c = 1.621$, $d_m = 1.61$ g/cm^3. The Fp group is similar to that found in related compounds and the Fe–C distances also compare well with those found in analogous complexes, see **Fig. 11** on p. 267 [150].

References:

[1] Hallam, B.F.; Pauson, P.L. (Chem. Ind. [London] **1955** 653).

[2] Piper, T.S.; Wilkinson, G. (Chem. Ind. [London] **1955** 1296).

[3] Hallam, B.F.; Pauson, P.L. (J. Chem. Soc. **1956** 3030/7).

[4] Piper, T.S.; Wilkinson, G. (J. Inorg. Nucl. Chem. **3** [1956] 104/24).

[5] Strohmeier, W.; Lemmon, R.M. (Z. Naturforsch. **14a** [1959] 109/12).

[6] Green, M.L.H.; Nagy, P.L.I. (Z. Naturforsch. **18b** [1963] 162).

[7] Green, M.L.H.; Ishaq, M.; Mole, T. (2nd Intern. Conf. Organometal. Chem., Madison 1965, Abstr. 91).

[8] Jolly, P.W.; Bruce, M.I.; Stone, F.G.A. (J. Chem. Soc. **1965** 5830/7).

[9] Bruce, M.I.; Jolly, P.W.; Stone, F.G.A. (J. Chem. Soc. A **1966** 1602/6).

[10] Bennett, M.J.; Cotton, F.A.; Davison, A.; Faller, J.W.; Lippard, S.J.; Morehouse, S.M. (J. Am. Chem. Soc. **88** [1966] 4371/6).

[11] Cotton, F.A.; Musco, A.; Yagupsky, G. (J. Am. Chem. Soc. **89** [1967] 6136/9).

[12] Denisovich, L.I.; Gubin, S.P.; Chapovskii, Yu.A. (Izv. Akad. Nauk SSSR Ser. Khim. **1967** 2378/84; Bull. Acad. Sci. USSR Div. Chem. Sci. **1967** 2271/5).

[13] Nesmeyanov, A.N.; Chapovskii, Yu.A. (Izv. Akad. Nauk SSSR Ser. Khim. **1967** 2075/7; Bull. Acad. Sci. USSR Div. Chem. Sci. **1967** 1988/90).

[14] Bruce, M.I. (Org. Mass Spectrochem. **1** [1968] 503/17).

[15] Booth, B.L.; Haszeldine, R.N.; Tucker, N.I. (J. Organometal. Chem. **11** [1968] P5/P6).

[16] Cook, D.J.; Green, M.; Mayne, N.; Stone, F.G.A. (J. Chem. Soc. A **1968** 1771/5).

[17] Fuchs, B.; Ishaq, M.; Rosenblum, M. (J. Am. Chem. Soc. **90** [1968] 5293/5).

[18] Green, M.; Taunton-Rigby, A.; Stone, F.G.A. (J. Chem. Soc. A **1968** 2762/5).

[19] King, R.B. (J. Am. Chem. Soc. **90** [1968] 1417/29).

[20] Bichler, R.E.J.; Clark, H.C. (4th Intern. Conf. Organometal. Chem., Bristol 1969, Abstr. O 4).

[21] Cotton, F.A.; Marks, T.J. (J. Am. Chem. Soc. **91** [1969] 7523/4).

[22] Ciappenelli, D.J.; Rosenblum, M. (J. Am. Chem. Soc. **91** [1969] 3673/4).

[23] Cotton, F.A.; Marks, T.J. (J. Am. Chem. Soc. **91** [1969] 3178/82).

[24] Bichler, R.E.J.; Booth, M.R.; Clark, H.C. (J. Organometal. Chem. **24** [1970] 145/58).

[25] Campbell, C.H.; Green, M.L.H. (J. Chem. Soc. A **1970** 1318/28).

[26] Blackmore, T.; Bruce, M.I.; Davidson, P.J.; Iqbal, M.Z.; Stone, F.G.A. (J. Chem. Soc. A **1970** 3153/8).

[27] Gubin, S.P. (Pure Appl. Chem. **23** [1970] 463/87, 481/4).

[28] Banks, R.E.; Haszeldine, R.N.; Lappin, M.; Lever, A.B.P. (J. Organometal. Chem. **29** [1971] 427/31).

[29] Giering, W.P.; Rosenblum, M. (J. Am. Chem. Soc. **93** [1971] 5299/301).

[30] Su, S.R.; Wojcicki, A. (J. Organometal. Chem. **31** [1971] C34/C36).

[31] Cutler, A.; Fish, R.W.; Giering, W.P.; Rosenblum, M. (J. Am. Chem. Soc. **94** [1972] 4354/5).

[32] Ciappenelli, D.J.; Cotton, F.A.; Kruczynski, L. (J. Organometal. Chem. **42** [1972] 159/62).

[33] Campbell, A.J.; Fyfe, C.A.; Goel, R.G.; Maslowsky, E.; Senoff, C.V. (J. Am. Chem. Soc. **94** [1972] 8387/91).

[34] Grishin, Yu.K.; Sergeyev, N.M.; Ustynyuk, Yu.A. (Org. Magn. Resonance **4** [1972] 377/90).

[35] King, R.B.; Efraty, A. (J. Fluorine Chem. **1** [1971/72] 283/94).

[36] Rosenblum, M. (Chem. Eng. News **50** [1972] 24).

[37] Giering, W.P.; Raghu, S.; Rosenblum, M.; Cutler, A.; Ehntholt, D.; Fish, R.W. (J. Am. Chem. Soc. **94** [1972] 8251/3).

[38] Giering, W.P.; Rosenblum, M.; Tancrede, J. (J. Am. Chem. Soc. **94** [1972] 7170/2).

[39] Grishin, Yu.K.; Sergeyev, N.M.; Ustynyuk, Yu.A. (J. Organometal. Chem. **34** [1972] 105/18).

[40] Churchill, M.R.; Ni Chang, S.W.Y. (J. Am. Chem. Soc. **95** [1973] 5931/8).

[41] Denisovich, L.I.; Gubin, S.P. (J. Organometal. Chem. **57** [1973] 109/19).

[42] Nicholas, K.M.; Rosenblum, M. (J. Am. Chem. Soc. **95** [1973] 4449/50).

[43] Raghu, S.; Rosenblum, M. (J. Am. Chem. Soc. **95** [1973] 3062/2).

[44] Sanders, A.; Cohen, L.; Giering, W.P.; Kenedy, D.; Magatti, C.V. (J. Am. Chem. Soc. **95** [1973] 5430/1).

[45] Giering, W.P.; Magatti, C.V.; Sanders, A. (6th Intern. Conf. Organometal. Chem., Amhurst, Mass., 1973; Abstr. 149).

[46] Rosan, A.; Rosenblum, M.; Tancrede, J. (J. Am. Chem. Soc. **95** [1973] 3062/4).

[47] Tancrede, J.M.P. (Diss. Brandeis Univ. 1973; Diss. Abstr. Intern. B **34** [1974] 3177).

[48] Rosenblum, M. (AD-774 324 [1974] 1/14; C.A. **81** [1974] No. 49714).

[49] Cohen, L.; Giering, W.P.; Kenedy, D.; Magatti, C.V.; Sanders, A. (J. Organometal. Chem. **65** [1974] C57/C60).

[50] Nicholas, N.; Raghu, S.; Rosenblum, M. (J. Organometal. Chem. **78** [1974] 133/7).

[51] Nesmeyanov, A.N.; Makarova, L.G.; Polovyanyuk, I.V. (Dokl. Akad. Nauk SSSR **217** [1974] 360/1; Dokl. Chem. Proc. Acad. Sci. USSR **214/219** [1974] 509/10).

[52] Su, S.R.; Wojcicki, A. (Inorg. Chim. Acta **8** [1974] 55/60).

[53] Sanders, A.; Giering, W.P. (J. Am. Chem. Soc. **96** [1974] 5247/8).

[54] Rosenblum, M. (Accounts Chem. Res. **7** [1974] 122/8).

[55] Sanders, A.; Magatti, C.V.; Giering, W.P. (J. Am. Chem. Soc. **96** [1974] 1610/1).

[56] Wilkinson, G. (Science **185** [1974] 109/12).

[57] Booth, B.L.; Haszeldine, R.N.; Tucker, N.I. (J. Chem. Soc. Dalton Trans. **1975** 1439/45).

[58] Cooke, M.; Russ, C.R.; Stone, F.G.A. (J. Chem. Soc. Dalton Trans. **1975** 256/9).

[59] Booth, B.L.; Haszeldine, R.N.; Tucker, N.I. (J. Chem. Soc. Dalton Trans. **1975** 1446/8).

[60] Cutler, A.R. (Diss. Brandeis Univ. 1975; Diss. Abstr. Intern. B **36** [1975] 233).

[61] Cutler, A.; Ehntholt, D.; Lennon, P.; Nicholas, K.; Marten, D.F.; Madhavarao, M.; Raghu, S.; Rosan, A.; Rosenblum, M. (J. Am. Chem. Soc. **97** [1975] 3149/57).

[62] Williams, J.P.; Wojcicki, A. (Inorg. Chim. Acta **15** [1975] L21/L22).

[63] Nicholas, K.M.; Rosan, A.M. (J. Organometal. Chem. **84** [1975] 351/6).

[64] Davidson, J.L.; Green, M.; Stone, F.G.A.; Welch, A.J. (J. Chem. Soc. Chem. Commun. **1975** 286/7).

[65] Sanders, A.; Giering, W.P. (J. Am. Chem. Soc. **97** [1975] 919/21).

[66] Bauch, T.; Sanders, A.; Magatti, C.V.; Waterman, P.; Judelson, D.; Giering, W.P. (J. Organometal. Chem. **99** [1975] 269/79).

[67] Chang, S.W.Y.N. (Diss. Univ. Illinois, Chicago 1975; Diss. Abstr. Intern. B **36** [1976] 5573).

[68] Moorhouse, S.; Wilkinson, G. (J. Organometal. Chem. **105** [1976] 349/55).

[69] Cutler, A.; Ehntholt, D.; Giering, W.P.; Lennon, P.; Raghu, S.; Rosan, A.; Rosenblum, M.; Tancrede, J.; Wells, D. (J. Am. Chem. Soc. **98** [1976] 3495/507).

[70] Florio, S.M.; Nicholas, K.M. (J. Organometal. Chem. **112** [1976] C17/C19).

[71] Sanders, A.; Bauch, T.; Magatti, C.V.; Lorenc, C.; Giering, W.P. (J. Organometal. Chem. **107** [1976] 359/75).

[72] Sanders, A.; Giering, W.P. (J. Organometal. Chem. **104** [1976] 67/78).

[73] Sanders, A.; Giering, W.P. (J. Organometal. Chem. **104** [1976] 49/65).

[74] Williams, J.P. (Diss. Ohio State Univ., Columbus 1975; Diss. Abstr. Intern. B **36** [1976] 3950).

[75] Davidson, J.L.; Green, M.; Stone, F.G.A.; Welch, A.J. (J. Chem. Soc. Dalton Trans. **1976** 2044/53).

[76] Sanders, A. (Diss. Boston Univ. Grad. School 1975; Diss. Abstr. Intern. B **36** [1976] 2814/5).

[77] Stone, F.G.A. (AD-AO 32905 [1976] 1/29; C.A. **87** [1977] No. 23353).

[78] Lennon, P.; Rosan, A.M.; Rosenblum, M. (J. Am. Chem. Soc. **99** [1977] 8426/39).

[79] Dizikes, L.J.; Wojcicki, A. (J. Am. Chem. Soc. **99** [1977] 5295/303).

[80] Chen, L.S.; Lichtenberg, D.W.; Robinson, P.W.; Yamamoto, Y.; Wojcicki, A. (Inorg. Chim. Acta **25** [1977] 165/72).

[81] Labinger, J.A. (J. Organometal. Chem. **136** [1977] C31/C36).

[82] Klemarczyk, P.; Price, T.; Priester, W.; Rosenblum, M. (J. Organometal. Chem. **139** [1977] C25/C28).

[83] Williams, J.P.; Wojcicki, A. (Inorg. Chem. **16** [1977] 2506/12).

[84] Williams, J.P.; Wojcicki, A. (Inorg. Chem. **16** [1977] 3116/24).

[85] Allison, N.T.; Kawada, Y.; Jones, W.M. (J. Am. Chem. Soc. **100** [1978] 5224/6).

[86] Gompper, R.; Bartmann, E. (Angew. Chem. **90** [1978] 490/1).

[87] Florio, S.M.; Nicholas, K.M. (J. Organometal. Chem. **144** [1978] 321/34).

[88] Johnson, M.D. (Angew. Chem. **90** [1978] 499).

[89] Abram, T.S.; Baker, R. (Syn. React. Inorg. Metal-Org. Chem. **9** [1979] 471/7).

[90] Abram, T.S.; Baker, R. (J. Chem. Soc. Chem. Commun. **1979** 267/8).

[91] Abram, T.S.; Baker, R.; Exon, C.M. (Tetrahedron Letters **1979** 4103/6).

[92] Davison, A.; Solar, J.P. (J. Organometal. Chem. **166** [1979] C13/C17).

[93] Kolobova, N.E.; Skripkin, V.V.; Rozantseva, T.V. (Izv. Akad. Nauk SSSR Ser. Khim. **1979** 1665; Bull. Acad. Sci. USSR Div. Chem. Sci. **1979** 1541).

[94] Kupletskaya, N.B.; Buyanovskaya, P.G.; Churanov, S.S. (Dokl. Akad. Nauk SSSR **248** [1979] 111/4; Dokl. Chem. Proc. Acad. Sci. USSR **244/249** [1979] 410/2).

[95] Fabian, B.D.; Labinger, J.A. (J. Am. Chem. Soc. **101** [1979] 2239/40).

[96] Rogers, W.N.; Baird, M.C. (J. Organometal. Chem. **182** [1979] C65/C68).

[97] Gompper, R.; Bartmann, E.; Nöth, H. (Chem. Ber. **112** [1979] 218/33).

[98] Laycock, D.E.; Hartgerink, J.; Baird, M.C. (J. Org. Chem. **45** [1980] 291/9).

[99] Lennon, P.J.; Rosan, A.; Rosenblum, M.; Tancrede, J.; Waterman, P. (J. Am. Chem. Soc. **102** [1980] 7033/8).

[100] McKinney, M.A.; Haworth, D.T. (J. Chem. Educ. **57** [1980] 110/2).

[101] Fabian, B.D.; Fehlner, T.P.; Hwang, L.-S.J.; Labinger, J.A. (J. Organometal. Chem. **191** [1980] 409/13).

[102] Riley, P.E.; Davis, R.E.; Allison, N.T.; Jones, W.M. (J. Am. Chem. Soc. **102** [1980] 2458/60).

[103] Slovokhotov, Yu.L.; Yanovskyi, A.I.; Andrianov, V.G.; Struchkov, Yu.T. (J. Organometal. Chem. **184** [1980] C57/C60).

[104] Rosenblum, M.; Waterman, P.S. (J. Organometal. Chem. **187** [1980] 267/75).

[105] Cotton, J.D.; Crisp, G.T. (J. Organometal. Chem. **186** [1980] 137/45).

[106] Herrmann, W.A.; Plank, J.; Bernal, I.; Creswick, M. (Z. Naturforsch. **35b** [1980] 680/8).

[107] Stone, J.A.; Laycock, D.E.; Lin, M.; Baird, M.C. (J. Chem. Soc. Dalton Trans. **1980** 2488/92).

[108] Rogers, W.N. (Diss. Queen's Univ., Kingston, Canada, 1980; Diss. Abstr. Intern. B **41** [1981] 3028).

[109] Hong, P.; Sonogashira, K.; Hagihara, N. (J. Organometal. Chem. **219** [1981] 363/9).

[110] Kupletskaya, N.B.; Buyanovskaya, P.G.; Churanov, S.S. (Zh. Obshch. Khim. **51** [1981] 1110/5; J. Gen. Chem. [USSR] **51** [1981] 929/34).

[111] Gompper, R.; Kottmair, E. (Tetrahedron Letters **1981** 2865/8).

[112] Fabian, B.D.; Labinger, J.A. (J. Organometal. Chem. **204** [1981] 387/92).

[113] Priester, W.; Rosenblum, M.; Samuels, S.B. (Syn. React. Inorg. Metal-Org. Chem. **11** [1981] 525/37).

[114] Abram, T.S.; Baker, R.; Exon, C.M.; Rao, V.B.; Turner, R.W. (J. Chem. Soc. Perkin Trans. I **1982** 301/6).

[115] Abram, T.S.; Baker, R.; Exon, C.M.; Rao, V.B. (J. Chem. Soc. Perkin Trans. I **1982** 285/94).

[116] Manganiello, F.J.; Christensen, L.W.; Jones, W.M. (J. Organometal. Chem. **235** [1982] 327/34).

[117] Manganiello, F.J.; Radcliffe, M.D.; Jones, W.M. (J. Organometal. Chem. **228** [1982] 273/9).

[118] Booth, B.L.; Casey, S.; Critchley, R.P.; Haszeldine, R.N. (J. Organometal. Chem. **226** [1982] 301/12).

[119] Riley, P.E.; Davis, R.E.; Allison, N.T.; Jones, W.M. (Inorg. Chem. **21** [1982] 1321/8).

[120] Baker, R.; Exon, C.M.; Rao, V.B.; Turner, R.W. (J. Chem. Soc. Perkin Trans. I **1982** 295/300).

[121] Bucheister, A.; Klemarczyk, P.; Rosenblum, M. (Organometallics **1** [1982] 1679/84).

[122] Manganiello, F.J. (Diss. Univ. Florida 1982; Diss. Abstr. Intern. B **44** [1983] 498).

[123] Lennon, P.; Rosenblum, M. (J. Am. Chem. Soc. **105** [1983] 1233/41).

[124] Zulu, S.J. (Diss. Boston Univ. Grad. School 1983; Diss. Abstr. Intern. B **44** [1983] 493).

[125] Radcliffe, M.D.; Jones, W.M. (Organometallics **2** [1983] 1053/5).

[126] Wright, M.E. (Organometallics **2** [1983] 558/60).

[127] Wright, M.E. (Diss. Univ. Arizona 1983; Diss. Abstr. Intern. B **44** [1983] 1833).

[128] Doig, S.J.; Hughes, R.P.; Patt, S.L.; Samkoff, D.E.; Smith, W.I. (J. Organometal. Chem. **250** [1983] C1/C4).

[129] Fabian, B.D.; Labinger, J.A. (Organometallics **2** [1983] 659/64).

[130] Wright, M.E.; Hoover, J.F.; Nelson, G.O.; Scott, C.P.; Glass, R.S. (J. Org. Chem. **49** [1984] 3059/63).

[131] Baker, R.; Keen, R.B.; Morris, M.D.; Turner, R.W. (J. Chem. Soc. Chem. Commun. **1984** 987/8).

[132] Glass, R.S.; McConnell, W.W. (Organometallics **3** [1984] 1630/2).

[133] Doig, S.J.; Hughes, R.P.; Davis, R.E.; Gadol, S.M.; Holland, K.D. (Organometallics **3** [1984] 1921/2).

[134] McConnell, W.W. (Diss. Univ. Arizona 1984; Diss. Abstr. Intern. B **46** [1985] 168).

[135] Silverman, G.S. (Diss. Univ. South Carolina 1984; Diss. Abstr. Intern. B **45** [1985] 2924/5).

[136] Manganiello, F.J.; Oon, S.M.; Radcliffe, M.D.; Jones, W.M. (Organometallics **4** [1985] 1069/72).

[137] Lisko, J.R.; Jones, W.M. (Organometallics **4** [1985] 944/6).

[138] Waterman, P.S.; Belmonte, J.E.; Bauch, T.E.; Belmonte, P.A.; Giering, W.P. (J. Organometal. Chem. **294** [1985] 235/50).

[139] Wright, M.E.; Nelson, G.O.; Glass, R.S. (Organometallics **4** [1985] 245/50).

[140] Hegedus, L.S.; Holden, M.S. (J. Org. Chem. **50** [1985] 3920/3).

[141] Bly, R.S.; Silverman, G.S.; Bly, R.K. (Organometallics **4** [1985] 374/83).

[142] Doig, S.J.; Hemond, R.C.; Hughes, R.P.; Stewart, L.A.; Whitman, D.W.; Davis, R.E.; Gadol, S.M.; Holland, C.D. (Abstr. Papers 189th Natl. Meeting Am. Chem. Soc., Miami Beach 1985, INOR 254).

[143] Gompper, R.; Bartmann, E. (Angew. Chem. **97** [1985] 207/8; Angew. Chem. Intern. Ed. Engl. **24** [1985] 209).

[144] Thomas, M.J.; Ahackleton, T.A.; Wright, S.C.; Gillis, D.J.; Colpa, J.P.; Baird, M.C. (J. Chem. Soc. Chem. Commun. **1986** 312/4).

[145] Maslowski, E. (Diss. Illinois Inst. Technol., Chicago 1970; Diss. Abstr. Intern. B **31** [1970] 617/8).

[146] Campbell, A.J. (Diss. Univ., Guelph, Canada, 1974; Diss. Abstr. Intern. B **35** [1974] 2058).

[147] Duncan, J.D. (Diss. Balliol College, Oxford 1969, pp. 1/157).

[148] Bruce, M.I.; Duffy, D.N.; Liddell, M.J.; Snow, M.R.; Tiekink, E.R.T. (J. Organometal. Chem. **335** [1987] 365/78).

[149] Bertani, R.; Traldi, P.; Cecchetti, W.; Polloni, R. (Inorg. Chim. Acta **134** [1987] 27/9).

[150] Bruce, M.I.; Liddell, M.J.; Snow, M.R.; Tieking, E.R.T. (Organometallics **7** [1988] 343/50).

Physical Constants and Conversion Factors

Avogadro constant N_A (or L) = 6.02214×10^{23} mol^{-1}

Faraday constant F = 9.64853×10^4 C/mol

molar gas constant R = 8.31451 J·mol^{-1}·K^{-1}

molar volume (ideal gas) V_m = 2.24141×10^1 L/mol
(273.15 K, 101325 Pa)

Planck constant h = 6.62608×10^{-34} J·s

elementary charge e = 1.60218×10^{-19} C

electron mass m_e = 9.10939×10^{-31} kg

proton mass m_p = 1.67262×10^{-27} kg

1 kg = 2.205 pounds

1 m = 3.937×10^1 inches = 3.281 feet

1 m^3 = 2.642×10^2 gallons (U.S.)

1 m^3 = 2.200×10^2 gallons (Imperial)

Force	N	dyn	kp
1 N	1	10^5	1.019716×10^{-1}
1 dyn	10^{-5}	1	1.019716×10^{-6}
1 kp	9.80665	9.80665×10^5	1

Pressure	Pa	bar	kp/m²	at	atm	Torr	lb/in²
1 Pa = 1 N/m²	1	10^{-5}	1.019716×10^{-1}	1.019716×10^{-5}	9.86923×10^{-6}	7.50062×10^{-3}	1.450378×10^{-4}
1 bar = 10^6 dyn/cm²	10^5	1	1.019716×10^4	1.019716	9.86923×10^{-1}	7.50062×10^2	1.450378×10^1
1 kp/m² = 1 mm H$_2$O	9.80665	9.80665×10^{-5}	1	10^{-4}	9.67841×10^{-5}	7.35559×10^{-2}	1.422335×10^{-3}
1 at (technical)	9.80665×10^4	9.80665×10^{-1}	10^4	1	9.67841×10^{-1}	7.35559×10^2	1.422335×10^1
1 atm = 760 Torr	1.01325×10^5	1.01325	1.033227×10^4	1.033227	1	7.60×10^2	1.469595×10^1
1 Torr = 1 mmHg	1.333224×10^2	1.333224×10^{-3}	1.359510×10^1	1.359510×10^{-3}	1.315789×10^{-3}	1	1.933678×10^{-2}
1 lb/in² = 1 psi	6.89476×10^3	6.89476×10^{-2}	7.03069×10^2	7.03069×10^{-2}	6.80460×10^{-2}	5.17149×10^1	1

Work, Energy, Heat	J	kW·h	kcal	Btu	eV
1 J = 1 W·s = 1 N·m = 10^7 erg	1	2.778×10^{-7}	2.39006×10^{-4}	9.4781×10^{-4}	6.242×10^{18}
1 kW·h	3.6×10^6	1	8.604×10^2	3.41214×10^3	2.247×10^{25}
1 kcal	4.1840×10^3	1.1622×10^{-3}	1	3.96566	2.6117×10^{22}
1 Btu (British thermal unit)	1.05506×10^3	2.93071×10^{-4}	2.5164×10^{-1}	1	6.5858×10^{21}
1 eV	1.602×10^{-7}	4.450×10^{-14}	3.8289×10^{-11}	1.51840×10^{-10}	1

$1 \, \mathrm{cm}^{-1} = 1.239842 \times 10^{-4}$ eV
1 hartree = 27.2114 eV
1 Hz = 4.135669×10^{-15} eV
1 eV ≙ 23.0578 kcal/mol

Power	kW	hp	kp·m·s^{-1}	kcal/s
1 kW = 10^3 J	1	1.35962	1.01972×10^2	2.39006×10^{-1}
1 hp (horsepower, metric)	7.3550×10^{-1}	1	7.5×10^1	1.7579×10^{-1}
1 kp·m·s^{-1}	9.80665×10^{-3}	1.333×10^{-2}	1	2.34384×10^{-3}
1 kcal/s	4.1840	5.6886	4.26650×10^2	1

References:

International Union of Pure and Applied Chemistry, Manual of Symbols and Terminology for Physicochemical Quantities and Units, Pergamon, London 1979; Pure Appl. Chem. **51** [1979] 1/41.

The International System of Units (SI), National Bureau of Standards Spec. Publ. 330 [1972];

Landolt-Börnstein, 6th Ed., Vol. II, Pt. 1, 1971, pp. 1/14.

ISO Standards Handbook 2, Units of Measurement, 2nd Ed., Geneva 1982.

Cohen, E. R., Taylor, B. N., Codata Bulletin No. 63, Pergamon, Oxford 1986.

Key to the Gmelin System of Elements and Compounds

System Number	Symbol	Element		System Number	Symbol	Element
1		Noble Gases		37	In	Indium
2	H	Hydrogen		38	Tl	Thallium
3	O	Oxygen		39	Sc, Y	Rare Earth
4	N	Nitrogen			La–Lu	Elements
5	F	Fluorine		40	Ac	Actinium
6	**Cl**	**Chlorine**		41	Ti	Titanium
7	Br	Bromine		42	Zr	Zirconium
8	I	Iodine		43	Hf	Hafnium
8a	At	Astatine		44	Th	Thorium
9	S	Sulfur		45	Ge	Germanium
10	Se	Selenium		46	Sn	Tin
11	Te	Tellurium		47	Pb	Lead
12	Po	Polonium		48	V	Vanadium
13	B	Boron		49	Nb	Niobium
14	C	Carbon		50	Ta	Tantalum
15	Si	Silicon		51	Pa	Protactinium
16	P	Phosphorus		**52**	**Cr**	**Chromium**
17	As	Arsenic		53	Mo	Molybdenum
18	Sb	Antimony		54	W	Tungsten
19	Bi	Bismuth		55	U	Uranium
20	Li	Lithium		56	Mn	Manganese
21	Na	Sodium		57	Ni	Nickel
22	K	Potassium		58	Co	Cobalt
23	NH_4	Ammonium		59	Fe	Iron
24	Rb	Rubidium		60	Cu	Copper
25	Cs	Caesium		61	Ag	Silver
25a	Fr	Francium		62	Au	Gold
26	Be	Beryllium		63	Ru	Ruthenium
27	Mg	Magnesium		64	Rh	Rhodium
28	Ca	Calcium		65	Pd	Palladium
29	Sr	Strontium		66	Os	Osmium
30	Ba	Barium		67	Ir	Iridium
31	Ra	Radium		68	Pt	Platinum
32	**Zn**	**Zinc**		69	Tc	Technetium[1]
33	Cd	Cadmium		70	Re	Rhenium
34	Hg	Mercury		71	Np,Pu...	Transuranium
35	Al	Aluminium				Elements
36	Ga	Gallium				

HCl

$CrCl_2$

$ZnCrO_4$

$ZnCl_2$

Material presented under each Gmelin System Number includes all information concerning the element(s) listed for that number plus the compounds with elements of lower System Number.

For example, zinc (System Number 32) as well as all zinc compounds with elements numbered from 1 to 31 are classified under number 32.

[1] A Gmelin volume titled "Masurium" was published with this System Number in 1941.

A Periodic Table of the Elements with the Gmelin System Numbers is given on the Inside Front Cover